An Introduction to Applied Statistical Thermodynamics

Stanley I. Sandler
University of Delaware

John Wiley & Sons, Inc.

VP & Executive Publisher:	Don Fowley
Acquisitions Editor:	Jennifer Welter
Editorial Assistant:	Alexandra Spicehandler
Marketing Manager:	Christopher Ruel
Designer:	Seng Ping Ngieng
Production Manager:	Janis Soo
Senior Production Editor:	Joyce Poh

This book was set in 10.5/12 Times Roman by Laserwords Private Limited and printed and bound by Hamilton Printing. The cover was printed by Hamilton Printing.

This book is printed on acid free paper.

Founded in 1807, John Wiley & Sons, Inc. has been a valued source of knowledge and understanding for more than 200 years, helping people around the world meet their needs and fulfill their aspirations. Our company is built on a foundation of principles that include responsibility to the communities we serve and where we live and work. In 2008, we launched a Corporate Citizenship Initiative, a global effort to address the environmental, social, economic, and ethical challenges we face in our business. Among the issues we are addressing are carbon impact, paper specifications and procurement, ethical conduct within our business and among our vendors, and community and charitable support. For more information, please visit our website: www.wiley.com/go/citizenship.

Copyright © 2011 John Wiley & Sons, Inc. All rights reserved. No part of this publication may be reproduced, stored in a retrieval system or transmitted in any form or by any means, electronic, mechanical, photocopying, recording, scanning or otherwise, except as permitted under Sections 107 or 108 of the 1976 United States Copyright Act, without either the prior written permission of the Publisher, or authorization through payment of the appropriate per-copy fee to the Copyright Clearance Center, Inc. 222 Rosewood Drive, Danvers, MA 01923, website www.copyright.com. Requests to the Publisher for permission should be addressed to the Permissions Department, John Wiley & Sons, Inc., 111 River Street, Hoboken, NJ 07030-5774, (201)748-6011, fax (201)748-6008, website http://www.wiley.com/go/permissions.

Evaluation copies are provided to qualified academics and professionals for review purposes only, for use in their courses during the next academic year. These copies are licensed and may not be sold or transferred to a third party. Upon completion of the review period, please return the evaluation copy to Wiley. Return instructions and a free of charge return shipping label are available at www.wiley.com/go/returnlabel. Outside of the United States, please contact your local representative.

Library of Congress Cataloging-in-Publication Data

Sandler, Stanley I., 1940-
 An introduction to applied statistical thermodynamics / Stanley I. Sandler.
    p. cm.
 Includes index.
 ISBN 978-0-470-91347-5 (pbk.)
 1. Statistical thermodynamics. 2. Thermodynamics–Industrial applications. I. Title.
 TP155.2.T45S36 2010
 621.402′1–dc22

2010034666

Printed in the United States of America
10 9 8 7 6 5 4 3 2 1

To Judith,
Catherine,
Joel,
And Michael

About the Author

Stanley I. Sandler is the H. B. du Pont Chair of Chemical Engineering and Professor of Chemistry and Biochemistry at the University of Delaware. He was department chair from 1982 to 1986 and Interim Dean of the College of Engineering in 1992. He earned the B. Ch. E degree in 1962 from the City College of New York, and the Ph. D. in chemical engineering from the University of Minnesota. He has been a visiting professor at Imperial College of London, the Technical University of Berlin, the University of California-Berkeley, the University of Queensland (Australia), the National University of Singapore and the University of Melbourne (Australia).

In addition to this book, Professor Sandler is the author of the textbook "Chemical, Biochemical and Engineering Thermodynamics" 4th edition published by Wiley, over 350 research papers, the author of one and editor of another book on thermodynamic modeling, and the editor of several conference proceedings. He is also the editor of the AIChE Journal. Among his many awards and honors are the Camille and Henry Dreyfus Faculty-Scholar Award (1971); a Research Fellowship (1980) and U. S. Senior Scientist Award (1988) from the Alexander von Humboldt Foundation (Germany); American Society for Engineering Education Lectureship Award (1988); the Professional Progress Award (1984); the Warren K. Lewis Award (1996) and Founders Award (2004) from the American Institute of Chemical Engineers; the E. V. Murphree Award (1996) from the American Chemical Society; the Rossini Lectureship Award (1997) from the International Union of Pure and Applied Chemistry; and election to the U. S. National Academy of Engineering (1996). He is a fellow of the American Institute of Chemical Engineers and the Institution of Chemical Engineers (Britain), and a Chartered Engineer (Britain).

Preface for Instructors

GOALS AND MOTIVATION

As the author of a widely used undergraduate textbook on thermodynamics (*Chemical, Biochemical and Engineering Thermodynamics*, 4th ed., by S. I. Sandler, John Wiley & Sons, Inc.) and a teacher of a graduate course in chemical engineering thermodynamics, I am frequently asked what I do teach in the graduate course. Its content has largely been influenced by our departmental policy of not accepting our own undergraduates into our graduate program. Consequently, I have found that the students in the graduate course have very varied backgrounds in thermodynamics. Some have had instruction in thermodynamics that has been axiomatic; others have had very applied courses; some have had only one course; others two; and still others have had very little thermodynamics as they studied disciplines other than chemical engineering. Therefore, my first goal in the graduate thermodynamics course is to bring everyone to approximately the same level. I do this by covering much of the material in my two-semester undergraduate textbook in the first half of the one-semester graduate course. My second goal is to introduce the first-year graduate students to the fundamental ideas and engineering uses of statistical thermodynamics, the equilibrium part of the statistical mechanics, in the remainder of the semester. It is for this part of the course that I use the material herein, developed and refined over years of teaching the course.

APPROACH

One-half of a semester is a short period in which to introduce statistical thermodynamics and some of its engineering and science applications, so compromises need to be made. The result, this book, will undoubtedly disappoint the true practitioners of statistical mechanics in its brevity; but I hope students will find it sufficiently interesting and useful that they may apply the insights and tools they gain to their own research, and perhaps pursue more rigorous courses devoted to the subject. Indeed, with the present emphasis on nano- and biotechnologies, molecular-level descriptions and understandings offered by statistical thermodynamics should be of increasing interest. I also hope that those who do not wish to pursue research using statistical thermodynamics will have gained an understanding of its utility and how they apply it in their work. What is not presented here is complete theoretical rigor in introducing statistical thermodynamics. My view, with which you may or may not agree, is that in a short introduction to statistical thermodynamics for students unfamiliar with the subject, there is greater value in showing how it can be useful than in dwelling on the fine details of derivations. These are best left to a full course on the subject.

Consequently, while terms such as *phase space* are mentioned in the text, I do not dwell on such concepts or rigorously derive the fundamentals of ensembles from such a basis. The reader is referred to the excellent (and much larger) textbooks by Tolman, Hill, McQuarrie, and others listed at the end of these prefaces for such rigorous presentations. Instead, the approach here is much more applied. This is perhaps best evidenced by, for example, the analysis of the virial equation of state first discussed in Chapter 7. There is found a development of the second virial coefficient that is similar to most other statistical mechanics books. However, in later chapters, I show how the exact composition dependence of the second virial coefficient derived from statistical thermodynamics has become the basis for mixing rules used with common equations of state such as the van der Waals, Peng-Robinson, and other cubic equations of state commonly used by chemical engineers and physical chemists. Then, in Chapter 12, the extension to the osmotic virial equation applicable to colloidal and protein solutions is introduced, and I discuss how the osmotic virial coefficient has been used to identify solution conditions for protein crystallization. Another example is the derivation of the Debye-Hückel limiting law for electrolyte solutions in Chapter 15. I then show how this has formed the basis for the development of electrolyte solution equations used in engineering. In the final chapter of this book is an analysis that allows the reader to understand the statistical thermodynamic assumptions that underlie the well-known equations of state and activity coefficient models commonly used in engineering and the sciences.

Computer simulation has become increasingly important in research and in obtaining insight into physical processes at the molecular level. This book provides an introduction to the simplest forms of Monte Carlo and molecular-dynamics simulation (albeit only for simple spherical molecules) and user-friendly MATLAB®[1] programs for doing such simulations, as well as some other calculations.

Only equilibrium properties are considered in this book, not dynamic or kinetic properties like the kinetic theory of gases or liquids. Therefore, statistical thermodynamics is the accurate description for the contents here, rather than the more general term of statistical mechanics.

In summary, the purpose of this book is to provide a readable introduction to statistical thermodynamics and show its utility—how the results obtained lead to useful generalizations for practical application. The book also illustrates the difficulties that arise in the statistical thermodynamics of dense fluids as seen in the discussion of liquids.

ACKNOWLEDGMENTS

There are many people whose assistance, direct and indirect, have contributed to this book. First and foremost is my wife, who over the years has put up with me busily working away in seclusion in my offices at home and abroad. Next are colleagues who have seen various versions of the manuscript for this book and have offered their comments and corrections, and especially Professor Shiang-Tai Lin of National Taiwan University. Zachery Ulissi, a University of Delaware student, converted a FORTRAN Monte Carlo program for the Lennard-Jones fluid made available to us by Professors Daan Frenkel (Cambridge University) and Berend Smit (University of California, Berkeley) into a user-friendly MATLAB® format. Zach also wrote the MATLAB®-based molecular dynamics for that potential. Professor Lester Woodcock of the University of Manchester, while a visiting professor at the University of Delaware, provided the excellent FORTRAN programs for the Monte Carlo simulations of the square-well, hard-sphere, and Lennard-Jones fluids, and a molecular

---

[1]MATLAB® is a registered trademark of The MathWorks, Inc.

dynamics program for the Lennard-Jones fluid that University of Delaware student Meghan McCabe adapted into the MATLAB® format. Professor Jaeeon Chang of the University of Seoul provided help in the development of the MATLAB® program for the solution of the Percus-Yevick equation for hard-spheres. Special thanks must go to my colleagues at the University of Delaware for their support while this book was being written, and at the University of Melbourne (Australia) for providing an office and other assistance for a month each year that allowed me to work undisturbed. I also need to acknowledge the forbearance of the students over the years who have put up with the early versions of the manuscript, which has largely been rewritten and improved based on their very helpful comments. Finally, I want to thank my editor Jennifer Welter and the production staff at John Wiley & Sons, Inc. for their continued help and encouragement.

WEBSITE

The following resources accompanying this book are available on the website www.wiley.com/college/sandler.

For students and instructors:

The collection of MATLAB® programs for some statistical thermodynamic calculations, including Monte Carlo and molecular-dynamics simulations described on page ix.

For instructors only:

Solutions Manual containing complete solutions to the problems in the text.

Image Gallery containing illustrations from the test in a format appropriate to include in lecture slides.

These last resources are only available to instructors adopting this text for a course. Visit the instruction section of the website www.wiley.com/college/sandler to register for password access to these resources.

Stanley I. Sandler
September 2010

Preface for Students

While classical thermodynamics can be used to describe many processes very well, such as phase behavior, chemical reaction equilibria, and interrelating heat and work flows on changes of state, it barely acknowledges the existence of molecules. In that sense, classical thermodynamics is not complete. Indeed, to apply classical thermodynamics, we need to know the properties of a substance or a mixture, such as its internal energy, enthalpy, Gibbs energy and heat of formation, and the parameters to be used in equation-of-state or activity coefficient models. However, classical thermodynamics does not provide us with a path to calculate the values of these parameters from knowledge of the molecules in the fluids of interest. Nor does classical thermodynamics provide any information on the underlying basis or assumptions for an equation of state such as that of van der Waals, or the activity coefficient models used by chemists and engineers.

Statistical thermodynamics, which starts with a description of individual molecules, can provide such information. For molecules that do not interact, which lead to an understanding of the ideal gas, the road is a straightforward one. However, as you will see in this book, the analysis for molecules that interact (especially in a dense fluid such as a liquid) is very much more complicated, and generally cannot be solved exactly. Nonetheless, we can obtain useful insights by starting with a molecular-level description and statistical thermodynamics. However, given the specific purpose for this book as an addition to a course on classical thermodynamics, only equilibrium properties are considered here, not dynamic or kinetic properties like the kinetic theory of gases or liquids; hence the use here of the term *statistical thermodynamics* rather than the more general term *statistical mechanics*.

Statistical thermodynamics is also not complete because some of the parameters we need, such as bond lengths, bond energies, interaction energies, etc., come from an even deeper look at molecules from various forms of spectroscopy and computational quantum mechanics. Those subjects are beyond the scope of this introductory text, so here we will assume that such information is available when needed.

With the present emphasis on nano- and biotechnologies, molecular-level descriptions, computer simulation, the understandings arising from statistical thermodynamics, and the ability to use molecular-level arguments to make useful predictions are all of increasing interest and importance in chemical engineering and physical chemistry. I hope the presentation here will be sufficiently interesting to students that some may be encouraged to apply it to their own research, and perhaps even to study the subject further. But I also hope that even those who do not wish to pursue further study of statistical thermodynamics will have gained some appreciation for the subject and its utility, and a better understanding of the limitations and nuances of the generalizations they apply in their work.

As computer simulation is of increasing importance in research and in obtaining insight into what is occurring at the molecular level, this book provides an introduction to the simplest forms of Monte Carlo and molecular-dynamics simulation, for simple spherical molecules. There are user-friendly MATLAB®[2] programs for doing such simulations and some other statistical thermodynamic calculations that can be downloaded from the website for this book www.wiley.com/college/sandler.

Stanley I. Sandler
September 2010

---

[2]MATLAB® is a registered trademark of The MathWorks, Inc.

MATLAB®[3] programs that accompany this book

1. **LJ_Virial**

   For computing the second virial coefficient and its first and second temperature derivatives for the Lennard-Jones 12-6 fluid. Program (M file) is in folder LJ_virial.

2. **MC_Squarewell**[4]

   A Monte Carlo simulation program for the square-well fluid. Data entered through a GUI, graphical output appears on screen and numerical output is in a spreadsheet. By setting the well width parameter R_{sw} equal to 1 on input this program does calculations for the hard-sphere fluid. M file is in folder MC_sqwell.

3. **MC_LJ**[4]

   A Monte Carlo simulation program for the Lennard-Jones 12-6 fluid. Data entered through a GUI, graphical output appears on screen and numerical output is in a spreadsheet. M file is in folder MC_LJ.

4. **MD_LJ**[4]

   An isothermal molecular-dynamics simulation program for the Lennard-Jones 12-6 fluid. Data entered through a GUI, graphical output appears on screen and numerical output is in a spreadsheet. M file is in folder MD_LJ.

5. **LJ_MD_MC**[5,6]

   A program for the Lennard-Jones 12-6 fluid that does both Monte Carlo and isothermal molecular-dynamics simulations for the same state conditions. Data entered through a GUI, graphical output appears on screen and numerical output is in a spreadsheet. M file is in folder LJ_MD_MC.

6. **MD_LJ2**[6]

   An isothermal molecular-dynamics program for the Lennard-Jones 12-6 fluid that also calculates the speed distribution. M file is in folder MD_LJ2.

7. **PYHS**

   A program for computing the radial distribution function for the hard-sphere fluid in the Percus-Yevick approximation. Graphical ouput appears on the screen and numerical output in a spreadsheet. M file is in folder Percus-Yevick HS.

8. **PYHSPMF**

   A program for computing the radial distribution function and potential of mean force for the hard-sphere fluid in the Percus-Yevick approximation. Graphical output appears on the screen and numerical ouput in a spreadsheet. M file is in folder PY HS PMF.

---

[3] MATLAB® is a registered trademark of The MathWorks, Inc.

[4] This MATLAB® program is based on a FORTRAN program of Professor Leslie Woodcock and is being used with his permission. The program conversion to MATLAB® has been done by the University of Delaware undergraduate Meghan McCabe.

[5] These programs in MATLAB® are based on the FORTRAN programs on the Website http://molsim.chem.uva.nl/frenkel_smit that accompanies the excellent book by D. Frenkel and B. Smit, *Understanding Molecular Simulation, From Algorithms to Applications*, 2nd ed. Academic Press, London, 2001. The programs are being used with the permission of Daan Frenkel and Berend Smit. The program conversions to MATLAB® have been done by University of Delaware undergraduate Zachery Ulissi.

[6] This molecular-dynamics MATLAB® code for the Lennard-Jones fluid was developed by University of Delaware undegraduate Zachery Ulissi.

A Partial List of the Many Books on Statistical Thermodynamics and Statistical Mechanics

This text focuses on the applications and practical utility of statistical thermodynamics.

For further study, including detailed derivations of the fundamental concepts statistical mechanics, you may want to consult the following books, among others. As these books have generally been written by chemists, phyicists and mathematicians, these books focus on the science, and do not include the practical applications that are the focus of this book.

D. A. McQuarrie, *Statistical Mechanics*, University Science Books, Sausalito, CA, 2000.

R. C. Tolman, *Statistical Mechanics*, Oxford University Press, London, 1938.

N. Davidson, *Statistical Mechanics*, McGraw-Hill, New York, 1962.

T. L. Hill, *Statistical Mechanics*, McGraw-Hill, New York, 1956.

T. L. Hill, *An Introduction to Statistical Thermodynamics*, Addison-Wesley, Reading, MA, 1960.

J. Kestin and J. R. Dorfman, *A Course in Statistical Mechanics*, Academic Press, New York, 1971.

J. E. Mayer and M. G. Mayer, *Statistical Mechanics*, Wiley, New York, 1940.

E. Schrödinger, *Statistical Thermodynamics*, Cambridge University Press, Cambridge, 1952.

L. D. Landau and E. M. Lifshitz, *Statistical Physics*, Pergamon Press, London, 1958.

G. S. Rushbrooke, *Statistical Mechanics*, Oxford University Press, London, 1949.

H. Eyring, D. Henderson, B. J. Stover, and E. M. Eyring, *Statistical Mechanics and Dynamics*, Wiley, New York, 1964.

R. H. Fowler and E. A. Guggenheim, *Statistical Thermodynamics*, Cambridge University Press, Cambridge, 1956.

E. A. Mason and T. H. Spurling, *The Virial Equation of State*, Pergamon, New York, 1969.

J. S. Rowlinson and F. L. Swinton, *Liquids and Liquid Mixtures*, 3rd ed., Butterworths, Oxford, 1982.

R. A. Robinson and R. H. Stokes, *Electrolyte Solutions*, 2nd ed., Academic Press, New York, 1965.

K. A. Dill and S. Bromberg, *Statistical Thermodynamics in Chemistry and Biology*, Garland Science, New York, 2003.

D. Chandler, *Introduction to Modern Statistical Mechanics*, Oxford University Press, London, 1987.

M. P. Allen and D. J. Tildesley, *Computer Simulation of Liquids*, Clarendon Press, Oxford, 1989.

D. Frenkel and B. Smit, *Understanding Molecular Simulation, From Algorithms to Applications*, 2nd ed., Academic Press, London, 2001.

A. R. Leach, *Molecular Modeling: Principles and Applications*, 2nd ed., Prentice-Hall, 2001.

Contents

PREFACE FOR INSTRUCTORS v

PREFACE FOR STUDENTS ix

CHAPTER 1   INTRODUCTION TO STATISTICAL THERMODYNAMICS 1
    1.1   Probabilistic Description   1
    1.2   Macroscopic States and Microscopic States   2
    1.3   Quantum Mechanical Description of Microstates   3
    1.4   The Postulates of Statistical Mechanics   5
    1.5   The Boltzmann Energy Distribution   6

CHAPTER 2   THE CANONICAL PARTITION FUNCTION 9
    2.1   Some Properties of the Canonical Partition Function   9
    2.2   Relationship of the Canonical Partition Function to Thermodynamic Properties   11
    2.3   Canonical Partition Function for a Molecule with Several Independent Energy Modes   12
    2.4   Canonical Partition Function for a Collection of Noninteracting Identical Atoms   13
    Chapter 2 Problems   15

CHAPTER 3   THE IDEAL MONATOMIC GAS 16
    3.1   Canonical Partition Function for the Ideal Monatomic Gas   16
    3.2   Identification of β as $1/kT$ 18
    3.3   General Relationships of the Canonical Partition Function to Other Thermodynamic Quantities   19
    3.4   The Thermodynamic Properties of the Ideal Monatomic Gas   22
    3.5   Energy Fluctuations in the Canonical Ensemble   29
    3.6   The Gibbs Entropy Equation   33
    3.7   Translational State Degeneracy   35
    3.8   Distinguishability, Indistinguishability, and the Gibbs' Paradox   37
    3.9   A Classical Mechanics–Quantum Mechanics Comparison: The Maxwell-Boltzmann Distribution of Velocities   39
    Chapter 3 Problems   42

CHAPTER 4   THE IDEAL DIATOMIC AND POLYATOMIC GASES 44
    4.1   The Partition Function for an Ideal Diatomic Gas   44
        4.1a   The Translational and Nuclear Partition Functions   45
        4.1b   The Rotational Partition Function   45
        4.1c   The Vibrational Partition Function   47
        4.1d   The Electronic Partition Function   48
    4.2   The Thermodynamic Properties of the Ideal Diatomic Gas   49
    4.3   The Partition Function for an Ideal Polyatomic Gas   53

4.4 The Thermodynamic Properties of an Ideal Polyatomic Gas   55
4.5 The Heat Capacities of Ideal Gases   58
4.6 Normal Mode Analysis: The Vibrations of a Linear Triatomic Molecule   59
Chapter 4 Problems   62

CHAPTER 5   CHEMICAL REACTIONS IN IDEAL GASES   64

5.1 The Nonreacting Ideal Gas Mixture   64
5.2 Partition Function of a Reacting Ideal Chemical Mixture   65
5.3 Three Different Derivations of the Chemical Equilibrium Constant in an Ideal Gas Mixture   67
5.4 Fluctuations in a Chemically Reacting System   70
5.5 The Chemically Reacting Gas Mixture: The General Case   73
5.6 Two Illustrations   80
Appendix: The Binomial Expansion   83
Chapter 5 Problems   85

CHAPTER 6   OTHER PARTITION FUNCTIONS   87

6.1 The Microcanonical Ensemble for a Pure Fluid   87
6.2 The Grand Canonical Ensemble for a Pure Fluid   89
6.3 The Isobaric-Isothermal Ensemble   92
6.4 The Restricted Grand or Semi-Grand Canonical Ensemble   93
6.5 Comments on the Use of Different Ensembles   94
Chapter 6 Problems   96

CHAPTER 7   INTERACTING MOLECULES IN A GAS   98

7.1 The Configuration Integral   98
7.2 Thermodynamic Properties from the Configuration Integral   100
7.3 The Pairwise Additivity Assumption   101
7.4 Mayer Cluster Function and Irreducible Integrals   102
7.5 The Virial Equation of State   109
7.6 Virial Equation of State for Polyatomic Molecules   114
7.7 Thermodynamic Properties from the Virial Equation of State   116
7.8 Derivation of Virial Coefficient Formulae from the Grand Canonical Ensemble   118
7.9 Range of Applicability of the Virial Equation   123
Chapter 7 Problems   124

CHAPTER 8   INTERMOLECULAR POTENTIALS AND THE EVALUATION OF THE SECOND VIRIAL COEFFICIENT   125

8.1 Interaction Potentials for Spherical Molecules   125
8.2 The Second Virial Coefficient in a Mixture: Interaction Potentials Between Unlike Atoms   136
8.3 Interaction Potentials for Multiatom, Nonspherical Molecules, Proteins, and Colloids   137
8.4 Engineering Applications and Implications of the Virial Equation of State   140
Chapter 8 Problems   144

CHAPTER 9   MONATOMIC CRYSTALS   147

    9.1   The Einstein Model of a Crystal   147
    9.2   The Debye Model of a Crystal   150
    9.3   Test of the Einstein and Debye Heat Capacity Models for a Crystal   157
    9.4   Sublimation Pressure and Enthalpy of Crystals   159
    9.5   A Comment on the Third Law of Thermodynamics   161
    Chapter 9 Problems   161

CHAPTER 10   SIMPLE LATTICE MODELS FOR FLUIDS   163

    10.1   Introduction   164
    10.2   Development of Equations of State from Lattice Theory   165
    10.3   Activity Coefficient Models for Similar-Size Molecules from Lattice Theory   168
    10.4   The Flory-Huggins and Other Models for Polymer Systems   172
    10.5   The Ising Model   178
    Chapter 10 Problems   184

CHAPTER 11   INTERACTING MOLECULES IN A DENSE FLUID. CONFIGURATIONAL DISTRIBUTION FUNCTIONS   185

    11.1   Reduced Spatial Probability Density Functions   185
    11.2   Thermodynamic Properties from the Pair Correlation Function   190
    11.3   The Pair Correlation Function (Radial Distribution Function) at Low Density   194
    11.4   Methods of Determination of the Pair Correlation Function at High Density   197
    11.5   Fluctuations in the Number of Particles and the Compressibility Equation   199
    11.6   Determination of the Radial Distribution Function of Fluids using Coherent X-ray or Neutron Diffraction   202
    11.7   Determination of the Radial Distribution Functions of Molecular Liquids   210
    11.8   Determination of the Coordination Number from the Radial Distribution Function   211
    11.9   Determination of the Radial Distribution Function of Colloids and Proteins   213
    Chapter 11 Problems   214

CHAPTER 12   INTEGRAL EQUATION THEORIES FOR THE RADIAL DISTRIBUTION FUNCTION   216

    12.1   The Yvon-Born-Green (YBG) Equation   216
    12.2   The Kirkwood Superposition Approximation   219
    12.3   The Ornstein-Zernike Equation   220
    12.4   Closures for the Ornstein-Zernike Equation   222
    12.5   The Percus-Yevick Hard-Sphere Equation of State   227
    12.6   The Radial Distribution Functions and Thermodynamic Properties of Mixtures   228
    12.7   The Potential of Mean Force   230
    12.8   Osmotic Pressure and the Potential of Mean Force for Protein and Colloidal Solutions   237
    Chapter 12 Problems   239

CHAPTER 13 DETERMINATION OF THE RADIAL DISTRIBUTION FUNCTION AND FLUID PROPERTIES BY COMPUTER SIMULATION 241

13.1 Introduction to Molecular Level Computer Simulation 242
13.2 Thermodynamic Properties from Molecular Simulation 245
13.3 Monte Carlo Simulation 249
13.4 Molecular-Dynamics Simulation 253
Chapter 13 Problems 255

CHAPTER 14 PERTURBATION THEORY 257

14.1 Perturbation Theory for the Square-Well Potential 257
14.2 First Order Barker-Henderson Perturbation Theory 262
14.3 Second-Order Perturbation Theory 265
14.4 Perturbation Theory Using Other Reference Potentials 269
14.5 Engineering Applications of Perturbation Theory 272
Chapter 14 Problems 274

CHAPTER 15 A THEORY OF DILUTE ELECTROLYTE SOLUTIONS AND IONIZED GASES 276

15.1 Solutions Containing Ions (and Electrons) 276
15.2 Debye-Hückel Theory 280
15.3 The Mean Ionic Activity Coefficient 291
Chapter 15 Problems 296

CHAPTER 16 THE DERIVATION OF THERMODYNAMIC MODELS FROM THE GENERALIZED VAN DER WAALS PARTITION FUNCTION 297

16.1 The Statistical-Mechanical Background 298
16.2 Application of the Generalized van der Waals Partition Function to Pure Fluids 301
16.3 Equation of State for Mixtures from the Generalized van der Waals Partition Function 310
16.4 Activity Coefficient Models from the Generalized van der Waals Partition Function 318
16.5 Chain Molecules and Polymers 329
16.6 Hydrogen-Bonding and Associating Fluids 332
Chapter 16 Problems 334

INDEX 335

Chapter 1

Introduction to Statistical Thermodynamics

INSTRUCTIONAL OBJECTIVES FOR CHAPTER 1

The goals of this chapter are for the student to:
- Understand the probabilistic description used in statistical thermodynamics
- Understand the distinction between macrostates and microstates
- Understand the quantum mechanics description that will be used
- Understand the postulates of statistical thermodynamics
- Understand the derivation of the Boltzmann energy distribution

1.1 PROBABILISTIC DESCRIPTION

The goal of statistical thermodynamics is to allow one to make predictions about the macroscopic properties of a system, such as its heat capacity, chemical equilibrium constant, equation of state, etc., using information only about the microscopic (or molecular) nature of the system. The methods used take advantage of the fact that the large numbers of molecules in any system of interest allows the use of statistics.

An example of what we are trying to do, and some of the difficulties inherent in any molecular description, is evident by considering the pressure of a gas on its container. This pressure is the result of collisions of gas molecules with the container walls. The force exerted on the wall by any one collision is almost infinitesimal; there are, however, about 10^{24} molecule-wall collisions per second for each square centimeter of surface for a gas at standard conditions. The result of so many collisions is a finite force or pressure. The pressure we measure, then, is an average over many, many molecular collisions. In fact, the measured pressure is a longtime average (on a molecular scale) of many molecular events. In a similar fashion, other macroscopic properties of a system can be related to longtime averages of the corresponding molecular processes.

A direct and deterministic way to proceed with the development of a microscopic theory would be to use a calculational scheme based on following the trajectories (position and velocity) of each molecule in the system. At each molecule-molecule or molecule-wall collision, which occurs about every 10^{-11} seconds for each molecule, new trajectories would have to be computed. Any macroscopic property could then be computed by calculating the appropriate longtime average of the appropriate microscopic property. For example, the gas pressure could be related to the average over time of the force on the container due to molecular collisions. Such calculations have been done for limited numbers of molecules and for short periods of time using computers; this technique is discussed in Chapter 13, but is not yet practical for routine engineering calculations. Furthermore, such calculations yield much more information than we actually need or, for that matter, want. One has little need for the location and velocity of each of the 10^{22} molecules in a liter of gas, with constant updating each time a collision occurs, when one's interest is merely with a small number of average macroscopic properties such as the pressure, temperature, internal energy, heat capacity, Gibbs and Helmholtz energies, etc. In fact, to compute these average properties from the inestimable reams of computer output corresponding to, say, one second in the history of a real gas is an impossible task. Instead, a way to proceed would be to first reduce all the exact information into a suitably compact statistical form. A thermodynamic property of the system could then be computed as an appropriate statistical average. For example, the information about particle velocities could be compactly presented in terms of the probability distribution for particle velocities or kinetic energies. The temperature of a monatomic gas could then be computed from the average kinetic energy.

An alternative approach to the above for the development of a microscopic theory is to start directly with a statistical or probabilistic description. That is, we no longer inquire about the velocity of each molecule, but only about the probability distribution of the velocities of all molecules. This is the procedure we shall follow here.

What we are trying to determine are probability distributions and average values of properties when considering all possible states of the molecules consistent with the constraints on the overall system—for example fixed temperature, volume, and number of molecules; or fixed pressure, volume, and number of molecules, etc. In the language of statistical thermodynamics, the collection of possible states consistent with the constraints is referred to as *the ensemble of states*. Special names are given to these ensembles, depending on the constraints. For example, the canonical ensemble refers to all states consistent with fixed temperature, volume, and number of molecules. The microcanonical ensemble refers to all states consistent with fixed total energy, volume, and number of molecules, while the constraints for the grand canonical ensemble are fixed volume, temperature, and chemical potential (partial molar Gibbs energy).

1.2 MACROSCOPIC STATES AND MICROSCOPIC STATES

The macroscopic state of a gas can be completely specified by giving the numerical values for a small number of parameters, such as the temperature (or energy), volume, and number of molecules or moles. The classical mechanical description of the microscopic state of the fluid is much more detailed, in that the position vector and velocity vector of each particle would be specified—that is, the microscopic description would be a specification $(\underline{r}_1, \underline{v}_1, \underline{r}_2, \underline{v}_2, \cdots, \underline{r}_n, \underline{v}_n)$ where $\underline{r}_i$ and $\underline{v}_i$ are the position and velocity vectors of the ith molecule in some suitable frame of reference. (When considering an "ideal gas," that is, a gas of noninteracting particles in the absence of external fields such as gravity, the location of the molecules is unimportant, and therefore the description of a microstate of noninteracting molecules need not include position vectors.)

In order for the microstate of a gas to be consistent with the observed macroscopic state, the following criteria must be met:

(a) the number of molecules in the microstate must be the same as the number of molecules in the macroscopic state;
(b) all the position vectors, $\underline{r}_i$, must be constrained to be within the volume V; and
(c) the energy of the microscopic state must be equal to the energy of the macroscopic state.

Clearly there are a very large number of microstates consistent with any macroscopic state of the system. That is, there are a very large number of position and velocity assignments for the molecules that are consistent with restrictions (a)–(c).

Any counting of microstates following only the prescriptions outlined above would overcount the number of microstates, in that we would be differentiating between microstates that are indistinguishable. For example, consider the two microstates $(\underline{v}_1, \underline{v}_2, \underline{v}_3 \cdots, \underline{v}_n)$ and $(\underline{v}_2, \underline{v}_1, \underline{v}_3 \cdots, \underline{v}_n)$ for a system of identical particles. The only difference between these two states is that in the second state particle 2 has the velocity that particle 1 had in the first state, and vice versa. In fact, there are $n!$ ways this set of velocities could be assigned to the n identical particles. The question then arises as to whether one can really distinguish between these $n!$ different states. If the particles were of macroscopic size, we could identify each particle by, for example, painting a number on it, and therefore be able to distinguish between the different velocity assignments. For microscopic particles, however, the $n!$ states must be considered identical and indistinguishable, and therefore should not be counted as separate microstates. This concept of indistinguishability of identical molecules is in accord with the Heisenberg uncertainty principle—that is, since we do not exactly know the position and velocity of any molecule at any one time as a result of the amplification of initial uncertainties in solving the laws of mechanics, at any later time we would not know which molecule was which.

1.3 QUANTUM MECHANICAL DESCRIPTION OF MICROSTATES

In the quantum mechanical description of molecular states not all values of the energy are allowed, only certain discreet values. For example, a single molecule or a particle in a cubic box of volume V and side $L = V^{1/3}$ cannot have any possible energy, as is allowed by classical mechanics, but only values of the energy ε given by

$$\varepsilon(l_x, l_y, l_z) = \frac{h^2}{8mV^{2/3}} \left(l_x^2 + l_y^2 + l_z^2\right) \tag{1.3-1}$$

where l_x, l_y, and l_z are integers that can take on the values 0, 1, 2, etc.; m is the particle mass; and $h = 6.62517 \times 10^{-27}$ erg-sec is Planck's constant. It is important to distinguish between an *energy state* and an *energy level*. An energy state is a particular specification of quantum numbers. For a single particle in a box, the three almost equivalent quantum number assignments given below represent three distinguishable states.

l_x	l_y	l_z
2	1	1
1	2	1
1	1	2

However, each of these energy states has the same value of the energy, or the same energy level

$$\varepsilon = \frac{h^2}{8mV}\left(2^2 + 1^2 + 1^2\right) = \frac{h^2}{8mV}\left(1^2 + 2^2 + 1^2\right) = \frac{h^2}{8mV}\left(1^2 + 1^2 + 2^2\right) = \frac{3h^2}{4mV}$$

Therefore, we refer to this energy level as being threefold *degenerate*, that is, there are three energy states consistent with this energy level.

In what follows, it will be necessary to enumerate the states of a system, for example, in order of increasing energy. This can be done in either of two equivalent ways. The first is to number the energy states, starting with the lowest state. In such a numbering system, there would be some adjacent states that have the same value of the energy; these correspond to microstates of the same energy level. An alternative procedure is to only number the energy levels. In this case each energy value would be distinct, but one would also have to specify the degeneracy of each energy level—that is, the number of states that have this value of the energy. The degeneracy of the jth molecular energy level will be denoted by ω_j.

Our interest will usually be in the energy states of a large assembly of molecules rather than in that of a single molecule. The energy of an assembly of noninteracting molecules is simply the sum of the energies of the individual molecules. To account for the indistinguishability of identical molecules, an energy state of the collection of molecules is specified by giving a set of occupation numbers—that is, a set of numbers $\underline{n} = (n_1, n_2, n_3, \ldots)$ where n_i is the number of molecules in the ith molecular state. An important feature of this method of state identification is that the indistinguishability of identical particles has been incorporated into this description in that we have not specified which molecule is in which energy state, but only how many molecules are in each of the energy states. Each occupation number can only take on the integer values 0, 1, 2, 3, etc. However, as we shall see shortly, our interest will be with those systems and macroscopic conditions where the number of energy states available to each molecule greatly exceeds the number of molecules present. Therefore, the number of energy states of an assembly of molecules in which any particular molecular energy state has higher than single occupancy is very much smaller than the number of energy states in which each occupation number is either zero or one.

Each distinct energy state of an assembly of molecules will be given by a set of occupation numbers; the ith energy state will be specified by the vector $\underline{n}^i = \left(n_1^i, n_2^i, \ldots\right)$, where n_j^i is the occupation number of the jth single molecule energy state in the ith energy state of the assembly of molecules. Thus, each $\underline{n}^i$ represents a distinct microstate of the system and the energy of this microstate is

$$E_i = \text{energy of the } i^{th} \text{ microstate} = \sum_{\substack{\text{all molecular} \\ \text{energy states} \\ j}} n_j^i \varepsilon_j \qquad (1.3\text{-}2)$$

where ε_j is the energy of the jth energy state of a single molecule. There will, of course, be many different microstates (sets of occupation numbers) consistent with any specific value of the energy E. Therefore, as before, we can shift our attention from energy states to energy levels, provided that for each energy level of the macroscopic system we also specify its degeneracy—that is, the number of microstates that have this energy level. We will denote the degeneracy of an energy level of an assembly of molecules by Ω, to distinguish it from the degeneracy of the energy level of a single molecule, ω.

1.4 THE POSTULATES OF STATISTICAL MECHANICS

The rigorous development of the principles of statistical mechanics is a very elegant and beyond the scope of this introduction to the subject. The reader is referred to the many excellent textbooks on the subject for such presentations. In contrast, the presentation here will be quite inelegant, but also very simple; it is based upon two postulates.

The first postulate:

> All microstates of the system of volume V that have the same energy and the same number of particles are equally probable.

This postulate is known as the *equal a priori probability* principle, and is a statement of complete ignorance. However, there is much to be said for this concept. First, it is the most minimalist statement that can be made. Any other assumption of probability assignment would require much more information about the system, information that, in fact, we do not have. (Think about this. How else would you assign probability to states of equal energy?) We can ultimately test this assumption by comparing the results of calculations for system properties based on this assumption with experimental measurements of these quantities. No evidence has been found to contradict the *equal a priori* assignment of probability.

The second postulate:

> The (long) time average of any mechanical property in a real macroscopic system is equal to the average value of that property over all the microscopic states of the system, each state weighted with its probability of occurrence, provided that the microscopic states replicate the thermodynamic state and environment of the actual system.

This postulate, which is called the *ergodic hypothesis*, merely sets down in words a concept we alluded to earlier, namely that any experimental measurement is really a long time measurement on a molecular time scale. So long, in fact, with respect to the rate of transition between the microscopic states, that during the time necessary to perform the measurement, the assembly of molecules will have gone through a very large, statistically representative number of microstates. Therefore, we can replace the time average with a statistical average.

Taken together, postulates I and II represent a complete framework for the construction of a statistical theory of thermodynamic processes. The first postulate tells us how to choose a probability distribution, and the second postulate establishes that thermodynamic properties computed with this probability distribution will be equivalent to those that we would measure. There is, however, an important restriction embodied in this equivalence of statistical and time averages, namely the replication of the thermodynamic state and environment of the real system. In this introductory chapter to statistical thermodynamics we shall be concerned with a system of constant volume and number of particles, but which is free to exchange energy with its surroundings.

6 Chapter 1: Introduction to Statistical Thermodynamics

1.5 THE BOLTZMANN ENERGY DISTRIBUTION

In this section we establish the way of assigning probabilities to states of different energies for a system of fixed volume and number of particles in contact with a large heat bath.[1] (Note the *equal a priori probability* principle deals only with assigning probabilities to states of equal energy, not of different energies.) Consider the system shown in Fig. 1.5-1, in which the macroscopic subsystems A and B are in contact with an infinite heat bath of constant temperature.

Here, the heat bath is considered to be so large that the subsystems A and B are unaffected by the presence of one another. That is, a fluctuation of energy or temperature in system A has no effect on system B, and vice versa. The subsystems A and B are completely unspecified, and need not be identical. Let $p_A(E_n)$ be the probability that the subsystem A is in one particular microstate s_A whose energy in E_n. (Note that this is not the probability of finding system A with the energy E_n, since there may be many microscopic states consistent with this energy level. By the first postulate, all these states are equally likely, so that the probability of finding subsystem A in energy level E_n is proportional to $\Omega_A(E_n)p_A(E_n)$, where $\Omega_A(E_n)$ is the degeneracy of the level E_n). By the *equal a priori probability* postulate, this probability can only be a function of the energy level. Similarly, let $p_B(E_m)$ be the probability of finding subsystem B in a particular microstate s_B whose energy level is E_m. We now can ask what the probability is of simultaneously finding system A in the state s_A and subsystem B in the state s_B. Since A and B are completely independent, this probability is

$$p_A(E_n)p_B(E_m) \tag{1.5-1}$$

that is, the product of the two separate probabilities.

Now consider a composite system formed from both subsystems A and B. The probability of finding this composite system in a particular microstate s_{AB} can, by the *equal a priori probability* postulate, only be a function to the total energy of the composite system. This probability will be written as $p_{AB}(E_{AB})$, where E_{AB} is the energy of the microstate. Furthermore, all microstates of the composite system with the energy E_{AB} have the same probability of occurrence. Suppose, $E_{AB} = E_n + E_m$. One particular microstate (among many) of the system A + B that has this energy is when subsystem A is in state s_A and subsystem B is in state s_B. The probability of occurrence of such an event is given by Eq. 1.5-1. Therefore, the probability of

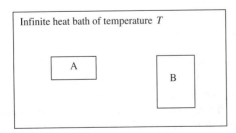

Figure 1.5-1 Systems A and B in an infinite heat bath of constant temperature.

---

[1]The simple argument here is based on one that appears in *Equilibrium Statistical Mechanics*, by E. A. Jackson, Prentice-Hall, Englewood Cliffs, NJ, 1968, and other books.

1.5 The Boltzmann Energy Distribution

occurrence of the state s_{AB} (or any other particular microstate of the system with the energy $E_n + E_m$) is

$$p_{AB}(E_n + E_m) = p_A(E_n)p_B(E_m) \tag{1.5-2}$$

Note that any other microstate with the same total energy also has this same probability of occurrence, for example $p_A(E_n + \delta)p_B(E_m - \delta)$ also equals $p_{AB}(E_n + E_m)$.

We can now inquire how these probabilities would change if we changed the value E_n without changing E_m. (Note that as far as the composite system is concerned, this is just one of many ways of changing the total energy.) In principle, the energy of the system is a discrete variable; however, if the energy levels are very closely spaced, we can treat the energy as a continuous variable and write this probability change in terms of derivatives with respect to energy. For the moment, we will assume that the energy levels are closely spaced; we will return to this question later. Taking the derivative of Eq. 1.5-2 with respect to E_n holding E_m constant, one obtains

$$\left.\frac{\partial}{\partial E_n}\right|_{E_m} p_{AB}(E_n + E_m) = \frac{dp_{AB}(E_n + E_m)}{d(E_n + E_m)} \left.\frac{d(E_n + E_m)}{d(E_n)}\right|_{E_m} = \frac{dp_{AB}(E_n + E_m)}{d(E_n + E_m)} \tag{1.5-3}$$

Now using Eq. 1.5-2, we also have

$$\left.\frac{\partial p_{AB}(E_n + E_m)}{\partial E_n}\right|_{E_m} = \left.\frac{\partial}{\partial E_n}\right|_{E_m} p_A(E_n)p_B(E_m) = \frac{dp_A(E_n)}{d(E_n)} p_B(E_m) \tag{1.5-4}$$

and therefore we have

$$\frac{dp_{AB}(E_n + E_m)}{d(E_n + E_m)} = \frac{dp_A(E_n)}{d(E_n)} p_B(E_m) \tag{1.5-5}$$

By a similar argument, one can show that changing E_m while holding E_n constant gives

$$\frac{dp_{AB}(E_n + E_m)}{d(E_n + E_m)} = p_A(E_n)\frac{dp_B(E_m)}{d(E_m)} \tag{1.5-6}$$

Now equating the results of Eqs. 1.5-5 and 1.5-6 we obtain

$$p_B(E_m)\frac{dp_A(E_n)}{dE_n} = p_A(E_n)\frac{dp_B(E_m)}{d(E_m)}$$

$$\text{or} \quad \frac{1}{p_A(E_n)}\frac{dp_A(E_n)}{dE_n} = \frac{1}{p_B(E_m)}\frac{dp_B(E_m)}{dE_m} \tag{1.5-7}$$

The interesting characteristic of Eq. 1.5-7 is that the left-hand side is independent of subsystem B, and the right-hand side of the equation is independent of subsystem A. Furthermore, as noted earlier, it is possible to make changes in subsystem A independent of any changes in subsystem B, and vice versa. One example would be to change the volume of subsystem A, which, as can be seen from Eq. 1.3-1 has the effect of changing all the energy levels in that subsystem, but no effect on subsystem B. That the relationship given in Eq. 1.5-7 must be maintained for all such changes means that each side of that equation must be independent of **both** subsystems A and B and can only depend on the properties of the reservoir, here characterized by

8 Chapter 1: Introduction to Statistical Thermodynamics

its temperature. This can be written as

$$\frac{d \ln p_A(E_n)}{dE_n} = \frac{d \ln p_B(E_m)}{dE_m} = -\beta \qquad (1.5\text{-}8)$$

(where we have introduced the negative sign for convenience, as will be evident later). From the discussion above, β cannot depend on the subsystems A and B, but may be some function of the character of the thermal reservoir or heat bath (such as its temperature) with which both subsystems are in contact. Indeed, one expects that changing the temperature of the reservoir would change the temperature of both subsystems, and that this would affect the energy probability distribution. Integrating Eq. 1.5-8 one obtains

$$p_A(E_n) = C_A e^{-\beta E_n} \quad \text{and} \quad p_B(E_m) = C_B e^{-\beta E_m} \qquad (1.5\text{-}9)$$

Each of the integration constants, C_A and C_B, are specific to the characteristics of their respective subsystems and can be determined from the normalization condition that each subsystem must be in one of its allowed energy states, that is,

$$\sum_{\substack{\text{states } n \text{ of} \\ \text{system A}}} p_A(E_n) = 1 \quad \text{and} \quad \sum_{\substack{\text{states } m \text{ of} \\ \text{system B}}} p_B(E_m) = 1 \qquad (1.5\text{-}10)$$

Therefore,

$$C_A = \frac{1}{\sum_{\substack{\text{states } n \text{ of} \\ \text{system A}}} e^{-\beta E_n}} \quad \text{and} \quad C_B = \frac{1}{\sum_{\substack{\text{states } m \text{ of} \\ \text{system B}}} e^{-\beta E_m}} \qquad (1.5\text{-}11)$$

We define the canonical partition function $Q(N, V, \beta)$ for any system to be

$$\boxed{Q(N, V, \beta) = \sum_{\substack{\text{states} \\ i}} e^{-\beta E_i(N, V)}} \qquad (1.5\text{-}12)$$

Note that the summation is over all the energy states of the system. With this definition of the canonical partition function, we have that the probability of occurrence of a particular microstate i with energy E_α is

$$\boxed{p_i(E_\alpha) = \frac{e^{-\beta E_\alpha}}{\sum_{\substack{\text{states} \\ i}} e^{-\beta E_i}} = \frac{e^{-\beta E_\alpha}}{Q(N, V, \beta)}} \qquad (1.5\text{-}13)$$

This is the important result of this introductory chapter. In what follows, it will be assumed that β is a positive number, and later we will show this to be true. The implication of $\beta > 0$ is that a state of higher energy has a lower probability of occurrence than a lower energy state.

Chapter 2

The Canonical Partition Function

INSTRUCTIONAL OBJECTIVES FOR THIS CHAPTER 2

The goals of this chapter are for the student to:

- Understand the derivation of the canonical partition function
- Understand the role of degeneracy in the probability distribution function
- Understand how thermodynamic properties are computed from the canonical partition function
- Understand the difference between the canonical partition function for a single molecule with several independent energy modes and the canonical partition function for a collection of identical molecules

2.1 SOME PROPERTIES OF THE CANONICAL PARTITION FUNCTION

In this section we consider some of the properties of the canonical partition function and its relation to thermodynamic properties. From Eq. 1.5-13 we have that the probability of occurrence of a particular microstate i with energy E_α is

$$p_i(E_\alpha) = \frac{e^{-\beta E_\alpha}}{\sum_{\substack{\text{states} \\ j}} e^{-\beta E_j}} = \frac{e^{-\beta E_\alpha}}{Q(N, V, \beta)} \quad \textbf{(1.5-13)}[1]$$

However, we can also ask what the probability is of finding the system in any microstate such that its energy is E_α. This would be the product of the probability that the system is in a particular microstate with energy E_α, and the degeneracy of that energy level (that is, the number of states having that energy), since by the *equal*

---

[1] Note that here and elsewhere, the notation that E_j is used for the energy of a particular microscopic state, while U is the average internal energy of the system, and it is U that appears in the equations of classical thermodynamics.

10 Chapter 2: The Canonical Partition Function

a priori probability assumption all states of the same energy have the same likelihood of occurrence. Therefore,

$$p(E_\alpha) = \sum_{\text{states } i \text{ whose energy is } E_\alpha} p_i(E_\alpha) = p_i(E_\alpha) \times \left\{\begin{array}{l}\text{number of states} \\ \text{with energy } E_\alpha\end{array}\right\} = p_i(E_\alpha) \times \Omega(E_\alpha)$$

where $p(E_\alpha)$ is the probability of finding the macrosystem in any microstate with the energy E_α, and $\Omega(E_\alpha)$ is the degeneracy.

(2.1-1)

Consequently, the probability of occurrence of a particular microstate i with energy E_α is

$$p_i(E_\alpha) = \frac{e^{-\beta E_\alpha}}{\sum_{\text{states } j} e^{-\beta E_j}} = \frac{e^{-\beta E_\alpha}}{Q(N, V, \beta)} \quad (2.1\text{-}2)$$

and the probability of occurrence of the energy level E_α, $P(E_\alpha)$ is

$$p(E_\alpha) = \omega(E_\alpha) p_i(E_\alpha) = \frac{\omega(E_\alpha) e^{-\beta E_\alpha}}{Q(N, V, \beta)} \quad (2.1\text{-}3)$$

Also, for later reference we note that the canonical partition function can be written either in terms of states or of levels, as follows:

$$Q(N, V, \beta) = \sum_{\substack{\text{states} \\ i}} e^{-\beta E_i} = \sum_{\substack{\text{levels} \\ j}} \omega(E_j) e^{-\beta E_j} \quad (2.1\text{-}4)$$

Since the probability of occurrence of any one microstate is proportional to $\exp(-\beta E)$, a particular state with a lower energy is more probable than a particular state with a higher energy. In fact, the state with the lowest energy is most probable. However, the degeneracy $\omega(E)$ of an energy level is an increasing function of the energy level. That is, generally there are more states possible having higher energy than a lower energy. For example, the kinetic energy of a particle in classical mechanics is $\frac{m}{2}(v_x^2 + v_y^2 + v_z^2)$, where m is the mass and v_i is the velocity in the ith coordinate direction. If, for demonstration, the velocities are restricted to be integers, the degeneracy of the energy level $m/2$ is 3 (one of the three velocities is 1 and the other two are 0), while the degeneracy of the level $10^8 m/2$ is very large. Therefore, the probability that the macroscopic system will have an energy level E is the product of an exponential term that is decreasing with increasing energy and a degeneracy that is increasing with increasing energy, as shown schematically in Fig. 2.1-1. (The probability distribution for collections of molecules is considered in Section 3.5.)

Therefore, the most probable energy level (not energy state!) is a balance between these two factors, as indicated in the last of the figures on the following page.

2.2 Relationship of the Canonical Partition Function to Thermodynamic Properties 11

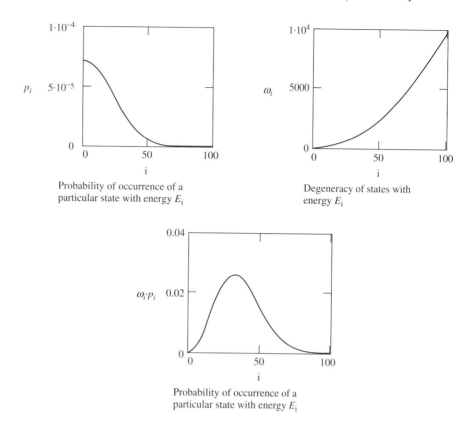

Figure 2.1-1 Probability Distribution for the Translational Energy States of a Single Molecule.

2.2 RELATIONSHIP OF THE CANONICAL PARTITION FUNCTION TO THERMODYNAMIC PROPERTIES

The internal energy U in thermodynamics is equal to the average value of the energy of the system $\overline{E}$. That is,

$$U = \sum_{\substack{\text{states}\\k}} E_k p(E_k) = \frac{\sum_{\substack{\text{states}\\k}} E_k e^{-\beta E_k}}{Q} \quad \text{where} \quad Q = \sum_{\substack{\text{states}\\k}} e^{-\beta E_k} \quad (2.2\text{-}1)$$

However,

$$\left(\frac{\partial Q}{\partial \beta}\right)_{N,V} = -\sum_{\substack{\text{states}\\k}} E_k e^{-\beta E_k} \quad (2.2\text{-}2)$$

so that

$$U = \overline{E} = -\left(\frac{\partial \ln Q}{\partial \beta}\right)_{N,V} = \frac{\sum\limits_{\substack{\text{states} \\ k}} E_k e^{-\beta E_k}}{Q} \qquad (2.2\text{-}3)$$

So we see that the "sum over states" or partition function can be used, rather than probability $p(E)$, to obtain an average value of a thermodynamic quantity. We will see in Chapter 3 that many thermodynamic properties can be obtained from the canonical partition function and its derivatives. However, before we can do this, we also have to consider some other general properties of this function.

2.3 CANONICAL PARTITION FUNCTION FOR A MOLECULE WITH SEVERAL INDEPENDENT ENERGY MODES

Consider a single molecule with, for simplicity, two completely independent energy modes. For example, translational energy that depends on the mass of the molecules and the center of mass velocity, and rotational energy that depends on the moment of inertia and the rotational velocity of the molecule. For simplicity of illustration, assume only three translational energy states exist—$\varepsilon_{\text{trans},1}$, $\varepsilon_{\text{trans},2}$ and $\varepsilon_{\text{trans},3}$—and that only three rotational states exist—$\varepsilon_{\text{rot},1}$, $\varepsilon_{\text{rot},2}$ and $\varepsilon_{\text{rot},3}$. All possible energy states of this highly simplified system are

$$\varepsilon_{\text{trans},1} + \varepsilon_{\text{rot},1},\ \varepsilon_{\text{trans},1} + \varepsilon_{\text{rot},2},\ \varepsilon_{\text{trans},1} + \varepsilon_{\text{rot},3},\ \varepsilon_{\text{trans},2} + \varepsilon_{\text{rot},1},\ \varepsilon_{\text{trans},2}$$
$$+ \varepsilon_{\text{rot},2},\ \varepsilon_{\text{trans},2} + \varepsilon_{\text{rot},3},\ \varepsilon_{\text{trans},3} + \varepsilon_{\text{rot},1},\ \varepsilon_{\text{trans},3} + \varepsilon_{\text{rot},2},\ \text{and}\ \varepsilon_{\text{trans},3} + \varepsilon_{\text{rot},3}$$

The single-particle canonical partition function, or sum over states, which we denote by q for this system is

$$\begin{aligned}
q &= e^{-\beta(\varepsilon_{\text{trans},1}+\varepsilon_{\text{rot},1})} + e^{-\beta(\varepsilon_{\text{trans},1}+\varepsilon_{\text{rot},2})} + e^{-\beta(\varepsilon_{\text{trans},1}+\varepsilon_{\text{rot},3})} \\
&\quad + e^{-\beta(\varepsilon_{\text{trans},2}+\varepsilon_{\text{rot},1})} + e^{-\beta(\varepsilon_{\text{trans},2}+\varepsilon_{\text{rot},2})} + e^{-\beta(\varepsilon_{\text{trans},2}+\varepsilon_{\text{rot},3})} \\
&\quad + e^{-\beta(\varepsilon_{\text{trans},3}+\varepsilon_{\text{rot},1})} + e^{-\beta(\varepsilon_{\text{trans},3}+\varepsilon_{\text{rot},2})} + e^{-\beta(\varepsilon_{\text{trans},3}+\varepsilon_{\text{rot},3})}
\end{aligned} \qquad (2.3\text{-}1)$$

which can be rewritten as

$$\begin{aligned}
q &= e^{-\beta\varepsilon_{\text{trans},1}}e^{-\beta\varepsilon_{\text{rot},1}} + e^{-\beta\varepsilon_{\text{trans},1}}e^{-\beta\varepsilon_{\text{rot},2}} + e^{-\beta\varepsilon_{\text{trans},1}}e^{-\beta\varepsilon_{\text{rot},3}} \\
&\quad + e^{-\beta\varepsilon_{\text{trans},2}}e^{-\beta\varepsilon_{\text{rot},1}} + e^{-\beta\varepsilon_{\text{trans},2}}e^{-\beta\varepsilon_{\text{rot},2}} + e^{-\beta\varepsilon_{\text{trans},2}}e^{-\beta\varepsilon_{\text{rot},3}} \\
&\quad + e^{-\beta\varepsilon_{\text{trans},3}}e^{-\beta\varepsilon_{\text{rot},1}} + e^{-\beta\varepsilon_{\text{trans},3}}e^{-\beta\varepsilon_{\text{rot},2}} + e^{-\beta\varepsilon_{\text{trans},3}}e^{-\beta\varepsilon_{\text{rot},3}}
\end{aligned} \qquad (2.3\text{-}2)$$

or

$$\begin{aligned}
q &= e^{-\beta\varepsilon_{\text{trans},1}}\left(e^{-\beta\varepsilon_{\text{rot},1}} + e^{-\beta\varepsilon_{\text{rot},2}} + e^{-\beta\varepsilon_{\text{rot},3}}\right) + e^{-\beta\varepsilon_{\text{trans},2}}\left(e^{-\beta\varepsilon_{\text{rot},1}}\right. \\
&\quad \left. + e^{-\beta\varepsilon_{\text{rot},2}} + e^{-\beta\varepsilon_{\text{rot},3}}\right) + e^{-\beta\varepsilon_{\text{trans},3}}\left(e^{-\beta\varepsilon_{\text{rot},1}} + e^{-\beta\varepsilon_{\text{rot},2}} + e^{-\beta\varepsilon_{\text{rot},3}}\right) \\
&= \left(e^{-\beta\varepsilon_{\text{trans},1}} + e^{-\beta\varepsilon_{\text{trans},2}} + e^{-\beta\varepsilon_{\text{trans},3}}\right)\left(e^{-\beta\varepsilon_{\text{rot},1}} + e^{-\beta\varepsilon_{\text{rot},2}} + e^{-\beta\varepsilon_{\text{rot},3}}\right) \\
&= q_{\text{trans}} q_{\text{rot}}
\end{aligned} \qquad (2.3\text{-}3)$$

Here, $q_{\text{trans}} = e^{-\beta\varepsilon_{\text{trans},1}} + e^{-\beta\varepsilon_{\text{trans},2}} + e^{-\beta\varepsilon_{\text{trans},3}}$ is the canonical partition function for the translational energy, and $q_{\text{rot}} = e^{-\beta\varepsilon_{\text{rot},1}} + e^{-\beta\varepsilon_{\text{rot},2}} + e^{-\beta\varepsilon_{\text{rot},3}}$ is the canonical partition function for the rotational energy.

While the model used here was a simple one, the important result is that if there are several independent energy modes $\varepsilon_A, \varepsilon_B, \varepsilon_C, \ldots$, then the canonical partition function is the product of the partition functions for the individual modes

$$\boxed{q = q_A q_B q_C \ldots} \quad \text{for independent energy modes} \quad (2.3\text{-}4)$$

2.4 CANONICAL PARTITION FUNCTION FOR A COLLECTION OF NONINTERACTING IDENTICAL ATOMS

Consider first a single atom with a collection of accessible energy states $\varepsilon_1, \varepsilon_2, \varepsilon_3, \ldots$. The canonical partition function, or sum over states, for this one-atom system is

$$q = e^{-\beta\varepsilon_1} + e^{-\beta\varepsilon_2} + e^{-\beta\varepsilon_3} \ldots \quad (2.4\text{-}1)$$

Now consider a collection of N such identical atoms, and assume that the atoms are sufficiently far apart that we can neglect their potential energy of interaction. Consequently, the atoms, for the present, will be considered to have only translational energy. To continue, it would be useful if we could specify the energy state of each atom. However, this is prohibited by the Heisenberg uncertainty principle. That is, since the position and velocity of each atom at an instant are imperfectly known (by the uncertainty principle), at some later time we could no longer be sure which atom had which velocity. (The situation would be simpler if we could paint a number on each atom, which is, of course, not possible.) Therefore, rather than specifying the energy state of each atom, we instead can define a state of the system by specifying the number of atoms in each energy state, but not indicating which atoms are in that state. In particular, we will use the notation that

n_j^i is the occupation number, or the number of atoms in the jth atomic state of a single atom in the ith macroscopic state of the collection of atoms.

Therefore, the ith macroscopic state of the collection of atoms is given by the collection of numbers $(n_1^i, n_2^i, n_3^i, \ldots)$ that specify the number of atoms having the energy $\varepsilon_1, \varepsilon_2, \varepsilon_3, \ldots$ etc. Clearly, with this definition

$$\sum_{\substack{\text{states } j \\ \text{of a} \\ \text{single atom}}} n_j^i = N = \text{the number of atoms in the system} \quad (2.4\text{-}2)$$

since each atom must be in one of the states of the possible atomic states. Also

$$E^i = \text{energy of the } i^{\text{th}} \text{ state of system} = \sum_{\substack{\text{states } j \\ \text{of a} \\ \text{single atom}}}^{N} \varepsilon_j n_j^i \quad (2.4\text{-}3)$$

Chapter 2: The Canonical Partition Function

With this notation, we will now consider a system with only two atoms. A listing of all possible microstates of the system, recognizing that the atoms are indistinguishable, would then be a list of all possible sets of two occupation numbers as shown below:

$(2, 0, 0, 0, \ldots), (1, 1, 0, 0, 0, \ldots), (1, 0, 1, 0, 0, \ldots), (1, 0, 0, 1, 0, \ldots),$

$(1, 0, 0, 0, \ldots, 1, \ldots 0), (0, 2, 0, 0, 0, \ldots), (0, 1, 1, 0, 0, \ldots), (0, 1, 0, 1, 0, \ldots),$

$(0, 1, 0, 0, 1, \ldots), (0, 1, 0, 0, 0, \ldots, 1, \ldots 0), (0, 0, 1, 1, 0, \ldots)$ etc.

The canonical partition function, or sum over states, for this system is then

$$Q = e^{-2\beta\varepsilon_1} + e^{-\beta(\varepsilon_1+\varepsilon_2)} + e^{-\beta(\varepsilon_1+\varepsilon_3)} + e^{-\beta(\varepsilon_1+\varepsilon_4)} + \cdots + e^{-2\beta\varepsilon_2} + e^{-\beta(\varepsilon_2+\varepsilon_3)}$$
$$+ e^{-\beta(\varepsilon_2+\varepsilon_4)} + e^{-\beta(\varepsilon_2+\varepsilon_5)} + \cdots + e^{-\beta(\varepsilon_3+\varepsilon_4)} + e^{-\beta(\varepsilon_3+\varepsilon_5)} + \cdots \text{etc.} \quad (2.4\text{-}4)$$

Compare this with the square of the single particle partition function q

$$q^2 = \left(e^{-\beta\varepsilon_1} + e^{-\beta\varepsilon_2} + e^{-\beta\varepsilon_3} \ldots\right)^2 = e^{-\beta(2\varepsilon_1)} + 2e^{-\beta(\varepsilon_1+\varepsilon_2)} + 2e^{-\beta(\varepsilon_1+\varepsilon_3)} + \cdots$$
$$+ e^{-\beta(2\varepsilon_2)} + 2e^{-\beta(\varepsilon_2+\varepsilon_3)} + \cdots \text{etc.} \quad (2.4\text{-}5)$$

So if we accept a small error in counting of the states with an occupation number of 2, i.e., $(2, 0, 0, 0, \ldots)$ then

$$Q = \frac{1}{2}q^2 \quad (2.4\text{-}6)$$

This assumption is satisfactory if the number of possible states of a single atom is very much larger than the number of atoms and the energy states are closely spaced, so that the probability of two atoms being in the same energy state is very small. In fact, that will always be the case for systems of interest to us. Also, this result can be generalized for the N particle system to obtain for N identical atoms

$$\boxed{Q = \frac{1}{N!}q^N} \quad (2.4\text{-}7)$$

where the factor $N!$ arises from the indistinguishability of N identical atoms.

It is important to compare and understand the difference between Eq. 2.3-4 and Eq. 2.4-7. In Eq. 2.3-4, the energy modes are independent, but distinguishable (that is, we can tell the difference between a translational motion and a rotation). In that case, the total partition function is the product of the partition functions for each of the energy modes. However, in Eq. 2.4-7, the individual atoms are independent of, but indistinguishable from, each other. The total partition function then is the product of the individual atom partition functions, but now divided by the factor $N!$ as a result of the indistinguishability of the atoms.

Extending this argument to a mixture of N_1 atoms of species 1, N_2 atoms of species 2, etc., we obtain the following general result for a (nonreacting) mixture

$$\boxed{Q(N_1, N_2, \ldots, V, \beta) = \prod_{i=1}^{} \frac{q_i^{N_i}}{N_i!}} \quad \text{nonreacting mixture} \quad (2.4\text{-}8)$$

Since atoms of any one species are indistinguishable from each other, the partition function for each species is $q_i^{N_i}/N_i!$; however, since the atoms of different species are distinguishable, the system partition function is the product of the partition functions for each species. Note also that in Eqs. (2.4-6 to 2.4-8) each of the individual atom partition functions depend upon volume and the still-unknown parameter β. This parameter, which is only a function of the temperature bath and not the system, will be evaluated in the next chapter by considering an especially simple system, the ideal monatomic gas.

CHAPTER 2 PROBLEMS

2.1 For a gas of N-like particles

$$Q = q^N/N!$$

where q is the partition function (sum over states) for one particle, and Q is the N-particle partition function. Show that if the gas consisted of N_1 particles of species 1, and N_2 particles of species 2, the appropriate expression for Q would be

$$Q = \frac{q_1^{N_1} q_2^{N_2}}{N_1! \, N_2!}$$

Here, q_1 and q_2 are the single particle partition functions for species 1 and 2, respectively.

Chapter 3

The Ideal Monatomic Gas

INSTRUCTIONAL OBJECTIVES FOR CHAPTER 3

The goals of this chapter are for the student to:

- Understand the generality of the identification of β with $(kT)^{-1}$
- Understand the ideal gas partition function for a monatomic gas
- Be able to compute the thermodynamic properties of an ideal monatomic gas
- Understand energy fluctuations in the canonical ensemble
- Understand the Gibbs entropy equation
- Understand the origin of the Gibbs paradox and its resolution

3.1 CANONICAL PARTITION FUNCTION FOR THE IDEAL MONATOMIC GAS

The canonical ensemble that we have so far been considering is shown in Fig. 3.1-1. Expressions for the canonical partition function for a system of N identical noninteracting particles are

$$Q = \frac{q^N}{N!} \tag{3.1-1}$$

and

$$q = \sum_{\substack{\text{states } i \text{ of a} \\ \text{single molecule}}} e^{-\beta \varepsilon_i} = \sum_{\substack{\text{energy levels } j \text{ of a} \\ \text{single molecule}}} \omega_j e^{-\beta \varepsilon_j} \tag{3.1-2}$$

The value of the parameter β has yet to be established. From the derivation in Chapter 1, it is clear that β is independent of the macroscopic system and is only a function of the characteristic of the thermal reservoir, which is its temperature. We will now establish the functional relationship between β and T by evaluating the partition function for one particularly simple system—a collection of noninteracting particles in a box, that is, an ideal monatomic gas. Most importantly, since β depends only on the reservoir, and not the system being considered, once its value is established using one system, it is applicable to all other systems. We will use the ideal monatomic gas as the test system to evaluate β since the properties of the ideal gas are known.

The simplest model for an ideal gas is N atoms contained in a cubic box of volume V. The allowed particle energy states are, from quantum mechanics, given by

$$\varepsilon(l_x, l_y, l_z) = \frac{h^2}{8mV^{2/3}}(l_x^2 + l_y^2 + l_z^2) \tag{3.1-3}$$

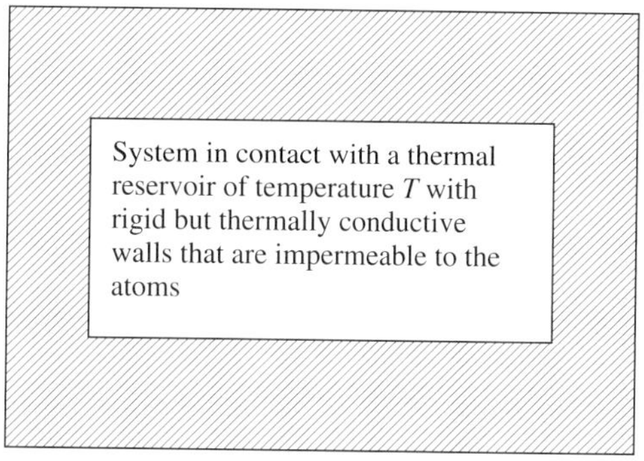

3.1 Canonical Partition Function for the Ideal Monatomic Gas

Figure 3.1-1 The Canonical Ensemble. A system with walls that are rigid (no volume change) and thermally conductive with fixed number of particles in contact with a thermal reservoir of temperature T. Therefore, N, V and T are fixed.

System in contact with a thermal reservoir of temperature T with rigid but thermally conductive walls that are impermeable to the atoms

where l_x, l_y, and l_z are the translational quantum numbers (each of which can take only positive integer values) and h is Planck's constant, which has a value of 6.62517×10^{-27} erg-sec. With these energy states, the single particle partition function is

$$q = \sum_{l_x}\sum_{l_y}\sum_{l_z} e^{-\frac{\beta h^2 (l_x^2+l_y^2+l_z^2)}{8mV^{2/3}}}$$

$$= \left(\sum_{l_x} e^{-\frac{\beta h^2 l_x^2}{8mV^{2/3}}}\right)\left(\sum_{l_y} e^{-\frac{\beta h^2 l_y^2}{8mV^{2/3}}}\right)\left(\sum_{l_z} e^{-\frac{\beta h^2 l_z^2}{8mV^{2/3}}}\right) = q_x q_y q_z \qquad (3.1\text{-}4)$$

Consider for the moment one of these sums,

$$\sum_{l_x=0}^{\infty} e^{-\Delta l_x^2} \qquad (3.1\text{-}5)$$

where $\Delta = \beta h^2/8mV^{2/3}$. The mass of a particle or atom m is of the order of 10^{-22} grams and suppose the system volume is 1 liter, then $V^{2/3}$ is of the order of 10^2 cm^2, and

$$\frac{h^2}{mV^{2/3}} = \mathcal{O}(10^{-33})$$

which has units of ergs, and the symbol $\mathcal{O}$ is used to indicate the order of magnitude. Therefore, if β is not very large—say of the order of 10^{16}, so that Δ is of the order of 10^{-17} or less—the sums in Eq. 3.1-5 can be simply evaluated by replacing them with integrals. To see this note that with Δ about 10^{-17}, for small values of l_x, say l_x less than 10^5, the summand changes very little in value between l_x and $l_x + 1$, so that the summand is almost a continuous variable. For large l_x, say l_x of 10^7 and larger, the relative or fractional change in going from l_x to l_x+1 is very small:

$$\frac{\ell_x+1}{\ell_x} = 1 + \frac{1}{\ell_x} = 1 + 10^{-7}$$

Chapter 3: The Ideal Monatomic Gas

so that the summand again behaves like a continuous variable. Therefore, it is an excellent approximation to write

$$\sum_{l_x} e^{-\beta h^2 l_x^2 / 8mV^{2/3}} = \int_0^\infty e^{-\beta h^2 x^2 / 8mV^{2/3}} \, dx = \sqrt{\frac{2m\pi V^{2/3}}{h^2 \beta}} \quad (3.1\text{-}6)$$

The remaining sums in Eq. 3.1-4 may be evaluated in precisely the same manner to obtain

$$q = \left(\sqrt{\frac{2\pi m V^{2/3}}{h^2 \beta}}\right)^3 = \left(\frac{2m\pi}{h^2 \beta}\right)^{\frac{3}{2}} V \quad (3.1\text{-}7)$$

therefore

$$Q(N, V, \beta) = \frac{q^N}{N!} = \frac{\left(\frac{2\pi m}{h^2 \beta}\right)^{3N/2}}{N!} V^N \quad (3.1\text{-}8)$$

and

$$\ln Q(N, V, \beta) = \frac{3N}{2} \ln\left(\frac{2\pi m}{h^2 \beta}\right) + N \ln V - \ln N! \quad (3.1\text{-}9)$$

The internal energy of the N-particle monatomic gas can then be computed from

$$U = -\left(\frac{\partial \ln Q}{\partial \beta}\right)_{V,N} = \frac{3N}{2\beta} \quad (3.1\text{-}10)$$

3.2 IDENTIFICATION OF β AS $1/kT$

By the ergodic hypothesis, the internal energy of Eq. 3.1-10 must be equal to that which is measured for an ideal gas. Since the internal energy of an ideal monatomic gas is known to be equal to $\frac{3}{2}NkT$, it follows that

$$\boxed{\beta = \frac{1}{kT}} \quad (3.2\text{-}1)$$

Furthermore, β is a universal parameter and only a function of the properties of the reservoir. Since β is only a function of the reservoir, and not the ideal gas system that we have used to determine its value, this identification of β with $(kT)^{-1}$ is always valid regardless of the system being considered. Therefore, Eq. 3.1-2 then becomes

$$\boxed{q = \sum_{\substack{\text{states } j \text{ of a} \\ \text{single molecule}}} e^{-\varepsilon_j / kT}} \quad (3.2\text{-}2)$$

The term $e^{-\varepsilon_j / kT}$ is referred to as the *Boltzmann factor*, and the partition function q is the Boltzmann factor weighted sum over the states available to the system.

Before proceeding further, it is worthwhile checking several of the assumptions we made earlier. One approximation was that the summation in Eq. 3.1-5 could be replaced with an integration. In order for this step to be justified, it was necessary that β not be too large. The Boltzmann constant is $k = 1.38044 \times 10^{-16}$ erg/deg, and T can, in principle, range from 0 K to infinity. Therefore, it is clear that, except at very low temperatures (less than 1 K), β will be less than 10^{16}, and the replacement of the summation with an integration was valid.

Another assumption that was made was that the number of molecular energy states was very much greater than the number of molecules, so that the likelihood of two molecules being in the same energy state was small. This was used in obtaining Eqs. 2.4-6 and 7. One liter of gas at standard temperature and pressure contains about 10^{22} molecules. As a rough estimate, the quantity

$$\frac{h^2}{mV^{2/3}} \sim 10^{-33} \text{ ergs}$$

may be taken as the spacing between energy levels; furthermore, each energy level has a very large degeneracy. Therefore, the number of energy states available to any one of the 10^{22} molecules is much greater than the number of molecules. We return to this question in Section 3.7.

With β now being identified as being $1/kT$, it then follows that the general expression for the canonical partition function is

$$Q(N, V, T) = \sum_{\substack{\text{states} \\ i}} e^{-E_i/kT} = \sum_{\substack{\text{levels} \\ j}} \Omega(E_j) e^{-E_j/kT} \quad \text{(3.2-3)}$$

3.3 GENERAL RELATIONSHIPS OF THE CANONICAL PARTITION FUNCTION TO OTHER THERMODYNAMIC QUANTITIES

We now have the partition function as a function of β or T, the volume, and the number of particles. (Notice that the partition function is a function of volume because the energy levels are a function of volume as seen in Eq. 3.1-3). Also, from Eq. 2.2-3 we have that

$$U = -\left(\frac{\partial \ln Q}{\partial \beta}\right)_V \quad \text{(2.2-3)}$$

Replacing β with $(kT)^{-1}$, it is easily shown that

$$\boxed{U = kT^2 \left(\frac{\partial \ln Q}{\partial T}\right)_{V,N}} \quad \text{(3.3-1a)}$$

or

$$U = \frac{kT^2}{Q} \left(\frac{\partial Q}{\partial T}\right)_{V,N} \quad \text{(3.3-1b)}$$

Since the partition function Q for a fixed number of particles N is a function at T and V, one can write

$$d \ln Q = \left(\frac{\partial \ln Q}{\partial T}\right)_{V,N} dT + \left(\frac{\partial \ln Q}{\partial V}\right)_{N,T} dV$$

Chapter 3: The Ideal Monatomic Gas

or

$$kT^2 \, d\ln Q = kT^2 \left(\frac{\partial \ln Q}{\partial T}\right)_{V,N} dT + kT^2 \left(\frac{\partial \ln Q}{\partial V}\right)_{N,T} dV$$

$$= U\,dT + kT^2 \left(\frac{\partial \ln Q}{\partial V}\right)_{N,T} dV \qquad (3.3\text{-}2)$$

Now using

$$U\,dT = -T^2 d\left(\frac{U}{T}\right) + T\,dU \qquad (3.3\text{-}3)$$

and rearranging Eqs. 3.3-2 and 3.3-3, we have

$$T\,dU = kT^2 \, d\left(\ln Q + \frac{U}{kT}\right) - kT^2 \left(\frac{\partial \ln Q}{\partial V}\right)_{T,N} dV$$

$$dU = kT\,d\left(\ln Q + \frac{U}{kT}\right) - kT \left(\frac{\partial \ln Q}{\partial V}\right)_{T,N} dV$$

$$= kT\,d\left(\ln Q + T\left(\frac{\partial \ln Q}{\partial T}\right)_{V,N}\right) - kT \left(\frac{\partial \ln Q}{\partial V}\right)_{T,N} dV \qquad (3.3\text{-}4)$$

But from the first and second laws of thermodynamics for a closed system, we know that

$$dU = T\,dS - P\,dV \qquad (3.3\text{-}5)$$

Comparing these two equations, we have

$$\boxed{P = kT \left(\frac{\partial \ln Q}{\partial V}\right)_{T,N}} \qquad (3.3\text{-}6)$$

and

$$dS = k\,d\left(\ln Q + \frac{U}{kT}\right) = k\,d\left(\ln Q + T\left(\frac{\partial \ln Q}{\partial T}\right)_{V,N}\right) \qquad (3.3\text{-}7)$$

On integrating this last equation we get

$$S = k\left(\ln Q + T\left(\frac{\partial \ln Q}{\partial T}\right)_{V,N}\right) + C \qquad (3.3\text{-}8a)$$

The constant of integration, C, may be set equal to zero by the third law of thermodynamics, which requires that the entropy of a system go to zero at 0 K. Therefore

$$\boxed{S = k\ln Q + kT \left(\frac{\partial \ln Q}{\partial T}\right)_{V,N}} \qquad (3.3\text{-}8b)$$

3.3 General Relationships of the Canonical Partition Function

Other thermodynamic properties can now be related to the partition function by using the usual equations of thermodynamics together with the relations for U, P, and S obtained above. For example,

$A(N, V, T) =$ Helmholtz energy $= U - TS$

$$= kT^2 \left(\frac{\partial \ln Q}{\partial T}\right)_{V,N} - kT \ln Q - kT^2 \left(\frac{\partial \ln Q}{\partial T}\right)_{V,N} = -kT \ln Q(N, V, T)$$

That is

$$\boxed{A(N, V, T) = -kT \ln Q(N, V, T)} \quad (3.3\text{-}9)$$

and if we have a mixture of particles at different species

$$\left(\frac{\partial A}{\partial N_i}\right)_{T,V,N_{j \neq i}} = \left(\frac{\partial G}{\partial N_i}\right)_{T,P,N_{j \neq i}} = -kT \left(\frac{\partial \ln Q}{\partial N_i}\right)_{T,V,N_{j \neq i}} = \mu_i$$

$$= \text{chemical potential} \quad (3.3\text{-}10)$$

where the chemical potential is the Gibbs energy per molecule. For future reference, we also note that

$$H = U + PV = kT^2 \left(\frac{\partial \ln Q}{\partial T}\right)_{V,N} + kTV \left(\frac{\partial \ln Q}{\partial V}\right)_{T,N} \quad (3.3\text{-}11)$$

$$C_V = \left(\frac{\partial U}{\partial T}\right)_V = 2kT \left(\frac{\partial \ln Q}{\partial T}\right)_{V,N} + kT^2 \left(\frac{\partial^2 \ln Q}{\partial T^2}\right)_{V,N}$$

$$= \frac{2kT}{Q} \left(\frac{\partial Q}{\partial T}\right)_{V,N} + \frac{kT^2}{Q} \left(\frac{\partial^2 Q}{\partial T^2}\right)_{V,N} - \frac{kT^2}{Q^2} \left(\frac{\partial Q}{\partial T}\right)_{V,N}^2 \quad (3.3\text{-}12)$$

and C_P can be obtained from the relation

$$C_P = C_V - T \left(\frac{\partial V}{\partial P}\right)_T \left(\frac{\partial P}{\partial T}\right)_V^2 = C_V - T \frac{\left(\frac{\partial P}{\partial T}\right)_V^2}{\left(\frac{\partial P}{\partial V}\right)_T}. \quad (3.3\text{-}13)$$

It is important to note that the relations of this section are always valid, independent of the system being considered. That is, even though we used the ideal monatomic gas to make the identification of β with $(kT)^{-1}$, that identification is always valid independent of the system being considered; thus, the equations in this section are valid for any system. In this regard, Eq. 3.3-9 provides the following interesting contrast between classical (or macroscopic) thermodynamics and statistical thermodynamics. The Helmholtz energy $A = U - TS$ as a function of N, V, and T is a fundamental equation of state[1] in the terminology of Gibbs in that if we have A as

---

[1] See, for example pp. 202–203 in *Chemical, Biochemical and Engineering Thermodynamics* by S. I. Sandler, John Wiley & Sons, Inc., 2006.

a function of N, V, and T, all other thermodynamic functions can be obtained from linear combinations of A and its derivatives with respect to N, V, and T. However, classical thermodynamics provides no guidance as to how to develop an equation for A as a function of N, V, and T, while statistical thermodynamics through the partition function Q and Eq. 3.3-9 provides the recipe, which is to enumerate all the energy states of the system and then do a Boltzmann factor weighted summation of all those states. This is easily accomplished for a system in which the molecules do not interact, as we show in the next section for the ideal (that is, noninteracting) monatomic gas and in the next chapter for the ideal diatomic and polyatomic gases.

3.4 THE THERMODYNAMIC PROPERTIES OF THE IDEAL MONATOMIC GAS

In the previous section the general equations relating the partition function and various thermodynamic functions were presented. Here we want to use these relationships to develop explicit expressions for the thermodynamic properties of an ideal monatomic gas. Before we can do this, however, we must refine our molecular model. So far, an atom has been considered to be a point mass in a cubic box. In fact, an atom is not merely a point mass, but an entity with a quite complicated electronic and nuclear structure, and there are numerous energy states associated with these internal degrees of freedom. The question that then arises is how these internal energy modes affect the partition function and thermodynamic properties of an ideal gas.

This question is answered by several observations and assumptions. The first of these is the Born-Oppenheimer approximation, which states that the translational (trans) energy states are independent of the electronic (elect) and nuclear (nuc) energy states. The next assumption is that the electronic and nuclear energy states of an atom may also be considered to be independent. Therefore, we can write the energy of an atom as the sum of three completely independent energy modes:

$$\varepsilon = \varepsilon_{\text{trans}} + \varepsilon_{\text{elect}} + \varepsilon_{\text{nuc}} \tag{3.4-1}$$

so that the partition function becomes

$$q = \sum_{\text{states of the atom}} e^{-(\varepsilon_{\text{trans}}+\varepsilon_{\text{elect}}+\varepsilon_{\text{nuc}})/kT} \tag{3.4-2}$$

Now using the independence of the energy states as was discussed in deriving Eq. 2.3-4, it is easily shown that

$$q = q_{\text{trans}} q_{\text{elect}} q_{\text{nuc}} \tag{3.4-3}$$

where

$$q_{\text{trans}} = \sum_{\substack{\text{translational} \\ \text{states i}}} e^{-\varepsilon_{\text{trans},i}/kT}, \quad q_{\text{elect}} = \sum_{\substack{\text{electronic} \\ \text{states i}}} e^{-\varepsilon_{\text{elect},i}/kT} \quad \text{and} \quad q_{\text{nuc}} = \sum_{\substack{\text{nuclear} \\ \text{states j}}} e^{-\varepsilon_{\text{nuc},j}/kT}$$

$$\tag{3.4-4}$$

The first of these partition functions is a sum over the translational energy states, which has already been evaluated in Eq. 3.1-7, using the particle-in-the-box model

3.4 The Thermodynamic Properties of the Ideal Monatomic Gas

for these energy levels, and replacing β with $(1/kT)$

$$q_{\text{trans}} = \left(\frac{2\pi m}{\beta h^2}\right)^{\frac{3}{2}} V = \left(\frac{2\pi m k T}{h^2}\right)^{\frac{3}{2}} V \quad (3.1\text{-}7)$$

The single particle translational partition function is frequently written as

$$q_{\text{trans}} = \frac{V}{\Lambda^3} \quad \text{where} \quad \Lambda = \sqrt{\frac{h^2}{2\pi m k T}} \quad (3.4\text{-}5)$$

is the de Broglie wavelength—that is, the wavelength equivalent of the momentum of a particle in the wave-particle duality theory of matter.[2] [Table 3.4-1 contains a list of the values of several constants used here and elsewhere in this book, several conversion factors, and a simplified formula for the calculation of the single particle translational partition function.]

The electronic partition function cannot be evaluated in such a general manner, since the electronic energy states depend on the electronic structure of the atoms, which is specific to each atomic species.[3] Therefore, the energy states used in the

Table 3.4-1 Constants and Conversion Factors in MKS Units

Constants

Avogadro's number N_{Av}	6.022×10^{23} molecules/mol
Boltzmann's constant k	1.38044×10^{-23} J/K $= 1.38044 \times 10^{-16}$ erg/K
Mass of an electron	9.1094×10^{-31} kg
Planck's constant h	6.6261×10^{-34} J·s
Speed of light (vacuum) c	2.9979×10^8 m/s
Gas constant $= N_{\text{Av}} \times k$	8.314×10^{-5} bar m^3/(mol K)

Conversion Factors

$1\,\text{J} = 1\,\text{kg·m}^2/\text{s}^2$
$1\,\text{eV} = 1.60206 \times 10^{-19}\,\text{J} = 1.60206 \times 10^{-12}\,\text{erg}$
$\quad = 23.0693\,\text{kcal/mol of electrons} = 96.49\,\text{kJ/mol of electrons}$
$1\,\text{Å} = 10^{-8}\,\text{cm}$

Translational partition function

$$\frac{q_{\text{trans}}}{V} = \left(\frac{2\pi m k T}{h^2}\right)^{\frac{3}{2}} = \Lambda^{-3} = \left(\frac{2\pi k}{h^2 N_{\text{Av}}}\right)^{\frac{3}{2}} (MT)^{3/2} = 1.88 \times 10^{20}(MT)^{3/2}\,\text{cm}^{-3}$$
$$= 1.88 \times 10^{26}(MT)^{3/2}\,\text{m}^{-3}$$

where m is the weight of a single atom $= M/N_{\text{Av}}$, M is the molecular weight in grams, T is the temperature (K), $V =$ volume in cm^3 or m^3 and $N_{\text{Av}} =$ Avogadro's number

---

[2] The analysis leading to this relation was developed in de Broghie's Ph.D. thesis in 1924, and for which he was awarded the Nobel Prize in Physics in 1929. See Problem 3.13.

[3] See, for example, the National Institute of Standards and Technology Chemistry WebBook, http://webbook.nist.gov/chemistry/, for spectroscopic data on electronic, rotational, and vibrational energy levels (the latter two needed in the study of polyatomic molecules). Much of the data are in terms of wavelengths or frequencies of emitted radiation and not explicitly in terms of energy levels. To convert a frequency to energy multiply by Planck's constant; also, (speed of light)/(wave length) gives the frequency.

24 Chapter 3: The Ideal Monatomic Gas

partition function calculation are obtained from tables of the electronic energy states. Since such tables usually list the energy levels and the degeneracy[4] of each level, rather than energy states, the partition function is computed as follows:

$$q_{\text{elect}} = \sum_{\substack{\text{electronic} \\ \text{states i}}} e^{-\varepsilon_{\text{elect},i}/kT} = \sum_{\substack{\text{electronic} \\ \text{levels j}}} \omega_{\text{elect},j} e^{-\varepsilon_{\text{elect},j}/kT}$$

$$= \omega_{\text{elect},1} e^{-\varepsilon_{\text{elect},1}/kT} + \omega_{\text{elect},2} e^{-\varepsilon_{\text{elect},2}/kT} + \cdots$$

$$= e^{-\varepsilon_{\text{elect},1}/kT} (\omega_{\text{elect},1} + \omega_{\text{elect},2} e^{-\Delta\varepsilon_{\text{elect},2}/kT} + \cdots) \quad (3.4\text{-}6)$$

where $\Delta\varepsilon_{\text{elect},2} = \varepsilon_{\text{elect},2} - \varepsilon_{\text{elect},1}$. Note that for the noble gases, the ground electronic state degeneracy is 1—that is, $\omega_{\text{elect},1} = 1$, while for alkali metal atoms it is 2.

The electronic energy levels mentioned above are determined from ultraviolet (UV) spectroscopic measurements. The principle of these measurements is that when an atom in the electronic ground state is subjected to UV radiation it may be excited to a higher electronic energy level. The energy difference between the ground state and the excited electronic state can then be determined by the frequency of the wavelength of the adsorbed radiation or the re-emitted radiation as the atom returns to its ground electronic energy level. In this manner, the energy levels of excited states relative to the ground state can be determined. However, the electronic energy of the ground state cannot be obtained by this technique. Furthermore, unless there are changes in the electronic structure (for example, by forming chemical bonds), there is no need to know the absolute energy content of the ground electronic state. (We will reconsider this question later when chemical reactions are studied in Chapter 5.) Therefore, by convention, we chose the energy level of the ground electronic state of an atom to be 0. With this convention, the electronic partition function is

$$q_{\text{elect}} = \omega_{\text{elect},1} + \omega_{\text{elect},2} e^{-\Delta\varepsilon_{\text{elect},2}/kT} + \cdots \quad (3.4\text{-}7a)$$

For most atoms, the energy level of the lowest excited energy level is rather high. For example, this value is 15.76 eV or 1521 kJ/mol for argon. Therefore, at room temperature

$$e^{-\Delta\varepsilon_{\text{elect},2}/kT} = e^{-450} \approx 0$$

Consequently, the degeneracy of the ground state alone is an excellent approximation to the electronic partition function. That is

$$q_{\text{elect}} = \omega_{\text{elect},1} \quad (3.4\text{-}7b)$$

The computation of the nuclear partition function is very similar to that of the electronic partition function, except that the nuclear energy levels are even much more widely spaced than the electronic energy levels, and do not change on chemical reaction. For example, $\Delta\varepsilon_{n2}$, the difference between the ground and first excited

---

[4] The ground state degeneracy of the electronic energy levels of an atom is equal to $2s+1$, where s is total electron spin angular momentum, which involves quantum mechanics that will not be considered here. Suffice it to say that the inert gases have an electronic degeneracy of 1, that of the alkali metals is 2, and oxygen is 3. Values for some atoms and molecules are available in M. Chase et al., JANAF Thermochemical Tables, *J. Phys. Chem. Ref. Data* **14**, Supplement 1 (1985).

3.4 The Thermodynamic Properties of the Ideal Monatomic Gas

nuclear energy states, is much larger than kT unless T is of the order of 10^{10} K. Therefore, for any situation of interest to us, the nuclear partition function can be written as

$$q_{nuc} = \omega_{nuc,1} \qquad (3.4\text{-}8)$$

so that the nuclear partition function is replaced with only the ground nuclear state degeneracy. Since the nuclear energy state of an atom is unchanged for any process we consider, including a chemical reaction, the nuclear partition function will appear only as a multiplicative factor in the total partition function, will not affect any measurable thermodynamic property, and will cancel out of most calculations. Therefore, generally we can set the nuclear partition function equal to unity.

Consequently, the partition function of a collection of N identical noninteracting atoms in a volume V can be written as

$$Q = \frac{\left[\left(\frac{2\pi m kT}{h^2}\right)^{\frac{3}{2}} V (\omega_{elect,1} + \omega_{elect,2} e^{-\Delta\varepsilon_{elect,2}/kT} + \cdots)\omega_{nuc,1}\right]^N}{N!}$$

$$= \frac{\left[\frac{V}{\Lambda^3}(\omega_{elect,1} + \omega_{elect,2} e^{-\Delta\varepsilon_{elect,2}/kT} + \cdots)\omega_{nuc,1}\right]^N}{N!} \qquad (3.4\text{-}9)$$

Using the relations of the previous section, we can now compute the thermodynamic functions for the assembly of noninteracting atoms as follows:

$$P = kT\left(\frac{\partial \ln Q}{\partial V}\right)_{N,T} = \frac{NkT}{V} \text{ (ideal gas law).} \qquad (3.4\text{-}10)$$

Also

$$U = kT^2\left(\frac{\partial \ln Q}{\partial T}\right)_{V,N} = \frac{3}{2}NkT \qquad (3.4\text{-}11)$$

if excited electronic states can be ignored and the electronic partition function can be written as $q_{elect} = \omega_{elect,1}$. However, if the first excited electronic energy level must be considered (generally, only at high temperatures) one obtains (see Problem 3.2)

$$U = \frac{3}{2}NkT + \frac{N\omega_{elect,2}\varepsilon_{elect,2} e^{-\varepsilon_{elect,2}/kT}}{q_{elect}} \qquad (3.4\text{-}12)$$

Using only Eq. 3.4-7b for the electronic partition function, we have

$$C_V = \left(\frac{\partial U}{\partial T}\right)_{N,V} = \frac{3}{2}Nk \qquad (3.4\text{-}13)$$

and

$$A = -kT \ln Q = -kT \ln \frac{q^N}{N!} = -NkT \ln q + kT \ln N! \qquad (3.4\text{-}14)$$

Now using Stirling's approximation (which will be done frequently in this book),

$$\ln N! = N \ln N - N = N \ln N - N \ln e \qquad (3.4\text{-}15)$$

we obtain (neglecting the nuclear partition function)

$$A = -NkT \ln q + NkT \ln N - NkT \ln e = -NkT \ln(qe/N)$$

$$= -NkT \ln \left[\left(\frac{2\pi mkT}{h^2} \right)^{\frac{3}{2}} \frac{Ve\omega_{\text{elect},1}}{N} \right] \quad (3.4\text{-}16)$$

and

$$S = k \ln Q + kT \left(\frac{\partial \ln Q}{\partial T} \right)_{V,N} = Nk \ln \left[\left(\frac{2\pi mkT}{h^2} \right)^{\frac{3}{2}} \frac{Ve^{5/2}\omega_{\text{elect},1}}{N} \right] \quad (3.4\text{-}17)$$

This last equation is referred to as the *Sackur-Tetrode equation*, and was originally derived based on the kinetic theory of gases.
Finally

$$\mu = -kT \left(\frac{\partial \ln Q}{\partial N} \right)_{V,T} = -kT \ln \left[\left(\frac{2\pi mkT}{h^2} \right)^{\frac{3}{2}} \frac{V}{N} \omega_{\text{elect},1} \right] = -kT \ln \left[\frac{q}{N} \right] = g$$
$$(3.4\text{-}18)^5$$

which is the Gibbs energy per molecule g. This expression for the chemical potential is of some interest. In particular, replacing V/N with kT/P using Eq. 3.4-10 and arbitrarily defining a pressure, P_o, to be the *standard state* pressure, then adding and subtracting $kT \ln P_o$ from the expression above and rearranging, we obtain

$$\mu(T, P) = -kT \ln \left\{ \left(\frac{2\pi mkT}{h^2} \right)^{\frac{3}{2}} \frac{kT}{P_o} \omega_{\text{elect},1} \right\} + kT \ln \left(\frac{P}{P_o} \right) \quad (3.4\text{-}19)$$

which is (in the form familiar to chemists and engineers)

$$\mu(T, P) = \mu_o(T, P_o) + kT \ln \left(\frac{P}{P_o} \right) \quad (3.4\text{-}20)$$

where

$$\mu_o(T, P_o) = -kT \ln \left[\left(\frac{2\pi mkT}{h^2} \right)^{\frac{3}{2}} \frac{kT}{P_o} \omega_{\text{elect},1} \right]$$

When using Eq. 3.4-20, one must remember that μ_o is a function of both temperature and the standard state pressure, and that this equation is only applicable to an ideal gas. For the case of the ideal monatomic gas considered here, from statistical mechanics we have obtained the temperature and pressure dependence of the chemical potential as given by Eq. 3.4-19. Furthermore, we have also obtained an explicit expression from which it is possible to calculate a numerical value for the standard

---

[5]This chemical potential is on a per-molecule basis. For a per-mole basis, multiply by Avogadro's number, or equivalently replace the Boltzmann constant with the gas constant.

3.4 The Thermodynamic Properties of the Ideal Monatomic Gas

state chemical potential:

$$\mu_o = -kT \ln \left[\left(\frac{2\pi mkT}{h^2} \right)^{\frac{3}{2}} \frac{kT}{P_o} \omega_{\text{elect},1} \right] \tag{3.4-21}$$

It is useful to discuss the units of the properties that are calculated using the equations above. The internal energy U will be depend on the units used for the Boltzmann constant and will be the total energy for N molecules. If Avogadro's number of molecules is used for N, then U will be energy per mole of molecules. A similar comment applies to the entropy, Helmholtz and Gibbs energies, the constant volume and constant pressure heat capacities, and the enthalpy. The chemical potential denoted by μ or g is on a per-molecule basis. These should be multiplied by Avogadro's number to obtain a value on a per-mole basis.

ILLUSTRATION 3.4-1

Compute the thermodynamic properties of 1 mole of argon at 300 K and 1 bar.

SOLUTION

Using the values of the parameters in Table 3.4-1 and the equations of this section we obtain

$$V = NRT/P = 1 \text{ mole} \times 8.314 \times 10^{-5} \frac{\text{bar} \cdot \text{m}^3}{\text{mole} \cdot \text{K}} \times 300 \text{ K}/1 \text{bar} = 0.025 \text{ m}^3$$

$$q = \left(\frac{2\pi mkT}{h^2} \right)^{\frac{3}{2}} V\omega_{\text{elect},1} = 1.88 \times 10^{26} (MT)^{3/2} \text{ m}^{-3} V \times 1$$

$$= 1.88 \times 10^{26} (39.945 \times 300)^{3/2} \times 0.025$$

$$= 6.146 \times 10^{30}$$

$$A = -NkT \ln \left[\frac{qe}{N} \right] = -RT \ln \left[\frac{qe}{N} \right] = -4.276 \times 10^4 \text{ joule/mol}$$

$$\mu = -kT \ln \left[\frac{q}{N} \right] = -4.207 \times 10^4 \text{ joule/mol}$$

$$S = Nk \ln \left[\left(\frac{2\pi mkT}{h^2} \right)^{\frac{3}{2}} \frac{Ve^{5/2} \omega_{\text{elect},1}}{N} \right] = Nk \ln \left[\frac{qe^{5/2}}{N} \right] = 155.01 \frac{\text{joule}}{\text{mol} \cdot \text{K}}$$

$$U = 3NkT/2 = 3.743 \times 10^3 \text{ joule/mol}$$

$$C_V = 3Nk/2 = 12.475 \frac{\text{joule}}{\text{mol} \cdot \text{K}}$$

$$H = U + PV = U + RT = 6.238 \times 10^3 \text{ joule/mol}$$

and

$$C_P = C_V + R = 20.792 \frac{\text{joule}}{\text{mol} \cdot \text{k}}$$

As a check, $S = \dfrac{U - A}{T} = \dfrac{3.743 \times 10^3 - (-4.276 \times 10^4) \text{ joule/mol}}{300\text{K}}$

$$= 155.01 \frac{\text{joule}}{\text{mol} \cdot \text{K}}$$

ILLUSTRATION 3.4-2

The following information is available about the first four electronic excited states of argon.

State	$\Delta\varepsilon$, eV	ω_{e_i}
$i = 1$	0	1
$i = 2$	11.548	5
$i = 3$	11.633	3
$i = 4$	11.723	1
$i = 5$	11.828	3

Although we cannot compute the absolute probability of occurrence of each of these excited levels (since we do not have information on all the excited states needed to compute the electronic partition function), we can compute the relative probabilities of occurrence. That is, to compute the absolute probability of energy level j, the following equation would be used:

$$p(\varepsilon_{\text{elect},i}) = \frac{\omega_{e_i} e^{-\varepsilon_{\text{elect},i}/kT}}{\sum_{\substack{\text{electronic} \\ \text{energy} \\ \text{states } j}} \omega_{e_j} e^{-\varepsilon_{\text{elect},j}/kT}}$$

We do not have the information needed to evaluate the electronic partition function in the denominator. However, we can compute the relative probability of occurrence of any two energy levels using

$$\frac{p(\varepsilon_{\text{elect},i})}{p(\varepsilon_{\text{elect},j})} = \frac{\omega_{\text{elect},i} e^{-\varepsilon_{\text{elect},i}/kT}}{\omega_{\text{elect},j} e^{-\varepsilon_{\text{elect},j}/kT}}$$

and in particular, the relative probability of occurrence of any energy level compared to the ground state is obtained from

$$\frac{p(\varepsilon_{\text{elect},i})}{p(\varepsilon_{\text{elect},1})} = \frac{\omega_{\varepsilon_{\text{elect},i}} e^{-\varepsilon_{\text{elect},i}/kT}}{\omega_{\varepsilon_{\text{elect},1}} e^{-\varepsilon_{\text{elect},0}/kT}} = \frac{\omega_{\varepsilon_{\text{elect},i}} e^{-\varepsilon_{\text{elect},i}/kT}}{1 e^{-0/kT}} = \omega_{\varepsilon_{\text{elect},i}} e^{-\varepsilon_{\text{elect},i}/kT}$$

The probability of occurrence of each of these energy levels relative to the ground state at 10000, 20000 and 30000 K are given below.

Level	$\Delta\varepsilon_{\text{elect}}$, eV	$\omega_{\varepsilon_{\text{elect},i}}$	$\omega_{\text{elect},i} e^{-\varepsilon_{\text{elect},i}/kT}$ $T = 10000$ K	$\omega_{\text{elect},i} e^{-\varepsilon_{\text{elect},i}/kT}$ $T = 20000$ K	$\omega_{\text{elect},i} e^{-\varepsilon_{\text{elect},i}/kT}$ $T = 30000$ K
$i = 1$	0	1	1	1	1
$i = 2$	11.548	5	7.520×10^{-6}	6.132×10^{-3}	5.729×10^{-2}
$i = 3$	11.633	3	4.088×10^{-6}	3.502×10^{-3}	3.326×10^{-2}
$i = 4$	11.723	1	1.227×10^{-6}	1.108×10^{-3}	1.071×10^{-2}
$i = 5$	11.828	3	3.260×10^{-6}	3.127×10^{-3}	3.082×10^{-2}
Degree of ionization at 1 atm			0.0120	0.941	~ 1.0

Also shown in the table is the degree of ionization at each temperature—that is the fraction of atoms for which an electron has jumped out of all possible orbitals to form an argon ion and free electron. The energy of ionization is 15.76 eV = 1520.6 kJ/mol, which is considerably higher than any of the excited states considered here. Therefore, looking at the results in the table, it may seem surprising that at the higher temperatures argon is either completely or almost completely ionized, even though the relative populations of the excited states are quite low. The explanation has to do with the degeneracy of the ionized state. Though we have only considered the electronic states here, each particle also has a range of translational states. As a result of ionization, there are now two sets of translational states available—those for the ion (which are essentially the same as for the atom, since the masses are almost identical), and also those for the electron, which are new. Since translational energy states are very closely spaced, the degeneracy of the ionized state, because of all the translational states available to both ion the electron, is enormous. Consequently, even though the likelihood of any one ionized state is small as a result of the large energy in the Boltzmann factor, the degeneracy multiplying this factor (which is the product of the electron and ion degeneracies) is so large that ionization is a likelier state than any of the excited atomic states.

3.5 ENERGY FLUCTUATIONS IN THE CANONICAL ENSEMBLE

In the calculations we have done so far, we have computed the average value of thermodynamic properties—for example, the average energy of a system in contact with a bath at fixed temperature. Since energy can be continually transferred between the system and the bath, it of interest to estimate the extent of the fluctuations of energy that are probable. To proceed, we will assume that the distribution of possible energy states around the average value is given by a normal or Gaussian distribution.

The Gaussian distribution is

$$f(x) = \frac{1}{\sigma\sqrt{2\pi}} e^{-\frac{1}{2}\left(\frac{x-\mu}{\sigma}\right)^2} \tag{3.5-1}$$

where

σ = standard deviation which is a measure of the breadth of the distribution

μ = mean of the distribution

x = dimensionless variable that can take on any value between $-\infty$ and $+\infty$

This distribution is normalized; that is, the integral overall values of x is unity as shown below:

$$\int_{-\infty}^{+\infty} \frac{1}{\sigma\sqrt{2\pi}} e^{-\frac{1}{2}\left(\frac{x-\mu}{\sigma}\right)^2} dx = \int_{-\infty}^{+\infty} \frac{1}{\sigma\sqrt{2\pi}} e^{-\frac{1}{2}\left(\frac{x-\mu}{\sigma}\right)^2} d(x-\mu)$$

$$= \frac{1}{\sqrt{\pi}} \int_{-\infty}^{+\infty} e^{-\frac{1}{2}\left(\frac{x-\mu}{\sigma}\right)^2} d\left\{\frac{1}{\sqrt{2}}\left(\frac{x-\mu}{\sigma}\right)\right\} \tag{3.5-2}$$

Chapter 3: The Ideal Monatomic Gas

Now letting $y = \frac{1}{\sqrt{2}}\left(\frac{x-\mu}{\sigma}\right)$ we obtain

$$\int_{-\infty}^{+\infty} \frac{1}{\sigma\sqrt{2\pi}} e^{-\frac{1}{2}\left(\frac{x-\mu}{\sigma}\right)^2} dx = \frac{1}{\sqrt{\pi}} \int_{-\infty}^{+\infty} e^{-y^2} dy$$

$$= \frac{2}{\sqrt{\pi}} \int_0^{\infty} e^{-y^2} dy = \frac{2}{\sqrt{\pi}} \cdot \frac{\sqrt{\pi}}{2} = 1 \quad (3.5\text{-}3)$$

Also, the average value of any function $G(x)$, represented as $\overline{G}$, is obtained from

$$\overline{G} = \int_{-\infty}^{+\infty} G(x) f(x)\, dx = \frac{1}{\sigma\sqrt{2\pi}} \int_{-\infty}^{+\infty} G(x) e^{-\frac{1}{2}\left(\frac{x-\mu}{\sigma}\right)^2} dx$$

Finally,

$$\sigma^2 = \overline{(\mu - x)^2} = \overline{(\overline{x} - x)^2} = \overline{x^2} - (\overline{x})^2 = \text{variance of the distribution} \quad (3.5\text{-}4)$$

where each overbar indicates the average value, here of the square of the differences (or fluctuations) of the instantaneous value of x from the average value $\mu = \overline{x}$. For reference, we note that for this distribution, the probability of finding a system in a state in which x is between $\mu - 0.674\sigma \leq x \leq \mu + 0.674\sigma$ is 50 percent, for $\mu - \sigma \leq x \leq \mu - \sigma$ is 68.27 percent, for $\mu - 3\sigma \leq x \leq \mu - 3\sigma$ is 99.73 percent, and for $\mu - 5\sigma \leq x \leq \mu - 5\sigma$ is 99.99994 percent.

With this as background, we can now inquire as to the extent of the fluctuations in energy in the canonical ensemble by assuming that the energy probability distribution is Gaussian. We will examine the case of energy fluctuations for the ideal monatomic gas (neglecting electronic and nuclear energy states) for which we know

$$\overline{E} = U = \frac{3}{2} NkT \quad \text{and} \quad C_V = \frac{3}{2} Nk$$

However, as the variable in the Gaussian distribution is dimensionless, we use as the variables

$$x = \frac{E}{3NkT/2} \quad \text{and} \quad \overline{x} = \frac{\overline{E}}{3NkT/2} = \frac{U}{3NkT/2} = 1$$

Now to obtain the standard deviation, which determines the extent of the energy fluctuations in energy in the canonical ensemble, we examine the quantity

$$\overline{\left(\frac{\overline{E} - E}{3NKT/2}\right)^2} = \frac{\overline{(\overline{E}^2 - 2E\overline{E} + E^2)}}{(3NkT/2)^2} = \frac{\overline{E}^2 - 2\overline{E}\cdot\overline{E} + \overline{E^2}}{(3NkT/2)^2}$$

$$= \frac{\overline{E}^2 - 2\overline{E}^2 + \overline{E^2}}{(3NkT/2)^2} = \frac{\overline{E^2} - \overline{E}^2}{(3NkT/2)^2} = \sigma^2$$

3.5 Energy Fluctuations in the Canonical Ensemble

$$\sigma^2 = \frac{1}{(3NkT/2)^2} \left[\frac{\sum\limits_{\substack{\text{states}\\i}} E_i^2 e^{-E_i/kT}}{Q(N,V,T)} - \frac{\left(\sum\limits_{\substack{\text{states}\\i}} E_i e^{-E_i/kT}\right)^2}{(Q(N,V,T))^2} \right] \tag{3.5-6}$$

$$= \frac{1}{(3NkT/2)^2} \left[\frac{(kT^2)^2}{Q} \left(\frac{\partial^2 Q}{\partial T^2}\right)_{N,V} + 2\frac{(kT)^2 T}{Q}\left(\frac{\partial Q}{\partial T}\right)_{N,V} - \frac{(kT^2)^2}{Q^2}\left(\frac{\partial Q}{\partial T}\right)^2_{N,V} \right]$$

This is to be compared with

$$C_V = \frac{2kT}{Q}\left(\frac{\partial Q}{\partial T}\right)_{V,N} + \frac{kT^2}{Q}\left(\frac{\partial^2 Q}{\partial T^2}\right)_{V,N} - \frac{kT^2}{Q^2}\left(\frac{\partial Q}{\partial T}\right)^2_{V,N}, \tag{3.5-7}$$

therefore,

$$\sigma^2 = \frac{kT^2 C_V}{(3NkT/2)^2} = \frac{kT^2(3Nk/2)}{(3NkT/2)^2} = \frac{2}{3N} \quad \text{and} \quad \sigma = \sqrt{\frac{2}{3N}}. \tag{3.5-8}$$

While the constant $\sqrt{2/3}$ is specific to the ideal monatomic gas, the inverse dependence on the square root of the number of particles is quite general for fluctuations. Consequently, the variance in energy as a fraction of total system energy decreases as the number of particles in the system increases. That is, the larger the number of molecules in the system, the smaller the percentage fluctuation in any of its properties. In fact, the dependence of the size of the fluctuations on particle number (or system size) is a quite general result.

To determine the likelihood of fluctuations in energy for the ideal monatomic gas, we can now write the probability distribution as

$$p(E) = \sqrt{\frac{3}{4\pi}N} e^{-\frac{1}{2}\left(\left(\frac{E}{\frac{3}{2}NkT}-1\right)\sqrt{\frac{3}{2}N}\right)^2} = \sqrt{\frac{3}{4\pi}N} e^{-\frac{1}{2}\left((\delta-1)\sqrt{\frac{3}{2}N}\right)^2} \tag{3.5-9}$$

where $\delta = \frac{E}{3NkT/2}$ and the probability density of the system having exactly the average energy $\overline{E} = \frac{3}{2}NkT$ is

$$p\left(E = \frac{3}{2}NkT\right) = \sqrt{\frac{3}{4\pi}N} \tag{3.5-10a}$$

so that

$$p(E) = p\left(E = \frac{3}{2}NkT\right) e^{-\frac{1}{2}\left((\delta-1)\sqrt{\frac{3}{2}N}\right)^2} \tag{3.5-10b}$$

32 Chapter 3: The Ideal Monatomic Gas

Now, consider the likelihood that in a system that consists of 1 mole (6.02×10^{23} molecules), the average energy of the system departs by 0.01 percent from the average value—that is, that the energy of the system is $1.0001 \times \frac{3}{2}NkT$ instead of $\frac{3}{2}NkT$, so that $\delta = 1.0001$. The probability of such a fluctuation relative to that of the system having the most likely value of the energy is

$$\frac{p\left(E = 1.0001 \frac{3}{2} NkT\right)}{p\left(E = \frac{3}{2} NkT\right)} = e^{-\frac{1}{2}\left(0.0001\sqrt{\frac{3}{2}N}\right)^2} = e^{-\frac{1}{2}(9.503 \times 10^7)^2}$$

$$= e^{-4.515 \times 10^{15}} \tag{3.5-11}$$

which is a very, very small number, even though the fluctuation considered is only 0.01 percent. This is because of the large number of molecules involved.

To understand the implication of the number of molecules in the system on the likely magnitude of the fluctuations, note that for the Gaussian distribution, there is a 50 percent probability that the system energy is within $\pm 0.674\sigma$ of its average value, and a 99.99 percent probability that the system energy is within $\pm 3.891\sigma$ of the average value. The percentage fluctuations at these levels of probability are shown in the table below for systems of various sizes. The observation that we can make from this table is that the possible magnitude of the energy fluctuations for a single particle is very large (indeed larger than the average value of the energy), while it is immeasurably small for a system consisting of one mole of molecules. In particular, if the average energy of one mole of an ideal gas system was 1 kJ, there is a 99.99 percent that the system would be in states such that its instantaneous energy was between $(1 - 4.09 \times 10^{-12})$ kJ and $(1 + 4.09 \times 10^{-12})$ kJ. One of the reasons statistical mechanics is useful is that for real systems with large numbers of molecules, the fluctuations from the average properties are so small as to be negligible.

	$\sigma_E / \overline{E}$			
Number of particles	1	10	100	6.022×10^{23} (1 mole)
50 percent likelihood	0.5503	0.1740	0.0550	7.09×10^{-14}
0.01 percent likelihood	3.177	1.005	0.3177	4.09×10^{-12}

Examples of how the probability distribution of energy fluctuations changes with the number of molecules is shown in Figures 3.5-1. Note the scale changes in the y-axis of these figures—and that δ is the fractional change from the average energy, that is, $\delta = \frac{E}{(3NkT/2)} - 1$.

It is interesting to note that, quite generally, the fluctuations or the statistical uncertainty of averages obtained from a sample of size N decreases as $1/\sqrt{N}$. The public sees this reported most frequently around election times, when the popular press reports the results of election polls. It is common to see a statement that 1000 voters were polled and that the reported accuracy of the poll is 3 percent. This estimate arises from $1/\sqrt{1000} = 1/31.6 = 0.0316$ or 3.16%, which is slightly over the reported 3 percent. However, the underlying accuracy of a statistical result also depends on there being no bias in the sampling. A good example of inaccuracy resulting from a biased sample was the predictions for the 1948 presidential election between the Democrat Harry S. Truman and the Republican Thomas E. Dewey. Most polls predicted Dewey would be the overwhelming winner, and one newspaper even incorrectly reported

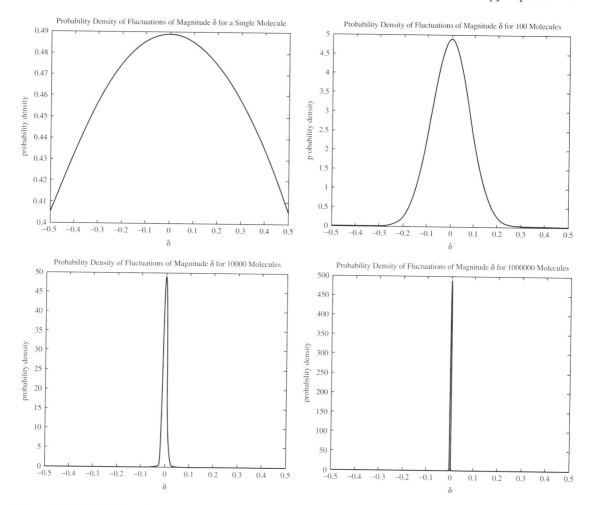

Figure 3.5-1 Probability of Energy Fluctuations as a Function of the Number of Molecules.

that as their initial post-election day headline. The problem was that the pre-election polling was the first to be largely done by telephone; but in 1948, only the wealthier people had telephones (hard to believe now), and weathier people tend to vote Republican. Consequently, the poll was biased in that Republican voters were more likely to be sampled than Democratic voters, which explains the incorrect prediction of the pollsters.

3.6 THE GIBBS ENTROPY EQUATION

It is of both interest and later utility to note that the expression for the entropy

$$S = k \left(\ln Q + T \left(\frac{\partial \ln Q}{\partial T} \right)_{N,V} \right) \qquad \text{(3.3-8b)}$$

can be written in another form. In particular, rewriting the relation between the probability of occurrence of a state and the partition function of Eq. 1.5-13 in terms

34   Chapter 3: The Ideal Monatomic Gas

of temperature instead of β, we have

$$p(N, V, T, E) = \frac{e^{-E/kT}}{Q(N, V, T)}$$

Notice that

$$\left(\frac{\partial \ln Q}{\partial T}\right)_{N,V} = \frac{1}{Q}\left(\frac{\partial Q}{\partial T}\right)_{N,V} = \frac{1}{Q}\left(\frac{\partial \sum_{\text{states } j} e^{-\frac{E_j}{kT}}}{\partial T}\right)_{N,V}$$

$$= \frac{1}{Q}\left(\sum_{\text{states } j} e^{-\frac{E_j}{kT}} \frac{E_j}{kT^2}\right)_{N,V} = \frac{U}{kT^2} \quad (3.6\text{-}1)$$

$$\sum_{\text{states } j} p(N, V, E_j) \ln p(N, V, E_j) = \sum_{\text{states } j} \frac{e^{-E_j/kT}(\ln e^{-E_j/kT} - \ln Q(N, V, T))}{Q(N, V, T)}$$

$$= \sum_{\text{states } j} \frac{e^{-E_j/kT}\left(\frac{-E_j}{kT} - \ln Q(N, V, T)\right)}{Q(N, V, T)} = \frac{-1}{kT}\sum_{\text{states } j} \frac{E_j e^{-E_j/kT}}{Q(N, V, T)} - \ln Q(N, V, T)$$

$$\quad (3.6\text{-}2)$$

$$= -\frac{U}{kT} - \ln Q(N, V, T) = -T\left(\frac{\partial \ln Q}{\partial T}\right)_{N,V} - \ln Q(N, V, T)$$

Consequently, comparing Eq. 3.3-8b with Eq. 3.6-2, we find that

$$S = k\left(\ln Q + T\left(\frac{\partial \ln Q}{\partial T}\right)_{N,V}\right) = -k\sum_{\text{states } j} p(N, V, E_j) \ln p(N, V, E_j) \quad (3.6\text{-}3)$$

This interesting result, known as the Gibbs entropy equation, relates the entropy to the probability of occurrence of the possible states of the system. For example, if there was only one state possible for the system, its probability of occurrence would be unity, and

$$S = -k\sum_{\text{states } j} p(N, V, E_j) \ln p(N, V, E_j) = -k(1 \times \ln(1)) = 0 \quad (3.6\text{-}4)$$

If there were two equally probable states, so that $p = 0.5$, then

$$S = -k(0.5 \times \ln(0.5) + 0.5 \times \ln(0.5)) = -k \ln(0.5) = 0.693\,k \quad (3.6\text{-}5)$$

if there are three equally probable states

$$S = -k\left(\frac{1}{3} \times \ln\left(\frac{1}{3}\right) + \frac{1}{3} \times \ln\left(\frac{1}{3}\right) + \frac{1}{3} \times \ln\left(\frac{1}{3}\right)\right) = -k \ln\left(\frac{1}{3}\right) = 1.0986\,k$$

$$\quad (3.6\text{-}6)$$

and finally, if there were ten equally likely states

$$S = -k \ln\left(\frac{1}{10}\right) = 2.303\,k \qquad (3.6\text{-}7)$$

In a similar manner, if there are more equally probable states possible, the entropy will increase further. (If the states are not equally probable, such as when considering states of different energy, the calculation of the entropy from the number of states available is slightly more complicated).

Generally, as the number of states available to a system increases, the uncertainty as to which state the system occupies increases and the entropy defined in terms of probability increases. Therefore, a statistical interpretation of entropy is that this function is related to the uncertainty of knowledge about the state of the system. This interpretation has been used in the science of information theory, where the entropy function is used as a measure of uncertainty contained in a communication that is imperfectly transmitted.

3.7 TRANSLATIONAL STATE DEGENERACY

Throughout the discussion so far, it has been assumed that the degeneracy of the translational states of a single molecule is very large, so there was only a small probability that two molecules occupied the same translational quantum state. Here we verify this assumption in a slightly approximate way. The starting point is the expression for the allowed translational energy states

$$\varepsilon(l_x, l_y, l_z) = \frac{h^2}{8mV^{2/3}}(l_x^2 + l_y^2 + l_z^2) = \frac{h^2}{8mV^{2/3}}R^2 \qquad (3.7\text{-}1)$$

where $R^2 = (l_x^2 + l_y^2 + l_z^2)$. Now note that the volume of a quadrant of a sphere with all quantum numbers l positive is

$$\text{volume} = \frac{\pi R^3}{6} = \frac{\pi}{6}(l_x^2 + l_y^2 + l_z^2)^{\frac{3}{2}} = \frac{\pi}{6}\left(\frac{8mV^{\frac{2}{3}}\varepsilon}{h^2}\right)^{\frac{3}{2}} = \frac{\pi}{6}\left(\frac{8m}{h^2}\right)^{\frac{3}{2}} V\varepsilon^{\frac{3}{2}} \qquad (3.7\text{-}2)$$

Here, each set of three quantum numbers represents a point within the quadrant of the sphere, and if the quantum numbers were continuous (which is how we will treat them here), this volume would be equal to the sum of all sets of quantum numbers representing translational energies between zero and ε. Now treating the volume and quantum numbers as continuous variables, we can (by differentiation) determine the change in volume with, for example, the quantum number l_x, and in this way estimate the degeneracy of that state. That is

$$\frac{d\,\text{volume}}{dl_x} = \frac{3}{2}\frac{\pi}{6}(l_x^2 + l_y^2 + l_z^2)^{\frac{1}{2}} \cdot 2l_x = \frac{\pi}{2}(l_x^2 + l_y^2 + l_z^2)^{\frac{1}{2}} \cdot l_x = \frac{\pi}{2}\left(\frac{8mV^{\frac{2}{3}}\varepsilon}{h^2}\right)^{\frac{1}{2}} \cdot l_x$$

$$(3.7\text{-}3)$$

Now, assuming approximate equipartition of energy between the three translational degrees of freedom

36 Chapter 3: The Ideal Monatomic Gas

$$\frac{1}{3}\varepsilon = \frac{h^2}{8mV^{\frac{2}{3}}}l_x^2 \quad \text{or} \quad l_x = \left(\frac{8mV^{\frac{2}{3}}\varepsilon}{3h^2}\right)^{\frac{1}{2}} \tag{3.7-4}$$

and

$$\frac{d\text{volume}}{dl_x} = \frac{\pi}{2}\left(\frac{8mV^{\frac{2}{3}}\varepsilon}{h^2}\right)^{\frac{1}{2}} \cdot l_x = \frac{\pi}{2}\left(\frac{8mV^{\frac{2}{3}}\varepsilon}{h^2}\right)^{\frac{1}{2}} \cdot \left(\frac{8mV^{\frac{2}{3}}\varepsilon}{3h^2}\right)^{\frac{1}{2}}$$

$$= \frac{\pi}{2\sqrt{3}}\left(\frac{8mV^{\frac{2}{3}}\varepsilon}{h^2}\right) \tag{3.7-5}$$

which is the degeneracy of the energy level l_x.

To obtain a numerical value for the single particle degeneracy, consider the example of one atom of argon in a one-liter box at 300 K. Using the expressions above, we find that the quantum number corresponding to the average energy is $l_x = 5.002 \times 10^9$ and that the degeneracy of this one quantum state corresponding to the average energy of the system is 6.808×10^{19}. These results establish two things. First, that for translations, the values of the quantum numbers involved are so large that we can—as we have done—treat the quantum numbers as continuous variables, since the fractional change in going from one quantum state to the next over the range of interest is so small as to be infinitesimal. Second, in addition to the very large number of quantum states available to a molecule (given the large values of the quantum numbers), the degeneracy of each state that would be of interest is so enormous that the probability of two molecules occupying the same quantum state is infinitesimal, as we have assumed.

It is also possible to estimate the number of spatial configurations available to molecules. This is most easily done by considering position to be a quantized variable and dividing the volume under consideration into a collection of n small, equal-size cubes rather than treating position as a continuous variable. (The number of available states will then depend on the size of each cube; however, as we are interested here in only ratios of numbers of states in different volumes, the size of the cube will cancel out of the final result.) To begin, first consider the number of positional states available to a single molecule in a volume V. The number of different positional states is equal to the number of (arbitrarily) chosen cubes. That is, if v is the volume of each small cube, then the number of possible locations of the molecule is just equal the number of cubes is $n = V/v$, since the likelihood of occupation of each cube is the same. For a system of the N noninteracting molecules, the number of different ways of distributing the molecules among the cubes allowing each cube to be occupied by any number of molecules (since we have assumed that there is no interaction among the molecules) is $n^N = (V/v)^N$.

With this as background, we consider the following question. Suppose there is a system of 1 mole noninteracting molecules in a volume V. What is the number states available to these molecules if they are restricted to one-half the volume V compared to the number of states if the molecules can be anywhere in the total volume?

Since the molecules do not interact, the energy of the system is the same whether the molecules are in the volume V or (somehow) clustered in $V/2$. Therefore, using the same cube volume v in both cases, the ratio of the number of possible states in the two cases is

$$\frac{\text{Number of positional states in volume } V/2}{\text{Number of positional states in volume } V} = \left(\frac{V/2v}{V/v}\right)^{N_{Av}} = \left(\frac{1}{2}\right)^{6.022 \times 10^{23}}$$

(3.7-6)

This is such an extremely small number that its numeric value is not easily computed (for reference, $0.5^{40} = 9 \times 10^{-13}$, and 40 is much less than 6.022×10^{23}). What this means is that, although it is theoretically possible for all the molecules to be clustered in one-half the volume V at some instant of time, the likelihood of that occurring is so remote that it will never be observed in the laboratory.

3.8 DISTINGUISHABILITY, INDISTINGUISHABILITY, AND THE GIBBS' PARADOX

There is an interesting paradox in classical thermodynamics that is resolved here using statistical thermodynamics. Consider a closed vessel separated into two compartments by a barrier. Both compartments contain ideal gases at the same temperature and pressure. Compartment A has a volume of V_A and contains $N_A = PV_A/kT$ atoms of species A, and compartment B has a volume of V_B and contains $N_B = PV_B/kT$ atoms of species B. When the barrier is removed and the species in the compartments have mixed, the entropy change for the process from classical thermodynamics is

$$S_f - S_i = (N_A + N_B)s_{\text{mix}} - N_A s_A - N_B s_B = N_A k \ln \frac{V_A + V_B}{V_A} + N_B k \ln \frac{V_A + V_B}{V_B}$$
$$= N_A k \ln \frac{N_A + N_B}{N_A} + N_B k \ln \frac{N_A + N_B}{N_B}$$

(3.8-1)

In this equation, the subscripts i and f refer the initial and final states, S is the total entropy, and s is the entropy per atom.

Now consider the case in which species A and B are identical. In this case, when the barrier is removed, there is no change in the entropy of the system; however, Eq. 3.8-1 predicts a positive entropy change. This is the paradox of Gibbs.

Statistical mechanics provides some insight into the origin of this paradox. The partition function for system of two different species separated by the impermeable membrane is (for simplicity, taking all the electronic degeneracies to be unity)

$$Q(N_A, N_B, V_A, V_B, T) = \frac{\left(\frac{2\pi m_A kT}{h^2}\right)^{\frac{3N_A}{2}} V_A^{N_A}}{N_A!} \frac{\left(\frac{2\pi m_B kT}{h^2}\right)^{\frac{3N_B}{2}} V_B^{N_B}}{N_B!}$$

(3.8-2)

and if A and B were identical (for example, species B transforms into species A) the partition function of Eq. 3.8-2, which we denote by Q', is

$$Q'(N_A, N_B, V_A, V_B, T) = \frac{\left(\frac{2\pi m_A kT}{h^2}\right)^{\frac{3(N_A+N_B)}{2}} V_A^{N_A} V_B^{N_B}}{N_A! N_B!}$$

(3.8-3)

38 Chapter 3: The Ideal Monatomic Gas

The partition function for the system with the partition removed is

$$Q(N_A, N_B, V_A + V_B, T)$$

$$= \frac{\left(\frac{2\pi m_A kT}{h^2}\right)^{\frac{3N_A}{2}} (V_A + V_B)^{N_A}}{N_A!} \frac{\left(\frac{2\pi m_B kT}{h^2}\right)^{\frac{3N_B}{2}} (V_A + V_B)^{N_B}}{N_B!} \quad (3.8\text{-}4)$$

Now, letting the atoms become identical (for example, by transforming B into A) the partition function becomes

$$Q'(N_A, N_B, V_A + V_B, T) = \frac{\left(\frac{2\pi m_A kT}{h^2}\right)^{\frac{3(N_A+N_B)}{2}} (V_A + V_B)^{N_A+N_B}}{N_A! N_B!} \quad (3.8\text{-}5)$$

But this is incorrect, since for $N_A + N_B$ identical atoms in a volume $V_A + V_B$, the correct partition function for this system is

$$Q''(N_A, N_B, V_A + V_B, T) = \frac{\left(\frac{2\pi m_A kT}{h^2}\right)^{\frac{3(N_A+N_B)}{2}} (V_A + V_B)^{N_A+N_B}}{(N_A + N_B)!} \quad (3.8\text{-}6)$$

Now note that

$$\frac{Q'(N_A, N_B, V_A + V_B, T)}{Q(N_A, N_B, V_A, V_B, T)} = \frac{\dfrac{\left(\frac{2\pi m_A kT}{h^2}\right)^{\frac{3(N_A+N_B)}{2}} (V_A + V_B)^{N_A+N_B}}{N_A! N_B!}}{\dfrac{\left(\frac{2\pi m_A kT}{h^2}\right)^{\frac{3(N_A+N_B)}{2}} V_A^{N_A} V_B^{N_B}}{N_A! N_B!}}$$

$$= \frac{(V_A + V_B)^{N_A+N_B}}{V_A^{N_A} V_B^{N_B}} \quad (3.8\text{-}7)$$

Using that

$$S = k\left(\ln Q + T\left(\frac{\partial \ln Q}{\partial T}\right)_{V,N}\right) \quad (3.8\text{-}8)$$

results in

$$S_f - S_i = k \ln\left[\frac{Q'(N_A, N_B, V_A + V_B, T)}{Q(N_A, N_B, V_A, V_B, T)}\right]$$
$$= (N_A + N_B) k \ln(V_A + V_B) - N_A k \ln V_A - N_B k \ln V_B$$
$$= N_A k \ln\left(\frac{V_A + V_B}{V_A}\right) + N_B k \ln\left(\frac{V_A + V_B}{V_B}\right) \quad (3.8\text{-}9)$$

which is the incorrect result of Eq. 3.8-1. If however we use the correct partition function

$$\frac{Q''(N_A, N_B, V_A + V_B, T)}{Q(N_A, N_B, V_A, V_B, T)} = \frac{\dfrac{\left(\dfrac{2\pi m_A kT}{h^2}\right)^{\frac{3(N_A+N_B)}{2}} (V_A + V_B)^{N_A+N_B}}{(N_A + N_B)!}}{\dfrac{\left(\dfrac{2\pi m_A kT}{h^2}\right)^{\frac{3(N_A+N_B)}{2}} V_A^{N_A} V_B^{N_B}}{N_A! N_B!}}$$

$$= \frac{(V_A + V_B)^{N_A+N_B} N_A! N_B!}{V_A^{N_A} V_B^{N_B} (N_A + N_B)!} \quad (3.8\text{-}10)$$

Then we obtain

$$S_f - S_i = k \ln \left[\frac{Q''(N_A, N_B, V_A + V_B, T)}{Q(N_A, N_B, V_A, V_B, T)}\right]$$

$$= k \left[\begin{array}{l}(N_A + N_B) \ln(V_A + V_B) - N_A \ln V_A - N_B \ln V_B + N_A \\ \ln N_A - N_A + N_B \ln N_B - N_B - (N_A + N_B)\ln(N_A + N_B) + (N_A + N_B)\end{array}\right]$$

$$= k \left[(N_A + N_B) \ln \frac{(V_A + V_B)}{(N_A + N_B)} - N_A \ln \frac{V_A}{N_A} - N_B \ln \frac{V_B}{N_B}\right] \quad (3.8\text{-}11)$$

$$= k \left[(N_A + N_B) \ln \frac{kT}{P} - N_A \ln \frac{kT}{P} - N_B \ln \frac{kT}{P}\right]$$

$$= k \ln \frac{kT}{P}[(N_A + N_B) - N_A - N_B] = 0$$

which is the correct result.

We see that the discrepancy between the two results, and the origin of the Gibbs paradox, is a because of the indistinguishability of identical molecules. In classical thermodynamics, all molecules are considered distinguishable, so that when the two types of molecules in the two compartments are mixed and then the molecules are made identical, the two types of molecules continue to be considered distinguishable, and there is no way to correct for this (other than including a correction factor as Gibbs did). However, in quantum mechanics, and therefore in statistical thermodynamics, we have to account for this indistinguishability of identical molecules by using the correct partition function Q'' (not the incorrect partition function Q') that results from changing B molecules into A molecules. That is we take into account that the A molecules resulting from changed B molecules are indistinguishable from the original A molecules.

3.9 A CLASSICAL MECHANICS–QUANTUM MECHANICS COMPARISON: THE MAXWELL-BOLTZMANN DISTRIBUTION OF VELOCITIES

The Boltzmann distribution of energy states and the concept of the partition function as the normalization constant can also be used in classical mechanics. In particular, consider the distribution of velocity of a particle. From the Boltzmann distribution, the probability of the velocity in the x direction being between v_x and $v_x + dv_x$, the velocity in the y direction being between v_y and $v_y + dv_y$ and the velocity in the z

40 Chapter 3: The Ideal Monatomic Gas

direction being between v_z and $v_z + dv_z$, is

$$p(v_x, v_y, v_z)dv_x dv_y dv_z = C e^{-m(v_x^2+v_y^2+v_z^2)/2kT} dv_x dv_y dv_z \quad (3.9\text{-}1)$$

Note that since any velocity is possible, there is an infinite choice of velocities, and the probability of any one particular velocity will be 0. Instead we examine the probability of occurrence of a velocity in an interval, for example between v_x and $v_x + dv_x$. Consequently, $p(v_x, v_y, v_z)$ is a probability density function, and C is the normalization parameter obtained from the condition that the integral of the probability density over all velocities (from $-\infty$ to $+\infty$) must equal 1, that is

$$1 = \int_{-\infty}^{\infty}\int_{-\infty}^{\infty}\int_{-\infty}^{\infty} p(v_x, v_y, v_z) dv_x dv_y dv_z = C \int_{-\infty}^{\infty}\int_{-\infty}^{\infty}\int_{-\infty}^{\infty} e^{-m(v_x^2+v_y^2+v_z^2)/2kT} dv_x dv_y dv_z$$

$$= C \int_{-\infty}^{\infty} e^{-mv_x^2/2kT} dv_x \int_{-\infty}^{\infty} e^{-mv_y^2/2kT} dv_y \int_{-\infty}^{\infty} e^{-mv_z^2/2kT} dv_z$$

or

$$C = \frac{1}{\int_{-\infty}^{\infty} e^{-mv_x^2/2kT} dv_x \int_{-\infty}^{\infty} e^{-mv_y^2/2kT} dv_y \int_{-\infty}^{\infty} e^{-mv_z^2/2kT} dv_z} \quad (3.9\text{-}2)$$

Now

$$\int_{-\infty}^{\infty} e^{-mv_x^2/2kT} dv_x = 2 \int_0^{\infty} e^{-mv_x^2/2kT} dv_x = \sqrt{\frac{2\pi kT}{m}}$$

and similarly for the integrals in the y and z directions. Therefore

$$C = \left(\frac{m}{2\pi kT}\right)^{3/2}$$

and $\quad p(v_x, v_y, v_z) dv_x dv_y dv_z = \left(\dfrac{m}{2\pi kT}\right)^{\frac{3}{2}} e^{-m(v_x^2+v_y^2+v_z^2)/2kT} dv_x dv_y dv_z \quad (3.9\text{-}3)$

This is the (classical mechanical) Maxwell-Boltzmann distribution of velocities.

Another form of the velocity distribution is to consider instead the scalar velocity $v^2 = v_x^2 + v_y^2 + v_z^2$ or "speed" and find the probability that the scalar velocity (which can only be positive) of the molecule is between v and $v + dv$. This is most easily obtained from the Maxwell-Boltzmann distribution of velocities and changing from the rectangular coordinates (v_x, v_y, v_z) to polar-spherical coordinates (v, θ, ϕ) so that Eq. 3.9-3 becomes

$$p(v, \theta, \varphi) v^2 \sin\theta \, dv d\theta \, d\phi = C e^{-mv^2/2kT} v^2 \sin\theta \, dv d\theta \, d\phi$$

with $\quad C = \dfrac{1}{\int_0^{\infty} e^{-mv^2/2kT} v^2 \, dv \int_0^{\pi} \sin\theta \, d\theta \int_0^{2\pi} d\phi} = \dfrac{1}{\dfrac{1}{4\pi}\left(\dfrac{2\pi kT}{m}\right)^{\frac{3}{2}} \cdot 2 \cdot 2\pi}$

$$= \left(\frac{m}{2\pi kT}\right)^{\frac{3}{2}}$$

so that

$$p(v, \theta, \varphi)v^2 \sin\theta \, dv \, d\theta \, d\phi = \left(\frac{m}{2\pi kT}\right)^{\frac{3}{2}} e^{-mv^2/2kT} v^2 \sin\theta \, dv \, d\theta \, d\phi \quad (3.9\text{-}4)$$

Now, integrating over the angles θ and ϕ, since our interest is only in the probability density for speed regardless of direction, we obtain

$$p(v)v^2 \, dv = \left[\int_0^\pi \sin\theta \, d\theta \, P(v, \theta, \varphi) \int_0^{2\pi} d\phi\right] v^2 \, dv$$

$$= \left(\frac{m}{2\pi kT}\right)^{\frac{3}{2}} e^{-mv^2/2kT} v^2 \, dv \int_0^\pi \sin\theta \, d\theta \int_0^{2\pi} d\phi$$

$$= 4\pi \left(\frac{m}{2\pi kT}\right)^{\frac{3}{2}} e^{-mv^2/2kT} v^2 \, dv \quad (3.9\text{-}5)$$

which is referred to as the *Maxwell distribution of scalar velocity or speed*.

It is interesting to compare the Maxwell-Boltzmann velocity distribution based on classical mechanics:

$$p(v_x, v_y, v_z) dv_x dv_y dv_z = \left(\frac{m}{2\pi kT}\right)^{\frac{3}{2}} e^{-m(v_x^2+v_y^2+v_z^2)/2kT} dv_x dv_y dv_z \quad (3.9\text{-}3)$$

with the quantum mechanical probability of finding a particle with quantum numbers l_x, l_y, and l_z based on Eqs. 3.1-4 and 3.1-7 with $\beta = 1/kT$, which results in

$$p(l_x, l_y, l_z) = \frac{e^{-h^2(l_x^2+l_y^2+l_z^2)/8mkTV^{2/3}}}{q_{\text{trans}}} = \frac{e^{-h^2(l_x^2+l_y^2+l_z^2)/8mkTV^{2/3}}}{\left(\frac{2\pi mkT}{h^2}\right)^{3/2} V}$$

$$= \frac{1}{V} \left(\frac{h^2}{2\pi mkT}\right)^{\frac{3}{2}} e^{-h^2(l_x^2+l_y^2+l_z^2)/8mkTV^{2/3}} \quad (3.9\text{-}6)$$

Clearly, there is a similarity between the two probability density functions. However, they could not have been expected to be identical, since the coordinates are different (velocities in classical mechanics and quantum numbers in quantum mechanics), and Planck's constant h does not appear in classical mechanics.

SUMMARY OF THE MOST IMPORTANT EQUATIONS IN THIS CHAPTER

General Results for Any System

$$q = \sum_{\substack{\text{states } j \text{ of a} \\ \text{single molecule}}} e^{-\varepsilon_j/kT}; \quad Q(N, V, T) = \sum_{\substack{\text{states} \\ i}} e^{-E_i/kT} = \sum_{\substack{\text{levels} \\ j}} \Omega(E_j) e^{-E_j/kT}$$

$$U = kT^2 \left(\frac{\partial \ln Q}{\partial T}\right)_{V,N}; \quad P = kT \left(\frac{\partial \ln Q}{\partial V}\right)_{T,N}; \quad A(N, V, T) = -kT \ln Q(N, V, T)$$

42 Chapter 3: The Ideal Monatomic Gas

$$S = k \ln Q + kT \left(\frac{\partial \ln Q}{\partial T}\right)_{V,N} ; \quad \mu = \left(\frac{\partial A}{\partial N}\right)_{V,T} = -kT \left(\frac{\partial \ln Q}{\partial N}\right)_{V,T}$$

$$C_V = \left(\frac{\partial U}{\partial T}\right)_V = 2kT \left(\frac{\partial \ln Q}{\partial T}\right)_{V,N} + kT^2 \left(\frac{\partial^2 \ln Q}{\partial T^2}\right)_{V,N}$$

Results Specific to the Ideal Monatomic Gas

$$q = \sum_{\substack{\text{states } j \text{ of a} \\ \text{single molecule}}} e^{-\varepsilon_j/kT}$$

$$= \left[\left(\frac{2\pi mkT}{h^2}\right)^{\frac{3}{2}} V (\omega_{\text{elect},1} + \omega_{\text{elect},2} e^{-\Delta \varepsilon_{\text{elect},2}/kT} + \cdots) \omega_{\text{nuc},1}\right]$$

$$Q(N,V,T) = \frac{\left[\left(\frac{2\pi mkT}{h^2}\right)^{\frac{3}{2}} V (\omega_{\text{elect},1} + \omega_{\text{elect},2} e^{-\Delta \varepsilon_{\text{elect},2}/kT} + \cdots) \omega_{\text{nuc},1}\right]^N}{N!}$$

$$S = Nk \ln \left[\left(\frac{2\pi mkT}{h^2}\right)^{3/2} \frac{V e^{5/2} \omega_{\text{elect},1}}{N}\right]; \quad \mu = -kT \ln \left[\left(\frac{2\pi mkT}{h^2}\right)^{3/2} \frac{V}{N} \omega_{\text{elect},1}\right]$$

$$U = \frac{3}{2} NkT; \quad C_V = \frac{3}{2} Nk \quad \text{and} \quad PV = NkT$$

CHAPTER 3 PROBLEMS

3.1 Compare the ratio of the number of F, Cl, Br, and I atoms in the first excited electronic state to that in the ground state at 1000 K and 1500 K.

Atom	$\Delta \varepsilon_{\text{elect}}$ from ground state (eV)	$\omega_{\text{elect},1}$	$\omega_{\text{elect},2}$
F	0.050	4	2
Cl	0.11	4	2
Br	0.46	4	2
I	0.94	4	2

3.2 For a monatomic gas with electronic energy spacing so large that only the ground state is important, it is easily shown that $C_V = 3Nk/2$. However, for a monatomic gas in which both the ground electronic state and first excited state are important, we must use

$$q_{\text{elect}} = \omega_{\text{elect},1} + \omega_{\text{elect},2} e^{-\varepsilon_{\text{elect},2}/kT}$$

Develop an expression for the internal energy and heat capacity for this case.

3.3 Compute and plot the constant-volume heat capacity for one mole of argon gas over the temperature range of 3000 K to 13000 K, assuming that the only electronic energy states that need to be considered are the ones in Illustration 3.4-2.

3.4 Calculate the numerical value of the canonical partition function for a single helium atom in a cubic box of edge 1 cm and the probability of finding the helium atom in a single energy level corresponding to the mean kinetic energy of the molecule at 300 K.

3.5 a. For Problem 3.4, calculate the probability of finding the helium atom within an energy range of ±1 percent of the mean value of the kinetic energy at 300 K.
  b. Repeat the for 100 helium atoms in the box.
  c. Repeat the calculation for 10,000 helium atoms.
  d. Repeat the calculation for 1,000,000 helium atoms.

3.6 Consider the mixing of N_A molecules of an ideal monatomic gas A and N_B molecules of a second ideal monatomic gas B at constant volume V and temperature T. Write the partition function for this system in terms of the partition functions q_A and q_B of molecules A and B. Develop expressions for the following:

a. The energy of the system, E.
b. The heat capacity of the system, C_V.
c. The total pressure of the system, P.
d. The entropy of the system in terms of the mole fractions x_A and x_B and the total pressure and temperature.
e. The increase in the entropy calculated in this way over that calculated as the sum of the entropies of the pure components, each at the same temperature and pressure. This is the entropy of mixing.

3.7 The translational energy levels for a particle in a rectangular box whose sides are of unequal lengths L_x, L_y, and L_z is

$$\varepsilon(l_x, l_y, l_z) = \frac{h^2}{8m}\left(\frac{l_x^2}{L_x^2} + \frac{l_y^2}{L_y^2} + \frac{l_z^2}{L_z^2}\right)$$

Develop the expression for the translational contribution to the partition function for a volume in this box, and comment on its difference from a molecule in a cube of the same total volume.

3.8 Consider an ideal gas of monatomic molecules on a line (for example in a nanopore) where they cannot pass each other, and where the translational motion is restricted to one dimension. The quantum mechanical translational energy levels in this case are

$$\varepsilon(l_x) = \frac{h^2}{8m}\left(\frac{l_x^2}{L_x^2}\right)$$

where L_x is the length of the one-dimensional line. What is the translational contribution to the partition function and thermodynamic properties of this gas?

3.9 A system can exist in two nondegenerate states: a ground state with energy ε_1, and a higher energy (excited) state ε_2. Develop expressions for the average energy, heat capacity, entropy, and Helmholtz energy of this system as a function of temperature.

3.10 A system has only three possible energy states: the ground state ε_1 with a degeneracy of ω_1, a first excited state ε_2 with a degeneracy of ω_2, and a second excited state ε_3 with a degeneracy of ω_3. Develop expressions for the average energy, heat capacity, entropy, and Helmholtz energy of this system as a function of temperature.

3.11 A large polymer molecule is made up of N monomer units, each of which can be in either a helix (H) or a coiled (C) state with energies of ε_H and ε_C, respectively. Assuming that the conformation of each monomer unit is independent of all other monomer units, determine the average fraction of monomers that are in the helix state as a function of a dimensionless temperature. How does the statistical degeneracy come in the result?

3.12 Show that if one had a fundamental equation of state for the Helmholtz energy—that is, an equation of the form, $A = f(N, V, T)$, where f is some unspecified function of N, V, and T—then all other thermodynamic functions can be obtained from f and its derivatives with respect to N, V, or T. In particular, find expressions for the internal energy $U(N, V, T)$, the entropy $S(N, V, T)$ the enthalpy $H(N, V, T)$, the pressure $P(N, V, T)$, and the constant volume and constant pressure heat capacities.

3.13 Compute the de Broglie wave length for a 60 kg mass (approximately the average weight of a US male) at 298 K.

3.14 Compute the chemical potential of Avogadro's number of molecules of argon at 298 K and 1 bar pressure.

Chapter 4

The Ideal Diatomic and Polyatomic Gases

INSTRUCTIONAL OBJECTIVES FOR CHAPTER 4

The goals of this chapter are for the student to:

- Understand the contributions of internal degrees of freedom (rotations and vibrations) to the ideal gas partition functions of diatomic and polyatomic molecules
- Be able to compute the thermodynamic properties of an ideal gas of diatomic molecules
- Be able to compute the thermodynamic properties of an ideal gas of polyatomic molecules

4.1 THE PARTITION FUNCTION FOR AN IDEAL DIATOMIC GAS

Here we extend the discussion of the previous chapter to the ideal diatomic gas here and to polyatomic gases in the next section. The main effect of the presence of additional atoms in each molecule is that the molecule now possesses internal energy modes associated with the rotational and vibrational motions, and also the bonding energy of forming a molecule from the separated atoms. For example, a diatomic molecule such as H_2, N_2, O_2 or CO has one vibrational energy mode (associated with the bond-stretching motion), two rotational energy modes (associated with rotations perpendicular to the line of molecular symmetry), three translational energy modes (associated with the center of mass motion), and energy due to its electronic state. (Rotation along the line of symmetry of a diatomic molecule is not a mode into which energy can be added or removed since the moment of inertia along the axis is 0 by the point-mass character of the atoms making up the molecule. So this energy mode is not considered.)

Evaluation of the partition function for a polyatomic gas requires that all the energy states of a molecule be identified and enumerated. As before, we expect the electronic and nuclear energy modes to be independent of each other and of the translational, rotational, and vibrational energy modes. Furthermore, since the translational energy states for a particle-in-a-box are dependent only on the mass of the particle and not its internal structure, the translational energy mode can be taken to be independent of the rotational and vibrational motions. However, the rotational and vibrational degrees of freedom may be coupled. The coupling arises because the rotational energy states depend upon the moment of inertia of the molecule, which varies with the interatomic

4.1 The Partition Function for an Ideal Diatomic Gas

separation that is constantly changing due to the vibrational motion. However, since the vibrational motions are very rapid and of small magnitude, an average moment of inertia is generally used to break this coupling. Then, the rotational energy states are approximated to be those of a rigid rotator with a fixed interatomic separation equal to the average or equilibrium separation distance, and the vibrational energy states are assumed to be those of a harmonic oscillator.

4.1a The Translational and Nuclear Partition Functions

With this decoupling of the energy modes of a diatomic molecule, we can write its partition function as

$$q = q_{\text{trans}} \, q_{\text{rot}} \, q_{\text{vib}} \, q_{\text{elect}} \, q_{\text{nuc}} \tag{4.1-1}$$

From the previous chapter, we can immediately write:

$$q_{\text{nuc}} = \omega_{\text{nuc},1} = 1 \tag{4.1-2}$$

unless otherwise specified, and

$$q_{\text{trans}} = \left(\frac{2\pi (m_1 + m_2)kT}{h^2} \right)^{\frac{3}{2}} V = \frac{V}{\Lambda^3} \tag{4.1-3}$$

where $m_1 + m_2$ is the total mass of the diatomic molecule.

4.1b The Rotational Partition Function

The energy levels for a rigid rotator obtained from quantum mechanics are

$$\varepsilon_{\text{rot},j} = \frac{j(j+1)h^2}{8\pi^2 I} \tag{4.1-4}$$

where I is the moment of inertia, which for a linear molecule is $I = m_1 d_1^2 + m_2 d_2^2$, ($d_i$ being the distance of the atom from the center of mass of the molecule). For a homonuclear molecule, that is a molecule composed of two identical atoms, $I = \frac{1}{2} m b^2$, where b is the bond length. For a linear molecule, there is a rotational degeneracy of $\omega_j = 2j + 1$ for the jth energy level that arises from the quantization in orientation of the angular momentum vector. Two cases must be distinguished when using Eq. 4.1-4 due to the symmetry properties of the wave function, the details of which are beyond the scope of our discussion. These symmetry properties result in a restriction on the allowed values of the quantum number j. For a diatomic molecule with distinguishable nuclei (i.e., a heteronuclear molecule such as CO, HD, etc.), all values of j are allowed. For a molecule with indistinguishable nuclei (i.e., the homonuclear molecules such as H_2, O_2, N_2, etc.), only the even or odd values of j are allowed, depending on whether the nuclear spin eigenstates are antisymmetric (parahydrogen) or symmetric (orthohydrogen), respectively. Furthermore, the nuclear degeneracy of the symmetric states (orthohydrogen) is 3, and that of antisymmetric state (parahydrogen) is unity.

For a heteronuclear diatomic molecule the rotational partition function is

$$q_{\text{rot}} = \sum_{j=0,1,2}^{\infty} (2j+1) e^{-j(j+1)h^2/8\pi^2 I kT} \tag{4.1-5}$$

46 Chapter 4: The Ideal Diatomic and Polyatomic Gases

or, using $\Theta_r = \frac{h^2}{8\pi^2 I k}$, which has units of temperature and will be referred to as the rotational temperature

$$q_{\text{rot}} = \sum_{j=0,1,2}^{\infty} (2j+1) e^{-j(j+1)\Theta_r/T} \tag{4.1-6}$$

This situation is similar to that for the translational partition function in that we have a difficult summation to evaluate. If we follow the same procedure as before, which is to replace the summation with an integration, then

$$q_{\text{rot}} = \int_0^{\infty} (2j+1) e^{-j(j+1)\Theta_r/T}\, dj = \frac{T}{\Theta_r} \int_0^{\infty} e^{-x}\, dx = \frac{T}{\Theta_r} \tag{4.1-7}$$

In this integration, the substitution $x = j(j+1)\Theta_r/T$ has been used. Of course, this replacement of a summation with an integration is valid only if $T \gg \Theta_r$. In fact, this replacement is satisfactory if $T > 5\Theta_r$. Table 4.1-1 contains a list of Θ_r values for a number of common diatomic molecules as determined from microwave spectroscopic measurements.

Before leaving the topic of the rotational partition function for diatomic molecules, two points must be resolved. The first concerns the evaluation of the rotational partition function if the temperature is not high enough to use Eq. 4.1-7. (Note, for example, that for hydrogen $\Theta_r = 87.5$ K, so that at room temperature $T/\Theta_r = 3.4$.) It can be shown that the summation Eq. 4.1-6 can be written as the following

Table 4.1-1 Rotational and Vibrational Temperatures and Dissociation Energies D_0 of Several Diatomic Molecules

Molecule	Θ_r (K)	Θ_v (K)	D_0 (ev)
H_2	87.5	6320	4.476
HD	65.8	5500	4.511
D_2	43.8	4490	4.553
HCl	15.2	4330	4.430
HBr	12.2	3820	3.75
N_2	2.89	3390	9.76
CO	2.78	3120	11.08
NO	2.45	2745	4.43
O_2	2.08	2278	5.08
Cl_2	0.351	814	2.48
Br_2	0.116	465	1.97
I_2	0.0537	309	1.54
HI	9.0	3200	2.75

Note: $D_0 = -(\varepsilon_{\text{elect},1} + \frac{1}{2}h\nu)$, and 1 ev = 1.60206×10^{-12} erg = 23.0693 kcal/mol of electrons = 96.49 kJ/mol of electrons.

series expansion:

$$q_{rot} = \frac{T}{\Theta_r}\left(1 + \frac{\Theta_r}{3T} + \frac{1}{15}\left(\frac{\Theta_r}{T}\right)^2 + \mathcal{O}\left(\frac{\Theta_r}{T}\right)^3\right) \quad (4.1\text{-}8)$$

which is used if T is not much greater than Θ_r.

The next point to be considered is the evaluation of the rotational partition function for homonuclear diatomic molecules—that is, the evaluation of the sum

$$\sum (2j+1)e^{-j(j+1)\Theta_r/T}$$

in which only odd or only even values of j are allowed. At high temperatures ($T \gg \Theta_r$), where up to high values of j must be considered, either of these summations will contain only half the terms that the summation over all j values has. Therefore, at high temperatures, to a good approximation, we have

$$\sum_{j=\text{odd}}(2j+1)e^{-j(j+1)\Theta_r/T} \approx \sum_{j=\text{even}}(2j+1)e^{-j(j+1)\Theta_r/T} = \frac{1}{2}\sum_{\text{all }j}(2j+1)e^{-j(j+1)\Theta_r/T}$$

or

$$q_{rot}(\text{even }j) = q_{rot}(\text{odd }j) = \frac{1}{2}q_{rot}(\text{all }j) = \frac{T}{2\Theta_r} \quad (4.1\text{-}9)$$

(Note that it is only at high temperatures that we can replace the summation by an integral; in that case, the adjacent terms in the summation are nearly equal, and the factor of one-half is correct. So Eq. 4.1-9 is valid when Eq. 4.1-7 is.) Consequently, the high-temperature rotational partition function for a general diatomic molecule is

$$q_{rot} = \frac{T}{\sigma\Theta_r} \quad (4.1\text{-}10)$$

where σ has a value of 1 for a heteronuclear molecule, and 2 for a homonuclear molecule. The value of σ, called the *symmetry number*, can be computed using the following simple device. Consider some spatial orientation of the molecule. The value of σ is the number of different orientations of the molecule indistinguishable from the first that can be obtained by all possible rotations of the molecule.

4.1c The Vibrational Partition Function

From quantum mechanics, the energy states for a harmonic oscillator are given by

$$\varepsilon_{\text{vib},j} = \left(j+\frac{1}{2}\right)h\nu, \quad \text{for } j = 0, 1, 2, \ldots \quad \text{where } \nu = \frac{1}{2\pi}\sqrt{\frac{K(m_1+m_2)}{m_1 m_2}} \quad (4.1\text{-}11)$$

where K is the force constant between the atoms. It should be noted that in the lowest quantum state, $j = 0$, the vibrational energy of the molecule is not zero, but rather is $\varepsilon_0 = h\nu/2$. (Although it may seem unusual that the lowest energy state is not zero, that is, that the molecule is still vibrating, this is in agreement with the Heisenberg uncertainty principle that states that both the position and velocity (or momentum) of any atom cannot be known precisely at the same time.) Each of the vibrational

48 Chapter 4: The Ideal Diatomic and Polyatomic Gases

energy levels for a diatomic molecule is nondegenerate (i.e., $\omega_{\text{vib},j} = 1$ for all j), so that the vibrational partition function is computed from

$$q_{\text{vib}} = \sum_{j=0}^{\infty} e^{-(j+\frac{1}{2})h\nu/kT} = e^{-\frac{h\nu}{2kT}} \sum_{j=0}^{\infty} e^{-jh\nu/kT} \tag{4.1-12}$$

Using $\Theta_v = h\nu/k$, which has units of temperature (referred to as the *vibrational temperature*), we get

$$\begin{aligned}
q_{\text{vib}} &= e^{-\Theta_v/2T} \sum_{j=0}^{\infty} e^{-j\Theta_v/T} = e^{-\Theta_v/2T} \sum_{j=0}^{\infty} \left(e^{-\Theta_v/T}\right)^j = \frac{e^{-\Theta_v/2T}}{1 - e^{-\Theta_v/T}} \\
&= \frac{1}{e^{\Theta_v/2T} - e^{-\Theta_v/2T}} = \frac{1}{2 \sinh\left(\dfrac{\Theta_v}{2T}\right)}
\end{aligned} \tag{4.1-13}$$

Table 4.1-1 also lists values of Θ_v for selected molecules.

It is interesting that for this partition function, the sum over states can be done exactly. This is fortunate, since the vibrational energy levels are sufficiently widely separated that the summation cannot be replaced by integration. It is not completely coincidental that the one summation that cannot be replaced by integration could be done exactly. Rather, it is for this reason that the harmonic oscillator approximation, which leads a simple analytic expression for the summation, was used to represent the vibrational energy states. Any inaccuracies that result from this approximation are kept small by treating Θ_v as an adjustable parameter to be fitted to the actual vibrational energy levels determined by infrared (IR) or Raman spectroscopic measurements.

Note that we have assumed that the vibrational and rotational energy states are decoupled. This is only an approximation, as there is some rotational-vibrational coupling, which can be seen in the fine structure of an IR spectrum. Although the inclusion of such a coupling is needed for high-accuracy predictions, it will not be considered here.

4.1d The Electronic Partition Function

The only part of the partition function of the diatomic molecule we have not yet discussed is that resulting from the electronic states. As before, this can be written as

$$q_{\text{elect}} = \sum_{\substack{\text{states} \\ i}} e^{-\varepsilon_{\text{elect},i}/kT} = \sum_{\substack{\text{levels} \\ j}} \omega_{\text{elect},j} e^{-\varepsilon_{\text{elect},j}/kT}$$

$$= \omega_{\text{elect},1} e^{-\varepsilon_{\text{elect},1}/kT} + \omega_{\text{elect},2} e^{-\varepsilon_{\text{elect},2}/kT} + \cdots \tag{4.1-14}$$

Usually, but not always, the spacing between the electronic energy states is so large (i.e., $\varepsilon_{\text{elect},2} \gg \varepsilon_{\text{elect},1}$) that only the first term in the sum needs to be retained. However, as a result of the previously chosen convention for electronic energy states (namely, that an atom in its lowest electronic energy state has zero energy), the ground electronic state $\varepsilon_{\text{elect},1}$ for a diatomic or polyatomic molecule is not equal to 0 as it is related to the energy of forming a chemical bond. In fact, the energy of chemical bond formation at 0 K is the sum of the ground-state electronic energy of the molecules $\varepsilon_{\text{elect},1}$ and the energy of the lowest vibrational state of the molecule $h\nu/2$ (c.f. Eq. 4.1-11). Therefore, the energy involved in a bond formation (or, alternatively, the negative of this), the bond dissociation energy at 0 K, referred to as D_0, is

$$D_0 = -\left(\varepsilon_{\text{elect},1} + \frac{1}{2}h\nu\right) \qquad (4.1\text{-}15)$$

The quantity D_o is also tabulated in Table 4.1-1. Values of the ground-state electronic degeneracy for some molecules have been tabulated.[1]

4.2 THE THERMODYNAMIC PROPERTIES OF THE IDEAL DIATOMIC GAS

Now that each of the partition function contributions has been evaluated, the total partition function for a single diatomic molecule (assuming $\omega_{\text{nuc}} = 1$) can be written as

$$q = \left(\frac{2\pi(m_1 + m_2)kT}{h^2}\right)^{\frac{3}{2}} V \left(\frac{T}{\sigma\Theta_r}\right) \left(\frac{e^{-\Theta_v/2T}}{1 - e^{-\Theta_v/T}}\right) \omega_{\text{elect},1} e^{-\varepsilon_{\text{elect},1}/kT}$$

$$= \left(\frac{2\pi(m_1 + m_2)kT}{h^2}\right)^{\frac{3}{2}} V \left(\frac{T}{\sigma\Theta_r}\right) \frac{\omega_{\text{elect},1} e^{D_0/kT}}{(1 - e^{-\Theta_v/T})} \qquad (4.2\text{-}1)^2$$

if the rotational energy mode is fully excited (i.e., $T > 5\Theta_r$). With this partition function for one diatomic molecule, it is relatively easy to show that for N identical diatomic molecules in a volume V at temperature T

$$\frac{-A}{NkT} = \ln\left[\left(\frac{2\pi(m_1 + m_2)kT}{h^2}\right)^{\frac{3}{2}} \frac{Ve}{N}\right] + \ln\left(\frac{T}{\sigma\Theta_r}\right) - \ln\left(1 - e^{-\Theta_v/T}\right)$$

$$+ \frac{D_0}{kT} + \ln\omega_{\text{elect},1} = \ln\left(\frac{qe}{N}\right) \qquad (4.2\text{-}2)$$

$$\frac{U}{NkT} = \frac{3}{2} + \frac{2}{2} + \frac{\frac{\Theta_v}{T}e^{-\Theta_v/T}}{(1 - e^{-\Theta_v/T})} - \frac{D_0}{kT} = \frac{5}{2} + \frac{\frac{\Theta_v}{T}e^{-\Theta_v/T}}{(1 - e^{-\Theta_v/T})} - \frac{D_0}{kT} \qquad (4.2\text{-}3)$$

$$\frac{C_v}{Nk} = \frac{3}{2} + \frac{2}{2} + \left(\frac{\Theta_v}{T}\right)^2 \frac{e^{-\Theta_v/T}}{(1 - e^{-\Theta_v/T})^2} = \frac{5}{2} + \left(\frac{\Theta_v}{T}\right)^2 \frac{e^{-\Theta_v/T}}{(1 - e^{-\Theta_v/T})^2}$$

$$= \frac{5}{2} + \left(\frac{\Theta_v}{2T}\right)^2 \frac{1}{\left(\sinh\left(\frac{\Theta_v}{2T}\right)\right)^2} \qquad (4.2\text{-}4)$$

---

[1]M. Chase et al., JANAF Thermochemical Tables, *J. Phys. Chem. Ref. Data* **14**, Supplement 1 (1985).

[2]Note that the numerical value of D_0/T can be so large, for example for N$_2$, that $e^{D_0/T}$ may be out of the allowable range of simple calculational programs. However, notice that it is value of $\ln q$ that is needed for the Helmholtz energy, and not the partition function q. So in some cases it is best to calculate $\ln q$ directly, rather than trying calculate q and then take its logarithm. In this way, the computational problem is avoided.

Chapter 4: The Ideal Diatomic and Polyatomic Gases

$$\frac{\mu}{kT} = -\ln\left[\left(\frac{2\pi(m_1+m_2)kT}{h^2}\right)^{\frac{3}{2}}\frac{V}{N}\right] - \ln\left(\frac{T}{\sigma\Theta_r}\right) + \ln\left(1 - e^{-\Theta_v/T}\right)$$

$$-\frac{D_0}{kT} - \ln\omega_{\text{elect},1} = -\ln\left[\frac{q}{N}\right] \quad (4.2\text{-}5)$$

$$\frac{S}{Nk} = \frac{7}{2} + \ln\left[\left(\frac{2\pi(m_1+m_2)kT}{h^2}\right)^{\frac{3}{2}}\frac{V}{N}\right] + \ln\left(\frac{T}{\sigma\Theta_r}\right) + \frac{\Theta_v}{T}\frac{e^{-\Theta_v/T}}{1 - e^{-\Theta_v/T}}$$

$$- \ln\left(1 - e^{-\Theta_v/T}\right) + \ln\omega_{\text{elect},1} \quad (4.2\text{-}6)$$

and

$$PV = NkT \quad (4.2\text{-}7)$$

Note that the ideal gas equation of state is again obtained. A conclusion we can reach from this is that departures from ideal gas behavior are not the result of the complicated structure of the molecules. As we shall see later, departures from ideal gas behavior instead result from the interactions between molecules.

It is useful to discuss the units of the properties calculated with the equations above. The internal energy U will be depend on the units used for the Boltzmann constant and is the total energy for N molecules. If Avogadro's number of molecules is used for N, then U will be energy per mole of molecules. It is also important to note that the internal energy of the diatomic molecule is calculated with respect to the separated atoms. This comment also applies to the entropy, Helmholtz and Gibbs energies, the constant volume and constant pressure heat capacities, and the enthalpy. The chemical potential denoted by μ or g is on a per-molecule basis, again with respect to the separated atoms. These should be multiplied by Avogadro's number to obtain a value on a per mole basis. Table 4.2-1 summarizes information on relating the contributions to the single particle partition function to mass, rotational and vibrational parameters in various units.

ILLUSTRATION 4.2-1

Calculate the thermodynamic properties of 1 mole of chlorine at 1 bar and 300 K.

SOLUTION

We have the following information for chlorine:

As an ideal gas at 1 bar and 300 K, $V = 0.025$ m^3 (see Illustration 3.4-1), $M = 71$ grams, $\Theta_r = 0.351$ K, $\sigma = 2$ (homonuclear molecule), $\Theta_v = 814$ K, $D_0 = 2.48$ ev, and $\omega_{\text{elect},1} = 2$.

$$q_{\text{trans}} = \left(\frac{2\pi mkT}{h^2}\right)^{\frac{3}{2}} V = 1.88 \times 10^{26}(MT)^{3/2}\text{m}^{-3}V$$

$$= 1.88 \times 10^{26}(71 \times 300)^{3/2} \times 0.025 = 1.461 \times 10^{31}$$

$$q_{\text{rot}} = \frac{T}{\sigma\Theta_r} = \frac{300}{2 \times 0.351} = 427.3;$$

4.2 The Thermodynamic Properties of the Ideal Diatomic Gas

$$q_{vib}\, q_{elect} = \omega_{elect,1} \frac{e^{D_0/kT}}{1 - e^{-\Theta_v/T}} = 2\frac{e^{(2.48\,\text{ev}\times 1.602\times 10^{-12}\,\text{erg/ev})/\left(1.38044\times 10^{-16}\,\frac{\text{erg}}{\text{K}} \times 300\,\text{K}\right)}}{1 - e^{-814/300}}$$

$$= 0.9879 \times 10^{42}$$

$$q = q_{trans}\, q_{rot}\, q_{vib}\, q_{elect} = 1.461 \times 10^{31} \times 427.3 \times 0.9914 \times 10^{42} = 6.168 \times 10^{75}$$

$$A = -NkT \ln\left[\frac{qe}{N}\right] = -RT \ln\left[\frac{qe}{N}\right] = -3.012 \times 10^5 \text{ joule/mol}$$

$$\frac{A}{NkT} = -\ln\left[\frac{qe}{N}\right] = -120.8$$

$$\mu = -kT \ln\left[\frac{q}{N}\right] = -2.987 \times 10^5 \text{ joule/mol}$$

$$\frac{U}{NkT} = \frac{5}{2} + \frac{\frac{\Theta_v}{T} e^{-\Theta_v/T}}{(1 - e^{-\Theta_v/T})} - \frac{D_0}{kT} = \frac{-2.292 \times 10^5 \text{ joule/mol}}{8.314 \times 300 \text{ joule/mol}} = -91.89$$

$$\frac{C_V}{Nk} = \frac{5}{2} + \left(\frac{\Theta_v}{T}\right)^2 \frac{e^{-\Theta_v/T}}{(1 - e^{-\Theta_v/T})^2} = \frac{25.45 \text{ joule/mol}}{8.314 \text{ joule/mol}} = 3.06$$

$$\frac{S}{Nk} = \frac{U}{NkT} - \frac{A}{NkT} = 28.87, \quad S = 240.1 \frac{\text{joule}}{\text{mol} \cdot \text{K}}$$

Note that for an ideal monatomic gas $U/NkT = 1.5$, the very much greater number (in magnitude) here is largely due to the energy contained in the chemical bond (i.e., D_0), though there are also small contributions from the rotational and vibrational motions. Similarly, the value of C_V/Nk for an ideal monatomic gas is 1.5; the value of 3.061 here is due to the additional contributions from the rotational and vibrational motions.

Problem 4.9 continues this illustration by asking for the calculation of the internal energy and the heat capacity of chlorine over the temperature range from 100 to 1000 K. The results are shown in Fig. 4.2-1 below.

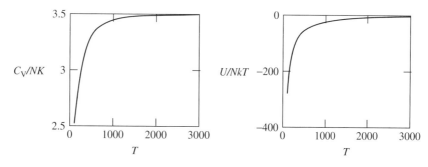

Figure 4.2-1 The reduced heat capacity C_V/Nk (left) and reduced internal energy U/NkT (right) for chlorine as a function of temperature.

52 Chapter 4: The Ideal Diatomic and Polyatomic Gases

Table 4.2-1 Useful Information for Computing Partition Functions

Translational partition function

$$\frac{q_{\text{trans}}}{V} = \left(\frac{2\pi \sum_i m_i kT}{h^2}\right)^{3/2} = \left(\frac{2\pi k}{h^2 N_{\text{Av}}}\right)^{3/2} (MT)^{3/2} = 1.88 \times 10^{20} (MT)^{3/2} \frac{1}{\text{cm}^3}$$

$$= 1.88 \times 10^{26} (MT)^{3/2} \frac{1}{\text{m}^3}$$

where m_i is the weight of atom i, M is the molecular weight of the molecule in grams, T is the temperature (K), V = volume in cm³ or m³, and N_{Av} = Avogadro's number.

Rotational partition function of a diatomic or linear polyatomic molecule (unitless)

$$q_{\text{rot}} = \frac{1}{\sigma}\left(\frac{8\pi^2 I kT}{h^2}\right) = \frac{T}{\sigma \Theta_r} = 0.02484 \times 10^{40} \frac{I_B T}{\sigma} = 0.6951 \frac{T}{B\sigma}$$

where I_B = moment of inertia (g cm²) and B = rotational constant (cm⁻¹)

Rotational partition function of a nonlinear polyatomic molecule (unitless)

$$q_{\text{rot}} = \frac{\sqrt{\pi}}{\sigma} \prod_{i=1}^{3}\left(\frac{8\pi^2 I_i kT}{h^2}\right)^{\frac{1}{2}} = \frac{\sqrt{\pi}}{\sigma}\left(\frac{T^3}{\Theta_A \Theta_B \Theta_C}\right)^{\frac{1}{2}} = \frac{0.006941 \times 10^{60}}{\sigma}(I_A I_B I_C T^3)^{1/2}$$

$$= \frac{1.02736}{\sigma}\left(\frac{T^3}{ABC}\right)^{\frac{1}{2}}$$

where I_A, I_B, I_C = moments of inertia (g-cm²) and A, B, C = rotational constants (cm⁻¹)

Vibrational temperature

Vibrational frequencies are determined spectroscopically and are usually reported as wave numbers ω in units of cm⁻¹. To convert a wave number to a frequency, multiply by the speed of light, c. To convert a wave number ω to a vibrational temperature, Θ_v use the following:

$$\Theta_v = \frac{h\nu}{k} = \frac{hc\omega}{k} = \frac{6.6261 \times 10^{-34}\, \text{J}\cdot\text{s} \times 2.9979 \times 10^{10}\, \text{cm}\cdot\text{s}^{-1} \times \omega\, \text{cm}^{-1}}{1.3807 \times 10^{-23}\, \text{J}\cdot\text{K}^{-1}} = 1.4387\, \omega\ (\text{K})$$

For the electronic partition function

$$1\, \text{eV} = 1.602 \times 10^{-19}\, \text{J} = 23.069\, \text{kcal/mol} = 96.49\, \text{kJ/mol}$$

4.3 THE PARTITION FUNCTION FOR AN IDEAL POLYATOMIC GAS

The analysis of the previous section is easily extended to molecules composed of more than two atoms. In particular, for a molecule composed of n atoms, there will be $3n$ kinetic energy modes in addition to energy in the electronic states of the molecule. Three of the energy modes of the molecule will be associated with the translational energy for which the partition function is

$$q_{\text{trans}} = \left(\frac{2\pi \sum_{\text{atoms } i} m_i kT}{h^2} \right)^{\frac{3}{2}} V = \frac{V}{\Lambda^3} \quad \textbf{(4.3-1)}$$

where $\sum m_i$ is the total mass of the molecule. There will also be two or three rotational energy modes, depending on whether the molecule is linear or nonlinear, respectively. For a linear molecule with two rotational degrees of freedom, such as CO_2, the rotational partition function (when $T \gg \Theta_r$) is

$$q_{\text{rot}} = \frac{T}{\sigma \Theta_r} \quad \textbf{(4.3-2a)}$$

while for a nonlinear molecule, for example water, the partition function is

$$q_{\text{rot}} = \frac{\sqrt{\pi}}{\sigma} \left(\frac{T^3}{\Theta_A \Theta_B \Theta_C} \right)^{\frac{1}{2}} \quad \textbf{(4.3-2b)}$$

Here Θ_A, Θ_B, and Θ_C are the three rotational temperatures computed from the three principal moments of inertia of the nonlinear molecule. The symmetry factor, σ, has the same meaning here as before; in particular, it is the number of ways that the molecule can be rotated without breaking any bonds to produce the same molecular orientation. For methane and carbon tetrachloride, for example, $\sigma = 12$.

The remaining $3n-5$ kinetic energy modes for linear molecules and $3n-6$ modes for nonlinear molecules are associated with the vibrational degrees of freedom. For a diatomic molecule, the vibration energy mode is a simple bond stretching motion. For a general polyatomic molecule, there is a collection of bond stretching and bond bending motions. All these complicated motions can be resolved into a linear combination of a number of independent motions by the procedure of normal coordinate analysis discussed in Section 4.6. The frequencies of oscillation for these normal vibrational modes can be determined by the techniques of infrared and Raman spectroscopy or quantum mechanical calculation, and a collection of vibrational temperatures (one for each normal mode), $\Theta_{v,i} = h\nu_i/k$, is obtained. The vibrational partition function for the molecule in the product of the vibrational partition functions for each of these independent normal vibrational modes. Therefore

$$q_{\text{vib}} = \prod_{i=1}^{l} \frac{e^{-\Theta_{v,i}/2T}}{1 - e^{-\Theta_{v,i}/T}}, \quad \textbf{(4.3-3)}$$

where $l = 3n-5$ for linear molecules and $3n-6$ for nonlinear molecules.

The electronic partition function can usually be written as

$$q_{\text{elect}} = \omega_{\text{elect},1} e^{-\varepsilon_{\text{elect},1}/kT} + \omega_{\text{elect},2} e^{-\varepsilon_{\text{elect},2}/kT} + \cdots \quad \textbf{(4.3-4a)}$$

54 Chapter 4: The Ideal Diatomic and Polyatomic Gases

though for many molecules this can be simplified to

$$q_{elect} = \omega_{elect,1} e^{-\varepsilon_{elect,1}/kT} \quad (4.3\text{-}4b)$$

since the electronic energy levels are usually, but not always, widely spaced. The ground electronic state degeneracy, $\omega_{elect,1}$, is unity for most chemically saturated molecules. The value of $\varepsilon_{elect,1}$, the lowest electronic energy level, is based on the separated atoms at rest as the zero energy state.

The total partition function for one molecule of an ideal polyatomic gas is

$$q = q_{trans} q_{rot} q_{vib} q_{elect} q_{nuc} \quad (4.3\text{-}5)$$

This is of the same general form as the partition function for the diatomic gas, except that the vibrational partition function, which contained only a single term corresponding to the vibrational frequency ν_i of a diatomic molecule, now contains a product of $3n-5$ or $3n-6$ terms, each corresponding to the frequency of one of the normal vibrational modes, and there may be two (linear molecule) or three (nonlinear molecule) rotational modes. This is shown in the equations below, and the following section shows how the thermodynamic properties change as a result.

$$q = \left(\frac{2\pi \sum_{\text{atoms } i} m_i kT}{h^2}\right)^{3/2} V \frac{T}{\sigma \Theta_r} \prod_{i=1}^{3n-5} \frac{e^{-\Theta_{v,i}/2T}}{1 - e^{-\Theta_{v,i}/T}} \omega_{elect,1} e^{-\varepsilon_{elect,1}/kT}$$

$$\text{linear polyatomic molecule}$$

$$= \left(\frac{2\pi \sum_{\text{atoms } i} m_i kT}{h^2}\right)^{3/2} V \frac{T}{\sigma \Theta_r} \prod_{i=1}^{3n-5} \left(\frac{1}{1 - e^{-\Theta_{v,i}/T}}\right) \omega_{elect,1} e^{D_0/kT} \quad (4.3\text{-}6)$$

and

$$q = \left(\frac{2\pi \sum_{\text{atoms } i} m_i kT}{h^2}\right)^{3/2} V \frac{\sqrt{\pi}}{\sigma} \left(\frac{T^3}{\Theta_A \Theta_B \Theta_C}\right)^{1/2} \prod_{i=1}^{3n-6} \frac{e^{-\Theta_{v,i}/2T}}{1 - e^{-\Theta_{v,i}/T}} \omega_{elect,1} e^{-\varepsilon_{elect,1}/kT}$$

$$\text{nonlinear molecule}$$

$$= \left(\frac{2\pi \sum_{\text{atoms } i} m_i kT}{h^2}\right)^{3/2} V \frac{\sqrt{\pi}}{\sigma} \left(\frac{T^3}{\Theta_A \Theta_B \Theta_C}\right)^{1/2} \prod_{i=1}^{3n-6} \left(\frac{1}{1 - e^{-\Theta_{v,i}/T}}\right) \omega_{elect,1} e^{D_0/kT}$$

$$(4.3\text{-}7)$$

where

$$D_0 = -\left(\varepsilon_{elect,1} + \frac{1}{2}\sum_i h\nu_i\right) \quad \text{and} \quad \frac{D_0}{kT} = -\left(\frac{\varepsilon_{elect,1}}{kT} + \frac{1}{2}\sum_i \frac{\Theta_{v,i}}{T}\right)$$

4.4 The Thermodynamic Properties of an Ideal Polyatomic Gas

Table 4.3-1 Characteristic Temperatures and Dissociation Energies D_0 for Several Triatomic and Polyatomic Molecules

Molecule	Θ_r (K)	Θ_v (K)	D_0 (ev)
Triatomic			
CO_2	0.561	3360, 954, 954, 1890	16.54
H_2O	40.1, 20.9, 13.4	5360, 5160, 2290	9.506
ClO_2	2.50, 0.478, 0.400	1360, 640, 1600	3.919
SO_2	2.92, 0.495, 0.422	1660, 750, 1960	11.01
N_2O	0.603	3200, 850, 850, 1840	11.44
NO_2	11.5, 0.624, 0.590	1900, 1980, 2330	9.615
Polyatomic			
NH_3	13.6, 13.6 8.92	4800, 1360, 4880, 4880, 2330, 2330	12.00
CH_4	7.54, 7.54, 7.54	4170, 2180, 2180, 4320, 4320, 4320, 1870, 1870, 1870	17.00
CH_3Cl	7.32, 0.637, 0.637	4270, 1950, 1050, 4380, 4380, 2140, 2140, 1460, 1460	16.07
CCl_4	0.0823, 0.0823, 0.0823	660, 310, 310, 1120, 1120, 1120, 450, 450, 450	13.39

Based on McQuarrie, *Statistical Mechanics*, Harper-Collins, 1976.

$$D_0 = -\left(\varepsilon_{\text{elect},1} + \frac{1}{2}\sum_i h\nu_i\right), \text{ also } 1 \text{ ev} = 1.602 \times 10^{-12} \text{ erg}$$

$$= 23.069 \text{ kcal/mol} = 96.50 \text{ kJ/mol}$$

The rotational and vibrational temperatures for a number of triatomic and polyatomic molecules are given in Table 4.3-1.

4.4 THE THERMODYNAMIC PROPERTIES OF AN IDEAL POLYATOMIC GAS

The thermodynamic functions for an ideal gas of N identical nonlinear polyatomic molecules in a volume V at temperature T is obtained by using Eqs. 4.3-1 to 4.3-5 in Eqs. 3.3-6 to 3.3-12 to obtain

$$\frac{A}{NkT} = -\ln\left[\left(\frac{2\pi \sum_{\text{atoms } i} m_i kT}{h^2}\right)^{\frac{3}{2}} \frac{Ve}{N}\right] - \ln\frac{\sqrt{\pi}}{\sigma}\left(\frac{T^3}{\Theta_A\Theta_B\Theta_C}\right)^{\frac{1}{2}}$$

$$+ \sum_{i=1}^{3n-6}\left[\frac{\Theta_{v,i}}{2T} + \ln\left(1 - e^{-\Theta_{v,i}/T}\right)\right] + \frac{\varepsilon_{\text{elect},1}}{kT} - \ln\omega_{\text{elect},1}$$

$$= -\ln\left[\left(\frac{2\pi \sum_{\text{atoms } i} m_i kT}{h^2}\right)^{\frac{3}{2}} \frac{Ve}{N}\right] - \ln\frac{\sqrt{\pi}}{\sigma}\left(\frac{T^3}{\Theta_A \Theta_B \Theta_C}\right)^{\frac{1}{2}}$$

$$+ \sum_{i=1}^{3n-6} \left[\ln\left(1 - e^{-\Theta_{v,i}/T}\right)\right] - \frac{D_0}{kT} - \ln\omega_{\text{elect},1} = -\ln\left[\frac{qe}{N}\right] \quad (4.4\text{-}1)$$

$$\frac{U}{NkT} = \frac{3}{2} + \frac{3}{2} + \sum_{i=1}^{3n-6}\left(\frac{\Theta_{v,i}}{2T} + \frac{\frac{\Theta_{v,i}}{T}e^{-\Theta_{v,i}/T}}{1 - e^{-\Theta_{v,i}/T}}\right) + \frac{\varepsilon_{\text{elect},1}}{kT}$$

$$= 3 + \sum_{i=1}^{3n-6}\left(\frac{\frac{\Theta_{v,i}}{T}e^{-\Theta_{v,i}/T}}{1 - e^{-\Theta_{v,i}/T}}\right) - \frac{D_0}{kT} \quad (4.4\text{-}2)$$

$$\frac{C_V}{Nk} = \frac{3}{2} + \frac{3}{2} + \sum_{i=1}^{3n-6}\left(\frac{\Theta_{v,i}}{T}\right)^2 \frac{e^{-\Theta_{v,i}/T}}{(1 - e^{-\Theta_{v,i}/T})^2} = \frac{3}{2} + \frac{3}{2}$$

$$+ \sum_{i=1}^{3n-6}\left(\frac{\Theta_{v,i}}{T}\right)^2 \frac{e^{-\Theta_{v,i}/T}}{(1 - e^{-\Theta_{v,i}/T})^2} = 3 + \sum_{i=1}^{3n-6}\left(\frac{\Theta_{v,i}}{2T}\right)^2 \frac{1}{\left(\sinh\left(\frac{\Theta_{v,i}}{2T}\right)\right)^2}$$

$$(4.4\text{-}3)$$

$$\frac{\mu}{kT} = -\ln\left[\left(\frac{2\pi \sum_{\text{atoms } i} m_i kT}{h^2}\right)^{\frac{3}{2}} \frac{V}{N}\right] - \ln\frac{\sqrt{\pi}}{\sigma}\left(\frac{T^3}{\Theta_A \Theta_B \Theta_C}\right)^{\frac{1}{2}}$$

$$+ \sum_{i=1}^{3n-6}\left[\frac{\Theta_{v,i}}{2T} + \ln(1 - e^{-\Theta_{v,i}/T})\right] + \frac{\varepsilon_{\text{elect},1}}{kT} - \ln\omega_{\text{elect},1}$$

$$= -\ln\left[\left(\frac{2\pi \sum_{\text{atoms } i} m_i kT}{h^2}\right)^{\frac{3}{2}} \frac{V}{N}\right] - \ln\frac{\sqrt{\pi}}{\sigma}\left(\frac{T^3}{\Theta_A \Theta_B \Theta_C}\right)^{\frac{1}{2}}$$

$$+ \sum_{i=1}^{3n-6}\left[\ln\left(1 - e^{-\Theta_{v,i}/T}\right)\right] - \frac{D_0}{kT} - \ln\omega_{\text{elect},1} = -\ln\left[\frac{q}{N}\right] \quad (4.4\text{-}4)$$

$$\frac{S}{Nk} = 4 + \ln\left[\left(\frac{2\pi \sum_{\text{atoms } i} m_i kT}{h^2}\right)^{\frac{3}{2}} \frac{V}{N}\right] + \ln\frac{\sqrt{\pi}}{\sigma}\left(\frac{T^3}{\Theta_A \Theta_B \Theta_C}\right)^{\frac{1}{2}}$$

$$+ \sum_{i=1}^{3n-6}\left[\frac{\Theta_{v_i}}{T}\frac{e^{-\Theta_{v,i}/T}}{1 - e^{-\Theta_{v,i}/T}} - \ln\left(1 - e^{-\Theta_{v,i}/T}\right)\right] + \ln\omega_{\text{elect},1} \quad (4.4\text{-}5)$$

4.4 The Thermodynamic Properties of an Ideal Polyatomic Gas

and
$$PV = NkT \quad (4.4\text{-}6)$$

The analogous equations for a linear molecule are:

$$\frac{A}{NkT} = -\ln\left[\left(\frac{2\pi \sum_{\text{atoms } i} m_i kT}{h^2}\right)^{\frac{3}{2}} \frac{Ve}{N}\right] - \ln\frac{T}{\sigma \Theta_r}$$

$$+ \sum_{i=1}^{3n-5}\left[\frac{\Theta_{v,i}}{2T} + \ln\left(1 - e^{-\Theta_{v,i}/T}\right)\right] + \frac{\varepsilon_{\text{elect},1}}{kT} - \ln\omega_{\text{elect},1}$$

$$= -\ln\left[\left(\frac{2\pi \sum_{\text{atoms } i} m_i kT}{h^2}\right)^{\frac{3}{2}} \frac{Ve}{N}\right] - \ln\frac{T}{\sigma \Theta_r}$$

$$+ \sum_{i=1}^{3n-5}\left[\ln\left(1 - e^{-\Theta_{v,i}/T}\right)\right] - \frac{D_0}{kT} - \ln\omega_{\text{elect},1} = -\ln\left(\frac{qe}{N}\right) \quad (4.4\text{-}7)$$

$$\frac{U}{NkT} = \frac{3}{2} + \frac{2}{2} + \sum_{i=1}^{3n-5}\left(\frac{\Theta_{v,i}}{2T} + \frac{\Theta_{v,i}}{T}\frac{e^{-\Theta_{v,i}/T}}{1 - e^{-\Theta_{v,i}/T}}\right) + \frac{\varepsilon_{\text{elect},1}}{kT}$$

$$= \frac{5}{2} + \sum_{i=1}^{3n-5}\left(\frac{\Theta_{v,i}}{T}\frac{e^{-\Theta_{v,i}/T}}{1 - e^{-\Theta_{v,i}/T}}\right) - \frac{D_0}{kT} \quad (4.4\text{-}8)$$

$$\frac{C_V}{Nk} = \frac{3}{2} + \frac{2}{2} + \sum_{i=1}^{3n-5}\left(\frac{\Theta_{v,i}}{T}\right)^2 \frac{e^{-\Theta_{v,i}/T}}{(1 - e^{-\Theta_{v,i}/T})^2} = \frac{5}{2} + \sum_{i=1}^{3n-5}\left(\frac{\Theta_{v,i}}{T}\right)^2 \frac{e^{-\Theta_{v,i}/T}}{(1 - e^{-\Theta_{v,i}/T})^2}$$

$$= \frac{5}{2} + \sum_{i=1}^{3n-5}\left(\frac{\Theta_{v,i}}{4T}\right)^2 \frac{1}{\left(\sinh\left(\frac{\Theta_{v,i}}{2T}\right)\right)^2} \quad (4.4\text{-}9)$$

$$\frac{\mu}{kT} = -\ln\left[\left(\frac{2\pi \sum_{\text{atoms } i} m_i kT}{h^2}\right)^{\frac{3}{2}} \frac{V}{N}\right] - \ln\frac{T}{\sigma \Theta_r}$$

$$+ \sum_{i=1}^{3n-5}\left[\frac{\Theta_{v,i}}{2T} + \ln(1 - e^{-\Theta_{v,i}/T})\right] + \frac{\varepsilon_{\text{elect},1}}{kT} - \ln\omega_{\text{elect},1}$$

$$= -\ln\left[\left(\frac{2\pi \sum_{\text{atoms } i} m_i kT}{h^2}\right)^{\frac{3}{2}} \frac{V}{N}\right] - \ln\frac{T}{\sigma \Theta_r}$$

$$+ \sum_{i=1}^{3n-5}\left[\ln\left(1 - e^{-\Theta_{v,i}/T}\right)\right] - \frac{D_0}{kT} - \ln\omega_{\text{elect},1} = -\ln\left(\frac{q}{N}\right) \quad (4.4\text{-}10)$$

58 Chapter 4: The Ideal Diatomic and Polyatomic Gases

$$\frac{S}{Nk} = \frac{7}{2} + \ln\left[\left(\frac{2\pi \sum_{\text{atoms } i} m_i kT}{h^2}\right)^{\frac{3}{2}} \frac{V}{N}\right] + \ln\frac{T}{\sigma \Theta_r}$$

$$+ \sum_{i=1}^{3n-5}\left[\frac{\Theta_{v,i}}{T} \frac{e^{-\Theta_{v,i}/T}}{1 - e^{-\Theta_{v,i}/T}} - \ln\left(1 - e^{-\Theta_{v,i}/T}\right)\right] + \ln \omega_{\text{elect},1} \quad \text{(4.4-11)}$$

and

$$PV = NkT \quad \text{(4.4-12)}$$

One interesting observation is that no matter how complicated the molecule is, as long as the fluid is composed of noninteracting molecules, the ideal gas equation of state results.

4.5 THE HEAT CAPACITIES OF IDEAL GASES

In chemistry and engineering, it is common to express the heat capacity (constant volume or constant pressure) of ideal gases in simple power-series form as a function of temperature. That is,

$$C_V = a + bT + cT^2 + dT^3 \quad \text{(4.5-1)}$$

However, we see from the results of this chapter and the previous one that there are exact expressions for the ideal gas heat capacity, and they are not of the power series form. There are advantages and disadvantages to the exact expressions. For example, for the diatomic gas only the single parameter Θ_v is needed, rather than the four adjustable parameters in the equation above. For a triatomic molecule, three vibrational temperatures are needed if the molecule is nonlinear, and four vibrational temperatures are needed for a linear molecule. As this number of parameters is approximately the same as in the power-series expansion, and the power series is algebraically simpler (no exponentials), it is generally simpler to use the power-series expansion. For molecules containing more than three atoms, the number of parameters needed for the statistical mechanical calculation of the heat capacity can greatly exceed the four parameters in the power-series expansion. It is for this reason that the power-series form is simpler to use for large molecules. For consistency, it is then common to use the power-series form for all molecules, even though an exact expression is available for small molecules. Also, by using the power series with parameters determined by fitting experimental data, there is no effect of the error that may have been introduced by the statistical thermodynamic assumption of the complete separation of the rotational and vibration degrees of freedom.

Recently the Design Institute for Physical Properties Research (DIPPR) of the American Institute of Chemical Engineers have correlated ideal gas constant-pressure heat capacity data using

$$C_P = k_1 + k_2 \left[\frac{(k_3/T)}{\sinh(k_3/T)}\right]^2 + k_4 \left[\frac{(k_5/T)}{\cosh(k_5/T)}\right]^2$$

where each of the k_i have been fit to experimental data. The justification for the use of hyperbolic functions is evident from the derivations in the previous section.

4.6 NORMAL MODE ANALYSIS: THE VIBRATIONS OF A LINEAR TRIATOMIC MOLECULE

In the discussion of this chapter, we have used that there are distinct vibrational modes of triatomic and larger molecules, but we have not considered what those modes are or how they could be identified. The identification of a set of vibrational motions that completely describe all possible vibrations of a molecule—that is, the normal modes—is discussed here. For simplicity of presentation of the ideas involved, consider the molecule shown below consisting of three identical atoms, each of mass m, and at equilibrium each is separated from the others by a distance b. To illustrate the concept of a normal model analysis for vibrations as simply as possible, only vibrations along the axis of the molecule will be considered here; the more complicated case of vibrations out of the axis of the molecule will be briefly mentioned later.

In modeling the vibrations, we assume that the triatomic molecule can be considered to consist of masses m connected by springs each with a spring constant K. This is shown below.

The potential energy of this system, PE, is

$$\text{PE} = \frac{K}{2}(x_2 - x_1 - b)^2 + \frac{K}{2}(x_3 - x_2 - b)^2 \tag{4.6-1}$$

and the kinetic energy, KE, is

$$\text{KE} = \frac{m}{2}(v_1^2 + v_2^2 + v_3^2) \quad \text{where} \quad v_i = \frac{dx_i}{dt} \tag{4.6-2}$$

To simplify these expressions, a new set of variables is introduced in which each distance variable is measured with respect to its equilibrium position, that is $\eta_i = x_i - x_{i0}$, where x_{i0} is the equilibrium position of the ith mass point. Now, using that in the equilibrium positions $x_{20} - x_{10} = x_{30} - x_{20} = b$, we obtain

$$\text{PE} = \frac{K}{2}(\eta_2 - \eta_1)^2 + \frac{K}{2}(\eta_3 - \eta_2)^2$$

and

$$\text{KE} = \frac{m}{2}\left(\dot{\eta}_1^2 + \dot{\eta}_2^2 + \dot{\eta}_3^2\right) \tag{4.6-3}$$

The classical equation of motion for any mass point is

$$m\ddot{\eta}_i = m\frac{d^2\eta_i}{dt^2} = F_i = -\frac{d(\text{PE})}{d\eta_i} \tag{4.6-4}$$

where F_i is the force on mass point i, which is obtained as the derivative of the potential energy with respect to the position of the mass point. Thus,

$$m\ddot{\eta}_1 = K(\eta_2 - \eta_1) \tag{4.6-5}$$

$$m\ddot{\eta}_2 = -K(\eta_2 - \eta_1) + K(\eta_3 - n_2) \tag{4.6-6}$$

$$m\ddot{\eta}_3 = -K(\eta_3 - \eta_2) \tag{4.6-7}$$

60 Chapter 4: The Ideal Diatomic and Polyatomic Gases

Now, to identify the normal modes for vibration, we note that for a periodic motion the equation of motion should be of the form

$$m\ddot{\zeta}_i = 4\pi m \nu^2 \zeta_i \tag{4.6-8}$$

which has the solution $\zeta_i = C_1 \sin(2\pi \nu_i t + C_2)$. Consequently, for the expected periodic motion, we should have that $m\ddot{\eta}_i = -\omega^2 m \eta_i$, where for simplicity the substitution $\omega^2 = 4\pi \nu^2$ has been made. With these substitutions, Eqs. 4.6-5 to 4.6-7 can be rewritten as

$$-m\omega^2 \eta_1 - K(\eta_2 - \eta_1) = 0 \tag{4.6-9}$$

$$-m\omega^2 \eta_2 + K(\eta_2 - \eta_1) - K(\eta_3 - \eta_2) = 0 \tag{4.6-10}$$

$$-m\omega^2 \eta_3 + K(\eta_3 - \eta_2) = 0 \tag{4.6-11}$$

or, in matrix form

$$\begin{bmatrix} K - m\omega^2 & -K & 0 \\ -K & 2K - m\omega^2 & -K \\ 0 & -K & K - m\omega^2 \end{bmatrix} \begin{bmatrix} \eta_1 \\ \eta_2 \\ \eta_3 \end{bmatrix} = \begin{bmatrix} 0 \\ 0 \\ 0 \end{bmatrix} \tag{4.6-12}$$

Since η_1, η_2, and η_3 can take on any arbitrary values, the only way that Eq. 4.6-12 can be satisfied at all times is for the determinate in the equation to be zero. That is, for

$$\begin{vmatrix} K - m\omega^2 & -K & 0 \\ -K & 2K - m\omega^2 & -K \\ 0 & -K & K - m\omega^2 \end{vmatrix} = 0 \tag{4.6-13}$$

or

$$\omega^2 (K^2 - \omega^2 m)(3mK - \omega^2 m^2) = 0 \tag{4.6-14}$$

This sixth order equation for ω has six solutions

$$\omega_1 = 0; \; \omega_2 = 0, \; \omega_{3\pm} = \pm\sqrt{\frac{K}{m}}; \quad \text{and} \quad \omega_{4\pm} = \pm\sqrt{\frac{3K}{m}} \tag{4.6-15}$$

The pair ω_{3+} and ω_{3-} refers to the two different phases of the same periodic motion, and ω_{4+} and ω_{4-} are the two different phases of another periodic motion; so of these only ω_{3+} (which we refer to as ω_3) and ω_{4+} (ω_4) are unique periodic motions that we consider further.

To understand what the normal mode motions are in the original coordinate system, we now substitute, in turn, each of the values of ω into Eqs. 4.6-9, 4.6-10 and 4.6-11. Using $\omega = 0$, we obtain

$$\eta_2 - \eta_1 = 0$$
$$2\eta_2 - \eta_1 - \eta_3 = 0 \tag{4.6-16}$$

and

$$\eta_3 - \eta_2 = 0$$

4.6 Normal Mode Analysis: The Vibrations of a Linear Triatomic Molecule   61

which has the solution that $\eta_1 = \eta_2 = \eta_3$. Thus, the mode with $\omega = 0$ is not a vibration, but rather a free translation of the molecule as a whole. That is, the motion is

$$\text{O}\!\!\sim\!\!\text{O}\!\!\sim\!\!\text{O}$$
$$\rightarrow \quad \rightarrow \quad \rightarrow$$

where the arrows, being of the same size and direction, indicate that the displacements of all the atoms are of the same magnitude and direction in this mode of motion.

Now, using $\omega = \sqrt{\frac{K}{m}}$, we obtain

$$\begin{aligned} -K\eta_1 - K(\eta_2 - \eta_1) &= 0; \quad K\eta_1 - K(\eta_2 - \eta_1) = 0 \\ -K\eta_2 + K(\eta_2 - \eta_1) - K(\eta_3 - \eta_2) &= 0 \\ -K\eta_3 + K(\eta_3 - \eta_2) &= 0 \end{aligned} \qquad (4.6\text{-}17)$$

which has the solution $\eta_1 = -\eta_3$ and $\eta_2 = 0$. Thus, in this mode, the central atom remains stationary, and atoms 1 and 3 vibrate with motions of equal magnitude and opposite direction, as indicated below:

$$\text{O}\!\!\sim\!\!\text{O}\!\!\sim\!\!\text{O}$$
$$\leftarrow \qquad\qquad \rightarrow$$

This vibrational mode is referred to as a symmetric stretching mode. As an example, this vibration in the linear molecule CO_2 has been reported from spectroscopy to be at a wave number of 1314 cm^{-1}, which corresponds to a vibrational frequency of

$$\omega_1 = \text{wave number} \times \text{speed of light} = 2.9979 \times 10^{10} \frac{\text{cm}}{\text{s}} \times 1314 \text{ cm}^{-1}$$
$$= 3.938 \times 10^{13} \text{ s}^{-1}$$

and a vibrational temperature Θ_v of 1890 K.

Finally, using $\omega = \sqrt{\frac{3K}{m}}$ we have

$$\begin{aligned} -3K\eta_1 - K(\eta_2 - \eta_1) &= 0 & \text{or} \quad \eta_2 &= -2\eta_1 \\ -3K\eta_2 + K(\eta_2 - \eta_1) - K(\eta_3 - \eta_2) &= 0 & \text{or} \quad \eta_2 &= \eta_1 - \eta_3 \qquad (4.6\text{-}18) \\ -3K\eta_3 + K(\eta_3 - \eta_2) &= 0 & \text{or} \quad \eta_2 &= -2\eta_3 \end{aligned}$$

Therefore, $\eta_2 = -2\eta_1 = -2\eta_3$, so that the end atoms move in the same direction and the center atom moves in the opposite direction with a displacement of twice the magnitude of each of the end atoms. This is shown below:

$$\text{O}\!\!\sim\!\!\text{O}\!\!\sim\!\!\text{O}$$
$$\rightarrow \quad \leftarrow \quad \rightarrow$$

This vibration is referred to as an asymmetric stretching mode. In CO_2, $\omega_2 = 2335$ cm^{-1} for this vibration, corresponding to a vibrational temperature Θ_v of 3360 K.

Therefore, for a linear molecule (atoms restricted to remain along a line) there are three modes of motion; one translational mode ($\omega = 0$), a symmetric stretching mode and an asymmetric stretching mode. Of course, the atoms in a real linear molecule are not restricted to move along a line. If we were to perform the same type of analysis for the three-dimensional motion of the molecule, we would find that there were 9 modes of motion or degrees of freedom for a linear molecule:

3 translational degrees of freedom

2 rotational degrees of freedom

4 vibrational degrees of freedom.

The two rotational motions are in directions perpendicular to the axis of the molecule. Two of the vibrational modes are as discussed above. The other two vibrations are two symmetric bending modes, one in a plane and the other perpendicular to it, of the form

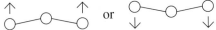

For carbon dioxide, these two identical vibration modes are $\omega_3 = 663$ cm^{-1} for this vibration, corresponding to a vibrational temperature Θ_v of 954 K.

In contrast, a nonlinear triatomic molecule also has 9 degrees of freedom, but distributed as follows:

3 translational degrees of freedom

3 rotational degrees of freedom

3 vibrational degrees of freedom

Two of the rotational degrees of freedom are perpendicular to the axis of the molecule, and the third is along the axis. The three vibrational modes—a symmetric stretch, a bending mode, and an asymmetric stretch—are schematically shown below:

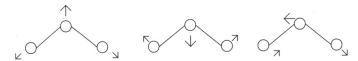

For water, the wave numbers for these vibrations are 3583 cm^{-1}, 1592 cm^{-1}, and 3725 cm^{-1}, respectively.

Similar analyses can be done for larger molecules. However, as the number of atoms in a molecule increases, the number of vibrational modes increases, and the algebra becomes more difficult. (In fact, each additional atom adds three additional vibrational degrees of freedom.) Also, very large molecules contain chains that are flexible, and it becomes increasingly difficult to make a clear distinction between rotations, vibrations, and, in some cases (i.e., polymers), even hindered translational motions.

CHAPTER 4 PROBLEMS

4.1 **a.** Derive Eqs. 4.2-2 to 4.2-7 for the linear diatomic molecule, starting from the single-particle partition function of Eq. 4.2-1.
  b. Also, show that for the linear diatomic molecule if $T \gg \Theta_v$ and $T \gg \Theta_r$
  $$\frac{E}{N} \to \frac{7}{2}kT \text{ and } \frac{C_V}{N} \to \frac{7}{2}k$$

4.2 Derive an expression for the chemical potential of an ideal diatomic gas.

4.3 Obtain the expressions for A and U for a diatomic gas when the second and third terms in Eq. 4.1-8 must be included.

4.4 Compute and plot U and C_V for CO as a function of temperature over the temperature range of 100 to 1500 K.

4.5 Calculate the entropy, heat capacity C_P, and chemical potential (Gibbs free energy) for nitrogen and hydrogen bromide at 25°C and 1 bar. The first electronic state is nondegenerate for both gases.

4.6 Calculate C_V in J/(mol K) and μ in J/mol for N_2 at 25°C and 1 bar pressure.

4.7 Calculate the fraction of CO_2 molecules in first four vibrational states at 200 K, 800 K, and 3000 K.

4.8 Calculate the constant-volume heat capacity C_V for H_2, HD, and D_2 at 150 K, 250 K, and 350 K, assuming that the atom-atom separation distance and bond force constant are the same in these species.

4.9 Calculate the constant volume heat capacity C_V (in joule/mol K) for N_2 and Cl_2 over the temperature range from 300 K to 2700 K. How well can these results be fit with a polynomial in T?

4.10 In this chapter, the harmonic oscillator approximation for the vibrational energy modes was used:

$$\varepsilon_{\text{vib},n} = \left(n + \frac{1}{2}\right) h\nu \quad n = 0, 1, 2, \ldots$$

A more realistic model is to include the first term in an expansion for anharmonicity:

$$\varepsilon_{\text{vib},n} = \left(n + \frac{1}{2}\right) h\nu - \chi \left(n + \frac{1}{2}\right)^2 h\nu$$
$$n = 0, 1, 2, \ldots$$

where χ is a small constant. What is the vibrational partition function for this case?

4.11 In this chapter, the rigid rotator approximation was used for the rotational energy modes

$$\frac{\varepsilon_{\text{rot},j}}{k} = j(j+1)\Theta_r \quad j = 0, 1, 2, \ldots$$

A more realistic model is to include the first term in an expansion to account for the fact that due to centrifugal forces, the molecules stretches slightly with increasing rotational motion. This is accounted for by including the first term in an expansion about the rigid rotator model that results in

$$\frac{\varepsilon_{\text{rot},j}}{k} = j(j+1)\Theta_r - \xi j^2 (j+1)^2 \quad j = 0, 1, 2, \ldots$$

where ξ is a small constant. What is the rotational partition function for this case?

4.12 Compute and plot U and C_V for H_2O as a function of temperature between 100 and 3000 K.

4.13 Compute and plot U and C_V for carbon dioxide at 1 bar and temperatures between 100 K and 1500 K.

4.14 Compute and plot U and C_V for methane at 1 bar and temperatures between 300 and 600 K.

4.15 Determine the constant volume and constant-pressure heat capacities for an ideal diatomic gas in the limit of $T \to 0$. A corollary to the third law of thermodynamics is that the heat capacity of a material in the perfect crystalline state should be 0 at absolute zero. Is the result you obtained consistent with the third law of thermodynamics? Explain.

4.16 Determine the constant volume and constant-pressure heat capacities for an ideal linear triatomic gas in the limit of $T \to 0$. A corollary to the third law of thermodynamics is that the heat capacity of a material in the perfect crystalline state should be 0 at absolute zero. Is the result you obtained consistent with the third law of thermodynamics? Explain.

4.17 When a normal mode analysis is done for a polyatomic molecule, one type of mode that may be found is a hindered rotation. For example, consider an ethane molecule. One internal motion of the molecule is a rotation around the carbon-carbon bond. However, the potential energy of the molecule is higher when the hydrogen atoms on two different carbons are aligned than when the hydrogens are in a staggered conformation so there is an energy barrier for this rotation. The form of the interaction energy for this rotation is periodic, and for ethane, with three hydrogens spaced 120° apart, is given by

$$u(\Phi) = \frac{1}{2} u(0)[1 - \cos(3\Phi)]$$

where $u(0)$ is the height of the energy barrier for the internal rotation of the molecule. At high temperatures ($kT \gg u(0)$), the barrier to rotation is small (or inconsequential) compared to the kinetic energy of the molecule, and the internal motion is essentially a free rotation. However, at low temperatures ($kT < u(0)$), the rotational energy barrier is large compared to the kinetic energy, and the motion around the carbon-carbon bond behaves as a vibration around the staggered conformation.
What is the contribution to the constant volume heat capacity of this energy mode at
a. $T = 0$ K, and at
b. at high temperatures (i.e., as $T \to \infty$)?

4.18 One assumption that is sometimes made is that for isotopic species, the bond length is the same in each of these species, as is the vibrational force constant k. Are the rotational and vibrational temperatures of H_2, D_2, and HD given in Table 4.1-1 consistent with this assumption?

Chapter 5

Chemical Reactions in Ideal Gases

In the previous chapters we saw how, from some simple assumptions, a whole framework could be developed permitting the calculation of the thermodynamic properties of dilute gases from the results of spectroscopic measurements. In this section, another use of statistical thermodynamics is developed. First, the concept of chemical equilibrium in an ideal gas mixture is discussed. Then it is shown how, from the same spectroscopic information used in the previous sections, the chemical equilibrium constant for an ideal gas phase reaction, the degree of ionization in a plasma (partially ionized gas), and the very large reactive contributions to the heat capacity can be calculated.

INSTRUCTIONAL OBJECTIVES FOR CHAPTER 5

The goals for this chapter are for the student to:

- Understand how the canonical partition function for an reacting ideal gas mixture differs from that for a nonreacting ideal gas system
- Be able to compute the chemical equilibrium constant for a reacting ideal gas mixture
- Be able to calculate the equilibrium compositions in a reacting ideal gas mixture
- Be able to compute the thermodynamic properties of a reacting ideal gas mixture

5.1 THE NONREACTING IDEAL GAS MIXTURE

Consider a system of N_A molecules of species A and N_B molecules of species B in a volume V and a temperature T. The partition function for this system is computed by evaluating the sum

$$Q(N_A, N_B, V, T) = \sum_{\substack{\text{all states } i \\ \text{of the system}}} e^{-E_i/kT} \tag{5.1-1}$$

Let $\underline{n}^i$ be the vector of occupation numbers for the ith state of the system; that is, $\underline{n}^i = (n^i_{A1}, n^i_{A2}, \ldots; n^i_{B1}, n^i_{B2}, \ldots)$, where n^i_{Aj} is the number of molecules of species A in the jth energy state of a species A molecule in the macroscopic state i of the system. The quantity n_{Bj}^i is similarly defined for species B. Each state occupation number vector must satisfy the following two restrictions:

$$\sum_j n_{Aj}^i = N_A \quad \text{and} \quad \sum_j n_{Bj}^i = N_B \tag{5.1-2}$$

Since a species A molecule is indistinguishable from other species A molecules, but distinguishable from a species B molecule (and vice versa), using the analysis developed in Chapter 2, it immediately follows that

$$Q(N_A, N_B, V, T) = Q_A(N_A, V, T) Q_B(N_B, V, T) \tag{5.1-3a}$$

Also, if the number of molecular energy states is much larger than the number of molecules (which is always the case for the systems we consider), then

$$Q(N_A, N_B, V, T) = Q_A(N_A, V, T) Q_B(N_B, V, T) = \frac{q_A^{N_A} q_B^{N_B}}{N_A! \, N_B!} \tag{5.1-3b}$$

and, more generally, for a mixture of S noninteracting species

$$Q(N_1, N_2, \ldots, N_S, V, T) = \prod_{i=1}^{S} \frac{q_i^{N_i}}{N_i!} \tag{5.1-4}$$

The thermodynamic properties of the S-component mixture are easily evaluated. For example

$$A = -kT \ln Q = -kT \ln \prod_{i=1}^{S} \frac{q_i^{N_i}}{N_i!} = -kT \sum_{i=1}^{S} \left(N_i \ln \frac{q_i e}{N_i} \right) \tag{5.1-5a}$$

Here, q_i is the partition function for species i, which may be an atom, a diatomic molecule, or a polyatomic molecule. Also,

$$P = kT \left(\frac{\partial \ln Q}{\partial V} \right)_{N_i, T} = \frac{kT}{V} \sum_i N_i \tag{5.1-5b}$$

which is the ideal equation of state. Earlier we saw that departure from ideal gas behavior was not the result of the internal structure of the molecules, and here we see it is not a result of forming a mixture. In a later chapter, we shall see that nonideal gas behavior results from the interactions between molecules.

5.2 PARTITION FUNCTION OF A REACTING IDEAL CHEMICAL MIXTURE

Above we showed that the partition function for a simple mixture of N_A molecules of species A and N_B molecules of species B is

$$Q(N_A, N_B, V, T) = \frac{q_A^{N_A} q_B^{N_B}}{N_A! \, N_B!} \tag{5.2-1}$$

66 Chapter 5: Chemical Reactions in Ideal Gases

that is to be compared with the partition function for a pure gas

$$Q(N, V, T) = \frac{q^N}{N!} \qquad (5.2\text{-}2)$$

In all cases, the single particle partition function is the sum over all the energy states accessible to a molecule.

Now, suppose a reversible reaction or isomerism can occur that transmutes species A into species B and vice versa,

$$A \leftrightarrow B$$

One can now ask what is the partition function appropriate to this chemically reacting mixture that initially contained N_A molecules of species A and N_B molecules of species B. The answer can immediately be written down from the observation that the effect of the chemical reaction is to make available to each molecule the energy states of *both* species. Therefore

$$Q = \frac{\left(\sum_{\substack{\text{accesible} \\ \text{energy states} \\ j}} e^{-\varepsilon_j/kT}\right)^N}{N!} = \frac{q^N}{N!} \qquad (5.2\text{-}3)$$

and for a single molecule

$$q = \sum_{\substack{\text{accessible} \\ \text{energy states}}} e^{-\varepsilon_j/kT} = \sum_{\substack{\text{energy states} \\ \text{available to} \\ \text{species A}}} e^{-\varepsilon_j^A/kT} + \sum_{\substack{\text{energy states} \\ \text{available to} \\ \text{species B}}} e^{-\varepsilon_j^B/kT}$$

$$= q_A + q_B$$

So that

$$Q = \frac{(q_A + q_B)^N}{N!} \qquad (5.2\text{-}4)$$

where

$$N = N_A + N_B$$

This expression can also be obtained by a less intuitive but somewhat more general procedure. Here we start with the observation that, due to the chemical reaction, the actual number of molecules of species A present at any time is not a known quantity, but may be 0, 1, 2, ..., N, where N_A^o and N_B^o are the number of molecules of species A and species B initially present, respectively, and $N = N_A^o + N_B^o$. Furthermore, for any particular set of values for N_A and N_B, Eq. 5.2-1 is valid. Therefore, the partition function for this chemically reacting mixture of A and B, for which all values of N_A

5.3 Three Different Derivations of the Chemical Equilibrium Constant in an Ideal Gas Mixture

and N_B are allowed subject to $N = N_A + N_B$, is

$$Q = \frac{q_A^N}{N!} + \frac{q_A^{N-1} q_B}{(N-1)!} + \frac{q_A^{N-2} q_B^2}{(N-2)! \, 2!} + \frac{q_A^{N-3} q_B^3}{(N-3)! \, 3!} + \cdots + \frac{q_A}{1} \frac{q_B^{N-1}}{(N-1)!} + \frac{q_B^N}{N!}$$

$$= \sum_{N_A=0}^{N} \sum_{N_B=0}^{N} \frac{q_A^{N_A} q_B^{N_B}}{N_A! \, N_B!} \qquad (5.2\text{-}5)$$

with the restriction on the double summation that

$$N_A + N_B = N_A^o + N_B^o = N \qquad (5.2\text{-}6)$$

The double summation of Eq. 5.2-5 with restriction Eq. 5.2-6 can be reduced to a single summation by eliminating N_B in terms of $N - N_A$, that is

$$Q = \sum_{N_A=0}^{N} \frac{q_A^{N_A} q_B^{N-N_A}}{N_A! (N-N_A)!} = \frac{1}{N!} \sum_{N_A=0}^{N} \frac{N!}{N_A! (N-N_A)!} q_A^{N_A} q_B^{N-N_A} = \frac{(q_A + q_B)^N}{N!}$$

$$(5.2\text{-}7)$$

The last term in the above equation is obtained from the term that precedes it by use of the binomial expansion, as discussed in the Appendix to this chapter, so that Eq. 5.2-3 is recovered.

5.3 THREE DIFFERENT DERIVATIONS OF THE CHEMICAL EQUILIBRIUM CONSTANT IN AN IDEAL GAS MIXTURE

Starting from Eq. 5.2-7, we will now obtain the chemical equilibrium constant for this reacting system by three different methods. Each of these methods leads to the same result, and the purpose of obtaining the same result by different methods is to illustrate the variety of methods that are used in statistical mechanics.

The first method is based upon the interrelationship between statistical and classical thermodynamics. From classical thermodynamics, we know that the criterion for equilibrium state in a closed system at constant T and V is the state of minimum Helmholtz energy A with respect to all possible variations consistent with the physical situation. Here, this implies that A should be a minimum with respect to N_A (or N_B), subject to the restriction of Eq. 5.2-6. Alternatively, since

$$A = -kT \ln Q \qquad (5.3\text{-}1)$$

where the partition function Q for any particular choice of N_A and N_B is given by Eq. 5.2-1, and the equilibrium requirement is that the $\ln Q$ have a maximum value subject to the constraints of constant V, T, and $N = N_A + N_B$. This maximum value can be found by the straightforward approach of eliminating N_B in terms to N_A and N, to obtain

$$\ln Q = N_A \ln q_A - N_A \ln N_A + (N - N_A) \ln q_B - (N - N_A) \ln(N - N_A) \qquad (5.3\text{-}2)$$

68 Chapter 5: Chemical Reactions in Ideal Gases

which is an unconstrained function of the single variable N_A. Now, setting[1]

$$\frac{d \ln Q}{d N_A} = 0 = \ln q_A - \ln N_A - \frac{N_A}{N_A} - \ln q_B + \ln(N - N_A) + \frac{N - N_A}{N - N_A}$$

$$= \ln q_A - \ln N_A - \ln q_B + \ln(N - N_A) = \ln q_A - \ln N_A - \ln q_B + \ln N_B$$

(5.3-3)

Therefore, the condition for equilibrium is that

$$\frac{q_A}{N_A} = \frac{q_B}{N_B} \quad \text{or} \quad \frac{N_B}{N_A} = \frac{q_B}{q_A} \quad (5.3\text{-}4)$$

This equation looks very much like an equilibrium constant relationship. It can be made to look more so by replacing particle numbers by number densities or number concentrations, that is

$$K_N \equiv \frac{(q_B/V)}{(q_A/V)} = \frac{(N_B/V)}{(N_A/V)} \quad \text{and} \quad \ln K_N = \ln\left(\frac{q_B}{V}\right) - \ln\left(\frac{q_A}{V}\right) \quad (5.3\text{-}5)^{[2]}$$

This is an interesting result, since it establishes that the equilibrium constant for a reacting ideal gas mixture can be computed from a ratio of single-particle partition functions.

The second method of computing the equilibrium concentrations in the simple reaction being considered is by computing the average value of the number of A molecules using the probability $p(N_A)$ of the occurrence of a state in which N_A molecules of species A are present, which is

$$p(N_A) = \frac{1}{Q} \frac{q_A^{N_A} q_B^{N-N_A}}{N_A!(N-N_A)!} = \frac{1}{Q} \frac{1}{N!}\left(\frac{N!}{N_A!(N-N_A)!} q_A^{N_A} q_B^{N-N_A}\right) \quad (5.3\text{-}6)$$

From this we obtain

$$\overline{N_A} = \sum_{N_A=0}^{N} N_A p(N_A) = \frac{1}{Q}\frac{1}{N!}\sum_{N_A=0}^{N} \frac{N_A N!}{N_A!(N-N_A)!} q_A^{N_A} q_B^{N-N_A}$$

$$= \frac{\frac{1}{N!}\sum_{N_A=0}^{N} \frac{N_A N!}{N_A!(N-N_A)!} q_A^{N_A} q_B^{N-N_A}}{\sum_{N_A=0}^{N} \frac{q_A^{N_A} q_B^{N-N_A}}{N_A!(N-N_A)!}}$$

$$= \frac{1}{\frac{(q_A+q_B)^N}{N!}} \frac{1}{N!}\sum_{N_A=0}^{N} \frac{N!}{(N_A-1)!(N-N_A)!} q_A^{N_A} q_B^{N-N_A} \quad (5.3\text{-}7)$$

---

[1] Note that N_A is an integer variable; however, as we are dealing with the order of 10^{23} molecules, we can consider it to be a continuous variable, as the difference between N_A and N_A+1 is so small. So we can take the derivative with respect to N_A.

[2] Because of the magnitudes of the individual partition functions, it is frequently desirable to compute $\ln q$ rather than q directly, and therefore $\ln K_N$ or $\ln K_C$.

5.3 Three Different Derivations of the Chemical Equilibrium Constant in an Ideal Gas Mixture

As shown in the Appendix to this chapter, by further analysis using the binomial expansion, this reduces to

$$\overline{N_A} = N \left(\frac{q_A}{q_A + q_B} \right) \tag{5.3-8}$$

and

$$\overline{N_B} = \overline{(N - N_A)} = \overline{N} - \overline{N_A} = N \left(1 - \frac{q_A}{q_A + q_B} \right) = \frac{q_B N}{q_A + q_B} \tag{5.3-9}$$

Therefore

$$\frac{\overline{N_A}}{\overline{N_B}} = \frac{\dfrac{q_A N}{(q_A + q_B)}}{\dfrac{q_B N}{(q_A + q_B)}} = \frac{q_A}{q_B} \tag{5.3-10}$$

Consequently, we again obtain the following for the molecular concentration-based equilibrium constant

$$K_N = \text{molecular concentration-based equilibrium constant} = \frac{(N_B/V)}{(N_A/V)} = \frac{(q_B/V)}{(q_A/V)} \tag{5.3-11}$$

The third method of obtaining the equilibrium constant is to use what is called the Maximum Term Method. We start by again noting that the probability that the system contains exactly N_A molecules of type A and $N - N_A$ molecules of type B is given by Eq. 5.3-6:

$$p(N_A) = \frac{q_A^{N_A} q_B^{N-N_A}}{N_A!(N-N_A)!} \frac{1}{Q} \tag{5.3-12}$$

where the partition function

$$Q = \frac{(q_A + q_B)^N}{N!} \tag{5.3-13}$$

does not depend on the value of N_A.

The most likely state of the system is that state for which $p(N_A)$ has a maximum value; we will use N_A^* to indicate this value of N_A. Thus, we want to find the value of N_A for which $p(N_A)$ or, equivalently, $\ln p(N_A)$, has a maximum value. To find this value, we assume that since the number of molecules is so large, we can treat N and N_A as continuous variables and set $d \ln p(N_A)/dN_A = 0$. So starting from

$$\ln p(N_A) = N_A \ln q_A + (N - N_A) \ln q_B - \ln N_A! - \ln(N - N_A)! - \ln Q$$

and using Stirling's approximation, we obtain

$$\ln p(N_A) = N_A \ln q_A + (N - N_A) \ln q_B - N_A \ln N_A + N_A - (N - N_A) \ln(N - N_A)$$
$$+ (N - N_A) - \ln Q$$

70 Chapter 5: Chemical Reactions in Ideal Gases

So that

$$\frac{d \ln p(N_A)}{dN_A} = \ln q_A - \ln q_B - \ln N_A - 1 + 1 + \ln(N - N_A) + (N - N_A)\frac{1}{(N-N_A)} - 1$$

$$= \ln q_A - \ln q_B - \ln N_A + \ln(N - N_A) = 0 \qquad (5.3\text{-}14)$$

Therefore, the most probable state is

$$\ln q_A - \ln q_B - \ln N_A^* + \ln(N - N_A^*) = 0$$

or

$$\frac{q_A}{q_B} = \frac{N_A^*}{N - N_A^*} = \frac{N_A^*}{N_B^*} \quad \text{and} \quad \frac{N_A^*/V}{N_B^*/V} = \frac{q_A/V}{q_B/V} = K_N \qquad (5.3\text{-}15)$$

Consequently, the most probable state of the system is that one in which the number ratio of the two species has the same numerical value as the partition function ratio. This is precisely the result obtained earlier.

The different procedures for obtaining the equilibrium constant are worthy of some further discussion. What was done first was to invoke the thermodynamic result that the state of equilibrium was a state of minimum Helmholtz energy subject to the restriction of constant total particle number, temperature and volume. The second method established that the same result could be obtained directly by determining the average number of particles of each species present. The third method established that this equilibrium state corresponds to the one state with the largest contribution to the partition function; that is, the most probable state of the system. This last procedure, referred to as the maximum-term method, has wide application in statistical thermodynamics, as will be seen. To understand why the maximum-term method gives the correct answer, we need to examine the magnitude of the fluctuations that are likely to occur. We do this in the following section.

5.4 FLUCTUATIONS IN A CHEMICALLY REACTING SYSTEM

An interesting question that arises is why these three different methods give the same result, and how large an error is incurred in computing the thermodynamic properties if one uses only the maximum term of the partition function for the calculation, rather than the complete partition function sum. This question can be answered by comparing logarithms of the partition functions. In particular, using

$$Q = \frac{(q_A + q_B)^N}{N!} \qquad (5.4\text{-}1)$$

and

$$Q^* = \frac{q_A^{N_A} q_B^{N_B}}{N_A! \, N_B!} \quad \text{where} \quad \frac{N_B}{N_A} = \frac{q_B}{q_A} \qquad (5.4\text{-}2)$$

Then

$$\ln Q = N \ln(q_A + q_B) - N \ln N + N \qquad (5.4\text{-}3)$$

and

$$\ln Q^* = N_A \ln q_A - N_A \ln N_A + N_A + N_B \ln q_B - N_B \ln N_B + N_B$$

$$= N_A \ln q_A - N_A \ln N_A + N_A + N_B \ln q_A + N_B \ln \left(\frac{N_B}{N_A}\right) - N_B \ln N_B + N_B$$

$$= (N_A + N_B) \ln(q_A/N_A) + N$$

But

$$\frac{q_A}{N_A} = \frac{q_B}{N_B} = \frac{q_A + q_B}{N_A + N_B} = \frac{q_A + q_B}{N}$$

so

$$\ln Q^* = N \ln \left(\frac{q_A + q_B}{N}\right) + N = N \ln(q_A + q_B) - N \ln N + N \quad \text{(5.4-4)}$$

Consequently, we are led to the conclusion that the value of the complete partition function (or its logarithm) and that obtained considering only the maximum term of the partition function are identical. In fact, this is not quite true. If we had retained the next term is Stirling's approximation

$$\ln N! = N \ln N - N + \frac{1}{2} \ln 2\pi N \quad \text{(5.4-5)}$$

we would have found that $\ln Q$ and $\ln Q^*$ differed, but only very slightly.

It is somewhat surprising that, for all practical purposes, the partition function for a chemically reacting gas is equal to the value of the partition function computed in just the equilibrium state. The implication is that the state of equilibrium is overwhelmingly probable compared to any of the other states of the system. We now demonstrate that this is true. The probability of occurrence of a state in which there are N_A molecules of species A and N_B molecules of species B is

$$p(N_A, N_B) = \frac{\frac{q_A^{N_A} q_B^{N_B}}{N_A! N_B!}}{Q} \quad \text{where} \quad Q = \frac{(q_A + q_B)^N}{N!} \quad \text{(5.4-6)}$$

or

$$\ln p(N_A, N_B) = N_A \ln q_A - \ln N_A! + N_B \ln q_B - \ln N_B! - \ln Q$$

and

$$\ln p(N_A, N) = N_A \ln q_A - \ln N_A! + (N - N_A) \ln q_B - \ln(N - N_A)! - \ln Q \quad \text{(5.4-7)}$$

Again using N_A^* to be the value at equilibrium—that is, when the values of N_A and N_B are related by Eq. 5.4-2—we now want to relate the probability of occurrence of some other state of the system to the probability of the equilibrium state. This can be done by a Taylor series expansion of the probability function if the state is not

72 Chapter 5: Chemical Reactions in Ideal Gases

too far removed from the equilibrium state, that is

$$\ln p(N_A, N) = \ln p(N_A, N)|_{N_A=N_A^*} + \left.\frac{\partial \ln p(N_A, N)}{\partial N_A}\right|_{N_A=N_A^*} (N_A - N_A^*)$$

$$+ \frac{1}{2} \left.\frac{\partial^2 \ln p(N_A, N)}{\partial N_A^2}\right|_{N_A=N_A^*} (N_A - N_A^*)^2 + \cdots \quad (5.4\text{-}8)$$

The linear term in this expansion vanishes by Eq. 5.3-14, and it can easily be shown that

$$\left.\frac{\partial^2 \ln p(N_A, N)}{\partial N_A^2}\right|_{N_A=N_A^*} = \frac{-N}{N_A^*(N - N_A^*)} \quad (5.4\text{-}9)$$

so

$$\ln p(N_A, N) = \ln p(N_A^*, N) - \frac{N(N_A - N_A^*)^2}{2(N_A^*)(N - N_A^*)} \quad (5.4\text{-}10)$$

or

$$p(N_A, N) = p(N_A^*, N) e^{-\frac{1}{2}\left(\frac{N_A^* - N_A}{\sigma}\right)^2} \quad \text{where} \quad \sigma^2 = \frac{N_A^*(N - N_A^*)}{N} \quad (5.4\text{-}11)$$

Equation 5.4-11 shows that the probability of occurrence of a state with N_A molecules of species A can be written, at least to first approximation, as a Gaussian distribution around the most probable state N_A^*. As was mentioned in Chapter 3, the quantity σ is the standard deviation of the distribution, and a measure of the breadth of the Gaussian distribution. Assuming that the equilibrium constant is finite, so that N_A and N_B are of the order of the number of molecules N, then σ^2 is of the order of N, and

$$\frac{\sigma}{N} = \frac{\mathcal{O}(N^{1/2})}{N} = \mathcal{O}\left(\frac{1}{\sqrt{N}}\right)$$

For example, for 1 mole (almost 10^{24} molecules), assuming that the reaction does not go completion, so that both N_A^* and $(N - N_A^*)$ are approximately of the order of 10^{24}, then

$$p(N_A, N) = p(N_A^*, N) e^{-\left(\frac{N_A^* - N_A}{\sqrt{\frac{N_A^*(N-N_A^*)}{N}}}\right)^2} \approx p(N_A^*, N) e^{-\left(\frac{N_A^* - N_A}{\sqrt{\frac{10^{24} \times 10^{24}}{10^{24}}}}\right)^2}$$

$$= p(N_A^*, N) e^{-\left(\frac{N_A^* - N_A}{10^{12}}\right)^2} \quad (5.4\text{-}12)$$

Since $(N_A^* - N_A)$ is of the order of 10^{24}, the exponential is of approximately e^{-24}, unless N_A is very, very close to the most likely (equilibrium) value of N_A^*. This result shows that the distribution of probability about the most likely distribution of particles is very sharply peaked, with a standard deviation of the distribution, σ/N, of the order of 10^{-12}. Consequently, the distribution is so sharply peaked that, in general, we could not measure the fluctuations around the most probable state. Therefore, the actual probability function for a large collection of molecules in chemical equilibrium is much more like a delta function than a broad Gaussian distribution. It is for this reason that the ratio of the number of molecules in the species A state to that in the B state at any instant can be obtained, to an excellent degree of accuracy, from looking only at the most probable state or the state of maximum value of $\ln p(N_A, N_B)$, for which

$$\frac{q_A}{q_B} = \frac{N_A}{N_B}$$

The result obtained above is an important one. First, it demonstrates that in a system with a large number of molecules the probability of occurrence of any state other than the equilibrium state is infinitesimally small—indeed, one part in a quadrillion—compared to that of the equilibrium state. Those departures of the macroscopic system from the equilibrium state that do have a finite probability of occurring, that is, those departures for which

$$\frac{N_A - N_A^*}{N_A^*} \sim 10^{-12}$$

are so infinitesimally small that they are beyond the range of most measurements. Also, the fact that the probability of occurrence of the equilibrium state is so very much greater than that of any other state explains why it is satisfactory to compute the partition function of the chemically reacting system by using the value of the partition function for only the equilibrium (most probable) state.

5.5 THE CHEMICALLY REACTING GAS MIXTURE: THE GENERAL CASE

Consider the general chemical reaction

$$\alpha A + \beta B + \cdots \leftrightarrow \rho R + \sigma S + \cdots \qquad (5.5\text{-}1)$$

which we will write as

$$\sum_i \nu_i I = 0 \qquad (5.5\text{-}2)$$

where ν_i is the stoichiometric coefficient for species i in the reaction. Of the three methods considered above for computing the equilibrium state, the simplest to extend to the general case is the first, based on classical thermodynamics. From classical thermodynamics we know that the condition for equilibrium is

$$\sum \nu_i \mu_i = 0 \qquad (5.5\text{-}3)$$

and from statistical mechanics for ideal gases (see Chapters 3 and 4) that

$$\mu_i = -kT \ln \frac{q_i}{N_i} \qquad (5.5\text{-}4)$$

74 Chapter 5: Chemical Reactions in Ideal Gases

Therefore, the general condition for chemical equilibrium in an ideal gas mixture is

$$\sum_i \nu_i \ln \frac{q_i}{N_i} = 0 = \sum_i \ln \left(\frac{q_i}{N_i}\right)^{\nu_i} = \ln \prod_i \left(\frac{q_i}{N_i}\right)^{\nu_i} \quad (5.5\text{-}5a)$$

or

$$\prod_i q_i^{\nu_i} = \prod_i N_i^{\nu_i} \quad (5.5\text{-}5b)$$

and

$$\prod_i \left(\frac{q_i}{V}\right)^{\nu_i} = \prod_i (q_i')^{\nu_i} = \prod_i \left(\frac{N_i}{V}\right)^{\nu_i} = K_N \quad (5.5\text{-}6a)$$

where, again, K_N is the equilibrium constant in terms of numbers of molecules. Using instead molar concentrations $C_i = N_i/(V N_{Av})$, where N_{Av} is Avogadro's number

$$\prod_i \left(\frac{q_i}{V}\right)^{\nu_i} = \prod_i (C_i N_{Av})^{\nu_i} = (N_{Av})^{\sum_i \nu_i} \prod_i (C_i)^{\nu_i} = K_N \quad \text{we have}$$

$$K_C = \prod_i (C_i)^{\nu_i} = \frac{K_N}{(N_{Av})^{\sum_i \nu_i}} = (N_{Av})^{-\sum_i \nu_i} \prod_i \left(\frac{q_i}{V}\right)^{\nu_i}$$

$$= (N_{Av})^{-\sum_i \nu_i} \prod_i (q_i')^{\nu_i} \quad (5.5\text{-}6b)$$

where $q_i' = q_i/V$. Equation 5.5-6 is valid for any chemical reaction in an ideal gas mixture, not only the unimolecular reaction considered earlier.

ILLUSTRATION

Nitric oxide, NO, forms at high temperatures from the reaction of nitrogen and oxygen, $N_2 + O_2 \rightleftarrows 2NO$; for example, in an internal combustion engine (this is one of the reasons modern gasoline cars have catalytic converters). Estimate the fraction of air (21 volume % oxygen and 79 volume % nitrogen) that will be converted into NO at 3000 K.

SOLUTION

Starting with one mole of air, using x to be the extent of reaction, we have the following:

Gas	Initial	Final
O_2	0.21	$0.21 - x$
N_2	0.79	$0.79 - x$
NO	0	$2x$

So the equilibrium relation is

$$K_N = \frac{(2x)^2}{(0.21-x)(0.79-x)} = \frac{4x^2}{(0.21-x)(0.79-x)}$$

5.5 The Chemically Reacting Gas Mixture: The General Case

To calculate x the value of the equilibrium constant at 3000 K is needed. This is calculated from

$$K_N = \prod_i \left(\frac{q_i}{V}\right)^{\nu_i} = \frac{(q_{NO}/V)^2}{(q_{N_2}/V)(q_{O_2}/V)}$$

$$= \frac{(q_{\text{trans,NO}} q_{\text{rot,NO}} q_{\text{vib,NO}} q_{\text{elect,NO}}/V)^2}{(q_{\text{trans,}N_2} q_{\text{rot,}N_2} q_{\text{vib,}N_2} q_{\text{elect,}N_2}/V)(q_{\text{trans,}O_2} q_{\text{rot,}O_2} q_{\text{vib,}O_2} q_{\text{elect,}O_2}/V)}$$

$$= \left(\frac{M_{NO}^2}{M_{N_2} M_{O_2}}\right)^{\frac{3}{2}} \left(\frac{\sigma_{N_2} \Theta_{r,N_2} \sigma_{O_2} \Theta_{r,O_2}}{(\sigma_{NO} \Theta_{r,NO})^2}\right) \left(\frac{\omega_{\text{elect,NO}}}{1 - e^{-\Theta_{v,NO}/T}}\right)^2$$

$$\times \left(\frac{1 - e^{-\Theta_{v,N_2}/T}}{\omega_{\text{elect,}N_2}}\right) \left(\frac{1 - e^{-\Theta_{v,O_2}/T}}{\omega_{\text{elect,}O_2}}\right) e^{(2D_{0,NO} - D_{0,N_2} - D_{0,O_2})/kT}$$

$$= \left[\left(\frac{30.01^2}{28.02 \times 32}\right)^{\frac{3}{2}} \left(\frac{2 \times 2.89 \times 2 \times 2.08}{(1 \times 2.45)^2}\right) \left(\frac{2}{1 - e^{-2745/3000}}\right)^2 \left(\frac{1 - e^{-3390/3000}}{1}\right) \right.$$

$$\left. \times \left(\frac{1 - e^{-2278/3000}}{2}\right) e^{(2 \times 4.43 - 9.76 - 5.08)1.602 \times 10^{-12}/(1.38044 \times 10^{16} \times 3000)}\right]$$

$$= 1.007 \times 4.006 \times 11.13 \times 0.677 \times 0.266 \times 8.987 \times 10^{-11} = 7.264 \times 10^{-10}$$

where each term has been calculated separately to show the dominance of the electronic energy difference in the calculation. Using this value of the equilibrium constant, $x = 5.489 \times 10^{-6}$ or 5.489 parts per million (ppm).

The thermodynamic properties of the reacting system can, based on the analysis in this section, be computed from the partition function

$$Q(T, V, \underline{N}) = \prod_{i=1}^{S} \frac{q_i^{N_i^*}}{N_i^*!} \tag{5.5-7}$$

However, to proceed, it should be noted that using the stoichiometric coefficient notation of Eq. 5.5-2, the number of molecules of species i present at any time (N_i) is related to the number of molecules of this species initially present, $N_{i,o}$, by the relation

$$N_i = N_{i,o} + \nu_i X \tag{5.5-8}$$

so that

$$\partial N_i = \nu_i \partial X \tag{5.5-9}$$

Chapter 5: Chemical Reactions in Ideal Gases

(Note that defined this way, X has units of number of molecules.) Using this notation, the pressure of the reacting system is computed from

$$P(\underline{N}, V, T) = kT \left(\frac{\partial \ln Q}{\partial V}\right)_T = kT \sum_{i=1}^{S} \frac{\partial}{\partial V}\bigg|_T (N_i^* \ln q_i - N_i^* \ln N_i^* + N_i^*)$$

$$= kT \sum_{i=1}^{S} \left[\ln\left(\frac{q_i}{N_i^*}\right)\left(\frac{\partial N_i^*}{\partial V}\right)_T + \frac{N_i^*}{V}\right]$$

$$= \frac{kT}{V} \sum_{i=1}^{S} N_i^* + kT \left(\frac{\partial X}{\partial V}\right)_T \sum_{i=1}^{S} \ln\left(\frac{q_i}{N_i^*}\right)^{\nu_i} \quad (5.5\text{-}10)$$

where the last term on the right-hand side vanishes due to Eq. 5.5-5a. Therefore

$$P(\underline{N}, V, T) = \frac{kT}{V} \sum_{i=1}^{S} N_i^* \quad (5.5\text{-}11)$$

Likewise, the internal energy of the reacting system is

$$U(T, V, \underline{N}) = kT^2 \left(\frac{\partial \ln Q}{\partial T}\right)_V$$

$$= kT^2 \sum_{i=1}^{S} \left[\ln\left(\frac{q_i}{N_i^*}\right)\left(\frac{\partial N_i^*}{\partial T}\right)_V + N_i^* \left(\frac{\partial \ln q_i}{\partial T}\right)_V\right]$$

$$= kT^2 \sum_{i=1}^{S} \left[\ln\left(\frac{q_i}{N_i^*}\right) \nu_i \left(\frac{\partial X}{\partial T}\right)_V + N_i^* \left(\frac{\partial \ln q_i}{\partial T}\right)_V\right]$$

$$= kT^2 \left(\frac{\partial X}{\partial T}\right)_V \ln \prod_{i=1}^{S} \left(\frac{q_i}{N_i^*}\right)^{\nu_i} + kT^2 \sum_{i=1}^{S} \left[N_i^* \left(\frac{\partial \ln q_i}{\partial T}\right)_V\right]$$

$$= kT^2 \sum_{i=1}^{S} N_i^* \left(\frac{\partial \ln q_i}{\partial T}\right)_V = \sum_{i=1}^{S} U_i(T, V, N_i^*) \quad (5.5\text{-}12)$$

where again Eq. 5.5-5a has been used, and U_i is the internal energy of N_i^* molecules of species i at the temperature and species equilibrium number, just as in an ideal gas mixture. Equations 5.5-11 and 5.5-12 are the result of the fact that this is an ideal gas mixture. However, remember when using these equations that each N_i^* changes with temperature, and therefore the internal energy of the reacting mixture has an additional temperature dependence above that of a nonreacting mixture, due to the change in the extent of the interconversion of some species into others with temperature.

5.5 The Chemically Reacting Gas Mixture: The General Case

A not-so-obvious result is obtained by looking at the constant volume heat capacity, C_V, of the reacting gas

$$C_V = \left(\frac{\partial U}{\partial T}\right)_V = \frac{\partial}{\partial T}\bigg|_V kT^2 \sum_{i=1}^{S} N_i^* \left(\frac{\partial \ln q_i}{\partial T}\right)_V$$

$$= 2kT \sum_{i=1}^{S} N_i^* \left(\frac{\partial \ln q_i}{\partial T}\right)_V + kT^2 \sum_{i=1}^{S} \left(\frac{\partial N_i^*}{\partial T}\right)_V \left(\frac{\partial \ln q_i}{\partial T}\right)_V$$

$$+ kT^2 \sum_{i=1}^{S} N_i^* \left(\frac{\partial^2 \ln q_i}{\partial T^2}\right)_V$$

$$= \sum_{i=1}^{S} C_{V,i} + kT^2 \sum_{i=1}^{S} \left(\frac{\partial N_i^*}{\partial T}\right)_V \left(\frac{\partial \ln q_i}{\partial T}\right)_V$$

$$= \sum_{i=1}^{S} C_{V,i} + \left(\frac{\partial X}{\partial T}\right)_V kT^2 \sum_{i=1}^{S} \nu_i \left(\frac{\partial \ln q_i}{\partial T}\right)_V$$

$$= \sum_{i=1}^{S} C_{V,i} + \left(\frac{\partial X}{\partial T}\right)_V \sum_{i=1}^{S} \nu_i kT^2 \left(\frac{\partial \ln q_i}{\partial T}\right)_V$$

$$= \sum_{i=1}^{S} C_{V,i} + \left(\frac{\partial X}{\partial T}\right)_V \sum_{i=1}^{S} \nu_i u_i \qquad (5.5\text{-}13)$$

where u_i is the internal energy per molecule of species i. Here, the first term after the last equal sign is the sum of the heat capacities of the pure components at the equilibrium composition; the second term is new and can be a very significant contribution to the heat capacity as a result of the chemical reaction. This can be seen using the notation of Eq. 5.5-1 so that

$$\Delta_{\text{rxn}} u = \sum_i \nu_i u_i \quad \text{and} \quad C_V = \sum_i C_{V,i} + \left(\frac{\partial X}{\partial T}\right)_V \Delta_{\text{rxn}} u \qquad (5.5\text{-}14)$$

where $\Delta_{\text{rxn}} u$ is the internal energy change of reaction on a molecular (not molar) basis for the stoichiometry of Eqs. 5.5-1 and 5.5-2. Consequently, the last term in the equation above is the contribution to the constant volume heat capacity that is a result of the internal energy change on reaction during the course of the reaction. For engineering purposes, this equation is more conveniently written using molar heat capacities. By using Avogadro's number N_{Av} this results in the following expression:

$$C_V^{\text{tot}} = \sum_i n_i^* C_{V,i} + \left(\frac{\partial x}{\partial T}\right)_V N_{\text{Av}} \Delta_{\text{rxn}} u = \sum_i n_i^* C_{V,i} + \left(\frac{\partial x}{\partial T}\right)_V \Delta_{\text{rxn}} U \qquad (5.5\text{-}15)$$

where now the species heat capacities are on a molar basis, n_i^* are the number of moles of species i present at equilibrium, x is the molar extent of reaction for the stoichiometry of Eq. 5.5-2, and $\Delta_{\text{rxn}} U$ is the molar internal energy change on reaction for this stoichiometry. This reaction contribution to the heat capacity can

78 Chapter 5: Chemical Reactions in Ideal Gases

be very large (as we show in an illustration that follows) for a system with a large energy change on reaction, but only over the temperature range where the extent of reaction is changing appreciably with temperature (that is, over the temperature range where $(\partial x/\partial T)_V$ is nonzero.)

Though we do not consider transport properties here, when the heat capacity is large due to reaction, the thermal conductivity of the gas is also much larger than usual. To see this, consider a dissociation reaction with a large heat release. The gas molecules would then dissociate at the high temperature surface, absorbing heat of reaction; migrate to the cooler surface, re-associate and release the heat of reaction resulting in a large rate of heat transfer and a high effective thermal conductivity. One situation where it is especially important to account for this is in the design of heat shields for spacecraft entering the atmosphere of earth or other planets. Because of frictional heating, high temperatures sufficient to ionize the gas in the boundary layer of the spacecraft can result. Engineers need to account for this in their design.

To illustrate how to account for the change in heat capacity, consider the problem of computing the degree of ionization of a gas of atoms as a function of temperature. The partition function for the mixture of atoms, ions, and electrons is

$$Q = \frac{q_A^{N_A} q_i^{N_i} q_e^{N_e}}{N_A! \, N_i! \, N_e!} \tag{5.5-16}$$

where the subscripts A, i, and e designate the atoms, ions, and electrons, respectively. The number densities of these species are interrelated by the equilibrium relationship

$$\frac{(q_i/V)(q_e/V)}{q_A/V} = \frac{(N_i/V)(N_e/V)}{(N_A/V)} \tag{5.5-17}$$

or, using the molecular extent of reaction, X, as the independent variable, we have

$$N_A = N_A^o - X, \quad N_e = X \quad \text{and} \quad N_i = X$$

$$\ln Q = (N_A^o - X) \ln\left(\frac{q_A}{N_A^o - X}\right) + X \ln\left(\frac{q_i}{X}\right) + X \ln\left(\frac{q_e}{X}\right) + (N_A^o + X) \tag{5.5-18}$$

and

$$\frac{(q_i/V)(q_e/V)}{q_A/V} = \frac{X^2}{V(N_A^o - X)} \tag{5.5-19}$$

The partition functions for each of the species are

$$\left(\frac{q_A}{V}\right) = \left(\frac{2\pi m_A kT}{h^2}\right)^{\frac{3}{2}} \omega_{A,\text{elect},1}; \quad \left(\frac{q_e}{V}\right) = \left(\frac{2\pi m_e kT}{h^2}\right)^{\frac{3}{2}} \omega_{e,\text{elect},1}$$

and

$$\left(\frac{q_i}{V}\right) = \left(\frac{2\pi m_i kT}{h^2}\right)^{\frac{3}{2}} \omega_{i,\text{elect},1} e^{-\varepsilon_{i,\text{elect},1}/kT} \tag{5.5-20}$$

Here, we have taken the state of zero energy as the atom at rest and have attributed the electronic energy difference (between the atoms and the ions) to the ions. Furthermore, the partition function for the electron has been computed from the particle-in-the-box model. (Also, we have neglected an important interaction term among charged species that results in the so-called ionization potential lowering, a concept that is beyond the current discussion.) The electronic degeneracy for the atom is unity, and it is equal to 2 for the ion and electron. Thus

$$K_N = \frac{(q_i/V)(q_e/V)}{(q_A/V)} = 4\left(\frac{2\pi m_e kT}{h^2}\right)^{\frac{3}{2}} e^{-\varepsilon_{i,\text{elect},1}/kT}$$

$$= \frac{(N_i/V)(N_e/V)}{(N_A/V)} = \frac{X^2}{V(N_A^o - X)} \quad (5.5\text{-}21)$$

where the masses of the ion and the atom have been taken to be identical. Now, from the ideal gas law for the reacting system (Eq. 5.5-11)

$$PV = \sum_j N_j kT = [N_A + N_i + N_e]kT = (N_A^o - X + X + X)kT = (N_A^o + X)kT$$

$$(5.5\text{-}22)$$

so

$$V = (N_A^o + X)kT/P$$

and combining Eqs. 5.1-19 and 5.1-21

$$\frac{X^2}{(N_A^{o2} - X^2)} = \frac{4kT}{P}\left(\frac{2\pi m_e kT}{h^2}\right)^{\frac{3}{2}} e^{-\varepsilon_{i,\text{elect},1}/kT} = F(T, P) \quad (5.5\text{-}23)$$

Now, defining $Y = X/N_A^o$ to be the degree of ionization, we have

$$Y = \sqrt{\frac{F}{1+F}} \quad (5.5\text{-}24)$$

Also

$$\left(\frac{\partial \ln q_A}{\partial T}\right)_V = \left(\frac{\partial \ln q_e}{\partial T}\right)_V = \frac{3}{2T}; \quad \text{and} \quad \left(\frac{\partial \ln q_i}{\partial T}\right)_V = \frac{3}{2T} + \frac{\varepsilon_{i,\text{elect},1}}{kT^2}$$

so that

$$U = \frac{3}{2}kT(N_A + N_i + N_e) + N_i \varepsilon_{i,\text{elect},1} = \frac{3}{2}kT(N_A^o + X) + X\varepsilon_{i,\text{elect},1}$$

or

$$U(\text{per mole of argon initially present}) = \frac{3}{2}RT(1+Y) + Y\varepsilon_{i,\text{elect},1} \quad (5.5\text{-}25)$$

and

$$C_V = \left(\frac{\partial U}{\partial T}\right)_V = \frac{3}{2}R(1+Y) + \frac{3}{2}RT\left(\frac{\partial Y}{\partial T}\right)_V + \varepsilon_{i,\text{elect},1}\left(\frac{\partial Y}{\partial T}\right)_V$$

$$= \frac{3}{2}R(1+Y) + \left(\frac{3}{2}RT + \varepsilon_{i,\text{elect},1}\right)\left(\frac{\partial Y}{\partial T}\right)_V \quad (5.5\text{-}26)$$

Chapter 5: Chemical Reactions in Ideal Gases

Also

H(per mole of atoms initially present) $= U + PV$

$$= \frac{3}{2}RT(1+Y) + Y\varepsilon_{i,\text{elect},1} + (1+Y)RT$$

$$= \frac{5}{2}RT(1+Y) + Y\varepsilon_{i,\text{elect},1} \qquad (5.5\text{-}27)$$

and finally

$$C_P = \left(\frac{\partial H}{\partial T}\right)_P = \frac{5}{2}R(1+Y) + \frac{5}{2}RT\left(\frac{\partial Y}{\partial T}\right)_P + \varepsilon_{i,\text{elect},1}\left(\frac{\partial Y}{\partial T}\right)_P$$

$$= \frac{5}{2}R(1+Y) + \left(\frac{5}{2}RT + \varepsilon_{i,\text{elect},1}\right)\left(\frac{\partial Y}{\partial T}\right)_P \qquad (5.5\text{-}28)$$

It is interesting to note that, at low temperatures, $C_P = 5R/2$, while at high temperatures $C_P = 5R$. Can you explain why this is so?[3]

5.6 TWO ILLUSTRATIONS

The Ionization of Argon

One mole of gaseous argon at 1 atmosphere is to be heated at constant pressure to very high temperatures. Using the equations above:

(a) Compute and plot the degree of ionization of argon as a function of temperature from 1000 to 30000 K at 0.01, 0.1, 1 and 33.6 bar. The ionization energy for the first ionization of argon

$$\text{Ar} \leftrightarrow \text{Ar}^+ + e$$

is 15.76 electron volts. The mass of an electron is 9.1083×10^{-28}g.

(b) Compute and plot the constant pressure heat capacity for this plasma over the same temperature and pressure range.

SOLUTION

The number-based equilibrium constant is calculated as follows:

$$K_N = 4\left(\frac{2\pi m_e kT}{h^2}\right)^{\frac{3}{2}} e^{-\varepsilon_{i,\text{elect},1}/kT}$$

$$= 4 \times 1.88 \times 10^{26} \times (9.1083 \times 10^{-28})^{3/2} T^{3/2} e^{-15.76 \times 1.602 \times 10^{-12}/(1.38044 \times 10^{-16} \times T)} \frac{\text{ions}}{\text{m}^3}$$

---

[3] The answer is that before dissociation, only the atoms are present; each atom has three translational motions. After dissociation is complete, each atom has been replaced by an ion and an electron, for a total of six translational motions.

The values are shown in the figure below. From Eq. 5.5-23, the quantity of more interest is

$$F(T,P) = \frac{4kT}{P}\left(\frac{2\pi m_e kT}{h^2}\right)^{\frac{3}{2}} e^{-\varepsilon_{i,\text{elect},1}/kT} = \frac{1.38044 \times 10^{-23}\left(\frac{J}{K}\right) \times T(K)}{P(\text{bar}) \times 10^5 \left(\frac{J/m^3}{\text{bar}}\right)} K_N(m^{-3})$$

$$= \frac{1.38044 \times 10^{-28} \times T(K)}{P(\text{bar})} K_N$$

To calculate the degree of ionization, Y, Eq. 5.5-24 is used. Finally, the heat capacity is computed using Eq. 5.5-28 after computing the degree of ionization $(\partial Y/\partial T)_V$. The results of all these calculations are shown in the figures below.

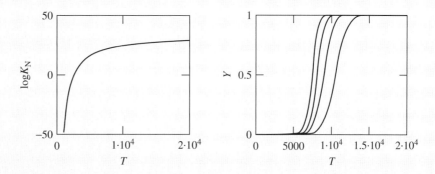

Log of the equilibrium constant K_N (left) and the degree of ionization Y (right) as a function of temperature. In the degree of ionization figure the lines are in order of the pressures are 0.01, 0.1, 1 and 33.6 bar, respectively.

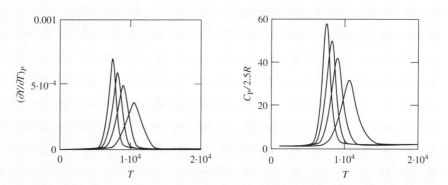

The derivative of degree of ionization Y with respect to temperature (left), $(\partial Y/\partial T)_P$ and the reduced constant pressure heat capacity $C_P/2.5R$ as a function of temperature. In these figures the lines are in order of the pressures are 0.01, 0.1, 1 and 33.6 bar, respectively.[4]

---

[4]Does the pressure dependence of the results shown in the figures agree with LeChatelier's principle?

The Adsorption of a Gas unto a Solid

Gas molecules at a pressure low enough that the gas can be considered ideal are in contact with a two-dimensional surface at which some of the molecules may be adsorbed. The partition function for a single gas molecule is q_{gas}, and the partition function for a single molecule adsorbed on the two-dimensional surface is q_{ad}. If there are M adsorption sites on a two-dimensional surface, and N identical molecules adsorbed, where $M \gg N$ so that the adsorbed molecules do not interact with each other, the partition function for the adsorbed molecules is

$$Q(N, M, T) = \frac{M!}{N!(M-N)!}[q_{ad}]^N$$

where the factorials arise from the number of ways of distributing N indistinguishable molecules over M distinguishable adsorption sites (since they are fixed on the surface of the graphite.)

Use this partition function to develop an expression for the fraction of the M adsorption sites that are occupied as a function of the pressure of an ideal gas, and q_{gas} and q_{ad}.

SOLUTION

The chemical potential of an ideal gas molecule is

$$\mu_{gas} = -kT\left(\frac{\partial \ln Q}{\partial N}\right)_T = -kT\frac{\partial}{\partial N}(N\ln q_{gas} - N\ln N + N)$$

$$= -kT\left(\ln q_{gas} - \ln N - \frac{N}{N} + 1\right) = -kT\ln\left(\frac{q_{gas}}{N}\right)$$

$$= -kT\ln\left(\frac{q'_{gas}V}{N}\right) = -kT\ln\left(\frac{q'_{gas}kT}{P}\right)$$

where as usual $q'_{gas} = q_{gas}/V$. The chemical potential of a gas molecule adsorbed on the surface is

$$\mu_{ad} = -kT\left(\frac{\partial \ln Q}{\partial N}\right)_T$$

$$= -kT\frac{\partial}{\partial N}[N\ln q_{ad} - N\ln N + N - (M-N)\ln(M-N) - (M-N)]$$

$$= -kT\left(\ln q_{ad} - \ln N - \frac{N}{N} + 1 + \ln(M-N) + 1 - 1\right) = -kT\ln\left(\frac{q_{ad}(M-N)}{N}\right)$$

At equilibrium, $\mu_{gas} = \mu_{ad}$, so that

$$\mu_{ad} = -kT\ln\left(\frac{q_{ad}(M-N)}{N}\right) = \mu_{gas} = -kT\ln\left(\frac{q'_{gas}kT}{P}\right)$$

$$\frac{q_{ad}(M-N)}{N} = \frac{q'_{gas}kT}{P} \quad \text{or} \quad M - N = N\frac{q'_{gas}kT}{q_{ad}P}$$

$$M = N\left(1 + \frac{q'_{gas}kT}{q_{ad}P}\right)$$

Therefore, the fractional coverage $\theta = N/M$ is

$$\theta = \frac{N}{M} = \frac{1}{1 + \frac{q'_{gas} kT}{q_{ad} P}} = \frac{\left(\frac{P}{kT}\right)}{\frac{q'_{gas}}{q_{ad}} + \frac{P}{kT}} = \frac{P}{P + kT \frac{q'_{gas}}{q_{ad}}} = \frac{P}{P + \lambda}$$

This equation shows that at low pressure $\lambda = \frac{kT q'_{gas}}{q_{ad}} \gg P$, the extent of adsorption (fractional coverage) increases linearly with pressure, while at high pressures $P \gg \frac{kT q'_{gas}}{q_{ad}} = \lambda$, the coverage saturates at complete coverage $\theta = 1$. The fractional coverage is an S-shaped curve between these two extremes and is referred to as the Langmuir adsorption isotherm; λ is the temperature-dependent Langmuir parameter.

APPENDIX: THE BINOMIAL EXPANSION[5]

The canonical partition function for a binary mixture of molecules A and B that undergo the chemical reaction A ↔ B is, as shown in the text

$$\sum_{N_A=0}^{N} \frac{q_A^{N_A} q_B^{N-N_A}}{N_A!(N-N_A)!} \tag{5.2-7}$$

We can evaluate this using the general form of the binomial expansion, written here as

$$\sum_{M=0}^{N} \frac{N!}{M!(N-M)!} x^M y^{N-M} = (x+y)^N \tag{A5-1}$$

or to match Eq. 5.2-7

$$\frac{1}{N!} \sum_{M=0}^{N} \frac{N!}{M!(N-M)!} x^M y^{N-M} = \frac{(x+y)^N}{N!} \tag{A5-2}$$

Therefore, the partition function for this reacting mixture is

$$Q = \sum_{M=0}^{N} \frac{q_A^M q_B^{N-M}}{M!(N-M)!} = \frac{1}{N!} \sum_{M=0}^{N} \frac{N! q_A^M q_B^{N-M}}{M!(N-M)!} = \frac{(q_A + q_B)^N}{N!} \tag{A5-3}$$

Another series summation of interest, based in Eq. 5.3-7, is

$$\sum_{N_A=0}^{N} \frac{N_A q_A^{N_A} q_B^{N-N_A}}{N_A!(N-N_A)!} \tag{A5-4}$$

---

[5] According to Wikipedia, the binomial expansion is attributed to the 17th-century mathematician Blaise Pascal, but was known to other mathematicians including Halayuhda in India in the 10th century, Omar Khayyám in Persia in the 11th century, and Yang Hui in China in the 13th century, all of whom derived similar results.

84 Chapter 5: Chemical Reactions in Ideal Gases

that arises in the equation

$$\overline{N_A} = \sum_{N_A=0}^{N} N_A p(N_A) = \frac{1}{Q} \sum_{N_A=0}^{N} \frac{N_A q_A^{N_A} q_B^{N-N_A}}{N_A!(N-N_A)!} \quad (5.3\text{-}7)$$

The summation of Eq. A5-4, written in generic form so that the result can also be used later, is

$$\sum_{M=0}^{N} \frac{M}{M!(N-M)!} x^M y^{N-M} = \sum_{M=1}^{N} \frac{M}{M!(N-M)!} x^M y^{N-M} \quad (\mathbf{A5\text{-}5})$$

Here, in the second form of this equation, we have changed the lower limit of the summation from $M = 0$ to $M = 1$, since the $M = 0$ term in the sum is 0 as a result of M in the numerator. The goal is to transform Eq. A5-5 into a form of Eq. A5-1. To proceed further, we define the new variables $MM = M - 1$ and $NN = N - 1$. Note that with these substitutions, the variable MM goes from $MM = 0$ ($M = 1$) to $M = N$; that is, equivalent to $MM = N - 1 = NN$. Therefore

$$\sum_{M=0}^{N} \frac{M}{M!(N-M)!} x^M y^{N-M} = \sum_{M=1}^{N} \frac{M}{M!(N-M)!} x^M y^{N-M}$$

$$= \sum_{MM=0}^{NN} \frac{1}{(M-1)!(NN-MM)!} x^{MM+1} y^{NN-MM}$$

$$= \frac{x}{NN!} \sum_{MM=0}^{NN} \frac{NN!}{MM!(NN-MM)!} x^{MM} y^{NN-MM}$$

$$= \frac{x}{(N-1)!} \sum_{MM=0}^{NN} \frac{NN!}{MM!(NN-MM)!} x^{MM} y^{NN-MM}$$

$$= \frac{x}{(N-1)!} (x+y)^{NN} = \frac{x}{(N-1)!} (x+y)^{N-1} \quad (\mathbf{A5\text{-}6})$$

by Eq. A5-1. Now

$$\overline{N_A} = \sum_{N_A=0}^{N} N_A p(N_A) = \frac{1}{Q} \sum_{N_A=0}^{N} \frac{N_A q_A^{N_A} q_B^{N-N_A}}{N_A!(N-N_A)!} \quad (5.3\text{-}7)$$

Therefore

$$\overline{N_A} = \frac{\sum_{N_A=0}^{N} \frac{N_A q_A^{N_A} q_B^{N-N_A}}{N_A!(N-N_A)!}}{\sum_{N_A=0}^{N} \frac{q_A^{N_A} q_B^{N-N_A}}{N_A!(N-N_A)!}} = \frac{\frac{q_A}{(N-1)!}(q_A+q_B)^{N-1}}{\frac{(q_A+q_B)^N}{N!}} = N \frac{q_A}{q_A + q_B} \quad (\mathbf{A5\text{-}7})$$

which is the desired result, Eq. 5.3-8.

CHAPTER 5 PROBLEMS

5.1 At very high temperatures, water may dissociate according to the following reaction:

$$H_2O \leftrightarrow H_2 + \frac{1}{2}O_2$$

If pure water is in a constant-volume container at an initial concentration of 20 mol/liter at 298 K, determine the equilibrium compositions over the temperature range of 800 to 1100 K. The degeneracy of the ground electronic state of hydrogen is 1.

5.2 An interesting problematic reaction in syngas production is

$$CO + 1/2O_2 \leftrightarrow CO_2$$

Calculate the equilibrium constants from 298 K to 3000 K. Note that O_2 has a triplet electronic ground state.

5.3 Compute the equilibrium constant for the water-gas reaction $CO_2 + H_2 \leftrightarrow CO + H_2O$ over the temperature range of 900 to 3000 K. For these molecules, $\omega_{elect,1} = \omega_{nuc,1} = 1$.

5.4 In many cases it is necessary to have information available on air at such high temperatures that this gas mixture dissociates. The four reactions that must be considered at temperatures up to about 9000 K are

$$O_2 \leftrightarrow O + O, \; N_2 \leftrightarrow N + N, \; NO \leftrightarrow N + O \quad \text{and}$$
$$O_2 + N_2 \leftrightarrow 2NO$$

For engineering purposes, it is frequently useful to develop expressions for the equilibrium constant in the form

$$K_C = CT^\eta e^{-\theta/T}$$

where C, η, and θ are constants. While such a form for the equilibrium constant is not exact, it can be a quite accurate and very useful approximation over limited temperature ranges.

By starting with the pertinent partition functions and restricting the temperature range to 2000 K to 8000 K, obtain numerical values for each of the constants for all of the reactions.

Data:

Species	Θ_r(K)	Θ_v(K)	D_o, eV	$\omega_{elect,1}$	$\omega_{elect,2}$	$(\varepsilon_{elect,2} - \varepsilon_{elect,1})/k$
N_2	2.89	3390	9.76	1	-	very large
NO	2.45	2745	4.43	2	2	178 K
O_2	2.08	2278	5.08	2	2	11,300 K

Low-temperature air is a mixture containing 79 mole percent nitrogen and 21 mole percent oxygen. Also, find the equilibrium composition of N_2, O_2, NO, N, and O using the equilibrium constants obtained above.

5.5 At high temperatures, hydrogen molecules are partially dissociated to hydrogen atoms

$$H_2 \leftrightarrow 2H \quad \omega_{H_2,elect,1} = 1 \quad \omega_{H,elect,1} = 2$$

Obtain an expression for the equilibrium constant for this dissociation $K = [H]^2/[H_2]$ as a function of temperature. (Data: See Table 4.1-1.)

5.6 Calculate the equilibrium constant for the hydrogen iodide pyrolysis $2HI \leftrightarrow H_2 + I_2$ at 700 K. Use the data in Table 4.1-1.

5.7 Gaseous hydrogen is really a mixture of parahydrogen (antisymmetric-spin eigenstates, even j rotational states, and a nuclear degeneracy of 1), and orthohydrogen (symmetric-spin eigenstates, odd j rotational states, and a nuclear degeneracy of 3).

a. Compute and plot the constant-volume heat capacity of parahydrogen, orthohydrogen, and normal hydrogen (a 1-to-3 mixture of para- and orthohydrogen) as a function of temperature from 0 K to 300 K.

b. Compute the equilibrium ratio of orthohydrogen to parahydrogen from 0 K to 300 K.

c. Compute and plot (on the same graph as part a) the constant-volume heat capacity for an equilibrium mixture of ortho- and parahydrogen from 0 K to 300 K.

5.8 Compute the equilibrium constant for the hydrogen-deuterium exchange reaction $H_2 + D_2 \leftrightarrow 2HD$ at 100 K and 1000 K. (See Table 4.1-1 for data.)

5.9 Consider the two reactions $H_2 \leftrightarrow 2H$ and $H \leftrightarrow H^+ + e^-$. The energy of ionization of atomic hydrogen is 13.53 eV; and the degeneracies of the electron and the hydrogen atom are both 2, while that of the proton (ionized hydrogen) is 1. It is conjectured that the dissociation of molecular hydrogen (H_2) into atomic hydrogen (H) is essentially complete before the ionization of the atomic hydrogen begins. Verify this by the following procedure:

a. At one atmosphere total pressure, find the temperature T at which $\frac{[H^+]}{[H]} = 0.01$, where [ ] denotes concentration.

b. At this temperature T show that $[H] \gg [H_2]$.

5.10 a. Derive an expression for the entropy change of a chemical reaction in terms of the partition functions of the reacting gases.

b. Calculate ΔS for $2H_2 + O_2 \leftrightarrow 2H_2O$ at 500 K at 1 atm starting with a stoichiometric ratio of H_2 and O_2.

5.11 For the reaction $N_2 + O_2 \leftrightarrow 2NO$, using partition functions, calculate the enthalpy change on reaction and the equilibrium constant at 1000 K. (See data in Table 4.1-1.)

5.12 At very high temperatures, alkali metals may associate in the vapor phase. Of interest here is the possibility that sodium will undergo vapor-phase dimerization $2\text{Na} \leftrightarrow \text{Na}_2$. Compute the equilibrium constant and the fraction of sodium associated for this reaction at constant pressure and temperatures of 800 and 1200 K. The following data are available:

	Na	Na$_2$
Molecular weight	23	46
Θ_r		0.221 K
Θ_v		229 K
D_0		17.3 kcal/mol
$\omega_{\text{elect},1}$	2	1

5.13 Compute the equilibrium constant for the following reaction over the temperature range of 300 to 3000 K.

$$H_2 + Cl_2 \leftrightarrow 2HCl$$

The electronic ground state for H$_2$ and HCl will be assumed to be nondegenerate; for Cl$_2$ it is 2. Other data for these molecules can be found in Table 4.1-1.

5.14 Calculate the fraction of hydrogen molecules dissociated into atoms over the temperature range of 1000 K to 5000 K. The dissociation is to take place in a constant volume container with an initial hydrogen molecule concentration of 0.01 mol/l $\omega_{H_2,\text{elect},1} = 1$; $\omega_{H,\text{elect},1}$. (Data: See Table 4.1-1.)

5.15 Calculate the fraction of hydrogen atoms that ionize at 10000 K and 15000 K in a constant volume container with an initial hydrogen atom concentration of 0.01 mol/l. Assess effect of higher electronic states of the H atom on the calculation. (Choosing the separated proton and electron as the state of 0 electronic energy, the electronic energy of ground state of the atom is -13.53 ev relative to the separated proton and electron at rest. Also, $\omega_{H,\text{elect},1} = 2$, $\omega_{e^-,\text{elect},1} = 2$ and $\omega_{H^+,\text{elect},1} = 1$. Higher electronic states of the atom are given by $\varepsilon_{\text{elect},n} = -\frac{13.53}{n^2}$ev. The number of states having the energy $\varepsilon_{\text{elect},n}$—that is $\omega_{\text{elect},n}$—is n^2 (so that all energy levels except the first are degenerate).

5.16 Consider the ideal gas bimolecular reaction:

$$A + B \leftrightarrow C + D$$

Starting with an equimolar mixture of A and B, develop the equations for the changes in internal energy, heat capacity C_V, and volume from the initial state to the equilibrium reaction state, at fixed temperature and pressure, in terms of the single-particle partition functions and their derivatives with respect to temperature.

5.17 The fundamental thermodynamic relation for a three-dimensional fluid is $dU = TdS - PdV$.
The analogous relation for a one-dimensional fiber (polymer chain, fibrous protein, muscle in the body, etc.) is $dU = TdS + \tau dL$, where L is the length of the fiber (replacing volume in the 3-D fluid) and τ is the tension (the tension is analogous to pressure in a 3-D fluid, but the sign is opposite since pressure is outward pointing and tension is inward pointing). Consider a material composed of a chain of M units, each of which has two states—an unstretched state α that has a length L_α and a partition function q_α, and a stretched state β that has a length L_β and a partition function q_β. The two states could be representative of the two states in a random coil-helix transition of a protein or nucleic acid in solution, or the unstretched and stretched states of a muscle fiber.
  a. Write the partition function for this system.
  b. Obtain an expression for the tension τ as a function of the total chain length L and $M, M_\alpha, L_\alpha, L_\beta, q_\alpha, q_\beta$, and kT.
  c. What is the ratio of chain units in each of the states at 0 tension ($\tau = 0$)?

5.18 Calculate the equilibrium compositions for the following reaction at 1 bar over the range of temperatures from 298.15 K to 1000 K, starting with a stoichiometric mixture of nitrogen and hydrogen with no other constituents:

$$\frac{1}{2}N_2 + \frac{3}{2}H_2 \leftrightarrow NH_3$$

5.19 In a high-temperature surface catalytic process, nitrogen is dissociated into nitrogen radicals according to the reaction $N_2 + M \leftrightarrow 2N + M$.
Calculate the equilibrium constant for this reaction at 4400 K, assuming nitrogen remains in its ground electronic state and all electronic degeneracies are 1. Starting with pure nitrogen, calculate the fraction of nitrogen that is dissociated at 1 bar and this temperature.

5.20 Consider this ideal gas reaction:

$$A + B \leftrightarrow C$$

Develop the equations for the changes in internal energy and entropy from the initial state to the equilibrium reaction state, at fixed temperature and volume, in terms of the single-particle partition functions and their derivatives with respect to temperature.

Chapter 6

Other Partition Functions

So far we have considered only the canonical partition function that describes a closed system of constant volume and constant temperature (since the system is in an infinite bath of constant temperature). As in classical thermodynamics, systems subject to other constraints are also of interest. Indeed, in statistical mechanics, it is sometimes more convenient to consider other system constraints, as we will show later in the derivation of the virial equation of state. There are two types of system constraints that are of special interest. The first is the closed, isolated, constant-volume system, which is a system of constant volume, and energy. The partition function describing such a system is referred to as the microcanonical partition function. The second system of interest is an open system of fixed volume in an infinite bath of temperature T and chemical potential (or molar Gibbs energy) μ for one or more species. The grand canonical partition function describes such a system. Each of these and two other partition functions are discussed in this chapter.

INSTRUCTIONAL OBJECTIVES FOR CHAPTER 6

The goals for this chapter are for the student to:

- Understand the microcanonical partition function
- Understand the grand canonical partition function
- Understand the isobaric-isothermal partition function
- Understand the semi-grand canonical partition function

6.1 THE MICROCANONICAL ENSEMBLE FOR A PURE FLUID

The system constraints on the microcanonical ensemble of constant number of particles (assuming no chemical reactions), volume, and energy are shown in Fig. 6.1-1, and the partition function for this system is easily derived as shown here. From the first postulate of statistical mechanics—that "all microstates of the system of volume V that have the same energy and the same number of particles are equally probably"—it follows that all microstates in the microcanonical ensemble have the same probability of occurrence. Therefore, if the degeneracy of the state N, V and E is $\Omega(N, V, E)$, it follows that the probability of occurrence of any microstate with fixed N, V, E is

$$p(N, V, E) = \frac{1}{\Omega(N, V, E)} \quad \text{and} \quad \sum_{\text{microstates}} p(N, V, E) = \sum_{\text{microstates}} \frac{1}{\Omega(N, V, E)} = 1$$

(6.1-1)

Chapter 6: Other Partition Functions

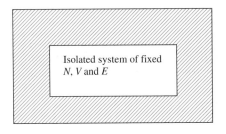

Figure 6.1-1 The Microcanonical Ensemble. A rigid system with no contact with its surroundings, and therefore of fixed number of particles, volume, and energy.

Now, using Eq. 3.6-4

$$S = -k \sum_{\text{states } j} p(N, V, E_j) \ln p(N, V, E_j) \qquad (3.6\text{-}4)$$

for the microcanonical ensemble, we have

$$S = -k \sum_{\text{states } j} \frac{1}{\Omega(N, V, E_j)} \ln\left(\frac{1}{\Omega(N, V, E_j)}\right)$$

$$= k\Omega(N, V, E_j) \left[\frac{1}{\Omega(N, V, E_j)} \ln \Omega(N, V, E_j)\right] = k \ln \Omega(N, V, E_j) \qquad (6.1\text{-}2)$$

since each of the Ω states has an equal likelihood of occurrence. To obtain the other thermodynamic properties for the microcanonical ensemble, we use known relationships from classical thermodynamics as follows:

$$dS = \left(\frac{\partial S}{\partial U}\right)_{V,N} dU + \left(\frac{\partial S}{\partial V}\right)_{U,N} dV + \left(\frac{\partial S}{\partial N}\right)_{U,V} dN$$

$$= \frac{1}{T} dU + \frac{P}{T} dV - \frac{G}{T} dN = \frac{1}{T} dU + \frac{P}{T} dV - \frac{\mu}{T} dN \qquad (6.1\text{-}3)$$

Therefore

$$\frac{1}{T} = \left(\frac{\partial S}{\partial U}\right)_{V,N} = k \left(\frac{\partial \ln \Omega(N, U, V)}{\partial U}\right)_{V,N} \qquad (6.1\text{-}4)$$

$$\frac{P}{T} = \left(\frac{\partial S}{\partial V}\right)_{U,N} = k \left(\frac{\partial \ln \Omega(N, U, V)}{\partial V}\right)_{U,N} \qquad (6.1\text{-}5)$$

and

$$\frac{G}{T} = \frac{\mu}{T} = -\left(\frac{\partial S}{\partial N}\right)_{U,V} = -k \left(\frac{\partial \ln \Omega(N, U, V)}{\partial N}\right)_{U,V} \qquad (6.1\text{-}6)$$

Consequently, if we could formulate an expression for the degeneracy of the system as a function of the number of particles, energy, and volume, we would have all of the thermodynamic properties of the system from simple calculations.

6.2 THE GRAND CANONICAL ENSEMBLE FOR A PURE FLUID

Another useful construct is the grand canonical ensemble and partition function for a system of fixed volume, fixed temperature (by being in contact with an infinite, constant temperature bath), and fixed chemical potential (by being in contact with an infinite reservoir of constant chemical potential of the species of interest) as indicated in Fig. 6.2-1. For such a system, it can be shown (Problem 6.1) by following the same argument as was used to derive the canonical partition function, that the grand canonical partition function is

$$\Xi(V, T, \mu) = \sum_N e^{N\mu/kT} \sum_{\substack{\text{energy states } i \\ \text{for } N \text{ molecules}}} e^{-E_i(N,V)/kT} = \sum_N e^{N\mu/kT} Q(N, V, T)$$

(6.2-1)

where $\Xi(V, T, \mu)$ is the grand canonical partition function, and $Q(N, V, T)$ is the canonical partition function discussed previously. In the grand canonical ensemble, the average number of particles $\overline{N}$ in the system is computed from

$$\overline{N} = \frac{\sum_N e^{N\mu/kT} \sum_{\substack{\text{energy states } i \\ \text{for } N \text{ molecules}}} N e^{-E_i(N,V)/kT}}{\Xi(V, T, \mu)} = \frac{\sum_N e^{N\mu/kT} N Q(N, V, T)}{\Xi(V, T, \mu)} \quad (6.2\text{-}2)$$

and the average energy U from

$$U = \overline{E} = \frac{\sum_N e^{N\mu/kT} \sum_{\substack{\text{energy states } i \\ \text{for } N \text{ molecules}}} E_i(N, V) e^{-E_i(N,V)/kT}}{\Xi(V, T, \mu)} \quad (6.2\text{-}3)$$

Note also that

$$\left(\frac{\partial \Xi}{\partial \mu}\right)_{V,T} = \frac{\partial}{\partial \mu}\bigg|_{V,T} \sum_N e^{N\mu/kT} \sum_{\substack{\text{energy states } i \\ \text{for } N \text{ molecules}}} e^{-E_i(N,V)/kT}$$

$$= \frac{1}{kT} \sum_N N e^{N\mu/kT} \sum_{\substack{\text{energy states } i \\ \text{for } N \text{ molecules}}} e^{-E_i(N,V)/kT}$$

so that

$$\overline{N} = kT \left(\frac{\partial \ln \Xi}{\partial \mu}\right)_{V,T} \quad (6.2\text{-}4)$$

System with rigid, thermally conductive walls that are permeable to the atoms and is in contact with a thermal reservoir of temperature T and an external system of particles of fixed chemical potential μ.

Figure 6.2-1 The Grand Canonical Ensemble. A system with rigid but thermally conductive walls that are permeable to the atoms in contact with a thermal reservoir of temperature T and particle reservoir of fixed chemical potential μ. Therefore, V, T, and μ are fixed.

Chapter 6: Other Partition Functions

and

$$\left(\frac{\partial \Xi}{\partial T}\right)_{V,\mu} = \frac{\partial}{\partial T}\bigg|_{V,\mu} \sum_N e^{N\mu/kT} \sum_{\substack{\text{energy states } i \\ \text{for } N \text{ molecules}}} e^{-E_i(N,V)/kT}$$

$$= \frac{1}{kT^2} \sum_N \sum_{\substack{\text{energy states } i \\ \text{for } N \text{ molecules}}} (E_i - \mu N) e^{-E_i(N,V)/kT} e^{N\mu/kT}$$

or

$$U - \mu \overline{N} = kT^2 \left(\frac{\partial \ln \Xi}{\partial T}\right)_{V,\mu}$$

so that

$$U = kT^2 \left(\frac{\partial \ln \Xi}{\partial T}\right)_{V,\mu} + \mu kT \left(\frac{\partial \ln \Xi}{\partial \mu}\right)_{V,T} \quad (6.2\text{-}5)$$

To proceed further, we note that since Ξ is a function of V, T and μ, we can write

$$d \ln \Xi(V, T, \mu) = \left(\frac{\partial \ln \Xi}{\partial T}\right)_{V,\mu} dT + \left(\frac{\partial \ln \Xi}{\partial \mu}\right)_{V,T} d\mu + \left(\frac{\partial \ln \Xi}{\partial V}\right)_{\mu,T} dV$$

$$= \frac{U - \mu \overline{N}}{kT^2} dT + \frac{\overline{N}}{kT} d\mu + \left(\frac{\partial \ln \Xi}{\partial V}\right)_{\mu,T} dV \quad (6.2\text{-}6)$$

Also

$$d\left(\frac{\overline{N}\mu}{kT}\right) = \frac{\overline{N}}{kT} d\mu + \frac{\mu}{kT} d\overline{N} - \frac{\overline{N}\mu}{kT^2} dT$$

so that

$$\frac{\overline{N}}{kT} d\mu - \frac{\overline{N}\mu}{kT^2} dT = d\left(\frac{\overline{N}\mu}{kT}\right) - \frac{\mu}{kT} d\overline{N}$$

Substituting this into Eq. 6.2-6 above results in

$$d \ln \Xi(V, T, \mu) = \frac{U - \mu \overline{N}}{kT^2} dT + \frac{\overline{N}}{kT} d\mu + \left(\frac{\partial \ln \Xi}{\partial V}\right)_{\mu,T} dV$$

$$= \frac{U}{kT^2} dT + d\left(\frac{\overline{N}\mu}{kT}\right) - \frac{\mu}{kT} d\overline{N} + \left(\frac{\partial \ln \Xi}{\partial V}\right)_{\mu,T} dV \quad (6.2\text{-}7)$$

Then, note that

$$\frac{U}{kT^2} dT = \frac{1}{k}\left[-d\left(\frac{U}{T}\right) + \frac{1}{T} dU\right]$$

6.2 The Grand Canonical Ensemble for a Pure Fluid

$$d \ln \Xi(V,T,\mu) = \frac{U}{kT^2}dT + d\left(\frac{\overline{N}\mu}{kT}\right) - \frac{\mu}{kT}d\overline{N} + \left(\frac{\partial \ln \Xi}{\partial V}\right)_{\mu,T}dV$$

$$= \frac{U}{kT}dT - \frac{1}{k}d\left(\frac{U}{T}\right) + d\left(\frac{\overline{N}\mu}{kT}\right) - \frac{\mu}{kT}d\overline{N} + \left(\frac{\partial \ln \Xi}{\partial V}\right)_{\mu,T}dV$$

$$dU = kTd\left(\ln \Xi + \frac{U}{kT} - \frac{\overline{N}\mu}{kT}\right) - kT\left(\frac{\partial \ln \Xi}{\partial V}\right)_{\mu,T}dV + \mu\, d\overline{N} \quad \textbf{(6.2-8)}$$

However, from classical thermodynamics, we know that

$$dU = TdS - PdV + \mu\, d\overline{N}$$

Consequently, we can make the following identifications:

$$P = kT\left(\frac{\partial \ln \Xi}{\partial V}\right)_{\mu,T} \quad \textbf{(6.2-9)}$$

and

$$S = k\left(\ln \Xi + \frac{U}{kT} - \frac{\overline{N}\mu}{kT}\right)$$

$$= k\left(\ln \Xi + T\left(\frac{\partial \ln \Xi}{\partial T}\right)_{V,\mu} + \mu\left(\frac{\partial \ln \Xi}{\partial \mu}\right)_{V,T} - \frac{\overline{N}\mu}{kT}\right)$$

$$= k\left(\ln \Xi + T\left(\frac{\partial \ln \Xi}{\partial T}\right)_{V,\mu} + \mu\frac{\overline{N}}{kT} - \frac{\overline{N}\mu}{kT}\right)$$

$$= k\ln \Xi + kT\left(\frac{\partial \ln \Xi}{\partial T}\right)_{V,\mu} \quad \textbf{(6.2-10)}$$

There is another relation that is especially useful. Starting from the classical thermodynamic relation

$$S = \frac{U}{T} + \frac{PV}{T} - \frac{\overline{N}\mu}{T}$$

we have

$$k\ln \Xi + kT\left(\frac{\partial \ln \Xi}{\partial T}\right)_{V,\mu} = kT\left(\frac{\partial \ln \Xi}{\partial T}\right)_{V,\mu} + \mu k\left(\frac{\partial \ln \Xi}{\partial \mu}\right)_{V,T}$$

$$+ kV\left(\frac{\partial \ln \Xi}{\partial V}\right)_{\mu,T} - \mu k\left(\frac{\partial \ln \Xi}{\partial \mu}\right)_{V,T}$$

that provides the following useful relationship for the volumetric equation of state:

$$\ln \Xi = V\left(\frac{\partial \ln \Xi}{\partial V}\right)_{\mu,T} \quad \text{so that} \quad P = kT\left(\frac{\partial \ln \Xi}{\partial V}\right)_{\mu,T} = \frac{kT}{V}\ln \Xi \quad \textbf{(6.2-11)}$$

or

$$\frac{PV}{kT} = \ln \Xi \quad \textbf{(6.2-12)}$$

92 Chapter 6: Other Partition Functions

Finally, we note that in the grand canonical ensemble, the probability that the system is in a state with N particles and energy E with a fixed system volume of V is

$$p(N, E, V) = \frac{e^{N\mu/kT} e^{-\frac{E_i(N,V)}{kT}}}{\Xi(V, T, \mu)} = \frac{e^{N\mu/kT} e^{-E_i(N,V)/kT}}{\sum_N e^{N\mu/kT} \sum_{\substack{\text{energy states } i \\ \text{for } N \text{ molecules}}} e^{-E_i(N,V)/kT}} \quad (6.2\text{-}13)$$

Now, starting from

$$\sum_{\substack{\text{states of energy } E, \\ \text{number of molecules } N, \\ \text{in volume } V}} p(N, E, V) \ln p(N, E, V)$$

$$= \sum_{\substack{\text{states of energy } E, \\ \text{number of molecules } N, \\ \text{in volume } V}} \frac{e^{N\mu/kT} e^{-E_i(N,V)/kT} \left(\frac{N\mu}{kT} + -\frac{E_i(N,V)}{kT} - \ln \Xi(V, T, \mu) \right)}{\Xi(V, T, \mu)}$$

$$= \frac{\overline{N}\mu}{kT} - \frac{U}{kT} - \ln \Xi = \frac{\overline{N}\mu}{kT} - \frac{U}{kT} - \frac{PV}{kT} = -\frac{S}{k}$$

we have

$$S = -k \sum_{\substack{\text{states of energy } E, \\ \text{number of molecules } N, \\ \text{in volume } V}} p(N, E, V) \ln p(N, E, V) \quad (6.2\text{-}14)$$

which establishes the validity of the Gibbs entropy equation when used with the grand canonical ensemble.

6.3 THE ISOBARIC-ISOTHERMAL ENSEMBLE

Another type of ensemble is a system of fixed number of molecules in a container with flexible walls that freely allow heat flows from an infinite bath of fixed pressure and fixed temperature. (See Fig. 6.3-1.) This set of variables is of interest as it generally occurs in chemical engineering. The isobaric-isothermal partition function denoted by $\Delta(N, T, P)$ is

$$\Delta(N, T, P) = \sum_{\text{volume } V} \sum_{\text{states } i} e^{-E_i(N,V)/kT} e^{-PV/kT} = \sum_{\text{volume } V} Q(N, V, T) e^{-\frac{PV}{kT}} \quad (6.3\text{-}1)$$

The probability of occurrence of a microstate of the isobaric-isothermal system is

$$p(N, E_i, P) = \frac{e^{-E_i(N,V)/kT} e^{-PV/kT}}{\Delta(N, T, P)} \quad (6.3\text{-}2)$$

and using Eq. 3.6-3, the Gibbs entropy equation

$$S(N, T, P) = -k \sum_{\text{volume } V} \sum_{\text{states } i} p(N, E_i, P) \ln p(N, E_i, P)$$

> System with completely flexible, thermally conductive walls that are impermeable to the atoms and in contact with a thermal reservoir of temperature T and subject to an external pressure P.

Figure 6.3-1 The Isobaric-Isothermal Ensemble. A system with completely flexible, thermally conductive walls that are impermeable to the atoms in contact with a thermal reservoir of temperature T and subject to an external pressure P. Therefore, N, P, and T are fixed.

it is easily shown that

$$G(N, T, P) = -kT \ln \Delta(N, T, P) \qquad (6.3\text{-}3)$$

from which all other thermodynamic properties can be obtained by differentiation. (See Problems 6-5 and 6-6.)

6.4 THE RESTRICTED GRAND OR SEMI-GRAND CANONICAL ENSEMBLE

Another type of ensemble is one that would represent, for example, an osmotic equilibrium system in which a mixture in a container is in contact with an infinite bath at fixed temperature and chemical potential of species 1, μ_1 (see Fig. 6.4-1). The walls of the container are a semi-permeable membranes such that heat can readily flow into or out of the system; however, only species 1 molecules can cross the membrane. Therefore, the chemical potential of species 1 in the system is fixed, as are temperature, volume, the numbers of molecules of species 2, species 3, etc. The semi-grand canonical partition function is

$$\Gamma(\mu_1, N_2, N_3, .., T, V) = \sum_{N_1} e^{-E_i(N_1, N_2, ..., V)/kT} e^{N_1 \mu_1 / kT}$$

$$= \sum_{N_1} Q(N_1, N_2, .., V, E_i) e^{N_1 \mu_1 / kT} \qquad (6.4\text{-}1)$$

It can then be shown that

$$A(T, V, N_2, \ldots, \mu_1) = \mu_1 N_1 - kT \ln \Gamma(T, V, N_2, \ldots, \mu_1) \qquad (6.4\text{-}2)$$

from which other thermodynamic properties can be derived.

> System with rigid, thermally conductive walls that are permeable to only atoms of species 1 in a mixture, and is in contact with a thermal reservoir of temperature T and an external system of species 1 atoms of fixed chemical potential μ_1.

Figure 6.4-1 The Restricted Grand or Semi-Grand Canonical Ensemble. A system with rigid but thermally conductive walls that are permeable only to the species 1 atoms that is in contact with a thermal reservoir of temperature T and particle reservoir of fixed chemical potential μ_1. Therefore, the temperature, the chemical potential of species 1 μ_1, and the particle numbers of all other species $(N_2, N_3, \ldots)$ are fixed.

6.5 COMMENTS ON THE USE OF DIFFERENT ENSEMBLES

Now that several different ensembles have been presented, it is useful to comment on their interrelationships and the most advantageous use of each. To start, we should point out that under identical state conditions, one should obtain the same calculated result for any thermodynamic property regardless of which ensemble is used,[1] provided the system contains a very large number of particles so that fluctuations are small. For example, the calculated value of the entropy using the microcanonical ensemble at fixed number of particles, volume, and energy should be the same as that obtained using the canonical ensemble at fixed number of particles, volume, and temperature if that temperature corresponds to the average energy used in the microcanonical ensemble calculation. The same is true for other properties, and also with other ensembles.

To see this, we first remember the discussion of fluctuations in energy in Chapter 3 and in chemical reactions in Chapter 5, and therefore we expect that fluctuations in thermodynamic properties will decrease as the number of particles increases proportional to $1/\sqrt{N}$. The so-called thermodynamic limit is $N \to \infty$; in this case, fluctuations can be ignored. This is the case we are interested in, as N is very large (of the order of Avogadro's number) for situations of interest to us here. Also, in this limit, as discussed in Section 5.3, not all states of the system need to be considered in the evaluation of any partition function, but only the most probable state. So, for example, starting with the canonical partition function

$$Q(N, V, T) = \sum_{\substack{\text{states} \\ i}} e^{-E_i/kT} = \sum_{\substack{\text{levels} \\ j}} \Omega(E_j) e^{-E_j/kT} \qquad (3.2\text{-}3)$$

and using only the most probable value of the energy level, which by the maximum term discussion of previous chapters is also the average value $\overline{E}$ and the macroscopically measureable internal energy U, in the canonical partition function, we have

$$Q(N, V, T) = \Omega(\overline{E}) e^{-\overline{E}/kT} \quad \text{or} \quad kT \ln Q(N, V, T) = -A = kT \ln \Omega(\overline{E}) - \overline{E}$$

or

$$A = U - TS = -kT \ln \Omega(\overline{E}) + U \qquad (6.5\text{-}1)$$

So then

$$S = k \ln \Omega(\overline{E})$$

which is the previously derived Eq. 6.1-2 for the entropy at constant N, V, and E in the microcanonical ensemble.

Similarly, starting from the grand canonical partition function

$$\Xi(V, T, \mu) = \sum_N e^{\frac{N\mu}{kT}} \sum_{\substack{\text{energy states } i \\ \text{for } N \text{ molecules}}} e^{-\frac{E_i(N,V)}{kT}} = \sum_N e^{\frac{N\mu}{kT}} Q(N, V, T) \qquad (6.2\text{-}1)$$

---

[1] This is a variation of the "Sandler rule"—that nature does not care how we choose to describe her; therefore, different descriptions or ways of proceeding, if done correctly, should give the same final answer. However, that does not mean that all ways of proceeding will be equally expeditious in obtaining that answer.

6.5 Comments on the Use of Different Ensembles

The value of this partition function for only the most probable (and average) number of particles, $\overline{N}$, corresponding to the fixed values of the chemical potential μ, V, and T is

$$\Xi(V, T, \mu) = e^{\frac{\overline{N}\mu}{kT}} \sum_{\substack{\text{energy states } i \\ \text{for } \overline{N} \text{ molecules}}} e^{-\frac{E_i(\overline{N}, V)}{kT}} = e^{\frac{\overline{N}\mu}{kT}} Q(\overline{N}, V, T)$$

or

$$\ln \Xi(V, T, \mu) = \frac{PV}{kT} = \frac{\overline{N}\mu}{kT} + \ln Q(\overline{N}, V, T) = \frac{G}{kT} + \ln Q(\overline{N}, V, T) \quad \text{(6.5-2)}$$

where the first equality results from Eq. 6.2-12. Continuing, we have

$$G - PV = A + PV - PV = A = -kT \ln Q(\overline{N}, V, T) \quad \text{(6.5-3)}$$

This last equality shows that if we start from the grand canonical ensemble partition function at fixed V, T, and μ and use only the most probable or average number of particles $N = \overline{N}$ corresponding to the fixed values of V, T, and μ, we recover the thermodynamic relation for the canonical ensemble partition function and the Helmholtz energy. Similarly, using the most probable state arguments, it is possible to show similar interrelationships between other statistical mechanical ensembles in the thermodynamic limit—that is, for $N \to \infty$.

So if there is an equivalence between the different ensembles in the thermodynamic limit, why have several different ensembles been introduced here? The answer is much like the situation in classical thermodynamics in that depending on the constraints on the system, different thermodynamic functions may be of interest or use. For example, for the completely isolated system, that is a system of fixed N, V, and E, we know that the entropy achieves a maximum; for such a system it is the microcanonical ensemble that is appropriate, and there is a direct relationship between the natural logarithm of the microcanonical partition function and the entropy. Similarly, for a system at fixed N, V, and T, we know from classical thermodynamics that the Helmholtz energy is a minimum at equilibrium. Under these constraints, it is the canonical ensemble that is appropriate, and the Helmholtz energy is directly related to the canonical ensemble partition function. In a similar fashion, the isobaric-isothermal ensemble is the appropriate one for a system constrained to fixed N, P, and T, and its partition function is related to the Gibbs energy.

Another reason different ensembles have been introduced is that some computations or analyses are easier in one ensemble than another. For example, in the next chapter we derive the virial equation of state, first using the canonical ensemble, and then again using the grand canonical ensemble. These two derivations show that it is simpler to start from the grand canonical ensemble. Also, there are situations in which the physical system is best described by one ensemble rather than another. For example, consider the case of trying to obtain the isotherm for the adsorption of a gas on a porous solid adsorbent. If the gas external to the adsorbent is at a fixed temperature and pressure, it is also at a fixed chemical potential μ. Then, since from classical thermodynamics we know that at equilibrium the chemical potential of the species as a gas and as an adsorbed phase must be the same, the average number of adsorbed molecules is most easily computed using the grand canonical ensemble. An example of when the restricted grand canonical partition function would be the appropriate one to use is in determining the equilibrium concentration on one side of

Chapter 6: Other Partition Functions

a membrane that is permeable only to that species, if its chemical potential (i.e., its concentration and solution conditions) is fixed on the other side of membrane (for example, by being in contact with an infinite bath of that species).

So a conclusion of this brief discussion is that in statistical mechanics it is useful to use different ensembles depending on the physical situation, just as in classical thermodynamics, where the equilibrium state of systems under different constraints gives rise to different functions that have extreme values (i.e., entropy for a system at fixed N, V, and E; Helmholtz energy for a system at fixed N, V, and T; and Gibbs energy for a system at fixed N, P, and T).

CHAPTER 6 PROBLEMS

6.1 Following a similar analysis to that leading to the canonical partition function, show that for a system at fixed volume V in contact with a bath at temperature T and chemical potential μ, the probability of occurrence a particular microstate with number of molecules N_i and energy E_i is

$$p(E_i, N_i) = \frac{e^{-E_i/kT} e^{N\mu/kT}}{\Xi(V, T, \mu)}$$

where

$$\Xi(V, T, \mu) = \sum_N \sum_{\substack{\text{energy} \\ \text{states}}} e^{-E_j(N,V)/kT} e^{N\mu/kT}$$

$$= \sum_N Q(N, V, T) e^{N\mu/kT}$$

Here, Ξ is the grand canonical partition function, and Q is the canonical partition function.

6.2 Show that

$$P = kT \left(\frac{d \ln \Xi}{dV}\right)_{\mu,T} = \frac{kT}{V} \ln \Xi \text{ or}$$

$$\ln \Xi = \frac{PV}{kT} = NZ$$

where $Z = PV/NkT$ is the compressibility factor, and

$$S = kT \left(\frac{d \ln \Xi}{dT}\right)_{V,\mu} + k \ln \Xi \text{ and}$$

$$N = kT \left(\frac{d \ln \Xi}{d\mu}\right)_{V,T}$$

6.3 Show that for two phases in equilibrium, say a and b, the grand canonical partition function for the total system of both phases is $\Xi = \Xi_a \Xi_b$.

What does this mean thermodynamically? Is it also true that the canonical partition function Q is $Q = Q_a Q_b$?

6.4 Derive the expression for the isothermal-isobaric partition function Eq. 6.3-1.

6.5 Relate the isothermal-isobaric partition function and its derivatives to the thermodynamic properties such as the internal energy, enthalpy, volume, entropy, Gibbs energy, Helmholtz energy, and the constant-pressure heat capacity.

6.6 Derive the isobaric-isothermal partition function for a binary mixture.

6.7 Derive the thermodynamic properties from the binary mixture isobaric-isothermal partition function.

6.8 Determine the magnitude of possible fluctuations in the number of particles in the grand canonical ensemble.

6.9 Determine the magnitude of possible fluctuations in energy in the isobaric-isothermal ensemble.

6.10 Determine the magnitude of possible fluctuations in volume in the isobaric-isothermal ensemble.

6.11 Combining Eqs. 6.2-9 and 6.2-12 we have

$$P = kT \left(\frac{\partial \ln \Xi}{\partial V}\right)_{\mu,T} = \frac{kT}{V} \ln \Xi$$

Mathematically explain how this is possible.

6.12 Write the grand canonical partition function for an ideal
 a. monatomic gas
 b. diatomic gas

6.13 Write the isobaric-isothermal partition function for an ideal
 a. monatomic gas
 b. diatomic gas

6.14 Derive the expression for the semi-grand canonical partition function, Eq. 6.4-1.

6.15 Relate the semi-grand canonical partition function and its derivatives to the thermodynamic properties such as the internal energy, enthalpy, volume, entropy, Gibbs energy, Helmholtz energy, and the constant-volume heat capacity.

6.16 A flat graphite surface is in contact with a reservoir of gas molecules at a fixed chemical potential. There are M sites on the graphite surface on which gas molecules can adsorb, each site can adsorb only a single molecule, and the adsorbed molecules do not interact with each other. Starting from the grand canonical partition function, develop an expression for the coverage—that is, the fraction of the adsorption sites that are occupied (contain an adsorbed gas molecule) as a function of the chemical potential of the gas molecules in the reservoir.

Hints:
1. The number of ways of distributing N indistinguishable molecules on M adsorption sites that are distinguishable because they are fixed in space is $M!/N!(M-N)!$
2. Using the binomial expansion discussed in the Appendix to Chapter 5 with q_A being the partition function for site containing an adsorbed molecule and $q_B = 1$ for a site without an adsorbed molecule.

Chapter 7

Interacting Molecules in a Gas

In this chapter we introduce the idea that molecules interact, so that the energy of interaction contributes to the total energy of the system and must to be taken into account in the partition function. We do that in this chapter for the case of the disordered molecules in a dilute gas using graph theory to derive the virial equation of state. In the following chapter we compute values for the second virial coefficient for specific molecule interaction models. In succeeding chapters, we will then consider interacting molecules in a crystal, which is a dense but well-ordered medium; and then a liquid, which is a dense but disordered substance.

INSTRUCTIONAL OBJECTIVES FOR CHAPTER 7

The goals for this chapter are for the student to:

- Understand the concept of the configuration integral
- Understand the pairwise additivity assumption for the intermolecular potential
- Understand the graph theoretic method used to derive the virial equation of state from the canonical partition function
- Understand the derivation of the virial equation of state from the grand canonical partition function

7.1 THE CONFIGURATION INTEGRAL

Consider a gas composed of N identical atoms. The spatial position of each of these N atoms is specified by the collection of N vectors $(\underline{r}_1, \underline{r}_2, \ldots, \underline{r}_N)$, where $\underline{r}_j$ has the components (x_j, y_j, z_j) and the kinetic energies by the N translational energy quantum number vectors $(\underline{l}_1, \underline{l}_2, \ldots, \underline{l}_N)$, where $\underline{l}_j$ has the components (l_{xj}, l_{yj}, l_{zj}). The potential energy of interaction for the N-particle system will be written as $u(\underline{r}_1, \underline{r}_2, \ldots, \underline{r}_N)$, where we have assumed that the particles are spherically symmetric, so that the interaction energy of the system is only a function of the particle location. Note also that we have treated the kinetic and internal (electronic) energies of an atom as quantized variables; but the potential energy is being treated as a classical variable, since very small changes in position (allowed by the Heisenberg uncertainty principle) produce very small interaction energy changes, so the spacing between the potential energy levels is infinitesimal. We will also assume that the kinetic energy states available to a particle are not affected by its interactions with other particles.

In the analysis of this chapter, we will limit our consideration to a system of N identical monatomic particles. At the expense of greater complexity in both the notation and physical description, these restrictions could be removed. For example, polyatomic molecules could be considered if we specify both the position $\underline{r}$ and orientation of each molecule, as well as its translational, rotational, and vibrational quantum numbers (or, classically, the translational and rotational velocity vectors), and realize that the potential energy is now a function of both the position and orientation of each of the molecules. Furthermore, the extension to mixtures of species can easily be made. However, our purpose here is to develop, in a quantitative fashion, the fundamental aspects of studying nonideal gases; and the basic ideas of the development will only be obscured by complexities discussed above. Therefore, our development will be restricted to monatomic gases.

Each of the atoms has a set of allowable kinetic and internal energy states, which we shall assume is unaffected by the presence of other atoms. Consequently, if E^i represents the ith state of the N particle system, then

$$E^i = \sum_{j=1}^{N} \varepsilon_{j,\text{int}}^i + \sum_{j=1}^{N} \frac{h^2}{8mV^{2/3}} \left[\left(l_{x,j}^i\right)^2 + \left(l_{y,j}^i\right)^2 + \left(l_{z,j}^i\right)^2 \right] + u\left(\underline{r}_1^i, \underline{r}_2^i, \ldots, \underline{r}_N^i\right)$$

(7.1-1)

where $\varepsilon_{j,\text{int}}^i$ is the internal energy of the jth molecule in the ith system state, which is the electronic energy for a monatomic molecule (or the electronic, rotational, and vibrational energy for a polyatomic molecule), $\underline{r}_j^i$ represents its position vector, and each l is a kinetic energy quantum number. Assuming that the electronic, kinetic, and potential energy states are independent of each other, the partition function for this system can immediately be written as

$$Q(N, V, T) = \frac{(q_{\text{int}})^N}{N!} \left(\frac{2\pi mkT}{h^2}\right)^{3N/2} V^N C \int_V \ldots \int_V e^{-u(\underline{r}_1, \ldots \underline{r}_N)/kT} \, d\underline{r}_1 \ldots d\underline{r}_N$$

(7.1-2)

where $\int_V [\ ]d\underline{r}_i = \iiint_V [\ ]dx_i dy_i dz_i$. The constant C has been introduced as a normalization factor, and to keep the partition function dimensionally consistent. In particular, from the discussions of the previous chapters, if there are no interactions between the molecules, then

$$Q(N, V, T) = \frac{(q_{\text{int}})^N}{N!} \left(\frac{2\pi mkT}{h^2}\right)^{3N/2} V^N \quad \text{so that}$$

$$C \int_V \ldots \int_V e^{-u(\underline{r}_1, \ldots \underline{r}_N)/kT} \, d\underline{r}_1 \ldots d\underline{r}_N = 1$$

However, when $u(\underline{r}_1,..\underline{r}_N) = 0$, $\int_V \ldots \int_V e^{-u(\underline{r}_1,...\underline{r}_N)/kT} d\underline{r}_1 \ldots d\underline{r}_N = V^N$. Therefore $CV^N = 1$ or $C = V^{-N}$, and

$$Q(N, V, T) = \frac{\left(\frac{2\pi mkT}{h^2}\right)^{\frac{3N}{2}} (q_{\text{int}})^N}{N!} \iint e^{-u(\underline{r}_1,...\underline{r}_N)/kT} d\underline{r}_1 \ldots d\underline{r}_N$$

$$= \frac{\left(\frac{2\pi mkT}{h^2}\right)^{\frac{3N}{2}} (q_{\text{int}})^N}{N!} Z(N, V, T)$$

Here $Z(N, V, T)$ is the so-called configuration integral,

$$Z(N, V, T) = \int_V \ldots \int_V e^{-u(\underline{r}_1,...\underline{r}_N)/kT} d\underline{r}_1 \ldots d\underline{r}_N \qquad (7.1\text{-}3)$$

which is a function of the number of particles N, the volume V, and temperature T. The number of integrals depends on the number of molecules N, the dependence on T is through the integrand, and the integration limits for each position vector depend on the shape of the volume V.

7.2 THERMODYNAMIC PROPERTIES FROM THE CONFIGURATION INTEGRAL

Clearly, once the configuration integral is known, all the thermodynamic properties of the system can be evaluated:

$$A = -kT \ln Q = -NkT \ln q_{\text{int}} + NkT \ln N - NkT + 3NkT \ln \Lambda - kT \ln Z(N, V, T) \qquad (7.2\text{-}1)$$

where for simplicity we have used the earlier De Broglie wave length notation

$$\Lambda = \left(\frac{h^2}{2\pi mkT}\right)^{\frac{1}{2}} \qquad (3.4\text{-}5)$$

Then, for example

$$P = -\left(\frac{\partial A}{\partial V}\right)_{N,T} = kT \left(\frac{\partial \ln Z(N, V, T)}{\partial V}\right)_{N,T} \qquad (7.2\text{-}2)$$

Note that if there is no potential energy of interaction between the molecules—that is, if $u(\underline{r}_1, \ldots, \underline{r}_N) = 0$, then $Z = V^N$, and the partition function reduces to that for the ideal gas, which implies that

$$P = \frac{NkT}{V}$$

Consequently, the ideal gas law is obtained only if there is no energy of interaction between the particles; all the nonideal gas effects in the equation of state are contained in the configuration integral $Z(N, V, T)$.

The remainder of this chapter will be concerned with the evaluation of this quantity for one case—a moderately dense gas. In particular, our interest is in developing a theoretical basis for the virial equation of state

$$\frac{P}{\rho kT} = 1 + B_2(T)\rho + B_3(T)\rho^2 + \cdots \qquad (7.2\text{-}3)$$

which is a Taylor series expansion in density around the ideal gas result, with

$$B_2(T) = \lim_{\rho \to 0}\left(\frac{\partial(P/\rho kT)}{\partial \rho}\right)_T; \quad B_3(T) = \frac{1}{2}\lim_{\rho \to 0}\left(\frac{\partial^2(P/\rho kT)}{\partial \rho^2}\right)_T; \cdots$$

$$B_n(T) = \frac{1}{(n-1)!}\lim_{\rho \to 0}\left(\frac{\partial^n(P/\rho kT)}{\partial \rho^n}\right)_T \qquad (7.2\text{-}4)$$

7.3 THE PAIRWISE ADDITIVITY ASSUMPTION

In order to evaluate the configuration integral, it is necessary to have an expression for the interaction energy between the molecules (actually, we consider only atoms here). Thus, before proceeding with the analysis of the configuration integral, a discussion of the general character of the interaction energy of an assembly of molecules is useful. The interaction energy between two molecules is written as $u(\underline{r}_1, \underline{r}_2)$ and is a function of the positions of molecule 1 and 2 at $\underline{r}_1$ and $\underline{r}_2$, respectively. However, by the assumption of spherical symmetry of the molecules (atoms), the interaction energy will only be a function of r_{12}, the distance between molecules 1 and 2, $r_{12} = |\underline{r}_1 - \underline{r}_2| = \left[(x_1 - x_2)^2 + (y_1 - y_2)^2 + (z_1 - z_2)^2\right]^{\frac{1}{2}}$.

Even though at this point we may not know the details of the interaction potential, we can specify two boundary conditions. First, since atoms cannot overlap, we can expect that $u(r_{12}) \to \infty$ as $r_{12} \to 0$. Second, at infinite separation, we expect that there is no energy of interaction between the molecules, that is $u(r_{12}) \to 0$ as $r_{12} \to \infty$.

The interaction energy between three particles can be written as

$$u(\underline{r}_1, \underline{r}_2, \underline{r}_3) = u(r_{12}) + u(r_{13}) + u(r_{23}) + u(r_{12}, r_{13}, r_{23}) \qquad (7.3\text{-}1)$$

The first three terms on the right-hand side give the total potential energy as a sum of three pairwise or two-body (two-atom) interaction terms. The last term represents the correction to this pairwise additivity assumption as a result of the distortion of the electron clouds of the atoms due to the presence of the other atoms in close proximity. We will, for the present, neglect this nonpairwise additivity term, since its contribution may be small, except for very dense fluids. Therefore, for the three-particle system we will assume

$$u(r_1, r_2, r_3) = u(r_{12}) + u(r_{13}) + u(r_{23}) = \sum_i \sum_{\substack{j \\ 1 \le i \le j \le 3}} u(r_{ij}) \qquad (7.3\text{-}2)$$

or for the general N particle system

$$u(\underline{r}_1, \ldots, \underline{r}_N) = \sum_i \sum_{\substack{j \\ 1 \le i < j \le N}} u(r_{ij}) \qquad (7.3\text{-}3)$$

Chapter 7: Interacting Molecules in a Gas

This is the assumption of pairwise additivity. Then the Boltzmann factor in the interaction energy can be written as

$$e^{-u(\underline{r}_1,\ldots \underline{r}_N)/kT} = e^{-\sum_i \sum_{j,\, 1\le i<j\le N} u(r_{ij})/kT} = \prod_i \prod_{\substack{j \\ 1\le i\le j\le N}} e^{-u(r_{ij})/kT} \quad (7.3\text{-}4)$$

and

$$Z(N,V,T) = \int_V \cdots \int_V \prod_i \prod_{\substack{j \\ 1\le i\le j\le N}} e^{-u(r_{ij})/kT}\, d\underline{r}_1 \ldots d\underline{r}_N \quad (7.3\text{-}5)$$

7.4 MAYER CLUSTER FUNCTION AND IRREDUCIBLE INTEGRALS

To proceed further and derive the virial equation of state, the following analysis will be used, based on the Mayer cluster function $f_{ij}(r_{ij})$, defined as follows:

$$f_{ij}(r_{ij}) = e^{-u(r_{ij})/kT} - 1 \quad (7.4\text{-}1)$$

The significance of the term "cluster function" will become evident later; for the moment, we notice that the function f_{ij} is well behaved, with the following limits:

$$\text{as } r_{ij} \to 0, \quad u(r_{ij}) \to \infty \quad \text{and} \quad f_{ij} = -1$$

and

$$\text{as } r_{ij} \to \infty, \quad u(r_{ij}) \to 0 \quad \text{and} \quad f_{ij} = 0 \quad (7.4\text{-}2)$$

Using Eq. 7.4-1 in Eq. 7.3-5 we obtain

$$Z(N,V,T) = \int_V \cdots \int_V \prod_i \prod_{\substack{j \\ 1\le i\le j\le N}} (1 + f_{ij})\, d\underline{r}_1 \ldots d\underline{r}_N \quad (7.4\text{-}3)$$

which can be expanded into a sum of products

$$Z(N,V,T) = \int_V \cdots \int_V \left(1 + \sum_i \sum_{\substack{j \\ j>i}} f_{ij} + \sum_i \sum_{\substack{j \\ j>i}} \sum_{i'} \sum_{\substack{j' \\ j'>i'}} f_{ij} f_{i'j'} + \ldots \right) d\underline{r}_1 \ldots d\underline{r}_N \quad (7.4\text{-}4)$$

where the complicated restrictions on the summations are necessary to insure that each cluster function is not counted more than once since, for example, f_{12} and f_{21} represent the same interaction.

There are now several similar methods that can be used for the evaluation of the configuration integral. We will use a technique that amounts to the reduction of Eq. 7.4-3 into a collection successively more complicated integrals. The simplest term in the configuration integral to evaluate is the first one, which can be evaluated exactly

$$\int_V \cdots \int_V 1\, d\underline{r}_1 \ldots d\underline{r}_N = V^N \quad (7.4\text{-}5)$$

7.4 Mayer Cluster Function and Irreducible Integrals

The next integral is

$$\int_V \cdots \int_V f_{ij} \, d\underline{r}_1 \ldots d\underline{r}_i \ldots d\underline{r}_j \ldots d\underline{r}_N \qquad (7.4\text{-}6)$$

Since the integrand is only a function of $\underline{r}_i$ and $\underline{r}_j$ (really r_{ij}), the integration over all other position coordinates can be performed to get

$$V^{N-2} \int_V \int_V f_{ij} \, d\underline{r}_i \, d\underline{r}_j = V^{N-2} \int \cdots \int f_{ij} dx_i dy_i dz_i dx_j dy_j dz_j \qquad (7.4\text{-}7a)$$

The choice of the origin of the coordinate system is arbitrary. To do the integration above, it is especially convenient to choose the origin to be at the location of particle j. Then we change the variables of integration from $(x_i, y_i, z_i, x_j, y_j, z_j)$ to $(x_{ij}, y_{ij}, z_{ij}, x_j, y_j, z_j)$, where $x_{ij} = x_i - x_j$, etc. It is a simple task to show that the Jacobian of this transformation is unity. Also, the cluster integral f_{ij} is only a function of the variables x_{ij}, y_{ij}, and z_{ij}, so that the integral over particle j can be done and we have

$$V^{N-2} \iint f_{ij} dx_i dy_i dz_i dx_j dy_j dz_j = V^{N-2} \iint f_{ij} dx_{ij} dy_{ij} dz_{ij} dx_j dy_j dz_j$$

$$= V^{N-1} \iiint f_{ij} dx_{ij} dy_{ij} dz_{ij} \qquad (7.4\text{-}7b)$$

Next, the variable of integration is changed from the position vector in rectangular coordinates to one in spherical coordinates, that is

$$\underline{r}_{ij} = (x_{ij}, y_{ij}, z_{ij}) \Rightarrow \underline{r}_{ij} = (r_{ij}, \theta, \phi)$$

and

$$dx_{ij} dy_{ij} dz_{ij} \Rightarrow r_{ij}^2 \sin\theta \, d\theta \, d\phi \, dr_{ij}$$

where r_{ij} is the scalar distance between molecule i and molecule j. Furthermore, since $u(\underline{r}_i, \underline{r}_j) = u(r_{ij})$, it follows that f_{ij} is a function of only r_{ij}, so that

$$V^{N-1} \iiint f_{ij} dx_{ij} dy_{ij} dz_{ij} = V^{N-1} \iiint f_{ij}(r_{ij}) r_{ij}^2 \sin\theta \, d\theta \, d\phi \, dr_{ij}$$

$$= V^{N-1} 4\pi \int_0^\infty f_{ij}(r_{ij}) r_{ij}^2 \, dr_{ij} \qquad (7.4\text{-}7c)$$

This is as far as we can go with the evaluation of this integral until the functional form of the interaction potential $u(r_{ij})$ is specified. We refer to this integral as an irreducible integral, which we indicate as β_1, that is

$$\beta_1 = 4\pi \int_0^\infty f_{ij}(r_{ij}) r_{ij}^2 \, dr_{ij} \qquad (7.4\text{-}8a)$$

and denoted graphically as

Here, the filled circles represent molecules whose positions are to be integrated over the volume, and the line shows that there is an interaction between these molecules. For later reference, we note that

$$\beta_1 = \frac{1}{V} \iint f_{ij} dr_i dr_j = 4\pi \int_0^\infty f_{ij}(r_{ij}) r_{ij}^2 dr_{ij} \qquad (7.4\text{-}8b)$$

In Eq. 7.4-4, the term

$$\int_V \cdots \int_V \sum_i \sum_{\substack{j \\ j>i}} f_{ij} d\underline{r}_1 \ldots d\underline{r}_N \qquad (7.4\text{-}9a)$$

results in $N(N-1)/2$ terms containing β_1, where $N(N-1)/2$ represents the number of distinct pairs that can be formed from N molecules. Since $N \gg 1$, we will neglect terms of order unity with respect to N and write this as $N^2/2$, so that

$$\int_V \cdots \int_V \sum_i \sum_{\substack{j \\ j>i}} f_{ij} d\underline{r}_1 \ldots d\underline{r}_N = V^{N-1} \frac{N^2}{2} \beta_1 \qquad (7.4\text{-}9b)$$

Thus, so far we have

$$Z(N, V, T) = V^N + V^N \left(\frac{N^2}{2}\right)\left(\frac{\beta_1}{V}\right) \qquad (7.4\text{-}10)$$

The next type of product that arises in the expansion of the configuration integral is

$$\int_V \cdots \int_V \sum_i \sum_{\substack{j \\ j>i}} \sum_{i'} \sum_{\substack{j' \\ j'>i'}} f_{ij} f_{i'j'} d\underline{r}_1 \ldots d\underline{r}_N \qquad (7.4\text{-}11)$$

Here two cases arise: (a) i, j, i', and j' are all different; and (b) either $i = i'$ or $j = j'$. A representative example of the first case is the product $f_{12} f_{34}$, which results in the integral

$$\int_V \cdots \int_V f_{12} f_{34} d\underline{r}_1 \ldots d\underline{r}_N = V^{N-4} \int_V \cdots \int_V f_{12} f_{34} d\underline{r}_1 d\underline{r}_2 d\underline{r}_3 d\underline{r}_4$$

$$= V^{N-4} \left(\iint_{V\ V} f_{12} d\underline{r}_1 d\underline{r}_2\right)\left(\iint_{V\ V} f_{34} d\underline{r}_3 d\underline{r}_4\right) \qquad (7.4\text{-}12a)$$

The first equality results from the fact that the integrand is only a function of the position vectors $\underline{r}_1, \underline{r}_2, \underline{r}_3$, and $\underline{r}_4$; the last relation arises because the integrand is a product of two factors, the first of which depends only on $\underline{r}_1$ and $\underline{r}_2$, while the

7.4 Mayer Cluster Function and Irreducible Integrals 105

second is a function of $\underline{r}_3$ and $\underline{r}_4$. Then, from Eq. 7.4-8b, we have

$$\int_V \cdots \int_V f_{12} f_{34} \, d\underline{r}_1 \ldots d\underline{r}_N$$

$$= \int_V \cdots \int_V d\underline{r}_5 \ldots d\underline{r}_N \int_V \int_V \int_V \int_V f_{12} f_{34} \, d\underline{r}_1 \, d\underline{r}_2 \, d\underline{r}_3 \, d\underline{r}_4$$

$$= V^{N-4} \int_V \int_V f_{12} \, d\underline{r}_1 \, d\underline{r}_2 \int_V \int_V f_{34} \, d\underline{r}_3 \, d\underline{r}_4 = V^{N-2} \beta_1^2 = V^N \left(\frac{\beta_1}{V}\right)^2 \quad \text{(7.4-12b)}$$

This type of integral is represented by the cluster diagram

Therefore, the contribution to the partition function of any term of the form with nonrepeated indices is

$$\int_V \cdots \int_V f_{ij} f_{i'j'} \, d\underline{r}_1 \ldots d\underline{r}_N \quad \text{is} \quad V^N \left(\frac{\beta_1}{V}\right)^2 \quad \text{(7.4-13)}$$

It remains to count the number of terms of this form that occur. The number of pairs (i, j), (i', j') with the restrictions that $j' > i'$, $i' > i$, $j > i$ is

$$\frac{1}{2!} \left(\frac{N(N-1)}{2}\right) \left(\frac{N(N-1)}{2} - 1\right) \approx \frac{N^4}{8}$$

where the term 2! is included so that the products $f_{12} f_{34}$ and $f_{34} f_{12}$ are not counted as two separate terms.

The next integral to be considered is

$$\int_V \cdots \int_V f_{12} f_{13} \, d\underline{r}_1 \ldots d\underline{r}_N \quad \text{(7.4-14)}$$

which diagrammatically can be represented as

This integral is representative of the second class of integrals appearing in the term in Eq. 7.4-4 that is quadratic in f. In integrals of this type, the integrations over all position vectors other than $\underline{r}_1$, $\underline{r}_2$, and $\underline{r}_3$ may be done to give

$$V^{N-3} \int_V \int_V \int_V f_{12} f_{13} \, d\underline{r}_1 \, d\underline{r}_2 \, d\underline{r}_3$$

Since f_{12} is a function of only r_{12}, and f_{13} is a function of only r_{13}, the obvious transformation is to a coordinate system that has as its origin the location of particle 1.

106 Chapter 7: Interacting Molecules in a Gas

Note that moving particle 2 does not change the interparticle separation distance r_{13}, and that moving particle 3 does not change the interparticle separation distance r_{12}; this is important for doing the integrations as these two variables are independent. Since the Jacobian of this transformation is unity, we have

$$V^{N-3} \int_V \int_V \int_V f_{12} f_{13} \, d\underline{r}_1 \, d\underline{r}_2 \, d\underline{r}_3 = V^{N-3} \int_V d\underline{r}_1 \int_V \int_V f_{12} f_{13} \, d\underline{r}_{12} \, d\underline{r}_{13}$$

$$= V^{N-2} \int_V \int_V f_{12} f_{13} \, d\underline{r}_{12} \, d\underline{r}_{13}$$

$$= V^{N-2} \int_V f_{12} \, d\underline{r}_{12} \int_V f_{13} \, d\underline{r}_{13} = V^N \left(\frac{\beta_1}{V}\right)^2 \quad (7.4\text{-}15)$$

The number of terms of this form that occur, neglecting terms of order unity with respect to N (the number of atoms), is

$$\frac{1}{2!} \left(\frac{N(N-1)}{2} \right) \left(\frac{N \cdot 2}{2} - 1 \right) \approx \frac{N^3}{8}$$

Consequently, if terms of order unity are neglected (since $N \gg 1$ and therefore $N^4 \gg N^3$), we have

$$\int_V \cdots \int_V \sum_{j' > i', i' > i} \sum_{j > i} f_{ij} f_{i'j'} \, d\underline{r}_1 \cdots d\underline{r}_N = \frac{1}{2} \left(\frac{N^2}{2} \right)^2 \left(\frac{\beta_1}{V} \right)^2 V^N \quad (7.4\text{-}16)$$

and

$$Z = V^N \left[1 + \left(\frac{N^2}{2} \right) \left(\frac{\beta_1}{V} \right) + \frac{1}{2} \left(\frac{N^2}{2} \right)^2 \left(\frac{\beta_1}{V} \right)^2 \right] \quad (7.4\text{-}17)$$

While the remaining terms in the series for the partition function may, at first glance, seem obvious, it is nonetheless useful to consider the next term in the series expansion of the configuration integral. This term will involve all possible integrals involving a triple product of Mayer functions $f_{ij} f_{i'j'} f_{i''j''}$. There are a number of different types of terms that contribute to this integral. The first is integrals in which none of the indices are repeated, corresponding to the cluster diagram

Integrals of this type are treated as follows:

$$\int_V \cdots \int_V f_{12} f_{34} f_{56} \, d\underline{r}_1 \cdots d\underline{r}_N = V^{N-6} \int_V \cdots \int_V f_{12} f_{34} f_{56} \, d\underline{r}_1 \cdots d\underline{r}_6$$

$$= V^{N-6} \int_V \int_V f_{12} \, d\underline{r}_1 \, d\underline{r}_2 \int_V \int_V f_{34} \, d\underline{r}_3 \, d\underline{r}_4 \int_V \int_V f_{56} \, d\underline{r}_5 \, d\underline{r}_6$$

$$= V^{N-6} \left[\int_V \int_V f_{12} \, d\underline{r}_1 \, d\underline{r}_2 \right]^3 = V^N \left(\frac{\beta_1}{V} \right)^3 \quad (7.4\text{-}18a)$$

7.4 Mayer Cluster Function and Irreducible Integrals 107

The number of terms of this type is

$$\frac{\left[\frac{N(N-1)}{2}\right]\left[\frac{(N-2)(N-3)}{2}\right]\left[\frac{(N-4)(N-5)}{2}\right]}{3!} \approx \frac{1}{3!}\left(\frac{N^2}{2}\right)^3 = \frac{1}{3!}\frac{N^6}{8} \quad (7.4\text{-}18\text{b})$$

(neglecting terms of order unity which are much smaller than the number of particles N.)

The next class of integrals or diagrams is that in which one of the indices is repeated, such as $f_{12}f_{23}f_{56}$, which is represented by

Note that each of these types of integrals has been evaluated above, and from that analysis we obtain

$$\int_V \cdots \int_V f_{12}f_{23}f_{56}\, d\underline{r}_1 \cdots d\underline{r}_N = V^N \left(\frac{\beta_1}{V}\right)^3 \quad (7.4\text{-}19\text{a})$$

The number of integrals of this type is

$$\frac{\left[\frac{N(N-1)}{2}\right]\left[\frac{2(N-2)}{2}\right]\left[\frac{(N-3)(N-4)}{2}\right]}{3!} \approx \frac{1}{3!}N\left(\frac{N^2}{2}\right)^2 = \frac{1}{3!}\frac{N^5}{4} \quad (7.4\text{-}19\text{b})$$

Another type of integral (or, equivalently, cluster diagram) that occurs is when two of the indices are repeated, such as $f_{12}f_{23}f_{34}$, represented by

It is straightforward (though a bit tedious) to show that this integral can be reduced to

$$\int_V \cdots \int_V f_{12}f_{23}f_{34}\, d\underline{r}_1 \cdots d\underline{r}_N = V^{N-3} \int_V \cdots \int_V f_{12}f_{23}f_{34}\, d\underline{r}_{12}\, d\underline{r}_{23}\, d\underline{r}_{34}$$

$$= V^N \left(\frac{\beta_1}{V}\right)^3 \quad (7.4\text{-}20\text{a})$$

(The procedure followed is first to change the origin of the coordinate system to the location of particle 3, and then do the r_{34} integration, leaving a double integral in r_{12} and r_{23}. The integral that remains is identical in form to the one in Eq. 7.4-8b and is evaluated in that way.) The number of integrals of this type is

$$\frac{\left[\frac{N(N-1)}{2}\right]\left[\frac{2(N-2)}{2}\right]\left[\frac{2(N-3)}{2}\right]}{3!} \approx \frac{1}{3!}N^2\frac{N^2}{2} = \frac{1}{3!}\frac{N^4}{2} \quad (7.4\text{-}20\text{b})$$

108 Chapter 7: Interacting Molecules in a Gas

Yet another type integral that is the product of three Mayer functions has three repeated indices, for example $f_{12}f_{13}f_{14}$ represented by the diagram

and corresponding to the integral

$$\int_V \cdots \int_V f_{12}f_{13}f_{14}\, d\underline{r}_1 \ldots d\underline{r}_N = V^{N-4} \int_V \cdots \int_V f_{12}f_{13}f_{14}\, d\underline{r}_1 \ldots d\underline{r}_4$$

$$= V^{N-4} \int_V d\underline{r}_1 \int_V f_{12}\, d\underline{r}_{12} \int_V f_{13}\, d\underline{r}_{13} \int_V f_{14}\, d\underline{r}_{14}$$

$$= V^{N-3} \left[\int_V f_{12}\, d\underline{r}_{12} \right]^3 = V^N \left(\frac{\beta_1}{V} \right)^3 \qquad (7.4\text{-}21a)$$

The number of integrals of this type is

$$\frac{\left[\dfrac{N(N-1)}{2}\right]\left[\dfrac{1(N-2)}{2}\right]\left[\dfrac{1(N-3)}{2}\right]}{3!} \approx \frac{1}{3!}N^2\frac{N^2}{8} = \frac{1}{3!}\frac{N^4}{8} \qquad (7.4\text{-}21b)$$

The final type of integral involving the product of three Mayer functions is that in which each of the subscripts is repeated twice, i.e., $f_{12}f_{23}f_{13}$ which corresponds to each particle interacting with two others. This is represented by the cluster diagram

This resulting integral can immediately be reduced to

$$V^{N-3} \int_V \int_V \int_V f_{12}f_{23}f_{13}\, d\underline{r}_1\, d\underline{r}_2\, d\underline{r}_3 \qquad (7.4\text{-}22)$$

However, one cannot make the transformation from $(\underline{r}_1, \underline{r}_2, \underline{r}_3)$ to $(\underline{r}_{12}, \underline{r}_{23}, \underline{r}_{13})$ and treat the integral as a product of three independent integrals. This is easily seen by noting that $\underline{r}_1, \underline{r}_2$, and $\underline{r}_3$ are each independent vectors, and each spans the volume V. However, $\underline{r}_{12}, \underline{r}_{23}$, and $\underline{r}_{13}$ are not independent vectors in that if, for example, $\underline{r}_{12}$ and $\underline{r}_{23}$ are fixed, this also fixes the vector $\underline{r}_{13}$. Therefore $\underline{r}_{13}$ is not independent of the other two vectors, and cannot span the volume V. Thus, the integral of Eq. 7.4-22 cannot be reduced to a product of β_1 integrals. Consequently, we now have a new class of irreducible integral that is defined as follows:

$$\beta_2 = \frac{1}{2!V} \int_V \int_V \int_V f_{ij}f_{jk}f_{ik}\, d\underline{r}_i\, d\underline{r}_j\, d\underline{r}_k \qquad (7.4\text{-}23a)$$

Note that there are

$$\frac{N(N-1)(N-2)}{3!} \approx \frac{N^3}{3!} \quad (7.4\text{-}23b)$$

terms in the general three f factor sum that are of the β_2 type.

By counting the number of triplets, it is easily established that the three f factor term contains a total of

$$\frac{1}{3!}\left[\frac{N(N-1)}{2}\right]\left[\frac{N(N-1)}{2}-1\right]\left[\frac{N(N-1)}{2}-2\right]$$

different terms, of which

$$\frac{1}{3!}\left[\frac{N(N-1)}{2}\right]\left[\frac{N(N-1)}{2}-1\right]\left[\frac{N(N-1)}{2}-2\right] - \frac{N(N-1)(N-2)}{3}$$

are of the β_1 type. Neglecting terms of order unity with respect to the number of molecules in the system, we get

$$\sum\sum\sum\int\cdots\int f_{ij}f_{i'j'}f_{i''j''}\,d\underline{r}_1\ldots d\underline{r}_N = \frac{1}{3!}\left(\frac{N^2}{2}\right)^3 V^N \left(\frac{\beta_1}{V}\right)^3 + \frac{N^3}{3!}V^N\left(\frac{2!\beta_2}{V^2}\right) \quad (7.4\text{-}24)$$

Since β_1 and β_2 are each of order V, the first of these two terms, which is of order N^6, is dominant over the second term of order N^3. Retaining only the dominant term (that is, neglecting the cyclic β_2 term), we have

$$Z = V^N \left\{1 + \left(\frac{N^2}{2}\right)\left(\frac{\beta_1}{V}\right) + \frac{1}{2!}\left(\frac{N^2}{2}\right)^2\left(\frac{\beta_1}{V}\right)^2 + \frac{1}{3!}\left(\frac{N^2}{2}\right)^3\left(\frac{\beta_1}{V}\right)^3\right\}$$

The expression on the right-hand side can be recognized as the first four terms in the series expansion of the exponential function $\exp\left(\frac{N^2}{2}\frac{\beta_1}{V}\right)$. By considering higher-order integrals (i.e., with four, five, or more Mayer functions), additional terms in this series expansion of the exponential function will be obtained.

7.5 THE VIRIAL EQUATION OF STATE

By neglecting all cyclic diagrams (or cyclic cluster integrals) in the expansion we obtain

$$Z(N, V, T) = V^N \exp\left(\frac{N^2\beta_1}{2V}\right) \quad (7.5\text{-}1)$$

We will return to the effect of the inclusion of cyclic diagrams shortly. However, it is of interest first to examine the results obtained based on this expression for the configuration integral. Using the relationship between the partition function and the pressure, we have

$$P = kT\left(\frac{\partial \ln Q}{\partial V}\right)_{N,T} = kT\left(\frac{\partial \ln Z}{\partial V}\right)_{N,T} = kT\frac{\partial}{\partial V}\left(N\ln V + \frac{N^2\beta_1}{2V}\right)_{N,T}$$

$$= \frac{NkT}{V}\left(1 - \frac{N\beta_1}{2V}\right) = \frac{NkT}{V}\left(1 - \frac{1}{2}\beta_1\rho\right) \quad \text{or} \quad \frac{PV}{NkT} = 1 - \frac{1}{2}\beta_1\rho \quad (7.5\text{-}2)$$

110 Chapter 7: Interacting Molecules in a Gas

This result is to be compared with one term virial equation of state

$$\frac{PV}{NkT} = \frac{P}{\rho kT} = 1 + B_2(T)\rho \qquad (7.5\text{-}3)$$

to obtain an expression for the second virial coefficient

$$B_2(T) = -\frac{1}{2}\beta_1 = -\frac{1}{2}\int_V \left(e^{-u(r)/kT} - 1\right) d\underline{r} = -2\pi \int_0^\infty \left(e^{-u(r)/kT} - 1\right) r^2 \, dr$$

$$= 2\pi \int_0^\infty \left(1 - e^{-u(r)/kT}\right) r^2 \, dr \qquad (7.5\text{-}4)$$

Equation 7.5-3 establishes that the departure from ideal gas behavior results from the interactions between molecules, and Eq. 7.5-4 provides an explicit relation between the second virial coefficient and the interaction energy between a pair of molecules. This is an interesting result, since we showed in Chapter 4 that molecular complexity (i.e., going from monatomic to polyatomic molecules) in the absence of intermolecular interactions did not result in departure from ideal gas behavior. It can be shown (though we will not do so here) that interacting polyatomic molecules are also described by Eq. 7.5-3, though in that case the relation comparable to Eq. 7.5-4 is more complicated, involving integrals over the relative orientations of the molecules as well as their intermolecular separation distance. This is discussed later in this chapter.

This series-type evaluation of the partition function can be extended to include higher-order terms—such as β_2 and other irreducible integrals—and thereby obtain higher-order virial coefficients. Before doing this, however, it is useful to reflect on the procedure so far used, and to make clear the ordering procedure to be used in computing the higher virial coefficients. The first class of integrals considered was, in diagrammatic form,

●——●

for the molecular interactions which occur between only two molecules. The next type of irreducible integral was the cyclic interaction between three molecules

$$\beta_2 = \frac{1}{2!V} \int_V \int_V \int_V f_{12} f_{23} f_{31} \, d\underline{r}_1 \, d\underline{r}_2 \, d\underline{r}_3 \qquad (7.5\text{-}5)$$

which is represented by

which is the only cyclic interaction that can occur between three molecules. Note that the four f-factor cluster diagram of the type

*

7.5 The Virial Equation of State

does not represent a new type of irreducible integral. In fact, it will result in a product of β_1 and β_2 integrals. One could anticipate this by noting that this cluster diagram has what is called an articulation point, which is a molecule (or, more correctly in graph theory, a node), indicated here by an asterisk, separating the diagram into two parts—one corresponding to a β_1 integral and the other a β_2 integral. If one chooses the location of the articulation molecule as the origin of the coordinate system for the initial integration, the two types of diagrams easily separate.

The next type of irreducible integral is in fact a collection of integrals representing all possible cyclic interactions between four molecules, of which there are 10 distinct types shown below. These interactions are represented by the irreducible integral

$$\beta_3 = \frac{1}{3!V} \int_V \int_V \int_V \int_V \left[3 f_{34} f_{23} f_{14} f_{12} + 6 f_{34} f_{23} f_{14} f_{13} f_{12} + f_{34} f_{24} f_{23} f_{14} f_{13} f_{12} \right]$$
$$\times d\underline{r}_1 \, d\underline{r}_2 \, d\underline{r}_3 \, d\underline{r}_4 \qquad (7.5\text{-}6)$$

The extension to higher order types of interactions is obvious, in principle, though very tedious in practice.

With this background, we now return to the problem of developing expressions for the higher virial coefficients. To obtain an expression for the expansion of the configuration integral (up to and including the third virial coefficient), we will retain all terms of order β_1 and β_2 while neglecting integrals of type β_3 and higher. If this is done for all integrals up to

$$\sum \sum \sum \int \cdots \int f_{ij} f_{i'j'} f_{i''j''} dr_i dr_j dr_{i'} dr_{j'} dr_{i''} dr_{j''} \qquad (7.5\text{-}7a)$$

one obtains

$$Z = V^N \left\{ 1 + \left(\frac{N^2}{2}\right)\left(\frac{\beta_1}{V}\right) + \frac{1}{2!}\left(\frac{N^2}{2}\right)^2 \left(\frac{\beta_1}{V}\right)^2 + \frac{1}{3!}\left(\frac{N^2}{2}\right)^3 \left(\frac{\beta_1}{V}\right)^3 + \frac{N^3}{3!} \left(\frac{2!\beta_2}{V^2}\right) \right\}$$
$$(7.5\text{-}7b)$$

The next term in the series arises from the integral

$$\sum \sum \sum \sum \int_V \cdots \int_V f_{ij} f_{i'j'} f_{i''j''} f_{i'''j'''} \, d\underline{r}_i \, d\underline{r}_j \, d\underline{r}_{i'} \, d\underline{r}_{j'} \, d\underline{r}_{i''} \, d\underline{r}_{j''} \, d\underline{r}_{i'''} \, d\underline{r}_{j'''}$$
$$(7.5\text{-}8a)$$

which contains the terms indicated in Table 7.5-1.

Chapter 7: Interacting Molecules in a Gas

Table 7.5-1 Different Four Bond Diagrams

Number of Different Indices	Example	Diagram	Contribution
8	$f_{12}f_{34}f_{56}f_{78}$		$V^N \left(\dfrac{\beta_1}{V}\right)^4$
7	$f_{12}f_{23}f_{45}f_{67}$		$V^N \left(\dfrac{\beta_1}{V}\right)^4$
6	$\begin{cases} f_{12}f_{23}f_{45}f_{56} \\ f_{12}f_{23}f_{34}f_{56} \end{cases}$		$V^N \left(\dfrac{\beta_1}{V}\right)^4$
5	$\begin{cases} f_{12}f_{23}f_{34}f_{45} \\ \\ f_{12}f_{13}f_{14}f_{15} \\ \\ f_{12}f_{23}f_{31}f_{45} \end{cases}$		$V^N \left(\dfrac{\beta_1}{V}\right)^4$ $\\$ $\\$ $V^N \left(\dfrac{\beta_1}{V}\right)\left(\dfrac{2!\beta_2}{V^2}\right)$
4	$f_{12}f_{23}f_{13}f_{24}$		$V^N \left(\dfrac{\beta_1}{V}\right)\left(\dfrac{2!\beta_2}{V^2}\right)$
	$f_{12}f_{23}f_{34}f_{14}$		Contribution to β_3 integral

Since there are $N^2/2$ ways to choose a pair of indices for a β_1 term, and $N^3/3!$ ways to choose a triplet of indices for a β_2 term, it is easy to show that for $N \gg 1$

$$\sum\sum\sum\sum \int_V \cdots \int_V f_{ij}f_{i'j'}f_{i''j''}f_{i'''j'''}\, d\underline{r}_1 \ldots d\underline{r}_N$$

$$= \frac{1}{4!}\left(\frac{N^2}{2}\right)^4 \left(\frac{\beta_1}{V}\right)^4 V^N + \frac{N^3}{3!}\left(\frac{N^2}{2}\right)\left(\frac{2!\beta_2}{V^2}\right)\left(\frac{\beta_1}{V}\right)V^N \quad (7.5\text{-}8b)$$

Here the β_3 term, which will lead to the fourth virial coefficient, has been neglected. By analogy, we find that, to the same order

$$\sum \int \cdots \int f_{ij}f_{i'j'}f_{i''j''}f_{i'''j'''}f_{i''''j''''}\, d\underline{r}_1 \ldots d\underline{r}_N$$

$$= \frac{1}{5!}\left(\frac{N^2}{2}\right)^5 \left(\frac{\beta_1}{V}\right)^5 V^N + \frac{1}{2!}\left(\frac{N^3}{3!}\right)\left(\frac{N^2}{2}\right)^2 \left(\frac{2!\beta_2}{V^2}\right)\left(\frac{\beta_1}{V}\right)^2 V^N \quad (7.5\text{-}9)$$

7.5 The Virial Equation of State

and

$$\sum \int \cdots \int f_{ij} f_{i'j'} f_{i''j''} f_{i'''j'''} f_{i''''j''''} f_{i'''''j'''''} d\underline{r}_1 \ldots d\underline{r}_N$$

$$= \frac{1}{6!} \left(\frac{N^2}{2}\right)^6 \left(\frac{\beta_1}{V}\right)^6 V^N + \frac{1}{3!} \left(\frac{N^3}{3!}\right) \left(\frac{N^2}{2}\right)^3 \left(\frac{2!\beta_2}{V^2}\right) \left(\frac{\beta_1}{V}\right)^3 V^N$$

$$+ \frac{1}{2!} \left(\frac{N^3}{3!}\right)^2 \left(\frac{2!\beta_2}{V^2}\right)^2 V^N \tag{7.5-10}$$

Therefore, by extension, we have

$$Z = V^N \left\{ 1 + \left(\frac{N^2}{2}\right)\left(\frac{\beta_1}{V}\right) + \frac{1}{2!}\left(\frac{N^2}{2}\right)^2\left(\frac{\beta_1}{V}\right)^2 + \frac{1}{3!}\left(\frac{N^2}{2}\right)^3\left(\frac{\beta_1}{V}\right)^3 \right.$$

$$+ \frac{1}{4!}\left(\frac{N^2}{2}\right)^4\left(\frac{\beta_1}{V}\right)^4 + \cdots \right\}$$

$$+ V^N \frac{N^3}{3!} \left(\frac{2!\beta_2}{V^2}\right) \left\{ 1 + \left(\frac{N^2}{2}\right)\left(\frac{\beta_1}{V}\right) + \frac{1}{2!}\left(\frac{N^2}{2}\right)^2\left(\frac{\beta_1}{V}\right)^2 + \cdots \right\}$$

$$+ V^N \frac{1}{2!} \left\{ \frac{N^3}{3!}\left(\frac{2!\beta_2}{V^2}\right) \right\}^2 \{1 + \cdots\} \tag{7.5-11}$$

which looks like the terms in the double power series expansion of

$$Z = V^N \exp\left\{\frac{N^2 \beta_1}{2V}\right\} \exp\left\{\frac{N^3 2!\beta_2}{3!V^2}\right\} = V^N \exp\left\{\frac{N^2 \beta_1}{2V}\right\} \exp\left\{\frac{N^3 \beta_2}{3V^2}\right\} \tag{7.5-12}$$

Now, by induction, we find that the general result is

$$Z = V^N \exp\left\{\frac{N^2 \beta_1}{2V}\right\} \exp\left\{\frac{N^3 \beta_2}{3V^2}\right\} \exp\left\{\frac{N^4 \beta_3}{4V^3}\right\} \cdots$$

$$= V^N \prod_{s \geq 1} \exp\left\{\frac{N^{s+1} \beta_s}{(s+1)V^s}\right\} = V^N \exp\left\{\sum_{s \geq 1} \frac{N^{s+1} \beta_s}{(s+1)V^s}\right\} \tag{7.5-13}$$

Consequently,

$$P = kT \left(\frac{\partial \ln Q}{\partial V}\right)_{T,N} = kT \left(\frac{\partial \ln Z}{\partial V}\right)_{T,N} = \rho kT \left\{ 1 - \sum_{s \geq 1} \frac{s\beta_s}{(s+1)} \rho^s \right\} \tag{7.5-14}$$

Comparing this result with the virial equation of state

$$P = \rho kT \{1 + B_2 \rho + B_3 \rho^2 + B_4 \rho^3 + \cdots\} \tag{7.5-15}$$

we have

$$B_2 = -\frac{\beta_1}{2}, \quad B_3 = -\frac{2\beta_2}{3}, \quad B_4 = -\frac{3\beta_3}{4}, \quad \text{etc.} \tag{7.5-16}$$

114 Chapter 7: Interacting Molecules in a Gas

Alternatively, if we write

$$P = \rho kT \left\{ 1 + \sum_{j=1} B_{j+1} \rho^j \right\} \quad (7.5\text{-}17)$$

then

$$B_{j+1} = -\frac{j}{j+1} \beta_j \quad (7.5\text{-}18)$$

(Note: This result for the configuration integral is not quite correct due to our inexact counting—that is, neglecting terms of order unity with respect to N. Had the counting been done in a more rigorous manner, the result would have been

$$Z = V^N \exp \left\{ \sum_{j=1}^{N-1} \frac{N!}{(N-j-1)!(j+1)} \frac{\beta_j}{V^j} \right\} \quad (7.5\text{-}19)$$

For our purpose, the difference between this result and that of Eq. 7.5-13 is negligible.)

We will not attempt to reduce the higher-order virial coefficients to a simple integral form as was done with the second virial coefficient, as this is a difficult, tedious task. It is useful, however, to notice that each virial coefficient arises from considering cyclic interactions of a specific class, and that the number of molecules in the closed cycle determines the order of the virial coefficient to which that cyclic interaction contributes. That is, the interaction between only two molecules results in the second virial coefficient, the interactions in a closed cycle of three molecules results in the third virial coefficient, etc.

We have now established that the configuration integral for a nonideal monatomic gas is

$$Z = V^N \exp \left\{ \sum_s \frac{N \beta_s}{s+1} \rho^s \right\} \quad \text{where} \quad \beta_1 = 4\pi \int_0^\infty \left(e^{-u(r)/kT} - 1 \right) r^2 \, dr \quad (7.5\text{-}20)$$

and that the other irreducible integrals are considerably more complicated. An important observation, however, is that each of the β integrals is a function of temperature. This configuration integral can now be used to evaluate the partition function for a real gas. For a monatomic gas, the result is

$$\ln Q(N, V, T) = \ln \left\{ \frac{\left(\frac{2\pi mkT}{h^2} \right)^{3N/2} Z(N, V, T)}{N!} \right\} \quad (7.5\text{-}21)$$

7.6 VIRIAL EQUATION OF STATE FOR POLYATOMIC MOLECULES

The analysis used here for monatomic particles, resulting in the family of β integrals, is also applicable to diatomic and polyatomic molecules. In fact, the only difference is that the cluster functions result in β integrals that are multidimensional over not only the separation distance between the particles, but also their relative orientations.

7.6 Virial Equation of State for Polyatomic Molecules 115

In particular, if we make the assumption that none of the internal energy modes of a molecule is affected by the interaction of the particles, the partition function is

$$\ln Q(N, V, T) = \ln \left\{ \frac{\left(\frac{2\pi mkT}{h^2}\right)^{3N/2} q_{\text{int}}^N(T) Z(N, V, T)}{N!} \right\} \quad (7.6\text{-}1)$$

For monatomic molecules, q_{int} is just the partition function for the electronic energy states of the atom; for more complicated molecules, q_{int} contains contributions from the rotational and vibrational energy modes as well. Also, for monatomic molecules, the interaction energy among the particles is only a function of the distance between them, and not their orientation. While polyatomic molecules are not spherically symmetric, the development of the partition function so far presented is still valid, except that interaction energy between to molecules is now a function of their relative orientation as well as their separation, and this must be included in the configuration integral. Consequently, each of the irreducible (β) integrals are then integrals over both orientation and separation distance and are more complicated than those presented in the previous section. For example, for diatomic or triatomic molecules the β_1 integral would be

$$\beta_1 = \frac{4\pi \iiint\limits_0^\infty \left(e^{-u(r,\underline{\omega}_1,\underline{\omega}_2)/kT} - 1\right) r^2 dr d\underline{\omega}_1 d\underline{\omega}_2}{\iint d\underline{\omega}_1 d\underline{\omega}_2} \quad (7.6\text{-}2)$$

where each $\underline{\omega}$ is the vector describing the orientation of the molecule in space (two angles for a linear molecule and three angles for a nonlinear molecule).

Using Eq. 7.6-1, it is easily shown that the volumetric equation of state for a polyatomic molecule is unchanged from that obtained previously for a monatomic gas,

$$P = kT \frac{\partial \ln Q}{\partial V} = \rho kT \left[1 - \sum_{j=1} \rho^j \frac{j}{j+1} \beta_j \right] = \rho kT \left[1 + \sum_{j=1} \rho^j B_{j+1} \right] \quad (7.6\text{-}3)$$

where

$$B_{j+1} = -\frac{j}{j+1} \beta_j$$

and, in particular

$$B_2(T) = -2\pi \frac{\iiint\limits_0^\infty \left(e^{-u(r,\underline{\omega}_1,\underline{\omega}_2)/kT} - 1\right) r^2 dr d\underline{\omega}_1 d\underline{\omega}_2}{\iint d\underline{\omega}_1 d\underline{\omega}_2} \quad (7.6\text{-}4)$$

Therefore, the partition function can be written in terms of the virial coefficients as follows:

$$\ln Q(N, V, T) = \frac{3N}{2} \ln \left(\frac{2\pi mkT}{h^2}\right) + N \ln q_{\text{int}} + \ln V^N - \sum_j N \frac{\rho^j}{j} B_{j+1} - \ln N \quad (7.6\text{-}5)$$

for both monatomic and polyatomic molecules (where $q_{\text{int}} = 1$ for a monatomic molecule).

7.7 THERMODYNAMIC PROPERTIES FROM THE VIRIAL EQUATION OF STATE

Once the canonical partition function is known, as is the case here for the slightly nonideal gas resulting in virial equation of state, all the thermodynamic properties can be obtained. For example

$$U(N, V, T) = kT^2 \left(\frac{\partial \ln Q}{\partial T}\right)_{N,V} = NkT^2 \left\{\frac{3}{2T} + \frac{d \ln q_{int}}{dT} - \sum_j \frac{\rho^j}{j} \frac{dB_{j+1}}{dT}\right\}$$

$$= \frac{3NkT}{2} + NkT^2 \frac{d \ln q_{int}}{dT} - NkT^2 \sum_j \frac{\rho^j}{j} \frac{dB_{j+1}}{dT} \qquad (7.7\text{-}1)$$

or

$$U(N, V, T) - U^{IG}(N, V, T) = -NkT^2 \sum_j \frac{\rho^j}{j} \frac{dB_{j+1}}{dT}$$

or

$$\frac{U(N, V, T) - U^{IG}(N, V, T)}{NkT} = -T \sum_j \frac{\rho^j}{j} \frac{dB_{j+1}}{dT} \qquad (7.7\text{-}2a)$$

where the superscript IG is used to indicate an ideal gas property at the same temperature, volume, and number of molecules. It is convenient to define a virial coefficient on a molar (rather than per-molecule) basis, as $\mathcal{B}_j = N_{Av} B_j$, where N_{Av} is Avogadro's number. Then, on a molar basis (indicated by an underbar)

$$\underline{U}(n, T) - \underline{U}^{IG}(n, T) = -RT^2 \sum_j \frac{n^j}{j} \frac{d\mathcal{B}_j}{dT} \qquad (7.7\text{-}2b)$$

where n is the molar density. The other thermodynamic properties are

$$S = k \ln Q + kT \left(\frac{\partial \ln Q}{\partial T}\right)_{V,N} = Nk \ln q_{int} - k \ln N! + Nk \ln \left(\frac{2\pi mkT}{h^2}\right)^{3/2}$$

$$+ Nk \ln V - Nk \sum_j \frac{\rho^j B_{j+1}}{j} + NkT \frac{d \ln q_{int}}{dT} + \frac{3k}{2} N - NkT \sum_j \frac{\rho^j}{j} \frac{dB_{j+1}}{dT}$$

$$(7.7\text{-}3a)$$

$$S(N, V, T) = S^{IG}(N, V, T) - NkT \sum_j \frac{\rho^j}{j} \frac{dB_{j+1}}{dT} \qquad (7.7\text{-}3b)$$

and

$$\frac{S(N, V, T) - S^{IG}(N, V, T)}{Nk} = -T \sum_j \frac{\rho^j}{j} \frac{dB_{j+1}}{dT} \qquad (7.7\text{-}3c)[1]$$

or

$$\frac{\underline{S}(n, T) - \underline{S}^{IG}(n, T)}{R} = -T \sum_j \frac{n^j}{j} \frac{d\mathcal{B}_{j+1}}{dT} \qquad (7.7\text{-}3d)$$

---

[1] Compare Eqs. 7.7-2a and 7.7-3c. Can you explain why the right-hand sides of these equations are identical?

7.7 Thermodynamic Properties from the Virial Equation of State

$$A(N, V, T) = U(N, V, T) - TS(N, V, T)$$

$$= U^{IG}(N, V, T) - NkT^2 \sum_j \frac{\rho^j}{j} \frac{dB_{j+1}}{dT}$$

$$- T\left(S^{IG}(N, V, T) - Nk \sum_j \frac{\rho^j B_{j+1}}{j} - NkT \sum_j \frac{\rho^j}{j} \frac{dB_{j+1}}{dT}\right)$$

(7.7-4a)

$$A(N, V, T) = A^{IG}(N, V, T) + NkT \sum_j \rho^j \frac{B_{j+1}}{j} \quad (7.7\text{-}4\text{b})$$

which can also be gotten directly from $A = kT \ln Q$, and

$$\frac{\underline{A}(n, T) - \underline{A}^{IG}(n, T)}{RT} = \sum_j n^j \frac{\mathcal{B}_{j+1}}{j} \quad (7.7\text{-}4\text{c})$$

Also

$$\mu(N, V, T) = \left(\frac{\partial A}{\partial N}\right)_{T,V} = \mu^{IG}(N, V, T) + \frac{\partial}{\partial N}\left\{kT \sum \frac{N^{j+1}}{V^j} \frac{B_{j+1}}{j}\right\}$$

$$= \mu^{IG}(N, V, T) + kT \sum (j+1) \frac{N^j}{V^j} \frac{B_{j+1}}{j} \quad (7.7\text{-}5)$$

For an ideal gas at any pressure P, we usually write

$$\mu^{IG}(T, P) = \mu^{IG}(T, P_0) + kT \ln \frac{P}{P_0} \quad (7.7\text{-}6\text{a})$$

where P_0 is a standard state pressure (frequently chosen to be 1 bar); and once P_0 is chosen, $\mu^{IG}(T, P_0)$ is a function of temperature only. With these definitions

$$\mu(T, P) = \mu^{IG}(T, P_0) + kT \ln \frac{P}{P_0} + kT \sum (j+1) \rho^j \frac{B_{j+1}}{j} \quad (7.7\text{-}6\text{b})$$

The fugacity $f(T, P)$ for a real gas is defined by the relation

$$\mu(T, P) = \mu^{IG}(T, P_0) + RT \ln \frac{f(T, P)}{P_0} \quad (7.7\text{-}7)$$

so that

$$f = Pe^{\sum(j+1)\rho^j \frac{B_{j+1}}{j}} = P\exp\left\{2\rho B_2 + \frac{3}{2}\rho^2 B_3 + \frac{4}{3}\rho^3 B_4 + \cdots\right\} \quad (7.7\text{-}8)$$

However, expressing fugacity as a power series in density is not the most useful relation for the engineer, since it is usually the pressure not the density that is known. Therefore, one first has to solve the volumetric equation of state for density for the given temperature and pressure, and then use this density in the equation above.

118 Chapter 7: Interacting Molecules in a Gas

The enthalpy of the nonideal gas is gotten from $H = U + PV$, so that

$$H(N, V, T) = U^{\text{IG}}(N, V, T) - NkT^2 \left\{ \rho \frac{dB_2}{dT} + \frac{1}{2}\rho^2 \frac{dB_3}{dT} + \frac{1}{3}\rho^3 \frac{dB_4}{dT} + \cdots \right\}$$

$$+ NkT\{1 + \rho B_2 + \rho^2 B_3 + \rho^3 B_4 + \cdots\}$$

$$= U^{\text{IG}}(N, V, T) + NkT$$

$$+ NkT \left\{ \rho \left(B_2 - T\frac{dB_2}{dT} \right) + \rho^2 \left(B_3 - T\frac{dB_3}{dT} \right) + \cdots \right\}$$

(7.7-9a)

$$H(N, V, T) = H^{\text{IG}}(N, V, T) + NkT \left\{ \rho \left(B_2 - T\frac{dB_2}{dT} \right) + \rho^2 \left(B_3 - T\frac{dB_3}{dT} \right) + \cdots \right\}$$

(7.7-9b)

and

$$\frac{H(N, V, T) - H^{\text{IG}}(N, V, T)}{NkT} = \rho \left(B_2 - T\frac{dB_2}{dT} \right) + \rho^2 \left(B_3 - T\frac{dB_3}{dT} \right) + \cdots$$

(7.7-9c)

7.8 DERIVATION OF VIRIAL COEFFICIENT FORMULAE FROM THE GRAND CANONICAL ENSEMBLE

Considerable effort was devoted to deriving expressions for the virial coefficients starting from the canonical ensemble. We will now re-derive the expressions for the virial coefficients starting from the grand canonical ensemble. The reasons for doing this are several:

(a) the derivation is easier, avoids the cumbersome apparatus of cluster integrals, and shows the advantage (in this case) of using the grand canonical ensemble;
(b) the derivation provides a simple method of obtaining expressions for all the higher virial coefficients;
(c) the derivation does not require pairwise additivity of the potential; and
(d) the derivation is equally applicable to classical or quantum fluids, and so is completely general.

We start with the definition for the grand canonical partition function

$$\Xi(\mu, V, T) = \sum_{E,N} e^{-E(N,V)/kT} e^{\mu N/kT} \qquad (7.8\text{-}1)$$

which can be written as

$$\Xi(\mu, V, T) = \sum_N e^{N\mu/kT} \sum_{\substack{E \\ \text{for fixed } N}} e^{-E(N,V)/kT} = \sum_N e^{\frac{\mu N}{kT}} Q(N, V, T) \qquad (7.8\text{-}2)$$

since

$$\sum_E e^{-E(N,V)/kT} = Q(N, V, T) = \text{canonical ensemble partition function.}$$

7.8 Derivation of Virial Coefficient Formulae 119

For convenience, we define the absolute activity as $\lambda = e^{\mu/kT}$, so that Eq. 7.8-2 can be written as

$$\Xi(V, T, \mu) = \Xi(V, T, \lambda) = \sum_N Q(N, V, T)\lambda^N \qquad (7.8\text{-}3)$$

This provides an expression for the grand canonical partition function in terms of a series expansion in the absolute activity, with coefficients that are canonical partition functions of increasing numbers of atoms (or molecules).

To proceed, we note that $Q(1, V, T)$ is the canonical partition function for a single particle in the volume V

$$Q(1, V, T) = Q_1 = q_{\text{int}}\left(\frac{2\pi m kT}{h^2}\right)^{\frac{3}{2}} V$$

since the configuration integral for a single particle is $Z(1, V, T) = V$. For simplicity we will use the notation that $z = \frac{Q_1 \lambda}{V} = q_{\text{int}}\left(\frac{2\pi m kT}{h^2}\right)^{3/2} \lambda$. The next term is

$$\lambda^2 Q(2, V, T) = \frac{1}{2!}\left[q_{\text{int}}\left(\frac{2\pi m kT}{h^2}\right)^{\frac{3}{2}}\right]^2 \lambda^2 Z(2, V, T) = \frac{z^2 Z(2, V, T)}{2!}$$

$$\underleftarrow{\qquad\qquad\text{Indistinguishability factor}\qquad\qquad}\underrightarrow{}$$

where

$$Z(2, V, T) = \iint e^{-u(\underline{r}_1, \underline{r}_2)/kT} \, d\underline{r}_1 \, d\underline{r}_2 = Z_2$$

In general

$$Z(M, V, T) = \int \ldots \int e^{-u(\underline{r}_1, \ldots \underline{r}_M)/kT} \, d\underline{r}_1 \ldots d\underline{r}_M = Z_M$$

$$\lambda^M Q(M, V, T) = \frac{1}{M!}\left[q_{\text{int}}\left(\frac{2\pi m kT}{h^2}\right)^{\frac{3}{2}}\right]^M \lambda^M Z(M, V, T) = \frac{z^M Z(M, V, T)}{M!}$$

so that

$$\Xi(\mu, V, T) = \sum_{N=0} Q(N, V, T)\lambda^N = \sum_{N=0} \frac{Z(N, T, V)z^N}{N!} = \sum_{N=0} \frac{Z_N z^N}{N!} \qquad (7.8\text{-}4)$$

We have shown previously (Eq. 6.2-12) that $PV = kT \ln \Xi$, which can be written as

$$\frac{PV}{kT} = \ln \Xi = \ln(1 + Y) \quad \text{where} \quad Y = Z_1 z + \frac{1}{2}Z_2 z^2 + \frac{1}{6}Z_3 z^3 + \cdots \qquad (7.8\text{-}5)$$

Next, expanding the logarithm

$$\ln(1 + Y) = Y - \frac{1}{2}Y^2 + \frac{1}{3}Y^3 - \cdots\cdots$$

120 Chapter 7: Interacting Molecules in a Gas

and grouping terms of similar power in z, we obtain

$$\frac{PV}{kT} = \ln \Xi = Z_1 z + \frac{z^2}{2}(Z_2 - Z_1^2) + \frac{z^3}{6}(Z_3 - 3Z_1 Z_2 + 2Z_1^3) + \cdots$$

or

$$\frac{PV}{kT} = V \sum_{j=1}^{\infty} b_j z^j = \ln \Xi \tag{7.8-6}$$

where

$$b_1 = \frac{Z_1}{V} = 1, \text{ since } Z_1 = V, \quad Vb_2 = \frac{1}{2}(Z_2 - Z_1^2), \quad Vb_3 = \frac{1}{6}(Z_3 - 3Z_1 Z_2 + 2Z_1^3)$$

and so on.

This gives PV/kT as a function of $z = Q_1 \lambda / V = q_{\text{int}} \left(2\pi mkT/h^2\right)^{3/2} \lambda$. However, what we really would like to have is a volumetric equation of state in the form of PV/kT as a function of N/V or ρ.

Now, note that from the definition of the partition function, the average number of particles in the system can be computed from the grand canonical ensemble as follows:

$$\overline{N} = \frac{\sum_N N Q(N, V, T) \lambda^N}{\Xi}$$

so

$$\overline{N} = \lambda \left(\frac{\partial \ln \Xi(T, V, \lambda)}{\partial \lambda}\right)_{T,V} = z \left(\frac{\partial \ln \Xi(T, V, z)}{\partial z}\right)_{T,V} \tag{7.8-7}$$

and therefore

$$\overline{N} = z \left(\frac{\partial \ln \Xi}{\partial z}\right)_{T,V} = V \sum_{j=1}^{\infty} j b_j z^j \tag{7.8-8a}$$

or

$$\frac{\overline{N}}{V} = \rho = z + 2b_2 z^2 + 3b_3 z^3 + \cdots \tag{7.8-8b}$$

This is an equation for ρ as a function of z, while what is needed is z as a function of ρ. To obtain such an expression, we must invert the series (called a series reversion)—that is, develop an expression for z as a function of density that can be used in Eq. 7.8-6. To do this we write

$$z = A_1 \left(\frac{\overline{N}}{V}\right) + A_2 \left(\frac{\overline{N}}{V}\right)^2 + A_3 \left(\frac{\overline{N}}{V}\right)^3 + \cdots = A_1 \rho + A_2 \rho^2 + A_3 \rho^3 + \cdots$$

$$\tag{7.8-9}$$

Using Eq. 7.8-8b in Eq. 7.8-9, we get

$$z = A_1 \left(z + 2b_2 z^2 + 3b_3 z^3 + \cdots\right) + A_2 \left(z + 2b_2 z^2 + 3b_3 z^3 + \cdots\right)^2$$
$$+ A_3 \left(z + 2b_2 z^2 + 3b_3 z^3 + \cdots\right)^3 + \cdots$$

7.8 Derivation of Virial Coefficient Formulae

and equating powers of z, we obtain

$$z^1: \quad 1 = A_1$$
$$z^2: \quad 2b_2 A_1 + A_2 = 0 \quad \Rightarrow \quad A_2 = -2b_2$$
$$z^3: \quad 3b_3 A_1 + 4b_2 A_2 + A_3 = 0 \Rightarrow A_3 = 8b_2^2 - 3b_3$$

etc.

so that

$$z = \left(\frac{N}{V}\right) - 2b_2 \left(\frac{N}{V}\right)^2 + \left(8b_2^2 - 3b_3\right)\left(\frac{N}{V}\right)^3 + \cdots \quad \text{(7.8-10)}$$

Using this expression in Eq. 7.8-6 gives

$$\frac{P}{kT} = \sum_{j=1}^{\infty} b_j \left[\rho - 2b_2 \rho^2 + \left(8b_2^2 - 3b_3\right)\rho^3 + \cdots \right]^j$$

and expanding the series and equating powers of ρ, we obtain

$$\frac{P}{kT} = \rho - 2b_2 \rho^2 + \left(8b_2^2 - 3b_3\right)\rho^3 + b_2 \rho^2 - 4b_2^2 \rho^3 + \cdots$$

$$\frac{P}{kT} = \rho - b_2 \rho^2 + \left(4b_2^2 - 2b_3\right)\rho^3 + \cdots \quad \text{(7.8-11)}$$

since $b_1 = 1$.

Comparing Eq. 7.8-11 with the virial equation of state

$$P/kT = \rho + B(T)\rho^2 + C(T)\rho^3 + \cdots$$

we have

$$B(T) = -b_2 = -\left(\frac{1}{2V}\right)\left(Z_2 - Z_1^2\right) = -\frac{1}{2V}\left(Z_2 - V^2\right)$$

$$= +\frac{1}{2V}\left\{V^2 - \iint_v e^{-u(\underline{r}_1,\underline{r}_2)/kT} d\underline{r}_1 \, d\underline{r}_2\right\}$$

$$= \frac{1}{2V} \iint \left(1 - e^{-u(\underline{r}_1,\underline{r}_2)/kT}\right) d\underline{r}_1 \, d\underline{r}_2 \quad \text{(7.8-12a)}$$

If the potential is spherically symmetric, i.e., $u(\underline{r}_1, \underline{r}_2) = u(r_{12})$, then

$$B(T) = \frac{1}{2V} V 4\pi \int_0^{\infty} \left(1 - e^{-u(r)/kT}\right) r^2 \, dr = 2\pi \int_0^{\infty} \left(1 - e^{-u(r)/kT}\right) r^2 \, dr \quad \text{(7.8-12b)}$$

which is precisely the expression we developed earlier in this chapter. In a similar fashion, we obtain

$$C(T) = \left(4b_2^2 - 2b_3\right) = \frac{Z_2(Z_2 - V^2)}{V^2} + \frac{V^3 - Z_3}{3V} + \cdots \quad \text{(7.8-13)}$$

Note that in this derivation, we never had to specify the form of the configuration integral or the interaction potential, whether or not the potential was pairwise additive, or even whether the system was described by classical or quantum mechanics (though both of these will be important in the evaluation of the configuration integrals). Therefore, this derivation is also applicable to more complicated configuration integrals—for example, when the intermolecular potential function is not spherically symmetric, so that more than just a single center-to-center position vector is necessary to specify the potential energy. This is the case in molecular fluids in which relative orientation vectors or several atom-atom distances are needed.

There is one part of this derivation that needs justification. We had the exact expression

$$\frac{PV}{kT} = \ln \Xi = \ln(1+Y) \quad \text{where} \quad Y = \sum_{N=1}^{\infty} \frac{z^N Z_N}{N!} = \sum_{N=1}^{\infty} Q_N \lambda^N \qquad (7.8\text{-}14)$$

and then used the series expansion

$$\ln(1+Y) = Y - \frac{1}{2}Y^2 + \frac{1}{3}Y^3 \ldots$$

For this series to converge rapidly, Y should be less than unity. In fact, Y is much greater than 1; it is of the order of the number of molecules in the system, N. Therefore, the series expansion would seem to be divergent. However, notice that Y is a sum of terms, each of which is a product of the form $Q(N)\lambda^N$. In particular, the first term in the series is $Q(1)\lambda = q_{\text{trans}}\lambda = q_{\text{trans}} e^{\mu/kT}$. To obtain a rough order of magnitude estimate for the terms in this product, consider the calculation of the argon as an ideal gas at 25°C:

$$q_{\text{trans}} = 0.245 \times 10^{27} \text{cm}^{-3} \times 2.24 \times 10^4 \text{cm}^3 = 0.548 \times 10^{23}$$

$$\mu = -9500 \text{ cal/mol} \quad \text{and} \quad \lambda = e^{\mu/kT} = e^{\frac{-9500}{2 \times 298}} = e^{-15.9}$$

So Y is a sum of terms, each of which is a product of the very large number $Q(N)$ and a very small number λ^N. In the limit of $\lambda = 0$, $Y = 0$, and $\ln(1+Y) = 0$. So if we expand $\ln(1+Y)$ about $\lambda = 0$, we have

$$\ln(1+Y) = \ln(1+Y)|_{\lambda=0} + \left.\frac{\partial \ln(1+Y)}{\partial \lambda}\right|_{\lambda=0} \lambda + \frac{1}{2!}\left.\frac{\partial^2 \ln(1+Y)}{\partial \lambda^2}\right|_{\lambda=0} \lambda^2$$

$$+ \frac{1}{3!}\left.\frac{\partial^3 \ln(1+Y)}{\partial \lambda^3}\right|_{\lambda=0} \lambda^3 + \cdots \qquad (7.8\text{-}15)$$

However, as mentioned above, $\ln(1+Y)|_{\lambda=0} = 0$, and

$$\left.\frac{\partial \ln(1+Y)}{\partial \lambda}\right|_{\lambda=0} = \frac{1}{\sum_{N=0}^{\infty} Q_N \lambda^N} \left.\sum_{N=0}^{\infty} N Q_N \lambda^{N-1}\right|_{\lambda=0} = Q_1 \qquad (7.8\text{-}16a)$$

7.9 Range of Applicability of the Virial Equation

$$\left.\frac{\partial^2 \ln(1+Y)}{\partial \lambda^2}\right|_{\lambda=0} = \frac{1}{\Xi}\sum N(N-1)Q_N\lambda^{N-2} - \frac{1}{\Xi^2}\left(\sum_{N=1}^{\infty} NQ_N\lambda^{N-1}\right)^2$$

$$= 2Q_2 - Q_1^2 \quad \text{and} \tag{7.8-16b}$$

$$\left.\frac{\partial^2 \ln(1+Y)}{\partial \lambda^3}\right|_{\lambda=0} = \frac{1}{\Xi}\sum (N-2)(N-1)(N)Q_N\lambda^{N-3}$$

$$- \frac{3}{\Xi^2}\left(\sum N(N-1)Q_N\lambda^{N-2}\right) \times \left(\sum NQ_N\lambda^{N-1}\right)$$

$$+ \frac{1}{\Xi^3}\left(\sum_{N=1}^{\infty} NQ_N\lambda^{N-1}\right)^3 = 3!Q_3 - 3Q_2Q_1 + Q_1^3 \quad (7.8\text{-}16\text{c})$$

So that

$$\ln(1+Y) \sim Q_1\lambda + \frac{1}{2!}\left(2Q_2 - Q_1^2\right)\lambda^2 + \frac{1}{3!}\left(3!Q_3 - 3Q_2Q_1 + Q_1^3\right)\lambda^3$$

$$= Q_1\lambda + Q_2\lambda^2 + Q_3\lambda^3 - \frac{Q_1^2\lambda^2}{2} - \frac{1}{2}Q_1Q_2\lambda^3 + \frac{1}{3!}Q_1^3\lambda^3 + \cdots$$

$$= \left(Q_1\lambda + Q_2\lambda^2 + Q_3\lambda^3 + \cdots\right) - \frac{1}{2}\left(Q_1^2\lambda^2 + Q_1Q_2\lambda^3 + \cdots\right)$$

$$+ \frac{1}{3}\left(Q_1^3\lambda^3 + \cdots\right)$$

$$= Y - \frac{1}{2}Y^2 + \frac{1}{3}Y^3 + \cdots \tag{7.8-17}$$

The important point is that we have obtained the desired result not by doing an expansion about the large term Y, but rather by expanding about the vanishingly small term λ.

7.9 RANGE OF APPLICABILITY OF THE VIRIAL EQUATION

In Table 7.9-1 are data for the compressibility factor of argon at 25°C at a collection of different pressures. We see that for argon at 25°C, which is well above its critical temperature of 150.87 K, there is an insignificant error in the compressibility factor when using the virial equation with only the second virial coefficient up to 10 atm; only about a 2 percent error at pressures up to 100 atm; and significant errors at higher pressures. In general, one can expect similar accuracies with other nonassociating gases well above their critical points. However, the error will be larger for gases below their critical points, and significantly greater for a gas in which association occurs by, for example, hydrogen bonding. Thus, there would be significant error when using a truncated virial expansion for a strong hydrogen-bonding fluid such as hydrogen fluoride, and to a lesser—but not negligible—extent for acetic acid, methanol, and water.

Table 7.9-1 The Compressibility Factor of Argon at 25°C and Predictions Using the Second and Third Virial Coefficients

P(atm)	$P/\rho kT$				error of using only B_2
	$1 + B_2\rho$	$+B_3\rho^2$	+remainder	Total	
1	1 −0.00064	+0.00000	+0.00000	0.99936	0%
10	1 −0.00648	+0.00020	−0.00007	0.99365	−0.013%
100	1 −0.06754	+0.02127	−0.00036	0.95337	−2.19%
1000	1 −0.38404	+0.68788	+0.37272	1.67616	−63.25%

Based on a table in E. A. Mason and T. Spurling, *Virial Equation of State*, Pergamon Press, New York, 1969.

CHAPTER 7 PROBLEMS

7.1 Show that the second virial coefficient for a mixture of species is given

$$B_{2,\text{mix}}(\underline{x}, T) = \sum_{i=1}^{C}\sum_{j=1}^{C} x_i x_j B_{2,ij}(T)$$

where

$$B_{2,ij}(T) = 2\pi \int_0^\infty \left(1 - e^{-u_{ij}(r)/kT}\right) r^2\, dr$$

and $u_{ij}(r)$ is the intermolecular potential for a species i-species j interaction.

7.2 Obtain expressions for the thermodynamic properties of a binary mixture described by the following equation of state:

$$P(\underline{x}, \rho, T) = \rho kT[1 + B_{2,\text{mix}}(\underline{x}, T)\rho]$$

and

$$B_{2,\text{mix}}(\underline{x}, T) = \sum_{i=1}^{C}\sum_{j=1}^{C} x_i x_j B_{2,ij}(T)$$

7.3 Show that the third virial coefficient for a mixture of species is given

$$B_{3,\text{mix}}(\underline{x}, T) = \sum_{i=1}^{C}\sum_{j=1}^{C}\sum_{k}^{C} x_i x_j x_k B_{3,ijk}(T)$$

7.4 Obtain expressions for the thermodynamic properties of a binary mixture described by the following equation of state:

$$P(\underline{x}, \rho, T) = \rho kT[1 + B_{2,\text{mix}}(\underline{x}, T)\rho + B_{3,\text{mix}}(\underline{x}, T)\rho^2]$$

and the mole fraction dependence of the second and third virial coefficients are as given in Problems 7.1 and 7.3.

7.5 Obtain expressions for the constant volume heat capacity for a monatomic gas that obeys the virial equation of state.

7.6 Since pressure is more easily measured than density, it is sometimes more convenient to use a virial expansion in terms of pressure as shown below

$$\frac{P}{\rho kT} = 1 + B_2^* P + B_3^* P^2 + B_4^* P^3 + \ldots$$

Relate the virial coefficients B_i^* to the coefficients B_i in the virial expansion considered in this chapter.

7.7 Obtain expressions for the thermodynamic properties of a gas using the virial expansion in pressure of the previous problem.

7.8 Develop the expressions for the $\underline{U} - \underline{U}^{\text{IG}}$ and $C_V - C_V^{\text{IG}}$ for a fluid described by the virial equation of state with only the second virial coefficient.

Chapter 8

Intermolecular Potentials and the Evaluation of the Second Virial Coefficient

In the previous chapter, we developed expressions for the equation of state and other thermodynamic properties of a nonideal gas in terms of the virial coefficients. However, to use these formulae, we need values for the virial coefficients as a function of temperature. The second virial coefficient for a monatomic species has been shown to be

$$B_2(T) = -\frac{1}{2}\beta_1 = 2\pi \int_0^\infty (1 - e^{-u(r)/kT})r^2\, dr \tag{7.5-4}$$

that can be explicitly evaluated once the form of the intermolecular potential is specified. Here we will consider a number of models for the intermolecular potential and examine the form of the virial coefficient for each of these models.

INSTRUCTIONAL OBJECTIVES FOR CHAPTER 8

The goals for this chapter are for the student to:

- Be able to compute values of the second virial coefficient as a function of temperature for different interaction potential models
- Be able to compute thermodynamic properties using the virial equation of state
- Be able to compute the second virial coefficients in mixtures
- Understand the engineering implications and applications of the virial equation of state

8.1 INTERACTION POTENTIALS FOR SPHERICAL MOLECULES

Hard-Sphere Potential

The simplest potential used to represent molecular interactions is the rigid or hard-sphere potential shown in Fig. 8.1-1 and is given by

$$u(r) = \begin{cases} \infty & r \leq \sigma \\ 0 & r > \sigma \end{cases} \tag{8.1-1}$$

126 Chapter 8: Intermolecular Potentials and the Evaluation of the Second Virial Coefficient

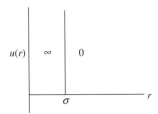

Figure 8.1-1 The Hard-Sphere Potential.

This very simple potential model accounts for the fact that molecules cannot overlap, but neglects the attractive forces between molecules. Using this potential in Eq. 7.5-4 gives us

$$B_2(T) = 2\pi \int_0^\infty (1 - e^{-u(r)/kT}) r^2 \, dr$$

$$= 2\pi \int_0^\sigma (1) r^2 \, dr + 2\pi \int_\sigma^\infty (1 - 1) r^2 \, dr = \frac{2\pi \sigma^3}{3} \quad \textbf{(8.1-2)}$$

There are two features of this expression for the virial coefficient that are interesting. First, the virial coefficient for this potential is always positive, which will also be true for any purely repulsive potential. By a repulsive potential, we mean a potential energy that is only positive, so that there is no negative or attractive energy between the molecules. Second, the virial coefficient for the hard-sphere fluid is temperature independent; this is only occurs for the rigid-sphere potential. Also note that the numerical value of the second virial coefficient for this interaction potential is exactly one-half the molecular volume.

Real molecules repel each other at short distances, since overlap cannot occur; but due to electrostatic forces (such as dipoles, induced dipoles, etc.), they attract each other at larger intermolecular separations. Consequently, the second virial coefficient is negative at low temperatures, where weak, long-range attractions are important; and positive at higher temperatures, where the kinetic energy of the molecules is sufficiently high that only the strong, short-range repulsions are important. This is

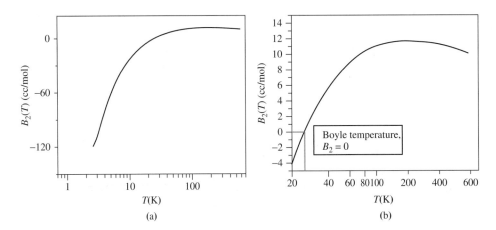

Figure 8.1-2 The Second Virial Coefficient for Helium.

shown in Fig. 8.1-2 for the second virial coefficient of helium. There is an intermediate temperature, referred to as the Boyle temperature, at which the second virial coefficient of a real fluid is equal to 0.

It is useful to note that if one had second virial coefficient data as a function of temperature, such as the data in Fig. 8.1-2, the form of Eq. 7.5-4 is such that it cannot be inverted in an exact manner to obtain the underlying intermolecular potential function $u(r)$. Consequently, what is usually done is to assume a simple model for the interaction between two atoms or molecules as we consider below, and then adjust the values of the parameters in the model to get the best possible fit of the experimental data. Because the true functional form of the interaction potential between atoms is not known—that very simple approximate models are used, and that there are experimental uncertainties in virial coefficient data—a potential determined in this way can be considered useful for correlation, but should not be considered a true representation of the actual interaction potential.

Point Centers of Repulsion Potential

In this model the interaction potential is also purely repulsive; however, the discontinuous nature of the rigid sphere potential is replaced with the smooth function of interatomic separation

$$u(r) = \frac{a}{r^n} \tag{8.1-3a}$$

shown in Fig. 8.1-3.

The second virial coefficient for this potential is

$$B_2(T) = 2\pi \int_0^\infty (1-e^{-a/kTr^n})r^2\,dr = \frac{2\pi}{3}\left(\frac{a}{kT}\right)^{3/n}\Gamma\left(1-\frac{3}{n}\right); \quad \text{provided } n \geq 4 \tag{8.1-4}$$

where $\Gamma(\ )$ is the gamma function. (If $n \leq 3$, the value of B_2 is infinite.) Though the virial coefficient is now temperature dependent, it is still only positive; it decreases in value with increasing temperature, unlike the behavior of the second virial coefficient shown in Fig. 8.1-2.

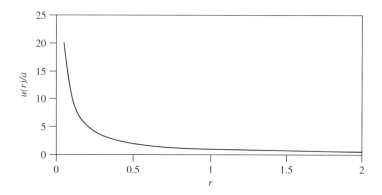

Figure 8.1-3 The Point Centers of Repulsion Potential.

128 Chapter 8: Intermolecular Potentials and the Evaluation of the Second Virial Coefficient

Coulomb Potential

Two point charges, q_1 and q_2, in a vacuum interact via the Coulomb potential

$$u(r) = \frac{q_1 q_2}{r} \tag{8.1-3b}$$

which is a special case of the Point Centers of Repulsion potential, and leads to an infinite virial coefficient (Problem 8.3). Charged particles must be treated in a different manner, and this is discussed in Chapter 15.

Potentials with Attraction

From experimental volumetric data for non-ideal gases, it is possible to obtain numerical values for the second virial coefficient. For many gases, particularly at low temperatures, the second virial coefficient is found to be negative—as has been shown earlier with the experimental data for helium. From Eq. 7.5-4 it is evident that the virial coefficient can only be negative if the intermolecular potential is negative (that is, attractive) for some intermolecular separations. The experimental observation that at higher temperatures the second virial coefficient becomes positive, as was shown earlier for helium, requires that any potential model must also have a repulsive part. Below are several simple potential models having both attractive and repulsive regions.

Square-Well Potential

The simplest analytical attractive-repulsive potential is the square-well model

$$u(r) = \begin{cases} \infty & r \leq \sigma \\ -\varepsilon & \sigma < r < R_{sw}\sigma \\ 0 & r \geq R_{sw}\sigma \end{cases} \tag{8.1-5}$$

which is shown in Fig. 8.1-4.

The second virial coefficient for the square-well potential is

$$B_2(T) = 2\pi \int_0^\sigma (1 - 0) r^2 \, dr + 2\pi \int_\sigma^{R_{sw}\sigma} (1 - e^{\varepsilon/kT}) r^2 \, dr$$

$$+ 2\pi \int_{R_{sw}\sigma}^\infty (1 - 1) r^2 \, dr$$

$$= \frac{2\pi \sigma^3}{3} + \frac{2\pi}{3}(1 - e^{\varepsilon/kT})(R_{sw}^3 - 1)\sigma^3 = \frac{2\pi \sigma^3}{3}[1 + (1 - e^{\varepsilon/kT})(R_{sw}^3 - 1)]$$

(8.1-6a)

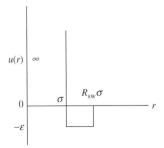

Figure 8.1-4 The Square-Well Potential.

or as a dimensionless quantity (using that B_2 and $2\pi\sigma^3/3$ both have units of volume)

$$\frac{B_2(T^*)}{\frac{2\pi\sigma^3}{3}} = B^*(T^*) = [1 + (1 - e^{1/T^*})(R_{sw}^3 - 1)] \quad (8.1\text{-}6b)$$

where $T^* = kT/\varepsilon$. The second virial coefficient computed with this expression has the property that at low temperatures, the exponential term dominates, and the second virial coefficient is negative. However, at high temperatures, the value of the second virial coefficient is positive. At very high temperatures, where the mean kinetic energy of the molecules is of much larger magnitude than the depth of the potential well ε, attractive forces are unimportant; and in this limit the virial coefficient becomes equal to that computed from the rigid-sphere model $2\pi\sigma^3/3$. This behavior of the second virial coefficient of the square-well potential for $R_{sw} = 1.5$ is shown in Fig. 8.1-5.

As the temperature increases, and B_2 goes from a negative to a positive quantity, there is a temperature at which the second virial coefficient is zero; this is known as the Boyle temperature, T_B. It is a simple exercise to show that for the square-well model:

$$T_B = \frac{\varepsilon/k}{\ln(R_{sw}^3/(R_{sw}^3 - 1))} \quad (8.1\text{-}7)$$

At temperatures below the Boyle temperature, the attractive part of the potential is clearly important, and the virial coefficient is negative. At temperatures higher than T_B, the repulsive part of the potential dominates, and the second virial coefficient is positive. Table 8.1-1 contains the square-well potential parameters for some simple fluids.

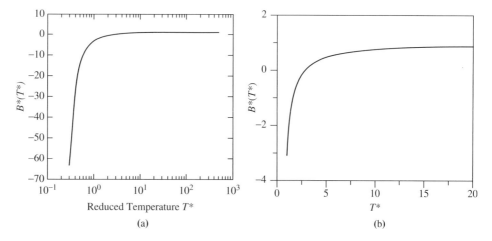

Figure 8.1-5 (a) The reduced second virial coefficient B^* for the square-well fluid with $R_{sw} = 1.5$ as a function of reduced temperature T^*; (b) high temperature range.

Table 8.1-1 Square-Well and Lennard-Jones 12-6 Potential Parameters Determined from Second Virial Coefficient Data Assuming the Molecules are Spheres[1]

Molecule	Square-Well Potential			Lennard-Jones 12-6 Potential	
	R_{sw}	σ (Å)	ε/k (K)	σ (Å)	ε/k (K)
argon	1.70	3.067	93.3	3.504	117.7
benzene	1.38	4.830	620.4	8.569	242.7
CF_4	1.48	4.103	191.1	4.744	151.5
CH_4	1.60	3.355	142.5	3.783	148.9
CO_2	1.44	3.571	283.6	4.328	198.2
krypton	1.68	3.278	136.6	3.827	164.0
n-pentane	1.36	4.668	612.3	8.497	219.5
neopentane	1.45	5.422	382.6	7.445	232.5
nitrogen	1.58	3.277	95.2	3.745	95.2
xenon	1.64	3.593	198.5	4.099	222.3

Mie and Lennard-Jones Potentials

Perhaps the most widely used potential for correlating experimental data on simple molecules is the Mie potential:

$$u(r) = \frac{d}{r^n} - \frac{c}{r^m} \tag{8.1-8}$$

This form of the potential model has the advantage of being a smooth, continuous function, and presumably more realistic than the interaction potentials considered so far. The evaluation of the virial coefficients for this potential requires numerical integration. The m parameter in the potential has been estimated from quantum-mechanical dispersion energy calculations to be equal to 6 for nonpolar molecules. It is the result of the instantaneous, coupled fluctuations of the distributions of the electrons around each atom resulting in a induced dipole-induced dipole net attraction, referred to as London dispersion forces. Largely for mathematical convenience, n is frequently chosen to be equal to 12, resulting in the commonly-used Lennard-Jones (12-6) potential for non-polar molecules shown in Fig 8.1-6.[2]

$$u(r) = 4\varepsilon\left[\left(\frac{\sigma}{r}\right)^{12} - \left(\frac{\sigma}{r}\right)^{6}\right] \tag{8.1-9}$$

It is easily shown that σ is the value of r for which $u(r)$ is equal to zero, and ε is the depth of the potential energy well which occurs at $r = 2^{1/6}\sigma$.

---

[1]From D. A. McQuarrie, *Statistical Mechanics*, HarperCollins, New York, 1976; original source is A. E. Sherwood and J. M. Prausnitz, *J. Chem. Phys.* **41**, 29 (1964).

[2]As shown in Section 8.3, this same potential can be used by rotationally averaging the permanent dipole–permanent dipole interactions among polar molecules, but this results in temperature dependent ε and σ parameters Eq. 8.3-7b.

8.1 Interaction Potentials for Spherical Molecules 131

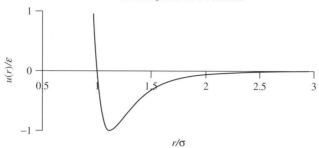

Figure 8.1-6 The Lennard-Jones 12-6 Potential.

The second virial coefficient is then computed from

$$B_2(T) = 2\pi \int_0^\infty \left(1 - \exp\left\{-\frac{4\varepsilon}{kT}\left(\left(\frac{\sigma}{r}\right)^{12} - \left(\frac{\sigma}{r}\right)^{6}\right)\right\}\right) r^2\, dr$$

$$= \frac{2\pi}{3}\sigma^3 \int_0^\infty \left(1 - \exp\left\{-\frac{4\varepsilon}{kT}\left(\frac{1}{y^4} - \frac{1}{y^2}\right)\right\}\right) dy \quad \text{(8.1-10)}$$

where $y = (r/\sigma)^3$. Defining a dimensionless reduced temperature $T^* = kT/\varepsilon$, the reduced (dimensionless) second virial coefficient can be written as

$$\frac{B_2(T^*)}{\frac{2\pi}{3}\sigma^3} = B^*(T^*) = \int_0^\infty \left(1 - \exp\left\{\frac{-4}{T^*}\left(\frac{1}{y^4} - \frac{1}{y^2}\right)\right\}\right) dy \quad \text{(8.1-11)}$$

This form of the equation indicates that at the same reduced temperature (T^*), all molecules represented by the Lennard-Jones 12-6 potential should have the same value of the reduced virial coefficient $B_2^*(T^*) = B_2/\frac{2\pi}{3}\sigma^3$. Thus we have a corresponding-states relation for the second virial coefficient of a Lennard-Jones fluid. The reducing parameters for this correlation are not the critical properties (as is commonly used in engineering), but rather the intermolecular potential parameters.

Figure 8.1-7 is a plot of $B^*(T^*) = B_2(T^*)/\frac{2\pi\sigma^3}{3}$ versus T^*. The values of $B^*(T^*)$ and its first two temperature derivatives ($B_1^* = T^*(dB^*/dT^*)$ and $B_2^* = (T^*)^2(d^2B^*/dT^{*2})$) needed to compute corrections to the ideal gas enthalpy and heat capacity, respectively, can be obtained using the MATLAB® program LJ_virial.m on the website www.wiley.com/college/sandler.[3] Table 8.1-1 contains the Lennard-Jones 12-6 potential parameters for some simple fluids. The following are estimates for the relations between the Lennard-Jones parameters and the critical constants obtained by fitting virial coefficient data:

$$\frac{kT_C}{\varepsilon} = 1.35; \quad \frac{N_{Av}\sigma^3}{V_C} = 0.35; \quad \frac{P_C\sigma^3}{\varepsilon} = 0.142 \quad \text{(8.1-12)}$$

We see from this figure that the second virial coefficient for the Lennard-Jones potential is negative at low temperatures and positive at high temperatures, as is

---

[3]This program is found in the folder MATLAB® programs in folder LJ virial. The readme.txt file provides information on its use.

Chapter 8: Intermolecular Potentials and the Evaluation of the Second Virial Coefficient

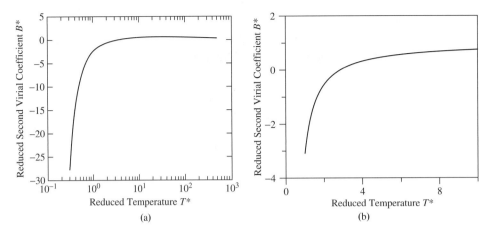

Figure 8.1-7 (a) Reduced Second Virial Coefficient of the Lennard-Jones 12-6 Potential; (b) high temperature range.

seen in experiments. An interesting feature of $B^*(T^*)$ is that—in agreement with measurements—at very high reduced temperatures, the second virial coefficient is a decreasing function of temperature, so that there is an intermediate temperature at which the second virial coefficient achieves a maximum. (This is difficult to see in Fig. 8.1-7 because of the scale.) That the second virial coefficient decreases with increasing temperature is easily explained. At high temperatures, it is only the repulsive part of the potential that is important in determining the value of the second virial coefficient. Since the repulsive part of the potential rises quite sharply with decreasing intermolecular separation distance r, this portion of the potential can almost, but not quite, be represented by a rigid-sphere potential. However, since the potential does have a finite—rather than infinite—slope, the effective hard-sphere diameter decreases as the average energy (or mean kinetic energy) of the system increases, as it does with increasing temperature. Thus, the effective hard-sphere diameter, and therefore the value of the second virial coefficient, decreases with increasing temperature at very high temperatures.

Exponential-6 (Modified Buckingham) Potential

This three parameter (ε, γ, and r_m) potential is given below.

$$u(r) = \begin{cases} \infty & r \leq r_m \\ = \dfrac{\varepsilon}{1 - 6/\gamma} \left[\dfrac{6}{\gamma} \exp\left(\gamma \left[1 - \dfrac{r}{r_m}\right]\right) - \left(\dfrac{r_m}{r}\right)^6 \right] & r > r_m \end{cases} \quad (8.1\text{-}13)$$

The advantage of this potential is that the three adjustable parameters allow greater flexibility in fitting experimental data. However, the disadvantages of this potential are that its derivatives are discontinuous at $r = r_m$ and the second virial coefficient must be evaluated numerically.

The Yukawa Potential

Another interaction potential that is frequently used, especially now for colloidal and protein solutions, is that of Yukawa:

$$u(r) = \begin{cases} \infty & 0 \leq r \leq \sigma \\ -\dfrac{\sigma \varepsilon}{r} \exp[-b(r-\sigma)] & r > \sigma \end{cases} \quad (8.1\text{-}14)$$

In this model, σ is the hard-sphere diameter, ε is the well depth, and b is a parameter that determines the range of the interaction. This potential is very similar to the shielded Coulomb potential for the interaction between charged particles that is mentioned in Section 8.4 and developed in Chapter 15, and is frequently used in place of it.

ILLUSTRATION 8.1-1

Below are experimental second virial coefficient data[4] for argon as a function of temperature.

	B(cc/mol)
T(K)	Ar
110	−156.0
125	−123.5
150	−84.7
200	−47.6
250	−28.0
300	−15.6
400	−0.9
500	+7.3
600	+12.5

Compare the experimental data with predictions using the Lennard-Jones 12-6 potential using the potential parameters for argon in Table 8.1-1.

SOLUTION

From Table 8.1-1,

$\sigma = 3.504$ Å $= 3.504 \times 10^{-8}$ cm and $\varepsilon/k = 117.7$ K. Therefore, $\dfrac{2\pi\sigma^3}{3} N_{Av} = 54.26$ cc/mol.

Now using the MATLAB® program LJ_VIRIAL to compute values of B^* we obtain

---

[4] From *The Virial Coefficients of Gases* by J. H. Dymond and E. B. Smith, Oxford University Press, 1969.

134 Chapter 8: Intermolecular Potentials and the Evaluation of the Second Virial Coefficient

$T(K)$	B cc/mol	$T^* = kT/\varepsilon$	B^*	B(calc) cc/mol
110	−156.0	0.9346	−2.855	−154.9
125	−123.5	1.0620	−2.284	−123.9
150	−84.7	1.2744	−1.644	−89.2
200	−47.6	1.6992	−0.924	−50.6
250	−28.0	2.1240	−0.533	−28.9
300	−15.6	2.5489	−0.289	−15.7
400	−0.9	3.3985	−0.0046	−0.25
500	+7.3	4.2481	0.154	+8.34
600	+12.5	5.0977	0.233	+13.7

The results are plotted below.

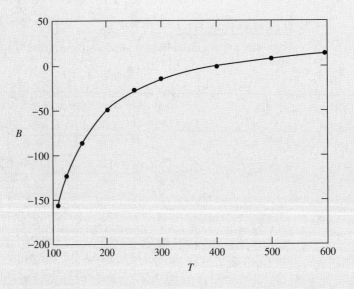

The points are the measured values and the line is the second virial coefficient calculated with the LJ 12-6 potential.

While the results are not in perfect agreement with experiment, as would be expected from an approximate potential such as the Lennard-Jones 12-6 potential, the results are very good over the temperature range considered.

ILLUSTRATION 8.1-2

Repeat Illustration 8.1-1 using the square-well potential.

SOLUTION

From Table 8.1-1,

$$\sigma = 3.067 \text{ Å} = 3.067 \times 10^{-8} \text{cm and } \varepsilon/k = 93.5 \text{ and } R_{SW} = 1.70. \text{ Therefore, } \frac{2\pi\sigma^3}{3} N_{Av}$$

$$= 36.39 \text{ cc/mol.}$$

Now using Eq. 8.1-6b to compute values of B^*, we obtain

$T(K)$	B cc/mol	$T^* = kT/\varepsilon$	B^*	B(calc) cc/mol
110	−156.0	1.179	−4.225	−153.7
125	−123.5	1.340	−3.341	−121.6
150	−84.7	1.608	−2.376	−86.4
200	−47.6	2.144	−1.326	−48.2
250	−28.0	2.680	−0.770	−28.0
300	−15.6	3.215	−0.427	−15.6
400	−0.9	4.287	−0.028	−1.02
500	+7.3	5.359	0.197	+7.18
600	+12.5	6.431	0.342	+12.4

The results of the calculation are plotted below.

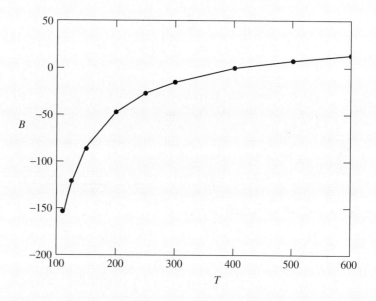

The points are the measured values and the line is the second virial coefficient calculated with the square-well potential.

What we see from the plots and tables of these two illustrations is that the results for the second virial coefficient of using the Lennard-Jones 12-6 and square-well potentials with appropriately fitted parameters are comparable. (The same will not be true when we consider high-density fluids in Chapters 11 and 12). A general implication of this is that second virial coefficient data generally cannot be used to "work backward" and uniquely determine the interaction potential between atoms, since different potentials with fitted parameters may give comparable results.

8.2 THE SECOND VIRIAL COEFFICIENT IN A MIXTURE: INTERACTION POTENTIALS BETWEEN UNLIKE ATOMS

In Problem 7.1 it was shown that

$$B_{2,\text{mix}}(\underline{x}, T) = \sum_{i=1}^{C} \sum_{j=1}^{C} x_i x_j B_{2,ij}(T) \qquad (8.2\text{-}1)$$

where

$$B_{2,ij}(T) = 2\pi \int_0^\infty (1 - e^{-u_{ij}(r)/kT}) r^2 \, dr \qquad (8.2\text{-}2)$$

Consequently, to use the virial equation in a mixture, we need to compute values of several virial coefficients at the same temperature. For example, in a binary mixture

$$B_{2,\text{mix}}(\underline{x}, T) = x_1^2 B_{2,11}(T) + 2x_1 x_2 B_{2,12}(T) + x_2^2 B_{2,22}(T) \qquad (8.2\text{-}3)$$

since $B_{2,12} = B_{2,21}$ as a result of $u_{12}(r) = u_{21}(r)$. To proceed further, we have to know not only the interaction potentials between like molecules (that is, $u_{11}(r)$ and $u_{22}(r)$), but also the cross or mixed interaction $u_{12}(r)$. A reasonable assumption is that if the $u_{11}(r)$ and $u_{22}(r)$ interaction potentials are of the same form—for example, both Lennard-Jones 12-6 potentials though with different parameters—then the same potential should also be used for the $u_{12}(r)$. That still leaves unresolved the choice of the appropriate potential parameters.

One way that potential parameter values are determined is by fitting experimental second virial coefficient data. As there are considerable experimental data for pure fluids, the pure-species potential parameters can be determined in this way. However, evaluating the potential parameters in the cross interaction $u_{12}(r)$ is more problematic for two reasons. First, there are only limited experimental data on mixture second virial coefficients, $B_{2,\text{mix}}$. Second, even when such data are available, since $u_{12}(r)$ must be computed after subtracting the contributions of $u_{11}(r)$ and $u_{22}(r)$, there is likely to be significant error in $u_{12}(r)$. Consequently, the usual procedure is to use a set of combining rules that relate the potential parameters of the 11 and 22 interaction potentials to those in the 12 potential. The most common combining rules are the following, the so-called Lorentz-Berthelot combining rules. For the distance parameter or core diameter (i.e., σ), the usual combining rule is

$$\sigma_{12} = \frac{1}{2}(\sigma_1 + \sigma_2) \qquad (8.2\text{-}4)$$

This equation is exact for the hard-sphere potential, and approximate for all other potential functions. The following combining rule is commonly used for the unlike energy parameter in a two-parameter potential (such as the Lennard-Jones potential):

$$\varepsilon_{12} = \sqrt{\varepsilon_1 \varepsilon_2}(1 - k_{12}) \qquad (8.2\text{-}5)$$

The use of the geometric average for energy parameters has an approximate basis from quantum chemistry calculations, but is not exact. The binary interaction parameter k_{12}, which is usually adjusted to fit mixture second virial coefficient or other experimental data, is meant to compensate for the approximate nature of the energy parameter combining rule. (Note that these combining rules are very much like the ones used for the parameters in cubic equations of state, as will be discussed in Section 8.4.) Additional combining rules are needed for interaction potentials with more than two parameters (for example, the modified Buckingham potential).

In closing this section, it is important to note that from Eq. 8.2-3, regardless of the interaction potentials used and their simplicity or complexity, the mixture second virial coefficient depends quadratically on composition. That is, the form of the interaction potentials used and the values of their parameters determine the numerical values of the like and unlike second virial coefficients; but theory leads to the exact result of a quadratic composition dependence.

8.3 INTERACTION POTENTIALS FOR MULTIATOM, NONSPHERICAL MOLECULES, PROTEINS, AND COLLOIDS

The intermolecular potential functions used for nonspherical molecules can be quite complicated because of the number of interaction sites present. One form of intermolecular potential is the complete atomistic approach in which the known geometry of the molecule is used; and each atom on one molecule is assumed to interact with each atom on other molecules using an atom-atom potential, such as the potentials discussed above, with parameters specific to each type of atom-atom interaction. In this model, the interaction between two molecules is the sum of all atom-atom interactions. For example, for a diatomic molecule (such as hydrogen chloride) shown in Fig. 8.3-1, for each center-of-mass separation and relative orientation of the two molecules, we would need the four interatomic distances shown to compute the total interaction energy for that configuration, and this calculation would have to be repeated for each configuration. As the number of atoms in a molecule increases, such calculations become increasingly difficult and further complicated by the fact that the configuration of larger molecules can change due to internal rotations around bonds and bending.

A slight generalization of the atomistic approach is the site-site model. In this model, each molecule is considered to have two or more interaction sites in which potential models such as those discussed in Section 8.2 are used represent each interaction; but each site does not necessarily have to be located at the center of each atom. Also, in some cases, additional sites are added that are not associated with any atom—as, for example, to represent point charges.

A simplification of these models is the united atom approximation in which a group of atoms are taken together as a single interaction site. For example, a $-CH_3$ group may be taken as a single interaction site. In such united atom models, hydrogen atoms are frequently lumped together with a larger atom (carbon in the example here) to make a single interaction site.

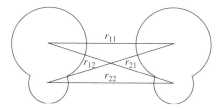

Figure 8.3-1 Interatomic Distances Needed to Calculate the Interaction Energy for One Configuration of a Pair of Diatomic Molecules.

Chapter 8: Intermolecular Potentials and the Evaluation of the Second Virial Coefficient

One problem that arises in the use of any of these models is that, generally, nonidentical site-site interactions must be considered. For example, if in the united atom approximation, we considered methanol to consist of a CH_3 group and an OH group, not only will we need the $CH_3 + CH_3$ and OH + OH interaction potential parameters, but also the parameters for the unlike CH_3+ OH interaction. This is the same problem of unlike site-site interaction arises that we considered in Section 8.2 when dealing with the monatomic interaction potentials for mixtures. For the same reason as discussed in Section 8.2, it is common to use the combining rules of Eqs. 8.2-4 and 8.2-5 to obtain the unlike site-site interaction parameters.

An extreme example of the site-site model is to treat a whole multiatom, nonspherical molecule as a single site. While not a realistic picture of a molecule, with a suitable adjustment of the potential parameters, this model can provide a reasonable description of the second virial coefficient. It is for this reason that square-well and Lennard-Jones 12-6 parameters were given in Table 8.1-1 for molecules such as CF_4, CH_4, nitrogen, carbon dioxide, n-pentane, and neopentane (2,2-dimethyl propane). However, there are several caveats in using these parameters. First, as expected, the use of a single-site model is more reasonable the closer the molecule is to being spherical. Thus, one expects the representation to be better for the almost spherical neopentane than for the more linear n-pentane. Second, potential parameter sets can also be regressed from other data—for example, from viscosity or thermal conductivity using relations obtained from the kinetic theory of gases. Both because the molecules discussed here consist of more than a single atom, and because the square-well and Lennard-Jones 12-6 potentials are just simple models of the interactions between molecules, different potential sets are obtained depending on the data used.

Another class of interaction potential models used for multiatom molecules is specific geometric shapes such as cylinders, spherocylinders (cylinders with hemispherical caps), ovalate spheroids, etc. We will not consider such models here. Still another class of potentials used for nonspherical molecules is the sum of a spherical potential with a nonspherical part, usually representing permanent dipoles (or multipoles) in the molecules. As an example, we will represent the relative orientation of two molecules as in Fig. 8.3-2 and use the following notation:

$$g(\theta_1, \theta_2, \phi_2 - \phi_1) = \sin\theta_1 \sin\theta_2 \cos\phi_{12} - 2\cos\phi_1 \cos\phi_2 \qquad (8.3\text{-}3)$$

where μ here is the permanent dipole moment of the molecule. With this notation, the following are some of the simplified interaction potentials used for multiatomic molecules.

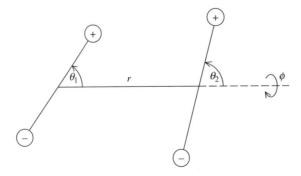

Figure 8.3-2 Angles Describing the Relative Orientation of Linear Molecules and/or Two Dipoles.

8.3 Interaction Potentials for Multiatom, Nonspherical Molecules, Proteins, and Colloids

Rigid Sphere Containing a Dipole

$$u(r, \theta_1, \theta_2, \phi_2 - \phi_1) = \begin{cases} \infty & r \leq \sigma \\ -\dfrac{\mu^2}{r^3} g(\theta_1, \theta_2, \phi_2 - \phi_1) & r > \sigma \end{cases} \quad (8.3\text{-}4)$$

Stockmayer Potential (Lennard-Jones 12-6 potential + dipole)

$$u(r, \theta_1, \theta_2, \phi_2 - \phi_1) = 4\varepsilon_o \left[\left(\frac{\sigma_o}{r}\right)^{12} - \left(\frac{\sigma_o}{r}\right)^6 \right] - \frac{\mu^2}{r^3} g(\theta_1, \theta_2, \phi_2 - \phi_1) \quad (8.3\text{-}5)$$

It has been shown[5] that by a Boltzmann-factor orientation averaging of this potential, one obtains

$$\langle u(r) \rangle = \frac{\int \int u(r, \theta_1, \theta_2, \phi_2 - \phi_1) e^{-u(r,\theta_1,\theta_2,\phi_2-\phi_1)/kT} \sin\theta_1 \sin\theta_2 \, d\theta_1 \, d\theta_2 \, d\phi_1 \, d\phi_2}{\int \int e^{-u(r,\theta_1,\theta_2,\phi_2-\phi_1)/kT} \sin\theta_1 \sin\theta_2 \, d\theta_1 \, d\theta_2 \, d\phi_1 \, d\phi_2}$$

$$= \varepsilon_o \left[\left(\frac{\sigma_o}{r}\right)^{12} - 2\left(\frac{\sigma_o}{r}\right)^6 \right] - \frac{2\mu^4}{3kTr^6} \quad (8.3\text{-}6)$$

which can be rewritten as

$$\langle u(r) \rangle = \varepsilon(T) \left[\left(\frac{\sigma(T)}{r}\right)^{12} - \left(\frac{\sigma(T)}{r}\right)^6 \right] \quad (8.3\text{-}7a)$$

with

$$\varepsilon(T) = \varepsilon_o \left(1 + \frac{\mu^4}{12kT\varepsilon_o\sigma_o^6} \right)^2 \quad \text{and} \quad \sigma(T) = \sigma_o \left[\frac{1}{1 + \frac{\mu^4}{12kT\varepsilon_o\sigma_o^6}} \right]^{\frac{1}{6}} \quad (8.3\text{-}7b)$$

That is, with this approximation, to compute the value of the second virial coefficient for the Stockmayer potential, the values of the second virial coefficient for the Lennard-Jones 12-6 fluid can be used, though the the parameters to be used—$\varepsilon(T)$ and $\sigma(T)$—are now functions of temperature.

The choice of intermolecular potentials that can be used for very large macromolecules, such as proteins and colloidal particles, is more problematic. While, in principle, atom-atom or site-site interactions can be used, there are so many atoms involved that summing over these in a large molecule is very difficult. In some cases, especially colloids or other macromolecules with no net charge, a simple hard-sphere model may be sufficient—but with a diameter characteristic of the macromolecule.[6] Another possibility is to use the square-well potential for neutral

---

[5] J. H. Bae and T. M. Reed III, *Ind. Eng. Chem.* **6**, 67 (1967).
[6] P. N. Pusey and W. van Megen, *Nature* **320**, 340 (1986).

140 Chapter 8: Intermolecular Potentials and the Evaluation of the Second Virial Coefficient

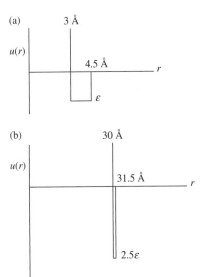

Figure 8.3-3 Typical Square-Well Potential for (a) an Atom and for (b) a Macromolecule Such As a Protein or Colloid.

macromolecules, though because of the size of the macromolecule and because so many site-site interactions are involved, the potential will look very different than for an atom-atom interaction. To be specific, for a 3 Å atom with the square-well parameter $R_{sw} = 1.5$, the range of the square well is 1.5 Å from (3 Å to 4.5 Å); for a 30 Å, because of the number of atom-atom interactions involved and the different separation distances, the well will be deeper and the center-to-center separation distance between the macromolecules greater; but the range of the square well will still be only about 1.5 Å. This is illustrated in Fig. 8.3-3 in which the separation distance is in Å. In Fig. 8.3-3b the macromolecule diameter is 30 Å; the range of the well remains that for atom-atom interactions of 1.5 Å; and, as an example, the well depth is 2.5 times the atom well depth ε. Thus, the interaction potential looks more like a sticky hard-sphere than the typical square-well potential.

There are many other potentials that have been also proposed for both spherical and nonspherical molecules; however, we will not consider them here.

8.4 ENGINEERING APPLICATIONS AND IMPLICATIONS OF THE VIRIAL EQUATION OF STATE

There are a number of applications and implications that arise from the virial equation of state. Some of these will be considered here.

Use of the Virial Equation of State as an Engineering Tool

In Chapter 7 we derived the first correction from ideal gas behavior and showed that the result is

$$\frac{P}{\rho kT} = 1 + B_2(T)\rho \tag{8.4-1}$$

In applications, this volumetric equation of state is only useful for a relatively dilute gas or vapor phase. In particular, it should not used for high pressure systems (see Table 7.9-1) or at temperatures and pressures close to where the fluid would condense to a liquid.

8.4 Engineering Applications and Implications of the Virial Equation of State

Including higher-order terms in the expansion can extend the range of applicability of the virial equation. For example, considering also all closed-cycle interactions between three molecules leads to

$$\frac{P}{\rho k T} = 1 + B_2(T)\rho + B_3(T)\rho^2 \qquad (8.4\text{-}2)$$

This will extend the range of the virial equation to somewhat higher densities. However, the third virial coefficient $B_3(T)$ is very difficult to evaluate for commonly used intermolecular interaction potentials. What is commonly done in application is to fit the second and third virial coefficients to experimental data.

Of course, if the virial coefficients are to be treated as fitted parameters, one need not stop at the third virial coefficient; in fact, one can include as many coefficients as can be justified by the quality of the experimental data. That is, one can use

$$\frac{P}{\rho k T} = 1 + B_2(T)\rho + B_3(T)\rho^2 + B_4(T)\rho^3 + B_5(T)\rho^4 + B_6(T)\rho^5$$
$$+ B_7(T)\rho^6 + \cdots \qquad (8.4\text{-}3)$$

At some point, the series in density is truncated, and the higher-order terms are neglected. There is a way to compensate for the terms that have been neglected. It is based on the idea that the exponential function has a series expansion with an infinite number of terms, that is

$$e^{\lambda \rho} = 1 + \lambda \rho + \frac{1}{2!}(\lambda \rho)^2 + \frac{1}{3!}(\lambda \rho)^3 + \frac{1}{4!}(\lambda \rho)^4 + \cdots \qquad (8.4\text{-}4)$$

Thus, for example, although the equation

$$\frac{P}{\rho k T} = 1 + B_2(T)\rho + B_3(T)\rho^2 + B_4(T)\rho^3 + B_5(T)\rho^4 e^{\lambda \rho} \qquad (8.4\text{-}5)$$

may not be very accurate, nonetheless it can be considered to have an infinite number of terms. This equation and the ones that immediately follow are referred to as *extended virial equations*, with the exponential term accounting for the neglected terms in the series.

While this last equation is not very useful, there are other volumetric equations of state of the extended virial form that have been used in engineering. These include, among many others, the equations of Benedict, Webb, and Rubin[7]

$$\frac{P}{\rho RT} = 1 + \left(B - \frac{A}{RT} - \frac{C}{RT^3}\right)\rho + \left(b - \frac{a}{RT}\right)\rho^2 + \frac{a\alpha}{RT}\rho^5$$
$$+ \frac{\beta \rho}{RT^3}(1 + \gamma \rho^2)\exp(-\gamma \rho^2) \qquad (8.4\text{-}6)$$

where here ρ is the molar density and R is the gas constant, and of Bender[8]

$$P = \rho T[R + B\rho + C\rho^2 + D\rho^3 + E\rho^4 + F\rho^5 + (G + H\rho^2)\exp(-a_{20}\rho^2)] \qquad (8.4\text{-}7)$$

---

[7]M. Benedict, G. B. Webb, and L. C. Rubin, *J. Chem. Phys.* **8**, 334 (1940) and later papers by the same authors.
[8]E. Bender, 5th *Symposium of Thermophysical Properties*, ASME, New York (1970), p. 227.

142 Chapter 8: Intermolecular Potentials and the Evaluation of the Second Virial Coefficient

with

$$B = a_1 - \frac{a_2}{T} - \frac{a_3}{T^2} - \frac{a_4}{T^3} - \frac{a_5}{T^4}; \quad C = a_6 + \frac{a_7}{T} + \frac{a_8}{T^2}; \quad D = a_9 + \frac{a_{10}}{T};$$

$$E = a_{11} + \frac{a_{12}}{T}; \quad F = \frac{a_{13}}{T} \quad G = \frac{a_{14}}{T^3} + \frac{a_{15}}{T^4} + \frac{a_{16}}{T^5}; \quad \text{and} \quad H = \frac{a_{17}}{T^3} + \frac{a_{18}}{T^4} + \frac{a_{19}}{T^5}$$

(8.4-8)

The virial equation origin of each of these equations of state is evident. Also, in these equations, the temperature dependent terms are meant to account for the temperature dependence of each of the virial coefficients.

For applications in which high accuracy is needed (for example, for the custody transfer of natural gas—essentially methane—or for steam in order to determine the efficiency of steam turbines), equations with many more adjustable parameters are used, and in some cases even different sets of parameters for different temperature-pressure ranges.

Mixing Rules for Simple Equations of State

In Problem 7.1, the exact composition dependence of the second virial coefficient for a mixture was found to be

$$B_{2,\text{mix}}(\underline{x}, T) = \sum_{i=1}^{C} \sum_{j=1}^{C} x_i x_j B_{2,ij}(T) \tag{8.4-9}$$

where $B_{2,ij}(T)$ is only a function of temperature, and not composition. Consequently, the second virial coefficient (and therefore the virial equation of state truncated at the second virial coefficient) is quadratic in composition.

In addition to the virial equation, other equations of state have been proposed and are used. For example, the van der Waals

$$P = \frac{RT}{\underline{V} - b} - \frac{a}{\underline{V}^2} \tag{8.4-10}$$

where $\underline{V}$ is the molar volume; other members of this class include the Soave[9] version of the Redlich-Kwong equation[10]

$$P = \frac{RT}{\underline{V} - b} - \frac{a(T)}{\underline{V}(\underline{V} + b)} \tag{8.4-11}$$

and the Peng-Robinson equation[11]

$$P = \frac{RT}{\underline{V} - b} - \frac{a(T)}{\underline{V}(\underline{V} + b) + b(\underline{V} - b)} \tag{8.4-12}$$

These equations have been developed by fitting experimental data and are not the results of exact theory. (In later chapters we will consider approximations that

---

[9]G. Soave, *Chem. Eng. Sci.* **27**, 1197 (1972).
[10]O. Redlich and J. N. S. Kwong, *Chem. Rev.* **44**, 233 (1949).
[11]D.-Y. Peng and D. B. Robinson, *IEC Fundam.* **15**, 59 (1976).

8.4 Engineering Applications and Implications of the Virial Equation of State

can be used to derive some of these equations.) Therefore, unlike the second virial coefficient, the composition dependence of the a and b parameters are not specified. However, the three equations above (and others of the so-called cubic form, since they can be written as a cubic equation in volume) can be expanded in an infinite series in virial form

$$\frac{P\underline{V}}{RT} = 1 + \left(b - \frac{a(T)}{RT}\right)\frac{1}{\underline{V}} + \cdots \qquad (8.4\text{-}13)$$

where the first two terms on the right-hand side of the equation are common to the family of cubic equations, and the further terms in the series are specific to each equation. Now, making the following identification with the virial equation of state

$$B_{2,\text{mix}}(\underline{x}, T) = \sum_{i=1}^{C}\sum_{j=1}^{C} x_i x_j B_{2,ij}(T) = b - \frac{a(T)}{RT} \qquad (8.4\text{-}14)$$

This identification suggests that the term on the right should also be quadratic in composition. One way to insure this is with the following mixing rules:

$$a(T) = \sum_{i=1}^{C}\sum_{j=1}^{C} x_i x_j a_{ij}(T) \quad \text{and} \quad b = \sum_{i=1}^{C}\sum_{j=1}^{C} x_i x_j b_{ij} \qquad (8.4\text{-}15)$$

This set of equations, originally used by van der Waals, is referred to as the *van der Waals one-fluid mixing rules*, since the mixture is described by the same equation as the pure fluids—though with composition-averaged parameters. However, these equations are incomplete in that, while the pure component parameters (i.e., a_{ii} and a_{jj}) can be determined from pure fluid properties, the cross terms (i.e., a_{ij}) are obtained from additional equations referred to as *combining rules*. (Compare these with the molecular level combining rules of Eqs. 8.2-4 and 8.2-5.)

The mixing rules of Eq. 8.4-15 are not exact. For example, the single relation of Eq. 8.4-14 has been used to constrain two functions, a and b. In fact, there are an infinite number of other mixing rules that could be developed that also satisfy Eq. 8.4-14—for example, one could add any composition dependent function $f(x)$ to a and then add $f(x)/RT$ to b, and satisfy Eq. 8.4-14.[12] Also, the composition of the third virial coefficient can be shown to be

$$B_{3,\text{mix}}(\underline{x}, T) = \sum_{i=1}^{C}\sum_{j=1}^{C}\sum_{k=1}^{C} x_i x_j x_k B_{3,ijk}(T) \qquad (8.4\text{-}16)$$

which is cubic in composition. However, from the van der Waals equation

$$B_{3,\text{mix}}(\underline{x}, T) = \frac{b^2}{2} = \frac{1}{2}\left[\sum_{i=1}^{C}\sum_{j=1}^{C} x_i x_j b_{ij}\right]^2 \qquad (8.4\text{-}17)$$

Thus, while the mixing rules of Eq. 8.4-14 give the correct composition dependence of the second virial coefficient, they are incorrect for any higher order virial coefficient.

---

[12] D. S. H. Wong and S. I. Sandler, *AIChE J.* **38**, 671 (1992).

144 Chapter 8: Intermolecular Potentials and the Evaluation of the Second Virial Coefficient

Nonetheless, the van der Waals one-fluid mixing rules have been in use for more than a century, and are satisfactory when dealing with species of similar chemical functionality, such as the family of hydrocarbons. There are more complicated and useful mixing rules that are used in place of the van der Waals one-fluid mixing rules for mixtures of dissimilar species.[13]

CHAPTER 8 PROBLEMS

8.1 The following second virial coefficient data are available for methane and CF_4

T (K)	B_{CH_4} (cm^3/mol)	T (K)	B_{CF_4} (cm^3/mol)
110	-344 ± 10	273.16	-111
200	-107 ± 2	373.16	-43.1
400	-15.5 ± 1	523.16	$+1.25$
500	-0.5 ± 1	673.16	$+23.6$
600	$+8.5 \pm 1$		

Use these data to estimate the Lennard-Jones parameters for CH_4 and CF_4.

8.2 Calculate the configurational contribution to the internal energy and heat capacity and the compressibility factor of argon vapor at its normal boiling point (87.2 K) using the Lennard-Jones 12-6 potential ($\sigma = 3.499$ Å and $\varepsilon/k = 118.13$ K).

8.3 Show that the second virial coefficient for the simple Coulomb potential of Eq. 8.4–22 diverges.

8.4 A gas chromatograph is to be used to analyze CO_2-CH_4 mixtures. To calibrate the response of the GC, a carefully prepared mixture of known composition is used. This mixture is prepared by starting with an evacuated steel cylinder and adding CO_2 until the pressure is exactly 2.500 atm 25°C. Then CH_4 is added until the pressure reaches exactly 5.000 atm at 25°C. Assuming that both CO_2 and CH_4 are represented by the L-J 12-6 potential with the parameters given below:

	σ(Å)	ε/k (K)
CH_4	4.010	142.87
CO_2	4.416	192.25

and that the following combining rule applies

$$u_{ij}(r) = 4\varepsilon_{ij}\left[\left(\frac{\sigma_{ij}}{r}\right)^{12} - \left(\frac{\sigma_{ij}}{r}\right)^{6}\right]$$

with $\sigma_{ij} = \frac{1}{2}[\sigma_{ii} + \sigma_{jj}]$ and $\varepsilon_{ij} = \sqrt{\varepsilon_{ii}\varepsilon_{jj}}$.
What is the exact composition of the gas in the cylinder?

8.5 For calibration of a gas chromatograph we need to prepare a gas mixture containing exactly 0.7 mole fraction methane and 0.3 mole fraction tetrafluoromethane at 300 K and 25 bar in a steel cylinder that is initially completely evacuated. Assume this mixture can be described by the virial equation using only the second virial coefficient, and the molecular interactions are described by the square-well potential and combining rules

$$u(r_{ij}) = \begin{cases} \infty & r \leq \sigma_{ij} \\ -\varepsilon_{ij} & \sigma_{ij} < r < R_{sw,ij}\sigma_{ij} \\ 0 & r \geq R_{sw,ij}\sigma_{ij} \end{cases}$$

with $\sigma_{ij} = \frac{1}{2}(\sigma_{ii} + \sigma_{jj})$, $\varepsilon_{ij} = \sqrt{\varepsilon_{ii}\varepsilon_{jj}}$ and $R_{sw,ij} = \frac{1}{2}(R_{sw,ii} + R_{sw,jj})$.

The potential parameters for CH_4 and CF_4 are as follows:

	σ(Å)	ε/k (K)	R_{sw}
CH_4	3.400	88.8	1.85
CF_4	4.103	191.1	1.48

The following two procedures will be considered for making the mixture of the desired composition at the specified conditions.

a. Procedure 1. CH_4 will be added isothermally to the initially evacuated cylinder until a pressure P_1 is obtained. Then CF_4 will be added isothermally until the pressure of 25 bar is obtained at 300 K. What should the pressure P_1 be to obtain exactly the desired composition?

b. Procedure 2. CF_4 will be added isothermally to the initially evacuated cylinder until a pressure P_2 is obtained. Then CH_4 will be added isothermally until the pressure of 25 bar is obtained at 300 K. What should the pressure P_2 be to obtain exactly the desired composition?

---

[13] See, for example, D. S. H. Wong and S. I. Sandler, *AIChE Journal* **38**, 671 (1992).

8.6 Compute the heat capacity of nitrogen at 150 K and 150 bar. At these conditions, nitrogen is described by the virial equation of state truncated at the second virial coefficient. Assume that the interaction potential between nitrogen molecules can be described by the spherically symmetric square-well potential. The following data are available:

$\varepsilon/k = 95.2 K; \quad \sigma = 3.277 \text{Å}; \quad R_{sw} = 1.58$

8.7 Repeat the calculation of Problem 8.6 but using an alternate set of square-well potential parameters that has been proposed for nitrogen:

$\varepsilon/k = 53.7 K; \quad \sigma = 3.299 \text{Å}; \quad R_{sw} = 1.87$

What can you say about the sensitivity of the predictions on the potential parameters based on the results of Problems 8.6 and 8.7?

8.8 Determine the importance of the attractive and repulsive parts of the intermolecular potential to the second virial coefficient by comparing values of the second virial coefficients for the Lennard-Jones 12-6 potential with those for an inverse 12th-power potential

$u(r) = 4\varepsilon(\sigma/r)^{12}$

having the same values of ε and σ over the temperature range of $T^* = kT/\varepsilon$ of 1 to 100.

8.9 The best estimates for the relations between the Lennard-Jones parameters and the critical constants were given in Eq. 8.1-12. Use these expressions to obtain the Lennard-Jones parameters for CH_4, CF_4, Ar, and CO_2, and then compute the second virial coefficients for these gases over the temperature range of 200 to 800 K.

8.10 Below are virial coefficient data[14] for Kr and N_2:

T(K)	B(cc/mol) Kr	B(cc/mol) N_2
110	−375	−132
125		−104
150	−200.7	−71.5
200	−116.9	−35.2
250	−75.7	−16.2
300	−50.5	−4.2
400	−22.0	+8.0
500	−8.1	+16.9
600	+2.0	+21.3
700		+24.0

The following additional data are available:

	Kr	N_2
T_C, K	209.41	126.2
P_C, MPa	5.50	3.394

a. Obtain the Lennard-Jones 12-6 parameters for these substances first by using the critical properties data, and second from the more difficult task of fitting the measured virial coefficient data given above.
b. Comment on the degree to which these substances satisfy the corresponding states principle for the second virial coefficient with the Lennard-Jones 12-6 potential based on the data above.

8.11 The partition function for a moderately dilute gas is

$$Q = \frac{(q'_{trans} q_{rot} q_{vib} q_{elect} q_{nuc})^N}{N!} V^N \exp\left(\frac{N^2 \beta}{2V}\right)$$

where $q'_{trans} = q_{trans}/V$ and $\beta = 4\pi \int f(r) r^2 dr$ with $f(r) = e^{-u(r)/kT} - 1$

a. Derive the equation of state for this gas.
b. For this gas develop expressions for the following in terms of β
   (i) departure of the internal energy from ideal gas behavior $\underline{U}(N,V,T) - \underline{U}^{IG}(N,V,T)$
   (ii) departure of the constant volume heat capacity from ideal gas behavior

   $C_V(N,V,T) - C_V^{IG}(N,V,T)$

   (iii) departure of the enthalpy from ideal gas behavior $\underline{H}(N,V,T) - \underline{H}^{IG}(N,V,T)$

c. Interactions in argon can be described reasonably well by the Lennard-Jones 12-6 potential with the following parameters $\varepsilon/k = 119.8$ K and $\sigma = 3.405$ Å. Using this information, at 150 K and 8 bar, compute the
   (i) volume of argon
   (ii) the internal energy, enthalpy and constant volume heat capacity deviations from ideal gas behavior

8.12 The series of books "Chemical Process Principles" by O. A. Hougen, K. M. Watson, and R. A. Ragatz (John Wiley & Sons, New York) contain many corresponding states charts that were of interest to engineers before computers were available. In particular, there

---

[14]From *The Virial Coefficients of Gases* by J. H. Dymond and E. B. Smith, Oxford University Press, 1969.

are charts of the fugacity for gases and the departures of enthalpy, entropy, and internal energy from ideal gas behavior. Here, such information will be computed for small segments of these charts using the virial equation of state.

Compute the fugacity and enthalpy, entropy and internal energy departures from ideal gas behavior as a function of $T_r = T/T_C$ and $P_r = P/P_C$ at $T_r = 1.00$, 2.00, and 3.00 using the virial equation of state. The results should be presented in graphical form. For these calculations, use the relationships between the critical properties and the Lennard-Jones 12-6 parameters given in Eqs. 8.1-12.

8.13 The triangular-well potential is

$$u(r) = \begin{cases} \infty & r \leq \sigma \\ -\varepsilon \dfrac{R_{tw}\sigma - r}{R_{tw}\sigma - \sigma} & \sigma < r < R_{tw}\sigma \\ 0 & r \geq R_{tw}\sigma \end{cases}$$

a. Obtain an expression for the second virial coefficient for this potential.
b. Does the second virial coefficient for the triangular-well potential have a maximum as a function of temperature?

8.14 The Sutherland potential is

$$u(r) = \begin{cases} \infty & r \leq \sigma \\ -\varepsilon \left(\dfrac{\sigma}{r}\right)^m & \sigma < r \end{cases}$$

a. Obtain an expression for the second virial coefficient for this potential. (Note that to obtain numerical values for the second virial coefficient for this potential one must either expand the exponential and integrate term-by-term, or evaluate the integral numerically.)
b. Sutherland suggested a value of $m = 3$. Show that the second virial coefficient does not converge for this value of m. (A better value from quantum mechanics is $m = 6$.)
c. Does the second virial coefficient for the Sutherland potential with $m = 6$ have a maximum as a function of temperature?

8.15 Assuming benzene ($T_C = 562.1$ K), ethane ($T_C = 305.4$ K), and the refrigerant R12 ($T_C = 385.0$ K) can be described by the Lennard-Jones 12-6 potential, what are the Boyle temperatures of these fluids?

8.16 Compare the values of the second virial coefficient as a function of reduced temperature for the two square-well potentials of Fig. 8.1-3.

8.17 Derive Eqs. 8.3-7.

Chapter 9

Monatomic Crystals

In this chapter we consider the statistical mechanics of monatomic crystals as the first example of a dense system of interacting molecules. A crystalline solid might seem like a difficult system to consider because of the small separations between atoms and their strong interactions. However, there is an important simplification, since the atoms are in a well-defined periodic structure or lattice and their locations are known. The only motions of the atoms are small vibrations around their equilibrium positions. Thus, the problems of the locations of the atoms and the thermodynamic properties of the crystal are separated. This is simpler than, for example, the statistical mechanics of liquids in which the average spatial arrangement of the molecules is unknown and must be found as part of the thermodynamic properties calculation.

Traditionally, since metals form crystalline structures, the thermodynamics of crystals was a field for metallurgists. The increased interest in crystalline materials among chemists and chemical engineers in recent years is a result of the importance of crystallization of electronic materials, proteins, colloids, biomolecules, polymers, and pharmaceuticals (the latter for both purification and drug delivery). In this chapter we only consider the simplest models of crystals, and only atomic crystals.

Each atom in a crystal is in close proximity to, and interacting with, other molecules. In particular, a molecule interacts strongly with its nearest neighbors and, to a lesser extent, with molecules that are increasingly further away. The sum of all these interactions results in a three-dimensional energy landscape in which the lowest energy state of each atom is at its equilibrium lattice site, and each atom vibrates about this lattice point in the three coordinate directions.

INSTRUCTIONAL OBJECTIVES FOR CHAPTER 9

The goals for this chapter are for the student to:

- Understand the Einstein model of a crystal
- Understand the Debye model of a crystal
- Understand the relation of these models to the third law of thermodynamics
- Understand the limitation of both these models by comparing the predicted heat capacities with experimental data

9.1 THE EINSTEIN MODEL OF A CRYSTAL

The simple Einstein model of a crystal results from considering one atom in the crystal, and assuming all the other atoms are fixed at their equilibrium lattice sites.

148 Chapter 9: Monatomic Crystals

Also, we will assume that the interaction energy landscape for this molecule is spherically symmetric, that is, the same in any direction away from its equilibrium lattice site. This would be rigorously correct if the all the other atoms in the crystal were smeared out on the surface of a sphere surrounding the lattice site of the molecule of interest. However, the other atoms are located at specific lattice sites, so the energy landscape is not symmetric.

In this model energy landscape, the interaction energy is very large if an atom gets very close to another atom in the lattice. Therefore, the motion of an atom is not the free translation as in an ideal gas, but rather a small wavelength vibration in a crystal. Consequently, the three translational motions of an atom in a gas are instead three vibrational motions in a crystal. As in a diatomic molecule, these three vibrational motions will be modeled as harmonic oscillators, and as a result of the spherically symmetric energy landscape assumption, these three vibrational motions are identical. In the three-dimensional Einstein model, all vibrations are at a single frequency denoted by ν_E, and each has the following energy levels:

$$\varepsilon_n = \left(n + \frac{1}{2}\right) h \nu_E \quad \text{with} \quad n = 0, 1, 2, \ldots \quad (9.1\text{-}1)$$

where ν_E is the classical vibrational frequency of the Einstein model in principle given by

$$\nu_E = \frac{1}{2\pi} \sqrt{\frac{f}{m}} \quad \text{where } f = \text{force constant} = \frac{d^2 u(r)}{dr^2}$$

m is the mass of an atom and $u(r)$ is the interaction energy landscape for an atom in the lattice.

The partition function of a crystal containing of N atoms in this model is

$$Q(N, V, T) = \sum_{\text{states } i} e^{-E_i(N,V)/kT} \quad (9.1\text{-}2)$$

where the energy of any state consists of two contributions. The first is the sum of the interaction energies of each atom with all others, all at their equilibrium lattice sites. This is a fixed number that we will write as E^{int}. The second contribution is as a result of the vibrational motions of each of the atoms. Therefore

$$Q(N, V, T) = \sum_{\text{states } i} e^{-E_i(N,V)/kT} = e^{-E^{\text{int}}/kT} \left(\sum_{\substack{\text{vibrational states } n \\ \text{of each atom}}} e^{-\varepsilon_n/kT} \right)^{3N}$$

$$= e^{-E^{\text{int}}/kT} (q_{\text{vib}})^{3N} \quad (9.1\text{-}3)$$

where

$$q_{\text{vib}} = \sum_{n=0}^{\infty} e^{-\varepsilon_n/kT} = \sum_{n=0}^{\infty} e^{-\left(n + \frac{1}{2}\right) \frac{h\nu_E}{kT}} = e^{-h\nu_E/2kT} \sum_{n=0}^{\infty} e^{-nh\nu_E/kT}$$

$$= \frac{e^{-h\nu_E/2kT}}{1 - e^{-h\nu_E/kT}} = \frac{e^{-\Theta_E/2T}}{1 - e^{-\Theta_E/T}} \quad (9.1\text{-}4)$$

with $\Theta_E = h\nu_E/k$ referred to as the Einstein vibrational temperature. For notational convenience we use $E^{int} = Nu$, where u is the interaction energy for each molecule at its equilibrium lattice site.

Therefore

$$Q(N, V, T) = e^{-Nu/kT} \left(\frac{e^{-\Theta_E/2T}}{1 - e^{-\Theta_E/T}} \right)^{3N} \tag{9.1-5}$$

$$A(N, V, T) = -kT \ln Q(N, V, T) = Nu - 3NkT \ln \left(\frac{e^{-\Theta_E/2T}}{1 - e^{-\Theta_E/T}} \right)$$

$$= Nu + \frac{3N}{2} h\nu_E + 3NkT \ln \left(1 - e^{-\Theta_E/T} \right)$$

$$= Nu_E^0 + 3NkT \ln \left(1 - e^{-\Theta_E/T} \right) \tag{9.1-6}$$

where $u_E^0 = u + \frac{3}{2}h\nu_E$ is the zero point energy (per atom) of the monatomic Einstein crystal, and

$$\mu(N, V, T) = \left(\frac{\partial A}{\partial N} \right)_{V,T} = \frac{A}{N} = u - 3kT \ln \left(\frac{e^{-\Theta_E/2T}}{1 - e^{-\Theta_E/T}} \right)$$

$$= u + \frac{3}{2}h\nu_E + 3kT \ln \left(1 - e^{-\Theta_E/T} \right) = u_E^0 + 3kT \ln \left(1 - e^{-\Theta_E/T} \right) \tag{9.1-7}$$

$$U(N, V, T) = kT^2 \left(\frac{\partial \ln Q(N, V, T)}{\partial T} \right)_{N,V}$$

$$= Nu + \frac{3N}{2}h\nu_E + 3NkT \left(\frac{\frac{\Theta_E}{T}}{e^{\Theta_E/T} - 1} \right) = Nu_E^0 + 3Nh\nu \frac{e^{-\Theta_E/T}}{1 - e^{-\Theta_E/T}} \tag{9.1-8}$$

$$S(N, V, T) = \frac{U(N, V, T) - A(N, V, T)}{T}$$

$$= 3Nk \left[\frac{\Theta_E}{T} \frac{e^{-\Theta_E/T}}{1 - e^{-\Theta_E/T}} - \ln \left(1 - e^{-\Theta_E/T} \right) \right] \tag{9.1-9}$$

and

$$C_V = \left(\frac{\partial U}{\partial T} \right)_{N,V} = 3Nk \left(\frac{\Theta_E}{T} \right)^2 \frac{e^{-\Theta_E/T}}{\left(1 - e^{-\Theta_E/T} \right)^2} \tag{9.1-10}$$

There are two limits of these equations that it are interesting to look at. The first is the limit of low temperature ($T \to 0$) for which $q_{vib} \to e^{-\Theta_E/T}$,

$$A(N, V, T \to 0) = Nu + \frac{3Nh\nu_E}{2} = Nu_E^0$$

$$U(N, V, T \to 0) = Nu_E^0$$

$$S(N, V, T \to 0) = 3Nk \frac{\Theta_E}{T} e^{-\Theta_E/T} \tag{9.1-11}$$

and

$$C_V(N, V, T \to 0) = \left(\frac{\partial U}{\partial T}\right)_{N,V} = 3Nk\left(\frac{\Theta_E}{T}\right)^2 e^{-\Theta_E/T}$$

Note that since $\lim_{x \to \infty} \frac{e^{-x}}{x} = \lim_{x \to \infty} \frac{e^{-x}}{x^2} = 0$, $S(N, V, T = 0) = 0$ for the Einstein crystal in accordance with the third law of thermodynamics that the entropy of a perfectly ordered crystal is 0 at the absolute zero of temperature. Also, that $C_V(N, V, T = 0) = 0$ is in accord with the experimental observation that the heat capacities of all substances go to zero at the absolute zero of temperature. However, experimentally, it is found that C_V goes to zero as T^3, which is not the prediction for the Einstein crystal.

The second interesting limit is that of high temperature ($T \to \infty$ or more correctly $T \gg \Theta_E$), in which case

$$q_{vib}(T > \Theta_E) = \frac{e^{-\Theta_E/2T}}{1 - e^{-\Theta_E/T}} = \frac{1 - \frac{\Theta_E}{2T} + \cdots}{1 - \left[1 - \frac{\Theta_E}{T} + \cdots\right]} = \frac{1}{\frac{\Theta_E}{T}} = \frac{T}{\Theta_E} \quad (9.1\text{-}12)$$

$$Q(N, V, T > \Theta_E) = e^{-\frac{Nu}{kT}}\left(\frac{kT}{h\nu_E}\right)^{3N} = e^{-\frac{Nu}{kT}}\left(\frac{T}{\Theta_E}\right)^{3N}$$

$$\ln Q(N, V, T > \Theta_E) = -\frac{Nu}{kT} + 3N \ln\left(\frac{kT}{h\nu_E}\right) = -\frac{Nu}{kT} + 3N \ln\left(\frac{T}{\Theta_E}\right)$$

$$A(N, V, T > \Theta_E) = -kT \ln Q(N, V, T > \Theta_E) = Nu - 3NkT \ln\left(\frac{T}{\Theta_E}\right) \quad (9.1\text{-}13)$$

$$\mu(N, V, T > \Theta_E) = \frac{A}{N} = u - 3kT \ln\left(\frac{T}{\Theta_E}\right)$$

$$U(N, V, T > \Theta_E) = Nu + 3NkT$$

$$S(N, V, T > \Theta_E) = 3Nk\left[1 - \ln\left(\frac{T}{\Theta_E}\right)\right]$$

and

$$C_V(T > \Theta_E) = 3Nk$$

Note that the high temperature limit of the constant volume heat capacity is $C_V(T > \Theta_E) = 3Nk$, which is known as the law of Dulong and Petit. This is to be compared with the constant-volume heat capacity of an ideal monatomic gas, which is $C_V = \frac{3}{2}Nk$. The difference arises because each atom of an ideal gas has three degrees of translational motion, each of which contributes $\frac{1}{2}k$ to the heat capacity, while each atom in a monatomic crystal has three degrees of vibrational motion, each of which—when fully excited (high temperature)—contributes k to the heat capacity.

9.2 THE DEBYE MODEL OF A CRYSTAL

The Einstein model of a monatomic crystal is a primitive one in that it is assumed that the motion of each atom is independent of all others. A better model is to allow all

9.2 The Debye Model of a Crystal

the molecules to vibrate simultaneously. For an N atom crystal, there is $3N$ degrees of freedom. To identify these, we could (in principle, but not in practice) do a normal mode analysis, as discussed in Section 4.6 for polyatomic molecules, and find that there are three translational motions corresponding to movement of the center of mass of the crystal, three rotational motions of the whole crystal, and $3N-6$ vibrational modes. The partition function of such a crystal, keeping the macroscopic crystal fixed in space so that there is no translational or rotational motion of the center of mass, is

$$Q(N, V, T) = e^{-u/kT} \prod_{i=1}^{3N-6} q_{\text{vib},i}\left(\frac{h\nu_i}{kT}\right) = e^{-u/kT} \prod_{i=1}^{3N-6} q_{\text{vib},i}\left(\frac{\Theta_{v,i}}{T}\right)$$

$$= e^{-u/kT} \prod_{i=1}^{3N-6} \frac{e^{-\Theta_{v,i}/2T}}{1 - e^{-\Theta_{v,i}/T}}$$

$$\text{or}\quad \ln Q(N, V, T) = -\frac{u}{kT} + \sum_{i=1}^{3N-6} \ln\left(\frac{e^{-\Theta_{v,i}/2T}}{1 - e^{-\Theta_{v,i}/T}}\right) \tag{9.2-1}$$

For the Einstein crystal, the vibrational frequency of each atom was assumed to be identical to ν_E, so the evaluation of the partition function was straightforward. Here, however, the independent vibrational modes have to be determined from a normal mode analysis, and involve $3N-6$ vibrational frequencies. One expects that the frequencies will range from high-frequency modes for the vibrational motion of a single atom, as considered in the Einstein model, to low-frequency (and therefore large wavelength) modes resulting from the concerted motion of large numbers of (in the limit, all) atoms in the crystal.

Instead of attempting to identify the $3N-6$ normal vibrational modes for the crystal, the Debye model uses a probability distribution of frequencies $g(\nu)$ defined such that the number of vibrational modes in the frequency range ν to $\nu+d\nu$ is $g(\nu)d\nu$. This probability distribution is normalized such that

$$\int_0^\infty g(\nu)\,d\nu = 3N - 6 \tag{9.2-2}$$

With this approximation

$$\ln Q(N, V, T) = -\frac{u}{kT} + \sum_{i=1}^{3N-6} \ln\left(\frac{e^{-\Theta_{v,i}/2T}}{1 - e^{-\Theta_{v,i}/T}}\right)$$

$$\rightarrow -\frac{u}{kT} + \int_0^\infty g(\nu) \ln\left(\frac{e^{-h\nu/2kT}}{1 - e^{-h\nu/kT}}\right) d\nu \tag{9.2-3}$$

The problem then becomes one of determining the probability distribution $g(\nu)$. In the Debye model, it is recognized that the low-frequency (large wavelength) collective motions make large contributions to the partition function. Such collective motions, which correspond to wavelengths of several or many lattice spacings, are rather insensitive to the specific atoms of the crystal. (This is an interesting and subtle idea. To understand the wave motion of a large amount of fluid, for example the ocean, we need only information about its macroscopic properties of viscosity and density and the laws of fluid mechanics to describe its motion. However, if we are interested in the

152 Chapter 9: Monatomic Crystals

vibration of a single molecule in ice, for example, we need detailed information on the mass of the atom and its interaction with surrounding molecules.) Thus, the general characteristics of the important large wavelength motions of atoms—in particular, the distribution of frequencies—in a crystal will be of a similar form for all crystals.

In the Debye model it is assumed that the frequency distribution of the large wavelength motions in a crystal is similar to that found in solid mechanics for the behavior of elastic waves in three-dimensional continuum, which is given by $g(\nu) = \alpha \nu^2$. This is taken to be valid for all frequencies up to the highest frequency (shortest wavelength) ν_D corresponding to the motion of a single atom with all other atoms fixed at their equilibrium positions. That is

$$g(\nu) = \begin{cases} \alpha \nu^2 & \text{for } 0 \leq \nu \leq \nu_D \\ 0 & \text{for } \nu_D \leq \nu \end{cases} \quad (9.2\text{-}4)$$

To find the parameter α the following normalization condition is used

$$\int_0^\infty g(\nu)\, d\nu = \int_0^{\nu_D} \alpha \nu^2\, d\nu = \frac{\alpha}{3}\left(\nu_D^3 - 0\right) = \frac{\alpha \nu_D^3}{3} = 3N - 6 \quad (9.2\text{-}5)$$

Since our interest is with macroscopic crystals of many atoms (such that $N \gg 1$), for here and what follows we will replace $3N-6$ with $3N$. Therefore

$$\alpha = \frac{9N}{\nu_D^3} \quad \text{and} \quad g(\nu) = \begin{cases} \dfrac{9N\nu^2}{\nu_D^3} & \text{for } 0 \leq \nu \leq \nu_D \\ 0 & \text{for } \nu_D \leq \nu \end{cases} \quad (9.2\text{-}6)$$

and

$$\ln Q(N, V, T) = -\frac{Nu}{kT} + \frac{9N}{\nu_D^3} \int_0^{\nu_D} \nu^2 \ln\left(\frac{e^{-h\nu/2kT}}{1 - e^{-h\nu/kT}}\right) d\nu$$

$$= -\frac{Nu}{kT} + \frac{9N}{\Theta_D^3} \int_0^{\Theta_D} \Theta^2 \ln\left(\frac{e^{-\Theta/2T}}{1 - e^{-\Theta/T}}\right) d\Theta$$

$$= -\frac{Nu}{kT} + \frac{9N}{\Theta_D^3} \int_0^{\Theta_D/T} \Theta^2 \left(\frac{\Theta}{2T}\right) d\Theta - \frac{9N}{\Theta_D^3} \int_0^{\Theta_D} \Theta^2 \ln\left(1 - e^{-\Theta/T}\right) d\Theta$$

$$= -\frac{Nu}{kT} - \frac{9N}{2\left(\frac{\Theta_D}{T}\right)^3} \frac{1}{4}\left(\frac{\Theta_D}{T}\right)^4 - 3NG\left(\frac{\Theta_D}{T}\right)$$

$$= -\frac{Nu}{kT} - \frac{9N}{8}\left(\frac{\Theta_D}{T}\right) - 3NG\left(\frac{\Theta_D}{T}\right)$$

where $G\left(\dfrac{\Theta_D}{T}\right) = 3\left(\dfrac{T}{\Theta_D}\right)^3 \displaystyle\int_0^{\Theta_D/T} x^2 \ln(1 - e^{-x})\, dx,$ and $x = \dfrac{h\nu}{kT} = \dfrac{\Theta}{T}$

$$(9.2\text{-}7)$$

9.2 The Debye Model of a Crystal

$$U(N, V, T) = kT^2 \left(\frac{\partial \ln Q(N, V, T)}{\partial T} \right)_V$$

$$= kT^2 \left[\frac{Nu}{kT^2} + \frac{9h\nu_D}{8kT^2} - \frac{9N}{\nu_D^3} \int_0^{\nu_D} \frac{1}{1 - e^{-h\nu/kT}} \frac{d}{dT}\left(1 - e^{-h\nu/kT}\right) \nu^2 \, d\nu \right]$$

$$= kT^2 \left[\frac{Nu}{kT^2} + \frac{9h\nu_D}{8kT^2} + \frac{9Nh}{\nu_D^3 kT^2} \int_0^{\nu_D} \frac{e^{-h\nu/kT}}{1 - e^{-h\nu/kT}} \nu^3 \, d\nu \right]$$

$$= Nu + \frac{9h\nu_D}{8} + \frac{9NkT}{\left(\frac{\Theta_D}{T}\right)^3} \int_0^{\frac{\Theta_D}{T}} \frac{e^{-x}}{1 - e^{-x}} x^3 \, dx$$

$$= Nu + \frac{9Nh\nu_D}{8} + 3NkTF\left(\frac{\Theta_D}{T}\right) = Nu_D^0 + 3NkTF\left(\frac{\Theta_D}{T}\right) \quad (9.2\text{-}8)$$

where

$$u_D^0 = u + \frac{9Nh\nu_D}{8}$$

is the zero point energy (per molecule) of the monatomic Debye crystal and

$$F\left(\frac{\Theta_D}{T}\right) = \frac{3}{\left(\frac{\Theta_D}{T}\right)^3} \int_0^{\frac{\Theta_D}{T}} \frac{e^{-x}}{1 - e^{-x}} x^3 \, dx = 3\left(\frac{T}{\Theta_D}\right)^3 \int_0^{\frac{\Theta_D}{T}} \frac{x^3 e^x}{e^x - 1} \, dx$$

$$= \frac{\Theta_D}{T} \frac{dG\left(\frac{\Theta_D}{T}\right)}{d\left(\frac{\Theta_D}{T}\right)} = -T \frac{dG\left(\frac{\Theta_D}{T}\right)}{d(T)}$$

Also $C_V = 3NkK\left(\frac{\Theta_D}{T}\right)$ where

$$K\left(\frac{\Theta_D}{T}\right) = \frac{3}{\left(\frac{\Theta_D}{T}\right)^3} \int_0^{\frac{\Theta_D}{T}} \frac{x^4 e^{-x}}{(1 - e^{-x})^2} \, dx = 3\left(\frac{T}{\Theta_D}\right)^3 \int_0^{\frac{\Theta_D}{T}} \frac{x^4 e^x}{(e^x - 1)^2} \, dx \quad (9.2\text{-}9)$$

$$\mu(N, V, T) = \left(\frac{\partial A}{\partial N}\right)_{V,T} = \frac{A}{N} = u + \frac{9h\nu_D}{8} + 3kTG\left(\frac{\Theta_D}{T}\right) \quad (9.2\text{-}10)$$

$$S(N, V, T) = \frac{U - A}{T} = \frac{1}{T}\left[Nu + \frac{9Nh\nu_D}{8} - 3NkTF\left(\frac{\Theta_D}{T}\right)\right.$$
$$\left. - Nu - \frac{9Nh\nu_D}{8} - 3NkTG\left(\frac{\Theta_D}{T}\right)\right]$$

$$= 3Nk\left(F\left(\frac{\Theta_D}{T}\right) - G\left(\frac{\Theta_D}{T}\right)\right) \quad (9.2\text{-}11)$$

Note that there is no closed, analytic form for the G, F, or K integrals. However, the following two limits are known:

as $T \to 0$,

$$F\left(\frac{\Theta_D}{T} \to \infty\right) \to \frac{\pi^4}{5\left(\frac{\Theta_D}{T}\right)^3} = \frac{\pi^4}{5}\left(\frac{T}{\Theta_D}\right)^3, \quad G\left(\frac{\Theta_D}{T} \to \infty\right) \to -\frac{\pi^4}{15}\left(\frac{T}{\Theta_D}\right)^3$$

and

$$K\left(\frac{\Theta_D}{T} \to \infty\right) \to \frac{12\pi^4}{15}\left(\frac{T}{\Theta_D}\right)^3 \qquad (9.2\text{-}12)$$

as $T \to \infty$

$$F\left(\frac{\Theta_D}{T} \to 0\right) = \frac{3}{\left(\frac{\Theta_D}{T}\right)^3} \int_0^{\Theta_D/T} \frac{x^3 e^{-x}}{1 - e^{-x}} dx = \frac{3}{\left(\frac{\Theta_D}{T}\right)^3} \int_0^{\Theta_D/T} \frac{x^3(1 - x + \cdots)}{1 - (1 - x + \cdots)} dx$$

$$= \frac{3}{\left(\frac{\Theta_D}{T}\right)^3} \frac{1}{3} \left(\frac{\Theta_D}{T}\right)^3 = 1 \qquad (9.2\text{-}13)$$

$$G\left(\frac{\Theta_D}{T} \to 0\right) = \frac{3}{\left(\frac{\Theta_D}{T}\right)^3} \int_0^{\Theta_D/T} x^2 \ln(1 - e^{-x}) dx = \ln\left(\frac{\Theta_D}{T}\right) - \frac{1}{3}, \text{ which diverges.}$$

$$(9.2\text{-}14)$$

and

$$K\left(\frac{\Theta_D}{T} \to 0\right) \to 1 \qquad (9.2\text{-}15)$$

More generally the behavior of the F function is given in Fig. 9.2-1 and Table 9.2-1 below. Also, values of the Debye temperature for some common elements are given in Table 9.2-2.

In the limiting case of $T \to 0$, we have the following:

$$U(N, V, T \to 0) = Nu + \frac{9N}{8}h\nu_D + 3NkT\frac{\pi^4}{5}\left(\frac{T}{\Theta_D}\right)^3$$

$$= Nu_D^0 + 3NkT\frac{\pi^4}{5}\left(\frac{T}{\Theta_D}\right)^3,$$

$$U(N, V, T = 0) = Nu + \frac{9Nk}{8}\Theta_D = Nu_D^0,$$

$$C_V(T \to 0) = \frac{12\pi^4 Nk}{5}\left(\frac{T}{\Theta_D}\right)^3, \quad C_V(T = 0) = 0 \qquad (9.2\text{-}16)$$

9.2 The Debye Model of a Crystal

$$A(N, V, T \to 0) = Nu + \frac{9N}{8}h\nu_D - \frac{\pi^4 NkT}{5}\left(\frac{T}{\Theta_D}\right)^3 = Nu_D^0 - \frac{\pi^4 NkT}{5}\left(\frac{T}{\Theta_D}\right)^3$$

$$\mu(N, V, T \to 0) = u + \frac{9}{8}h\nu_D - \frac{\pi^4 kT}{5}\left(\frac{T}{\Theta_D}\right)^3 = u_D^0 - \frac{\pi^4 kT}{5}\left(\frac{T}{\Theta_D}\right)^3$$

and

$$S(N, V, T \to 0) = \frac{U(N, V, T \to 0) - A(N, V, T \to 0)}{T} = \frac{4\pi^4 Nk}{5}\left(\frac{T}{\Theta_D}\right)^3$$

Finally

$$S(N, V, T = 0) = 0 \quad \text{and} \quad C_V(T = 0) = 0$$

so that for the Debye crystal (as for the Einstein model of a crystal), $S(N, V, T = 0) = 0$ in accordance with the third law of thermodynamics that the entropy of a perfect crystal is 0 at the absolute 0 of temperature. Also, that $C_V(N, V, T = 0) = 0$ agrees with the experimental observation that the heat capacity of all substances goes to 0 at the absolute 0 of temperature. Also, unlike the Einstein model, the heat

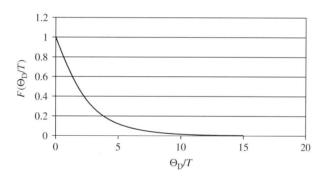

Figure 9.2-1 The function $F(\Theta_D/T)$ as a function of reciprocal temperature.

Table 9.2-1 $F(\Theta_D/T)$, $G(\Theta_D/T)$ and $K(\Theta_D/T)$ as a function Θ_D/T

Θ_D/T	$F(\Theta_D/T)$	$G(\Theta_D/T)$	$K(\Theta_D/T)$
0	1.0	−9.544	1.0
0.01	0.996	−4.942	1.0
0.10	0.963	−2.673	1.0
0.20	0.927	−2.017	0.998
0.50	0.825	−1.208	0.988
1.0	0.674	−0.683	0.952
3.0	0.284	−0.146	0.663
5.0	0.118	−0.046	0.369
10.0	0.019	-6.477×10^{-3}	0.076

Table 9.2-2 The Debye (and Einstein) Temperatures of Some Common Elements

Element	Θ_D (K)
Ag	215
Al	398
Au	185
Be	980
C	1860
Ca	230
Cd	165
Cr	405
Cu	315
Fe	453
Hg	90
K	99
Li	430
Mg	330
Mo	375
Na	160
Pb	86
Pt	225
W	315
Zn	240

From the theory of crystalline solids, the relationship between the Einstein and Debye temperatures is

$$\frac{\Theta_E}{\Theta_D} = \sqrt[3]{\frac{\pi}{6}} \quad \text{or} \quad \Theta_E = 0.806\Theta_D.$$

[Data in this table assembled from various sources.]

capacity of the Debye model goes to 0 as T^3, in agreement with the experimental data.

In the limiting case of $T \to \infty$, we have

$$U(N, V, T \to \infty) = Nu + \frac{9N}{8}h\nu_D + 3NkT$$

$$= Nu + \frac{9Nk}{8}\Theta_D + 3NkT = Nu_D^0 + 3NkT,$$

$$C_V(T \to \infty) = 3Nk$$

$$A(N, V, T \to \infty) = Nu + \frac{9N}{8}h\nu_D - 3NkT\left[\ln\left(\frac{\Theta_D}{T}\right) - \frac{1}{3}\right]$$

$$= Nu_D^0 - 3NkT\left[\ln\left(\frac{\Theta_D}{T}\right) - \frac{1}{3}\right], \tag{9.2-17}$$

$$\mu(N, V, T \to \infty) = u + \frac{9}{8}h\nu_D - 3kT\left[\ln\left(\frac{\Theta_D}{T}\right) - \frac{1}{3}\right]$$

$$= u_D^0 - 3kT\left[\ln\left(\frac{\Theta_D}{T}\right) - \frac{1}{3}\right]$$

and

$$S(N, V, T \to \infty) = \frac{U(N, V, T \to \infty) - A(N, V, T \to \infty)}{T}$$

$$= 3Nk + 3Nk\left[\ln\left(\frac{\Theta_D}{T}\right) - \frac{1}{3}\right] = Nk\left[3\ln\left(\frac{\Theta_D}{T}\right) + 2\right]$$

9.3 TEST OF THE EINSTEIN AND DEBYE HEAT CAPACITY MODELS FOR A CRYSTAL

As a test of the accuracy of the Einstein and Debye models, we consider the constant-volume heat capacity data for tungsten listed below.[1] The procedure that will be used is to fit the Einstein and Debye vibrational temperatures to the heat capacity data at one temperature, and then predict the heat capacity with both models at the other temperatures at which experimental data are available.

The following data are available for the constant volume heat capacity of tungsten

T (K)	1	10	20	30	40	50	60	70	80	90	100
C_v (J/mol K)	0.00104	0.0450	0.333	1.35	3.30	5.82	8.39	9.73	12.80	14.56	16.02

T (K)	120	140	180	240	400	800	1000	1200	1400	1600	1800
C_v (J/mol K)	18.24	19.82	21.78	23.35	24.74	26.14	26.70	27.25	27.82	28.42	29.06

T (K)	2000	2400	3000
C_v (J/mol K)	29.80	31.79	35.91

ILLUSTRATION

(a) Determine the Einstein temperature for tungsten from the heat capacity data at 140 K.
(b) Compute the heat capacity of tungsten at all temperatures using the value of the Einstein temperature computed in part (a).
(c) Determine the Debye temperature for tungsten from the heat capacity data at 140 K.
(d) Compute the heat capacity of tungsten at all temperatures using the value of the Debye temperature computed in part (c).

---

[1] G. K. White and S. J. Collocott, *J. Phys. Chem. Ref. Data*, **13**, 1251 (1984).

SOLUTION

(a) An Einstein vibrational temperature of 235.178 K reproduces the experimental heat capacity at 140 K.

(b) Using this value of the vibrational temperature, we compute the heat capacities shown below.

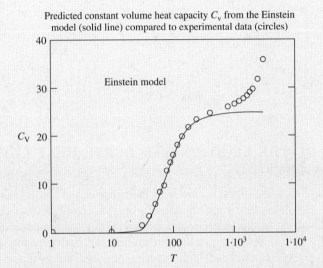

Predicted constant volume heat capacity C_v from the Einstein model (solid line) compared to experimental data (circles)

(c) A Debye vibrational temperature of 307.737 K reproduces the experimental heat capacity at 140 K.

(d) Using this value of the vibrational temperature, we compute the heat capacities shown below.

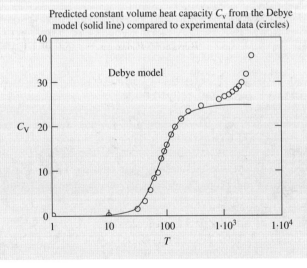

Predicted constant volume heat capacity C_v from the Debye model (solid line) compared to experimental data (circles)

There are several things to be noted from this illustration. First, the experimental heat capacity of tungsten at high temperatures exceeds the $3Nk = 24.942$ J mol^{-1} K^{-1}, which is the maximum value that is predicted from both the Einstein and Debye models. Second, to the scale of the graphs above, both the Einstein and Debye models predict similar heat capacities below about 500 K, though with different vibrational

9.4 Sublimation Pressure and Enthalpy of Crystals

temperatures. However, on closer examination, the two models give different predictions at very low temperatures. We see this in the table below.

T(K)		1	10	20	30
C_v (J/mol K)	experiment	1.04×10^{-3}	0.0450	0.333	1.35
C_v (J/mol K)	Einstein	8.45×10^{-7}	0.0270	0.604	2.43
C_v (J/mol K)	Debye	6.67×10^{-5}	0.0670	0.533	1.76

So it is evident that with an appropriately chosen vibrational temperature, the Einstein and Debye models give reasonable predictions of the constant volume heat capacity up to approximately twice the vibrational temperature. At higher temperatures, the predictions are increasingly in error. At such higher temperatures, the higher vibrational states are excited, and the vibrational motions are more complex. In this case, the simple Einstein and Debye models fail, and one must consider the more complicated subject of lattice dynamics—which is beyond the scope of this book. Nonetheless, the simple models of a crystal discussed here are useful for heat capacity predictions at moderate temperatures.

9.4 SUBLIMATION PRESSURE AND ENTHALPY OF CRYSTALS

Finally, even though we have just shown that the Einstein and Debye models of a monatomic crystal are of limited accuracy, it is of interest to estimate the sublimation pressure and the heat of sublimation of a monatomic crystal. We can do this for both the Einstein and Debye models of the crystal by equating the chemical potential of these crystals to that of ideal gas (since sublimation pressures are generally very low). That is

$$\mu_{\text{crystal}} = \mu_{\text{ideal gas}} = -kT \ln \left[\left(\frac{2\pi mkT}{h^2} \right)^{\frac{3}{2}} \frac{kT}{P} \right] \quad (9.4\text{-}1)$$

For the Einstein crystal, we have

$$\mu_{\text{crystal}} = u + \frac{3}{2} h \nu_E + 3kT \ln \left(1 - e^{-\frac{\Theta_E}{T}} \right) = \mu_{\text{ideal gas}}$$

$$= -kT \ln \left[\left(\frac{2\pi mkT}{h^2} \right)^{\frac{3}{2}} \frac{kT}{P} \right] \quad (9.4\text{-}2)$$

so that solving for the sublimation pressure P as a function of temperature, which is denoted by $P_E^{\text{sub}}(T)$, one obtains

$$kT \ln P_E^{\text{sub}}(T) = u + \frac{3}{2} h \nu_E + 3kT \ln \left(1 - e^{-\frac{\Theta_E}{T}} \right) + kT \ln \left[\left(\frac{2\pi mkT}{h^2} \right)^{\frac{3}{2}} kT \right]$$

$$= u_E^0 + 3kT \ln \left(1 - e^{-\frac{\Theta_E}{T}} \right) + kT \ln \left[\left(\frac{2\pi mkT}{h^2} \right)^{\frac{3}{2}} kT \right] \quad (9.4\text{-}3)$$

where $\Delta u_{E,sub}(T=0) = u + \frac{3}{2}h\nu_E = u_E^0$ is the sublimation energy at zero temperature and $P_E^{sub}(T)$ is the sublimation pressure for the Einstein crystal.[2] At low temperatures, this reduces to

$$kT \ln P_E^{sub}(T) = u_E^0 + 3kT \ln\left(1 - \left(1 - \frac{\Theta_E}{T}\right)\right) + kT \ln\left[\left(\frac{2\pi mkT}{h^2}\right)^{\frac{3}{2}} kT\right]$$

$$= u_E^0 + 3kT \ln\left(\frac{\Theta_E}{T}\right) + kT \ln\left[\left(\frac{2\pi mkT}{h^2}\right)^{\frac{3}{2}} kT\right] \quad (9.4\text{-}4)$$

The heat (enthalpy change) on sublimation is obtained from the classical thermodynamic relation

$$\left.\frac{d \ln P_E^{sub}}{dT}\right|_{\text{coexistence line}} = \frac{\Delta_{sub} H_E}{kT^2} \quad (9.4\text{-}5)$$

resulting in (see Problem 9.8)

$$\Delta_{sub} H_E = -u_E^0 - \frac{3h\nu_E}{1 - e^{-\Theta_E/T}} e^{-\Theta_E/T} + \frac{5}{2}kT$$

$$= -u_E^0 - \frac{3kT}{1 - e^{-\Theta_E/T}} \frac{\Theta_E}{T} e^{-\Theta_E/T} + \frac{5}{2}kT$$

Similarly, for the Debye crystal, we have

$$kT \ln P_D^{sub}(T) = u_D^0 + 3kTG\left(\frac{\Theta_D}{T}\right) + kT \ln\left[\left(\frac{2\pi mkT}{h^2}\right)^{\frac{3}{2}} kT\right] \quad (9.4\text{-}6)$$

where $u_D^0 = \Delta_{sub} u_D(T=0) = u + \frac{9}{8}h\nu_D$ is the sublimation energy at zero temperature for the Debye crystal. The low-temperature limit is

$$kT \ln P_D^{sub}(T) = \Delta_{sub} u_D(T=0) - \frac{\pi^4 kT}{5}\left(\frac{T}{\Theta_D}\right)^3 + kT \ln\left[\left(\frac{2\pi mkT}{h^2}\right)^{\frac{3}{2}} kT\right] \quad (9.4\text{-}7)$$

Also

$$\Delta_{sub} H_D = -u_D^0 + 3kT^2 \frac{dG\left(\frac{\Theta_D}{T}\right)}{dT} + \frac{5}{2}kT = -u_D^0 - 3kTF\left(\frac{\Theta_D}{T}\right) + \frac{5}{2}kT \quad (9.4\text{-}8)$$

In the equations of this section, the zero-point energies are not generally known. Therefore, the sublimation pressure equations are most useful if the sublimation pressure is known at one temperature. In such cases, this value can be used to obtain the zero-point energy (u_E^0 and/or u_D^0), and then predict the sublimation pressures at other temperatures (Problems 9.8 and 9.9).

---

[2] In the units generally used here, P has units of $J/m^3 = Pa$.

9.5 A COMMENT ON THE THIRD LAW OF THERMODYNAMICS

In this chapter we have shown that the entropies of perfectly ordered Einstein and Debye crystals are 0 at the absolute zero of temperature, which is the third law of thermodynamics. However, not all crystals are perfectly ordered at $T = 0$. For example, consider the case of a heteronuclear diatomic molecule, carbon monoxide, in which the energies of interaction in the two states shown below are so similar that in the process of forming a crystal, either of these states may occur:

```
C  C  C  C  C...        C  O  C  O  C.....
|  |  |  |  |    or     |  |  |  |  |
O  O  O  O  O           O  C  O  C  O
```

or, in fact, any other arrangement of the carbon monoxide molecules (note that the solid lines denote covalent bonds). Therefore, each molecule has a degeneracy in its orientation state of 2, and the orientational degeneracy Ω of a N molecule system is 2^N. Consequently, the entropy of an orientationally disordered carbon monoxide crystal is

$$S(N, V, T = 0) = k \ln \Omega = k \ln 2^N = Nk \ln 2 \tag{9.5-1}$$

This is referred to as the residual entropy of the carbon monoxide crystal and is a result of the fact that it is not perfectly ordered at absolute zero. Ice is another example of a crystal that has a residual entropy at the absolute zero of temperature that is a result of different orientations of covalently bonded and hydrogen-bonded $H \cdot O$.

CHAPTER 9 PROBLEMS

9.1 Consider a solid consisting of N atoms. Treat each atom as bound to a fixed center of force and behaving as independent harmonic oscillator of frequency ω. The binding energy of each atom in the ground state is ε_1, so that the energy states of a single oscillator are $(n\omega - \varepsilon_1)$ with $n = 1, 2, \ldots$ For simplicity, suppose that each atom can oscillate only in one dimension, and use the Einstein model of the crystal. Also consider the solid to be in equilibrium with a large volume of an ideal gas of the same atoms.
   a. What are the thermodynamic properties of this one-dimensional crystal?
   b. Discuss the behavior of this model equation at the limits of low and high temperature.
   c. Derive an expression for the sublimation pressure as a function of temperature for this simple model.

9.2 Repeat the calculations of Problem 9.1 for a two-dimensional Einstein crystal.

9.3 Use the Debye model to determine the equation of state for a solid (i.e., pressure as a function of temperature and volume). Express the results in terms of the Grüneisen parameter $\gamma = -\frac{V}{\Theta_D}\left(\frac{\partial \Theta_D}{\partial V}\right)_T$ and develop expressions for the limiting cases of (i) temperatures much greater than the Debye temperature

and (ii) temperatures much less than the Debye temperature. State and discuss all assumptions.

Also assume that the Grüneisen parameter, γ, is independent of temperature and express the coefficient of thermal expansion, $\alpha = \frac{1}{V}\left(\frac{\partial V}{\partial T}\right)_P$, in terms of γ, the heat capacity (C_V), and the bulk modulus (κ).

9.4 a. Derive the expression for the constant-volume heat capacity of a monatomic Einstein crystal in which the force constants for vibrations and therefore the vibrational frequencies in the x-, y-, and z-directions are different. That is, for a crystal in which the energy of each atom is given by

$$u(x, y, z) = u(x = y = z = 0) + \frac{a}{2}x^2 + \frac{b}{2}y^2 + \frac{c}{2}z^2$$

b. Graphite has a planar structure in which the atomic vibrations within the plane are equal and of high frequency:

$$\frac{h\nu_x}{k} = \Theta_x = \frac{h\nu_y}{k} = \Theta_y \gg \text{room temperature}$$

But the force constant—and, therefore, the atomic vibrational frequency—for motion out of the plane is small:

$$\frac{h v_z}{k} = \Theta_z \ll \text{room temperature}$$

Determine the temperature dependence of the low temperature constant-volume heat capacity of a monatomic Einstein model of graphite.

9.5 A simple model for the adsorption of a monatomic gas at low pressure (so molecule-molecule interactions can be ignored) on a crystal surface is that the energy of the adsorbed gas molecule is

$$u(x, y, z) = u(x_0, y_0, z_0)$$
$$+ \frac{\alpha}{2}\left[(x - x_0)^2 + (y - y_0)^2\right]$$
$$+ \frac{\beta}{2}(z - z_0)^2$$

where the vibrational motions within the plane of the crystal surface has a weaker force constant than the vibrational motion perpendicular to the crystal surface.
   a. What is the ground state energy of an adsorbed molecule?
   b. What is the canonical partition function for N identical adsorbed molecules on M lattice sites on the surface, assuming that the adsorbed molecules are indistinguishable?
   c. Develop an expression for the adsorption isotherm for this system assuming that the adsorbed molecules are in equilibrium with an ideal gas.

9.6 Use that the heat capacity C_v of copper at 100 K is 16.108 J/mol K and the Einstein model of the crystal to predict the heat capacity of copper at 1200 K (note that the experimental value is 30.635 J/mol K).

9.7 Using that the heat capacity C_v of silver at 25°C is 25.350 J/mol K, predict its heat capacity from 10 K to its melting point of 1234.9 K.

9.8 a. Derive Eq. 9.4-5.
   b. Derive Eq. 9.4-8.

9.9 The sublimation pressure of silver is 2.955×10^{-7} Pa at 830 K.
   a. Predict the sublimation pressure of silver from 830 K to its melting point of 1234.0 K using the Einstein model.
   b. Repeat the calculation using the Debye model.
   c. Compare the results of your predictions with the known sublimation pressure of 1.984×10^{-1} Pa at 1224 K, and with sublimation pressures at other temperatures that have been correlated[3] with the equation

$$\ln P^{\text{sub}}(\text{Pa}) = 29.635 - \frac{34974}{T} - 0.37677 \ln(T)$$

9.10 The sublimation pressure of zinc is 2.432×10^{-12} Pa at 298.15 K.
   a. Predict the sublimation pressure of zinc from 298.15 K to its melting point of 692.7 K using the Einstein model.
   b. Repeat the calculation using the Debye model.
   c. Compare the results of your predictions with the known sublimation pressure of 20.95 Pa at 692.7 K, and with sublimation pressures at other temperatures that have been correlated[3] with the equation

$$\ln P^{\text{sub}}(\text{Pa}) = 30.765 - \frac{15911}{T} - 0.7269 \ln(T)$$

9.11 Estimate the heat of sublimation of silver at 830 K:
   a. using the Einstein model;
   b. using the Debye model; and
   c. from the correlation in Problem 9.9.

9.12 Estimate the heat of sublimation of zinc at 298.15 K:
   a. using the Einstein model;
   b. using the Debye model; and
   c. from the correlation in Problem 9.10.

---

[3] These data are from the Design Institute for Physical Properties Research (DIPPR) of the American Institute of Chemical Engineers.

Chapter 10

Simple Lattice Models for Fluids

A crystal, as discussed in the previous chapter, is an ordered solid with molecules in fixed positions on a lattice. A gas is a low-density fluid with very little order among the molecules. A liquid is intermediate between these two limiting cases, as the molecules are in constant motion and there is some short-range order, in that there are one or more coordination shells depending on density. As will be seen in later chapters, the coordination shell in a liquid is more diffuse or spread out than in a solid in which there are fixed lattice positions. Liquids do exhibit increasing local order, that is, additional coordination shells with increasing numbers of molecules, as density increases.

We consider theoretically rigorous methods of describing liquids in subsequent chapters. Here we consider the enormous simplification of representing a liquid as a completely ordered system using a lattice model. While this is not especially useful for quantitative calculations, lattice models do provide some useful qualitative understandings not only of liquids, but also of the behavior of polymers and proteins in solution, and more generally for the development of models used in applied classical thermodynamics. Also, before the advent of computers (for computer simulation and the solution of integral equations, both of which are discussed in later chapters), lattice theories were commonly used for representing liquid state behavior.

In this chapter we will introduce some of the concepts in the simplest of lattice theories and briefly illustrate their utility.[1] For discussions of more complicated lattice theories (including cell models, decorated lattice theories, off-lattice models, and others) the reader should look elsewhere.

INSTRUCTIONAL OBJECTIVES FOR CHAPTER 10

The goals for this chapter are for the student to:

- Obtain an understanding of how simple models for interacting molecules can be obtained from lattice theories
- To understand how equation-of-state models used in applied thermodynamics can be derived from lattice theories

---

[1] Some of the ideas for this chapter came from the excellent book *Molecular Driving Forces: Statistical Thermodynamics in Chemistry and Biology* by K. A. Dill and S. Bromberg, Garland Science, New York, 2003.

- To understand how activity coefficient models used in applied thermodynamics can be developed from lattice models.

10.1 INTRODUCTION

Up to this point in our introduction to statistical mechanics, we have been considering reasonably exact statistical mechanics for a model representative of a real system, though we have made some assumptions in the solutions of the equations. For example, the choice of simple-interaction potential models used to calculate virial coefficients, the assumptions of harmoniticity of vibrational motions, the rigid rotator model for rotations, and the Einstein and Debye models for crystals. In this section we will go in another direction, using a very simplified model of a fluid—a perfect lattice—which we know to be incorrect, to obtain some insight into the underlying ideas of some of the models used in classical thermodynamics.

The model is a lattice in which each lattice site has z nearest neighbors; the value of z, which we leave unspecified, depends on the type of lattice being one-dimensional (a linear polymer chain), two-dimensional (a surface), or three-dimensional (a bulk fluid); and if two- or three-dimensional, whether it is a simple cubic, hexagonal close-packed, or another lattice structure. In the general analysis that follows, the exact value of the coordination number will be unimportant. The volume of each lattice point will be designated as b. The number of lattice points is M, so that the volume of the lattice is $V = Mb$ with N of these lattice sites occupied by atoms of species 1; and the remaining $M - N$ lattice sites, depending on the application, are either empty (when we are considering a volumetric or P-V-T equation of state) or occupied by species 2 atoms (when we are interested in the properties of mixtures).

We start by considering a system at fixed number of atoms N, volume V, and energy E—that is, the microcanonical ensemble. As we have shown in Chapter 6, in the microcanonical ensemble, the entropy is

$$S(N, V, T) = k \ln \Omega \qquad (6.1\text{-}2)$$

where Ω is the degeneracy of the system—here, the number of different ways of distributing N atoms of species 1 and $M - N$ atoms of species 2 (or vacancies) on the M lattice sites. Initially, we will assume that there are no molecule-molecule (or molecule-vacancy) interactions. Also, the lattice sites of the model are distinguishable by their fixed locations in space, and the atoms (or vacancies) are indistinguishable from each other; and, furthermore, since the energy (and consequently, the Boltzmann factors) of all configurations are equal, the atoms are considered to be randomly distributed. Therefore, we have that

$$\Omega = \frac{M!}{N!(M-N)!} \qquad (10.1\text{-}1)$$

so that

$$\begin{aligned} S = k \ln \Omega &= k[\ln M! - \ln N! - \ln(M-N)!] \\ &= k[M \ln M - M - (N \ln N - N + (M-N) \ln(M-N) - (M-N)] \quad (10.1\text{-}2) \\ &= k[M \ln M - N \ln N - (M-N) \ln(M-N)] \end{aligned}$$

This is the starting point for several of the analyses that follow.

ILLUSTRATION 10.1-1

The figure below shows one configuration in which 4 identical balls can be distributed over 9 lattice sites. How many different configurations are possible for this system?

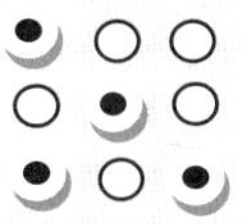

SOLUTION

Using Eq. 10.1.2, the number of possible different ways Ω of arranging the 4 identical balls on the 9 lattice sites is

$$\Omega = \frac{9!}{5!4!} = \frac{9 \cdot 8 \cdot 7 \cdot 6}{1 \cdot 2 \cdot 3 \cdot 4} = \frac{3024}{24} = 126$$

Note that while the balls are indistinguishable, the lattice sites are distinguishable since they are fixed in space. (If you are really bored, you can draw each of the configurations.)

10.2 DEVELOPMENT OF EQUATIONS OF STATE FROM LATTICE THEORY

From classical thermodynamics, the Helmholtz energy A is $A = U - TS$, and the pressure is

$$P = -\left(\frac{\partial A}{\partial V}\right)_{T,N} = -\left(\frac{\partial U}{\partial V}\right)_{T,N} + T\left(\frac{\partial S}{\partial V}\right)_{T,N} = P^{\text{eng}} + P^{\text{ent}} \quad (10.2\text{-}1)$$

where we have used P^{eng} and P^{ent} to represent the energetic and entropic contributions to the pressure, respectively. The simplest approximation is that the atoms and the vacancies (or holes) are randomly distributed on the lattice. Now, assuming that $V = Mb$, we have

$$P^{\text{ent}} = T\left(\frac{\partial S}{\partial V}\right)_{T,N} = \frac{T}{b}\left(\frac{\partial S}{\partial M}\right)_{T,N} = \frac{kT}{b}\left[\ln M + \frac{M}{M} - \ln(M-N) - \frac{M-N}{M-N}\right]$$

$$= \frac{kT}{b}\ln\frac{M}{(M-N)}$$

$$= \frac{kT}{b}\ln\frac{V}{(V-Nb)} = \frac{kT}{b}\ln\frac{1}{\left(1-\frac{Nb}{V}\right)} = -\frac{kT}{b}\ln\left(1-\frac{Nb}{V}\right) \quad (10.2\text{-}2)$$

Now using that $\ln(1+x) = x - \frac{1}{2}x^2 + \frac{1}{3}x^3 - \cdots$ we get

$$P^{\text{ent}} = -\frac{kT}{b}\ln\left(1-\frac{Nb}{V}\right) = +\frac{kT}{b}\left(\frac{Nb}{V} + \frac{1}{2}\left(\frac{Nb}{V}\right)^2 + \frac{1}{3}\left(\frac{Nb}{V}\right)^3 + \cdots\right) \quad \text{or}$$

$$\frac{P^{\text{ent}}V}{NkT} = 1 + \frac{1}{2}\frac{Nb}{V} + \frac{1}{3}\left(\frac{Nb}{V}\right)^2 + \cdots \quad (10.2\text{-}3)$$

166 Chapter 10: Simple Lattice Models for Fluids

This is in the form of the virial equation of state, but with temperature-independent virial coefficients—and with a very specific relationship between each of the virial coefficients.

Another interpretation is to consider only the first two terms on the right-hand side of this equation, and note that retaining only the first terms in a series expansion for the term on the right, we obtain

$$\frac{P^{ent}V}{NkT} = 1 + \frac{1}{2}\frac{Nb}{V} = \frac{1}{1 - \frac{1}{2}\frac{Nb}{V}} \quad \text{or} \quad P^{ent} = \frac{NkT}{V - N\frac{b}{2}} \quad (10.2\text{-}4)$$

which looks like the first term in the van der Waals equation of state.

To evaluate interaction energy among the atoms, we begin by noting that each molecule has z possible nearest neighbors and again use the assumption that the atoms and holes are randomly distributed. Since not all the nearest neighbor sites are occupied, each atom has zN/M occupied nearest neighbor sites and $z(M-N)/M$ unoccupied sites. Therefore, since there are a total of N atoms, using u to be the magnitude of each interaction (which is a negative number, since atoms attract at lattice separation distances) and dividing by the usual factor of 2 so as not to overcount the number of interactions, the total interaction energy is

$$U = \frac{N}{2}\frac{Nz}{M}u = \frac{N^2 z}{2M}u = \frac{N^2 zb}{2V}u \quad (10.2\text{-}5)$$

Then

$$P^{eng} = -\left(\frac{\partial U}{\partial V}\right)_{T,N} = \frac{N^2}{V^2}\frac{zbu}{2} = -\frac{aN^2}{V^2} \quad \text{where } a = -\frac{zbu}{2} \quad (10.2\text{-}6)$$

Therefore

$$P = P^{ent} + P^{eng} = \frac{NkT}{V - N\frac{b}{2}} - \frac{aN^2}{V^2} = \frac{RT}{\underline{V} - B} - \frac{a}{\underline{V}^2} \quad (10.2\text{-}7)$$

The last form of this equation has been changed to a molar basis by recognizing that $NkT = nN_{Av}kT = nRT$, where N_{Av} is Avogadro's number, $R = N_{Av}k$ is the gas constant, and $\underline{V} = V/n$. This result should be recognized as being the van der Waals equation. The derivation given here shows that the first term on the right-hand side of this equation of state is the entropic or free volume contribution to the pressure, and the second term is the energetic contribution to the pressure.

The van der Waals equation of state was an important contribution to thermodynamics, as it was the first equation to allow for a vapor-liquid phase transition. Indeed, his 1873 doctoral thesis was entitled *On the Continuity of the Gas and Liquid State* (English translation from the original Dutch). This equation was the first "cubic" equation of state in that, when written as

$$P(\underline{V} - B)\underline{V}^2 = RT\underline{V}^2 - a(\underline{V} - B) \quad \text{or} \quad P\underline{V}^3 - (PB + RT)\underline{V}^2 + a\underline{V} - B = 0 \quad (10.2\text{-}8)$$

it is cubic in volume.

Note that there is an inconsistency in the derivation of the van der Waals equation above. It arises as follows. For the energetic term, we have assumed that there is an interaction energy u between neighboring sites if they are occupied, but no interaction

10.2 Development of Equations of State from Lattice Theory

energy if one or both sites are unoccupied. This difference in interaction energies (and therefore Boltzmann factors) should result in the distribution of occupied and unoccupied sites being nonrandom. For example, if there is an attraction between occupied neighboring sites, we would expect more of these than based on a random distribution. Consequently, if there is an interaction energy difference, the expression for P^{ent} is incorrect.

This idea of nonrandomness also affects the enthalpic contribution to the equation of state. Following this idea that the likelihood of two adjacent sites being occupied is proportional to the Boltzmann factor in their energy of interaction, the probability that a lattice site next to an occupied site is filled would be $ce^{-u/kT}$, while the probability that it is vacant is c, where

$$c = \frac{1}{1 + e^{-u/kT}} \tag{10.2-9}$$

is the normalization constant—that is, the sum of the probabilities of the two states. In this case, the average interaction energy of the N atoms on the lattice is

$$U = \frac{N}{2}\frac{Nz}{M}u\frac{e^{-u/kT}}{1+e^{-u/kT}} = \frac{N^2 z}{2M}u\frac{e^{-u/kT}}{1+e^{-u/kT}} = \frac{N^2 zb}{2V}u\frac{e^{-u/kT}}{1+e^{-u/kT}} \tag{10.2-10}$$

resulting in

$$P^{\text{eng}} = -\left(\frac{\partial U}{\partial V}\right)_{T,N} = \frac{N^2}{V^2}\frac{zbu}{2}\frac{e^{-u/kT}}{1+e^{-u/kT}} = -\frac{a(T)N^2}{V^2}$$

where $\quad a(T) = -\frac{zbu}{2}\frac{e^{-u/kT}}{1+e^{-u/kT}} \tag{10.2-11}$

That the atoms in this model are no longer distributed randomly also means that the expression for the entropic contribution to the pressure needs to be modified. We will not proceed further with this, as the calculation becomes increasingly complicated and is based on a simple model where the only allowed locations of the atoms are on a lattice, which is not the case for a fluid (gas or liquid) in which molecules constantly move about.

For engineering use, the van der Waals equation of state is not very accurate, although modifications of this equation are. One of the earliest such modifications, as discussed in Section 8.4b (with a slightly different notation for the B parameter,) was the cubic Redlich-Kwong[2] equation

$$P = \frac{RT}{\underline{V} - B} - \frac{a}{\sqrt{T}\,\underline{V}(\underline{V} + B)} \tag{10.2-12}$$

and its modification by Soave[3] (Eq. 8.4-11) and the cubic Peng-Robinson[4] equation

$$P = \frac{RT}{\underline{V} - B} - \frac{a(T)}{\underline{V}(\underline{V} + B) + B(\underline{V} - B)} \tag{10.2-13}$$

---

[2] O. Redlich and J. N. S. Kwong, *Chem. Rev.* **44**, 233 (1949).
[3] G. Soave, *Chem. Eng. Sci.* **27**, 1197 (1972).
[4] D.-Y. Peng and D. B. Robinson, *IEC Fundam.* **15**, 59 (1976).

which is also cubic in volume. By appropriately choosing the volume parameter B and the temperature dependence of the $a(T)$ parameter, this cubic equation provides an adequate description of the vapor pressure and some other thermodynamic properties of nonpolar fluids (i.e., hydrocarbons, oxygen, nitrogen, etc.). A number of different temperature dependencies $a(T)$ have been proposed[5] to better correlate experimental data.

The three equations of state mentioned above (van der Waals, Redlich-Kwong, and Peng-Robinson) can all be expanded in virial equation form in powers of $B/\underline{V}$. The first terms in this expansion are

$$\frac{P\underline{V}}{RT} = 1 + \frac{B'(T)}{\underline{V}} + \frac{C'(T)}{\underline{V}^2} + \cdots$$

where

$$B' = B - \frac{a}{RT} \quad \text{van der Waals}$$

$$B' = B - \frac{a}{\sqrt{T}\,RT} \quad \text{Redlich-Kwong} \quad (10.2\text{-}14)$$

$$B' = B - \frac{a(T)}{RT} \quad \text{Peng-Robinson}$$

It is left as a problem (Problem 10.4) for the reader to obtain expressions for the third virial coefficient $C'(T)$ for these equations of state.

10.3 ACTIVITY COEFFICIENT MODELS FOR SIMILAR-SIZE MOLECULES FROM LATTICE THEORY

Before starting the discussion of activity coefficient models, it is useful to introduce the classical thermodynamic idea of an excess property of mixing. The formal definition of the excess property of mixing is that it is the difference between this property in the mixture and that which would be obtained upon forming an ideal mixture of the same components, each in the same state of aggregation of the mixture and at the same temperature and pressure as the mixture. Examples of several excess molar properties and the corresponding ideal mixture properties are given below.

$$\underline{V}^{\text{ex}}(x_1, x_2, \ldots, T, P) = \underline{V}_{\text{mix}}(x_1, x_2, \ldots, T, P) - \underline{V}_{\text{mix}}^{\text{IM}}(x_1, x_2, \ldots, T, P)$$
$$= \underline{V}_{\text{mix}}(x_1, x_2, \ldots, T, P) - \sum_i x_i \underline{V}_i(T, P)$$

$$\underline{U}^{\text{ex}}(x_1, x_2, \ldots, T, P) = \underline{U}_{\text{mix}}(x_1, x_2, \ldots, T, P) - \underline{U}_{\text{mix}}^{\text{IM}}(x_1, x_2, \ldots, T, P)$$
$$= \underline{U}_{\text{mix}}(x_1, x_2, \ldots, T, P) - \sum_i x_i \underline{U}_i(T, P)$$

$$\underline{S}^{\text{ex}}(x_1, x_2, \ldots, T, P) = \underline{S}_{\text{mix}}(x_1, x_2, \ldots, T, P) - \underline{S}_{\text{mix}}^{\text{IM}}(x_1, x_2, \ldots, T, P)$$
$$= \underline{S}_{\text{mix}}(x_1, x_2, \ldots, T, P) - \sum_i x_i [\underline{S}_i(T, P) - R \ln x_i]$$

---

[5] See, for example, Chapter 6 of S. I. Sandler, *Chemical, Biochemical and Engineering Thermodynamics*, 4th ed. John Wiley & Sons, Inc., Hoboken, 2006.

10.3 Activity Coefficient Models for Similar-Size Molecules from Lattice Theory

$$\begin{aligned}\underline{G}^{ex}(x_1, x_2, \ldots, T, P) &= \underline{G}_{mix}(x_1, x_2, \ldots, T, P) - \underline{G}_{mix}^{IM}(x_1, x_2, \ldots, T, P) \\ &= \underline{G}_{mix}(x_1, x_2, \ldots, T, P) - \sum_i x_i[\underline{G}_i(T, P) + RT \ln x_i]\end{aligned}$$

(10.3-1)

and

$$\begin{aligned}\underline{A}^{ex}(x_1, x_2, \ldots, T, P) &= \underline{A}_{mix}(x_1, x_2, \ldots, T, P) - \underline{A}_{mix}^{IM}(x_1, x_2, \ldots, T, P) \\ &= \underline{A}_{mix}(x_1, x_2, \ldots, T, P) - \sum_i x_i[\underline{A}_i(T, P) + RT \ln x_i]\end{aligned}$$

Also, the usual activity coefficient of species i in a mixture, γ_i, is related to the excess Gibbs energy of mixing and the Helmholtz energy of mixing as follows:

$$\left(\frac{\partial \underline{G}^{ex}}{\partial N_i}\right)_{T,P,N_{j \neq i}} = \left(\frac{\partial \underline{A}^{ex}}{\partial N_i}\right)_{T,V,N_{j \neq i}} = RT \ln \gamma_i(x_1, x_2, \ldots, T, P) \quad (10.3\text{-}2)$$

With this background, we now consider three lattices, each with the same spacing between lattice points and in which each lattice point has a coordination number of z and a volume of b. Each of the lattices has atoms at all its lattice points. The first lattice contains only M atoms of species 1, the second contains only $N - M$ atoms of species 2, and the third lattice of N lattice sites is formed by combining the first two lattices and mixing the species 1 and 2 atoms among the lattice sites. As the lattice spacings are the same in all three lattices, there is no volume change on forming a mixed atom lattice from the two pure atom lattices. That is,

$$\Delta_{mix} V = 0 = V^{ex} \quad (10.3\text{-}3)$$

Since there is only one way to distribute the M identical (and therefore indistinguishable) species 1 atoms on the first lattice, and $N - M$ identical species 2 atoms on the second lattice, the degeneracies of these lattices is 1, and this is represented as

$$\Omega_1(M; M) = 1 \text{ and } \Omega_2(N - M; N - M) = 1 \quad (10.3\text{-}4)$$

Here we have used the notation that $\Omega_1(M; M)$ is the number of different ways of putting M indistinguishable atoms of species 1 on M lattice sites, with a similar interpretation for $\Omega_2(N - M; N - M)$. Note that from $S(N, V, T) = k \ln \Omega$, the entropies of each of the pure lattices is zero.

We also use that $\Omega_{12}(M, N - M; N)$ is the number of ways of placing M atoms of species 1 and $N - M$ atoms of species 2 on the lattice of N sites. However, the number of different ways of distributing these atoms on the third lattice is given by Eq. 10.1-3, so that

$$\Omega_{12}(M, N - M; N) = \frac{N!}{M!(N - M)!} \quad (10.3\text{-}5)$$

Therefore the entropy of mixing, that is, forming the mixture lattice from the two pure component lattices is

170 Chapter 10: Simple Lattice Models for Fluids

$$\Delta_{mix} S = k \ln \frac{\Omega_{12}(M, N-M; N)}{\Omega_1(M; M)\Omega_2(N-M; N-M)}$$

$$= k[N \ln N - N - M \ln(M) + M - (N-M)\ln(N-M) + (N-M)]$$

$$= k\left[-M \ln \frac{M}{N} - (N-M)\ln \frac{(N-M)}{M}\right]$$

$$= -kN[x_1 \ln x_1 + x_2 \ln x_2] \tag{10.3-6a}$$

or, on a molar basis

$$\Delta_{mix}\underline{S} = -R[x_1 \ln x_1 + x_2 \ln x_2] \tag{10.3-6b}$$

where x_i is the mole fraction of species i. We recognize this to be the entropy of mixing an ideal solution from its pure components. Therefore, the excess entropy of mixing is 0:

$$\underline{S}^{ex} = 0 \tag{10.3-6c}$$

Next, consider the energy change on forming this mixture (that is, combining the lattices). In this calculation u_{11} is the average energy of a 1–1 interaction in the lattice, u_{22} is that of a 2–2 interaction, and $u_{12} = u_{21}$ is the average energy of a 1–2 interaction. The sum of the internal (or interaction) energies of the initial, unmixed lattices is

$$U_{unmix} = \frac{1}{2}Mzu_{11} + \frac{1}{2}(N-M)zu_{22} \quad \text{or} \quad \frac{U_{unmix}}{N} = \frac{1}{2}zx_1u_{11} + \frac{1}{2}zx_2u_{22} \tag{10.3-7}$$

To compute the energy of the mixed lattice, we note that the lattice contains M atoms of species 1 and $N - M$ atoms of species 2, and again make the assumption that all of the atoms are randomly distributed in the mixture. That is, each atom interacts with $Mz/N = x_1 z$ atoms of species 1 and $(N-M)z/N = x_2 z$ of species 2. The interaction energy of the mixed lattice is then

$$U_{mix} = M\left[\frac{1}{2}\frac{M}{N}zu_{11} + \frac{1}{2}\frac{(M-N)}{N}zu_{12}\right] + (N-M)\left[\frac{1}{2}\frac{M}{N}zu_{12} + \frac{1}{2}\frac{(N-M)}{N}zu_{22}\right] \tag{10.3-8a}$$

or

$$\frac{U_{mix}}{N} = x_1\left[\frac{1}{2}x_1 z u_{11} + \frac{1}{2}x_2 z u_{12}\right] + x_2\left[\frac{1}{2}x_1 z u_{12} + \frac{1}{2}x_2 z u_{22}\right]$$

$$= \frac{z}{2}[x_1^2 u_{11} + 2x_1 x_2 u_{12} + x_2^2 u_{22}] \tag{10.3-8b}$$

Therefore, the change in interaction energy on forming a mixture from the unmixed atoms is

$$\frac{U_{mix} - U_{unmix}}{N} = \frac{z}{2}[x_1^2 u_{11} + 2x_1 x_2 u_{12} + x_2^2 u_{22}] - \frac{1}{2}zx_1 u_{11} - \frac{1}{2}zx_2 u_{22}$$

$$= \frac{z}{2}[x_1(x_1-1)u_{11} + 2x_1 x_2 u_{12} + x_2(x_2-1)u_{22}] \tag{10.3-9a}$$

$$= \frac{z}{2}x_1 x_2 [2u_{12} - u_{11} - u_{22}] = \frac{z}{2}x_1 x_2 \psi = x_1 x_2 \chi$$

10.3 Activity Coefficient Models for Similar-Size Molecules from Lattice Theory 171

Consequently,

$$\frac{\Delta_{\text{mix}}U}{N} = \frac{U_{\text{mix}} - U_{\text{unmix}}}{N} = \frac{U^{\text{ex}}}{N} = x_1 x_2 \chi \qquad (10.3\text{-}9b)$$

In Eq. 10.3-9a $\psi = 2u_{12} - u_{11} - u_{22}$ is referred to as the exchange energy—that is, the interaction energy change as a result of breaking a 1–1 and a 2–2 interaction and forming two 1–2 interactions. Therefore, $\chi = z\psi/2$ is the energy change of simultaneously exchanging a species 1 atom and a species 2 atom in each of the pure-species lattices.

Summarizing, from this simple lattice model, we have that

$$V^{\text{ex}} = 0, \; S^{\text{ex}} = 0 \; \text{ and } \; U^{\text{ex}} = \chi N x_1 x_2 \; \text{ from which it follows that}$$
$$A^{\text{ex}} = G^{\text{ex}} = \chi N x_1 x_2 \qquad (10.3\text{-}10a)$$

This is the Regular Solution Model proposed by Scatchard,[6] based on the observations of J. H. Hildebrand.[7] This simple model is adequate for mixtures of molecules of approximately equal size and equal interaction energies. Note that this model also has the inconsistency of assuming a random placement of atoms on the lattice, but different energies of the different interactions that should lead to a nonrandom distribution of atoms on the lattice sites. Nonetheless, the model is a useful one in applied thermodynamics for slightly nonideal solutions of species that are of similar size and interaction energies.

In the above analysis, χ has units of energy on a per-molecule basis. Somewhat more useful is $\chi_m = \chi N_{\text{Av}}$, where N_{Av} is Avogadro's number and χ_m has units of energy per mole. Then

$$G^{\text{ex}} = \chi N x_1 x_2 = \chi N_{\text{Av}} \frac{N}{N_{\text{Av}}} x_1 x_2 = \chi_m n x_1 x_2 \quad \text{or} \quad \underline{G}^{\text{ex}} = \chi_m n x_1 x_2 \qquad (10.3\text{-}10b)$$

where $\underline{G}^{\text{ex}}$ is the excess Gibbs energy of mixing per mole of solution. Now using Eq. 10.3-2, we obtain[8]

$$RT \ln \gamma_1(x_1) = \chi_m (1 - x_1)^2 = \chi_m x_2^2 \quad \text{and} \quad \ln \gamma_2(x_2) = \chi_m (1 - x_2)^2 = \chi_m x_1^2$$
$$(10.3\text{-}11b)$$

There are a number of ways that this simple lattice model can be improved to yield activity coefficient models that are more widely applicable. One is to allow each of the lattices (pure species 1, pure species 2, and the mixed lattices) to have different lattice

---

[6] G. Scatchard, *Chem. Rev.* **8**, 321 (1931).

[7] J. H. Hildebrand, *J. Am. Chem. Soc.* **41**, 1067 (1919).

[8] It is important to note that the derivative is with respect to the number of molecules (or moles) of one species holding all other molecule numbers constant, and is not a derivative with respect to mole fractions. For a binary mixture, one can take the derivative with respect to x_1 by first replacing x_2 with $1 - x_1$, and the correct answer will be obtained. However, taking the derivative with respect to mole fraction in a multicomponent mixture will give an incorrect answer, since there are numerous ways of changing the mole fraction of species 1 by different adjustments of the mole fractions of the other species (i.e., changing the mole fraction of only other one species, changing the mole fractions of two other species together so the sum remains 1, etc.), and each will give a different result. Only by taking the derivative with respect to the number of one species holding all others constant is there no ambiguity, and the correct answer is obtained.

172 Chapter 10: Simple Lattice Models for Fluids

spacings; in this case V^{ex} and S^{ex} will both be nonzero. Another way is not to assume that the molecules mix randomly; for example, that the likelihood of the different interactions are proportional to the Boltzmann factors in their energies, that is

$$\frac{x_{12}}{x_{11}} = \frac{x_2}{x_1} \exp\left(-\frac{(u_{12} - u_{11})}{kT}\right) \quad \text{and} \quad \frac{x_{21}}{x_{22}} = \frac{x_1}{x_2} \exp\left(-\frac{(u_{21} - u_{22})}{kT}\right) \quad (10.3\text{-}10)$$

where x_{ij} is the local mole fraction (or local compositions) of species j around a central molecule of species i. This will result in a nonzero term for S^{ex} and a different expression than above for U^{ex}. The Wilson[9] local composition activity coefficient model has been developed based on such an assumption. This is discussed in Chapter 16.

10.4 THE FLORY-HUGGINS AND OTHER MODELS FOR POLYMER SYSTEMS

The simple lattice model just discussed has been extended to polymer (and protein) solutions by modeling these as a mixture of chain (multisite) molecules and solvent (single-site) molecules fully occupying the M sites of a lattice. In this model, there are considered to be p identical polymer molecules, each composed of c monomers in which each monomer occupies a single lattice site; and s solvent molecules, each occupying a single lattice site. Consequently, the total number of lattice sites in the mixed lattice is $N = p \times c + s$. Also, each lattice site has a coordination number z. Our interest in this section is in computing the entropy change on mixing when starting with one lattice of s sites fully occupied by solvent molecules and a second lattice of $p \times c$ sites fully occupied by polymer chains, and producing a lattice of N sites with both polymer chains and solvent molecules.

We will compute entropy of each of these systems using the microcanonical ensemble. In particular

$$S(s; s) = k \ln \Omega(s; s); \quad S(p; p \times c) = k \ln \Omega(p, p \times c) \quad \text{and} \quad (10.4\text{-}1)$$

$$S(p, s; p \times c + s) = k \ln \Omega(p, s; p \times c + s)$$

where, using the notation introduced earlier, $\Omega(p, s; p \times c + s)$ is the number of different ways that p identical chain molecules containing c monomer units and s identical single-site solvent molecules can be placed on the $p \times c + s$ lattice sites; the meanings of the other degeneracy terms (i.e., number of conformations) are as before.

The simplest degeneracy factor to start with is $\Omega(s; s)$, since there is only one way to put s indistinguishable solvent molecules on s lattice sites. Therefore, $\Omega(s; s) = 1$. The most difficult to compute is $\Omega(p, s; p \times c + s)$. We start by noting that there are M possible locations for placing the first unit of the first polymer. The second unit in the polymer chain, which is attached to the first, can be located at any of the z coordination sites of the first location. The third unit in the polymer chain can be located at any of the z coordination sites of the second unit except for the one site connected to the first unit—that is, there are only $z-1$ available sites. So a simple counting for the number of conformations (that is, ways of placing one polymer chain on the lattice) of a c-unit polymer chain on a N site lattice is $Nz(z-1)^{c-2}$.

---
[9]G. M. Wilson, J. Am. *Chem. Soc.*, **86**, 127 (1964).

10.4 The Flory-Huggins and Other Models for Polymer Systems

However, Flory[10] argued that this simple analysis, especially when continued to include many polymer chains on the lattice, overcounts the number of conformations possible since some of the coordination sites of a unit of the chain may be occupied either by a monomer unit of another polymer chain or by the present chain folding back upon itself. This is schematically illustrated in Fig. 10.4-1 for a simple two-dimensional square lattice, with the sites filled with polymer indicated in black and bonding between the monomers indicated by heavy black lines. There we see that monomer A has 3 vacant nearest neighbors (indicated by the thin connecting lines), while monomer B has only 2. Clearly, as the lattice becomes filled, each site has fewer vacant nearest-neighbor sites.

Flory continued by introducing the idea that an average way to account for this interference is to weight the likely number of available coordination sites with the fraction of the lattice that was still unoccupied. In this model the number of possible conformations for a 4-unit polymer would be

$$Nz\left(\frac{N-1}{N}\right)(z-1)\left(\frac{N-2}{N}\right)(z-1)\left(\frac{N-3}{N}\right)$$
$$=\left(\frac{z}{N}\right)\left(\frac{z-1}{N}\right)^2 N(N-1)(N-2)(N-3) \quad (10.4\text{-}2a)$$
$$=\left(\frac{z}{N}\right)\left(\frac{z-1}{N}\right)^2 \frac{N!}{(N-4)!}$$

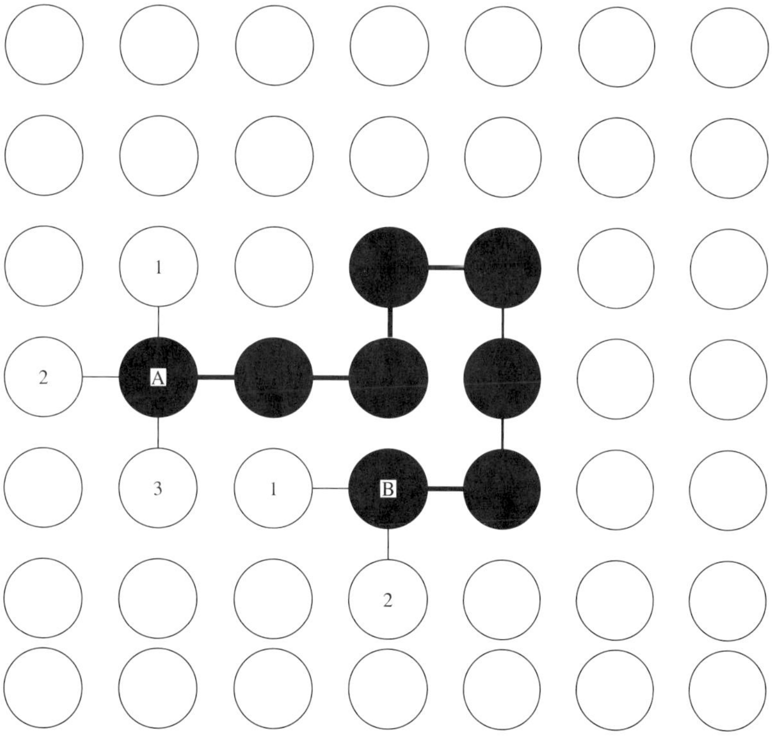

Figure 10.4-1 One arrangement of an 8-mer polymer on a square lattice.

---
[10]P. J. Flory, *J. Chem. Phys.* **9**, 660 (1941).

174 Chapter 10: Simple Lattice Models for Fluids

The number of conformations then available to the second 4-unit chain would then be

$$(N-4)z\left(\frac{N-5}{N}\right)(z-1)\left(\frac{N-6}{N}\right)(z-1)\left(\frac{N-7}{N}\right)$$

$$= \left(\frac{z}{N}\right)\left(\frac{z-1}{N}\right)^2 (N-4)(N-5)(N-6)(N-7) \quad \text{(10.4-2b)}$$

$$= \left(\frac{z}{N}\right)\left(\frac{z-1}{N}\right)^2 \frac{(N-4)!}{(N-8)!}$$

Therefore, the number of configurations available to both polymer chains is the product of the two:

$$\left(\frac{z}{N}\right)\left(\frac{z-1}{N}\right)^2 \frac{N!}{(N-4)!} \times \left(\frac{z}{N}\right)\left(\frac{z-1}{N}\right)^2 \frac{(N-4)!}{(N-8)!}$$

$$= \left(\frac{z}{N}\right)^2 \left(\frac{z-1}{N}\right)^{2\times 2} \frac{N!}{(N-8)!} \quad \text{(10.4-2c)}$$

By continuing and generalizing this analysis (Problem 10.9), it is easily shown that for p polymer chains (each composed of c monomer units) on a lattice of N sites, the number of possible conformations is

$$\left(\frac{z}{N}\right)^p \left(\frac{z-1}{N}\right)^{p(c-2)} \frac{N!}{(N-p\times c)!} \quad \text{(10.4-3)}$$

In each of these conformations, the s solvent molecules fill the remaining $p \times c + s$ lattice sites. Since the solvent molecules are identical, and therefore indistinguishable, their placement adds no additional distinguishable conformations, so that

$$\Omega(p, s; pc + s) = \left(\frac{z}{N}\right)^p \left(\frac{z-1}{N}\right)^{p(c-2)} \frac{N!}{(N-p\times c)!}$$

$$= \left(\frac{z}{p\times c + s}\right)^p \left(\frac{z-1}{p\times c + s}\right)^{p(c-2)} \frac{(p\times c + s)!}{s!} \quad \text{(10.4-4)}$$

The remaining degeneracy factor for the lattice of pure polymer is easily obtained from the expression above by setting $s = 0$. That is

$$\Omega(p; pc) = \left(\frac{z}{p\times c}\right)^p \left(\frac{z-1}{p\times c}\right)^{p(c-2)} (p\times c)! \quad \text{(10.4-5)}$$

Now using

$$\Delta_{\text{mix}} S = S(p, s; p \times c + s) - S(p, ; p \times c) - S(s; s)$$

$$= k \ln \Omega(p, s; p \times c + s) - k \ln \Omega(p; p \times c) - k \ln \Omega(s; s)$$

$$= k \ln \frac{\Omega(p, s; p \times c + s)}{\Omega(p; p \times c)\Omega(s; s)} \quad \text{(10.4-6)}$$

10.4 The Flory-Huggins and Other Models for Polymer Systems

we have

$$\Delta_{mix}S = k \ln \frac{\Omega(p, s; p \times c + s)}{\Omega(p; p \times c)\Omega(s; s)}$$

$$= k \ln \frac{\left(\dfrac{z}{p \times c + s}\right)^p \left(\dfrac{z-1}{p \times c + s}\right)^{p(c-2)} \dfrac{(p \times c + s)!}{s!}}{\left(\dfrac{z}{p \times c}\right)^p \left(\dfrac{z-1}{p \times c}\right)^{p(c-2)} (p \times c)!}$$

$$= k \ln \frac{\left(\dfrac{p \times c}{p \times c + s}\right)^{p(c-1)} (p \times c + s)!}{s!(p \times c)!}$$

$$= kp(c-1) \ln \left(\frac{p \times c}{p \times c + s}\right) + k[(p \times c + s)\ln(p \times c + s) - (p \times c + s)$$
$$- s \ln s + s - (p \times c)\ln(p \times c) + (p \times c)]$$

$$= kp(c-1) \ln \left(\frac{p \times c}{p \times c + s}\right) + k[(p \times c + s)\ln(p \times c + s) - s \ln s$$
$$- (p \times c)\ln(p \times c)]$$

$$= k[p \times c - p - p \times c]\ln(p \times c) + k[-p \times c - p + p \times c + s]$$
$$\ln(p \times c + s) - s \ln s$$

$$= k[(p + s)\ln(p \times c + s) - p\ln(p \times c) - s\ln s] \qquad (10.4\text{-}7)$$

Finally, on a per-molecule basis

$$\frac{\Delta_{mix}S}{k \times (p + s)} = -\frac{p}{p+s} \ln \frac{p \times c}{p \times c + s} - \frac{s}{p+s} \ln \frac{s}{p \times c + s} = -x_P \ln \phi_P - x_S \ln \phi_S \qquad (10.4\text{-}8a)$$

where $x_S = \frac{s}{p+s}$ and $x_P = \frac{p}{p+s}$ are the mole fractions of polymer and solvent, respectively, and $\phi_S = \frac{s}{p \times c + s}$ and $\phi_P = \frac{p \times c}{p \times c + s}$ are their volume fractions. On a molar basis, this equation is

$$\frac{\Delta_{mix}S}{R} = -x_P \ln \phi_P - x_S \ln \phi_S \qquad (10.4\text{-}8b)$$

Note that if the polymer is the same size as the solvent (so that $c = 1$), then each volume fraction becomes a mole fraction, and

$$\frac{\Delta_{mix}S(c=1)}{R} = -x_P \ln x_P - x_S \ln x_S \qquad (10.4\text{-}9)$$

which is the entropy change on forming an ideal mixture. Therefore, the excess entropy of mixing for forming a polymer solution from pure polymer and solvent is

$$\frac{\underline{S}^{ex}}{R} = \frac{\Delta_{mix}\underline{S}}{R} - \frac{\Delta_{mix}\underline{S}(c=1)}{R}$$

$$= -x_P \ln \phi_P - x_S \ln \phi_S + x_P \ln x_P + x_S \ln x_S = -x_P \ln \frac{\phi_P}{x_P} - x_S \ln \frac{\phi_S}{x_S} \quad \textbf{(10.4-10)}$$

The magnitudes of the entropy of mixing for two mixtures are shown in the figure below. One curve is for the case when the polymer has only a single monomer unit of the same size as the solvent (i.e., $c = 1$), and the second case is for $c = 1000$.

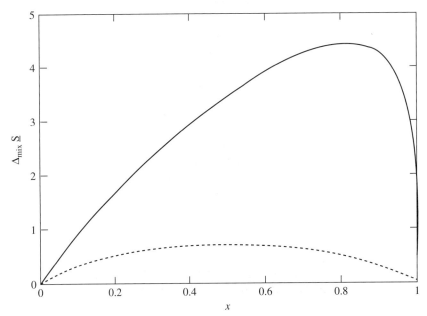

Entropy of mixing for a polymer and solvent of equal size (dashed line), and for a polymer that has 1000 monometer units, each of the size of a solvent molecule (solid line)

The Flory-Huggins model[11,12] is frequently used in polymer solution thermodynamics. It is obtained by adding to the excess entropy of mixing calculated above the following expression for the excess enthalpy of mixing

$$\underline{H}^{ex} = \chi RT(x_S + cx_P)\phi_S \phi_P \quad \textbf{(10.4-11)}$$

where χ is the so-called Flory parameter. This term can be thought of as an exchange energy—that is, to account for the energy differences between the solvent-solvent, monomer-monomer, and solvent-monomer interactions. With the inclusion of this term, we have

$$\frac{\underline{G}^{ex}}{RT} = \frac{\underline{H}^{ex} - T\underline{S}^{ex}}{RT} = \chi(x_S + cx_P)\phi_S \phi_P + x_P \ln \frac{\phi_P}{x_P} + x_S \ln \frac{\phi_S}{x_S} \quad \textbf{(10.4-12)}$$

---

[11] See reference 10.
[12] M. L. Huggins, *J. Chem. Phys.* **9**, 440 (1941).

10.4 The Flory-Huggins and Other Models for Polymer Systems

We then leave it to the reader (Problem 10.8) to show that for this model, the activity coefficients for the solvent and the polymer are

$$\ln \gamma_S = \ln \frac{\phi_S}{x_S} + \left(1 - \frac{1}{c}\right)\phi_P + \chi \phi_P^2 \text{ and}$$

$$\ln \gamma_P = \ln \frac{\phi_P}{x_P} - (c-1)\phi_S + c\chi \phi_S^2 \quad (10.4\text{-}13)$$

The Flory-Huggins model is widely used and a successful model for polymer solutions. However, from the derivation here, we see it also is not internally consistent. In particular, in computing the entropy of mixing, we assumed that, except for overlap and chain connectivity constraints, the polymer and solvent molecules could be located anywhere on the lattice with equal probability—that is, equal energies and therefore equal Boltzmann factors. However, by adding the excess enthalpy-of-mixing term, we recognize that different configurations can have different energies, and therefore are not equally likely. In spite of this internal inconsistency, the Flory-Huggins model has been very useful for polymer solutions.

There is another shortcoming of the Flory-Huggins model that is exposed in the derivation above. In particular, the polymer molecule is considered to be a linear chain. However, real polymers can have branches or rings, either of which makes the counting of lattice site occupancy and nearest neighbors more difficult. An additional inherent problem in the counting is that if a lattice site is initially vacant and available for the addition of the first segment of the polymer chain, there is a somewhat greater likelihood that this site is within a region of vacant lattice sites; so the probability of empty adjacent sites is somewhat higher than would be expected with the Flory model based on the overall fraction of vacant sites.

Various corrections to the Flory model have been proposed, and their derivations involved greater complexity than is justified here. Instead, we mention only one widely used equation: the Guggenheim-Staverman[13] model, generalized to a multicomponent mixture

$$\frac{S^{ex}}{R} = -\sum_{i=1}^{C} x_i \ln \frac{\phi_i}{x_i} - \frac{z}{2} \sum_{i=1}^{C} x_i q_i \ln \frac{\theta_i}{\phi_i} \quad (10.4\text{-}14)$$

In this equation, C is the number of components, z is the coordination number (usually set to 10), ϕ_i is the volume fraction

$$\phi_i = \frac{x_i r_i}{\sum_{j=1}^{C} x_j r_j} \quad (10.4\text{-}15a)$$

in which r_i is some measure of the volume of molecule i, and

$$\theta_i = \frac{x_i q_i}{\sum_{j=1}^{C} x_j q_j} \quad (10.4\text{-}15b)$$

---

[13]This model was developed independently by E. A. Guggenheim (see *Mixtures*, Clarendon Press, Oxford, 1952) and A. J. Staverman, *Rev. Trav. Chim. Pays-Bas*, **69**, 163 (1950).

178 Chapter 10: Simple Lattice Models for Fluids

is the surface area fraction of species i, in which q_i is some measure of the surface area of the molecule.

Note that Eqs. 10.4-14 and 10.4-15 are frequently used for all mixtures, not only polymer-solvent systems. In such cases, the surface area and volume parameters are those for the whole molecules. One common method, which is used in the correlative UNIQUAC[14] and predictive UNIFAC[15] activity coefficient models, is to assign a relative surface area and volume to each type of functional group and then obtain the r and q parameters for a molecule by summing the contributions of its functional groups. In this analysis, subgroupings within a molecule with specific chemical properties, such as $-CH_3$, $-COOH$, $-CNH_2$, etc., are designated functional groups.

10.5 THE ISING MODEL

The Ising[16] model is a very simple one-dimensional lattice model in which only neighboring sites interact. This is a model for magnetization of, for example, a ferromagnet. In this model of magnetism, there are two possible spin states for each lattice point. If two adjacent lattice points are in the same spin state, which we indicate by ↑↑ (though we could equally well indicate that by ↓↓ since the choice of up and down is arbitrary; all that matters is whether the neighboring states are parallel or antiparallel), the interaction energy is χ, where χ is the energy or coupling parameter that may be positive or negative in value. If adjacent spin states are opposite or antiparallel, indicated by ↑↓ or ↓↑, the energy is $-\chi$. Thus the energy change in going from, for example, the ↑↑ state to the ↑↓ is -2χ. The energy of this system of $N+1$ lattice points in any one configuration is

$$E = \chi \sum_{\substack{i>j \\ \text{adjacent to } i}}^{N+1} \sum_{j=1}^{N+1} \zeta_i \zeta_j = \chi \sum_{i=1}^{N} \zeta_i \zeta_{i+1} \qquad (10.5\text{-}1)$$

where ζ has values of either $+1$ (↑) or -1 (↓), and the canonical partition function for this system is

$$Q(N, V, T) = \sum_{\text{configurations}} e^{-\frac{\chi}{kT} \sum_{i=1}^{N} \zeta_i \zeta_{i+1}} \qquad (10.5\text{-}2)$$

where the outer sum is over all configurations, that is, combinations of the spin states of the chain of lattice points.

This sum, and therefore the partition function, is easily evaluated in this case. For example, for a lattice consisting of only two sites, the energy states of the system are χ and $-\chi$, so that the partition function is

---

[14]The UNIQUAC model was introduced by D. S. Abrams and J. M. Prausnitz, *AIChE J.*, **21**, 116 (1975).

[15]The UNIFAC model, which is a functional group contribution version of the UNIQUAC model, was developed by A. Fredenslund, R. L. Jones, and J. M. Prausnitz, *AIChE J.*, **21**, 1086 (1975), and has been refined many times since. A brief description of these models, including a recent version of UNIFAC, and programs for its use is available in S. I. Sandler, *Chemical, Biochemical and Engineering Thermodynamics*, 4th ed., John Wiley & Sons, Inc., Hoboken, 2006, Sections 9.5 and 9.6.

[16]E. Ising, *Z. Physik* **31**, 253 (1925).

10.5 The Ising Model

$$Q = e^{-\chi/kT} + e^{+\chi/kT} = 2\left[\frac{1}{2}(e^{-\chi/kT} + e^{+\chi/kT})\right] = 2\cosh\left(\frac{\chi}{kT}\right) \quad (10.5\text{-}3)$$

Next, consider a chain of three lattice sites. The possible energy states are as shown below.[17] Energy:

↑↑↑ (or equivalently ↓↓↓)	$+2\chi$
↑↑↓ (or equivalently ↓↓↑)	0χ
↑↓↑ (or equivalently ↓↑↓)	-2χ
↑↓↓ (or equivalently ↓↑↑)	0χ

Note that in this counting, the lattice sites are distinguishable by their location—that is ↑↑↓ is different from ↓↑↑; however, our choice of representing as state as ↑ or ↓ was arbitrary, as the energy only depends on whether adjacent states are parallel or antiparallel.

With this enumeration of the energy states, the partition function for this system is

$$Q = [e^{2\chi/kT} + 2e^0 + e^{-2\chi/kT}] = (e^{\chi/kT} + e^{-\chi/kT})^2$$

$$= \left[2\left(\frac{e^{\chi/kT}}{2} + \frac{e^{-\chi/kT}}{2}\right)\right]^2 = \left[2\cosh\left(\frac{\chi}{kT}\right)\right]^2 \quad (10.5\text{-}4)$$

and more generally, for a one-dimensional, $N+1$ lattice point system the partition function is

$$Q = \left[2\cosh\left(\frac{\chi}{kT}\right)\right]^N \quad (10.5\text{-}5)$$

We can now ask questions such as what is the likelihood of finding a five lattice point system in the state ↑↑↑↓↑. As the energy of this state is 0 (two parallel states and two anti-parallel states), so the probability of occurrence of this particular state is

$$p(\uparrow\uparrow\uparrow\downarrow\uparrow, \chi/kT) = \frac{1}{[2\cosh(\chi/kT)]^4} \quad (10.5\text{-}6)$$

We can also ask what the probability is of finding the system in any state with 4 ↑ and ↓—that is, in one of the following states:

↑↑↑↑↓, ↓↑↑↑↑, ↑↓↑↑↑, ↑↑↓↑↑ or ↑↑↑↓↑

Note that because of edge effects (the first two configurations above have only one sign reversal, and the remaining three have two sign reversals), the energies of these states are 2χ, 2χ, 0, 0, and 0, respectively. So the probability of finding the system in any state with 4 ↑ and ↓ is

$$p(4\uparrow \text{ and } \downarrow, \chi/kT) = \frac{2e^{-2\chi/kT} + 3}{[2\cosh(\chi/kT)]^4} \quad (10.5\text{-}7)$$

---

[17] There is a simple recipe for computing each of the weight factors in the table above and for larger systems. For a system of $N+1$ lattice points, the energy is $\chi[N - 2 \times$ (number of pairs of oppositely – oriented adjacent arrows)].

Chapter 10: Simple Lattice Models for Fluids

The average energy of this system is

$$\overline{E}(4\uparrow,\downarrow,\chi/kT) = \frac{4\chi e^{-2\chi/kT}}{[2\cosh(\chi/kT)]^4} \quad \text{or} \quad \frac{\overline{E}(4\uparrow,\downarrow,\chi/kT)}{\chi} = \frac{4e^{-2\chi/kT}}{[2\cosh(\chi/kT)]^4}$$

(10.5-8)

Now, consider a very large system ($N \to \infty$) so that edge effects in the lattice are not important and can be ignored. Suppose in this case we have a chain of $N+1$ lattice sites, of which m are $\uparrow$ and the remaining $N+1-m$ are $\downarrow$. The number of different arrangements of the two types of sites is the same counting problem as distributing m sites among $N+1$ lattice sites, for which the answer is

$$\frac{(N+1)!}{m!(N+1-m)!}$$

(10.5-9)

However, evaluating the energies for each of these states is quite complicated, since it depends on the particular conformation of up and down arrows. For example, if these were of equal number, the conformation $\uparrow\uparrow\uparrow\ldots\uparrow\downarrow\downarrow\ldots\downarrow$ with only a single sign reversal has an energy of $(N-2)\chi$ or $N\chi$ as $N\to\infty$, while the configuration $\uparrow\downarrow\uparrow\downarrow\ldots\uparrow\downarrow\uparrow\downarrow$ has N sign reversals and an energy of $-N\chi$. Configurations different from two above will have intermediate energies.

Even though the enumeration of states is extremely difficult, we can still evaluate the thermodynamic properties of the one-dimensional Ising model, since we know its partition function from Eq. 10.5-5. From the canonical ensemble, we find that for a system of $N+1$ sites

$$A(N+1, T, \chi) = -kT \ln Q = -kT \ln \left[2\cosh\left(\frac{\chi}{kT}\right) \right]^N$$

$$= -NkT \ln \left[2\cosh\left(\frac{\chi}{kT}\right) \right]$$

$$= -NkT \ln \left[e^{\frac{\chi}{kT}} + e^{-\frac{\chi}{kT}} \right]$$

(10.5-10)

Note that

$$A(N+1, T\to\infty, \chi) = -NkT \ln[e^0 + e^{-0}] = -NkT \ln 2$$

and

(10.5-11)

$$A(N+1, T\to 0, \chi) = -NkT \ln[e^{\chi/kT} + 0] = -N\chi$$

Also from

$$U(N+1, T, \chi) = kT^2 \left(\frac{\partial \ln Q}{\partial T}\right)_{N,V} = -N\chi \left(\frac{e^{\chi/kT} - e^{-\chi/kT}}{e^{\chi/kT} + e^{-\chi/kT}}\right)$$

so that

$$U(N+1, T\to\infty, \chi) = -\lim_{T\to\infty} N\chi \left(\frac{e^{\chi/kT} - e^{-\chi/kT}}{e^{\chi/kT} + e^{-\chi/kT}}\right) = 0 \quad (10.5\text{-}12)$$

$$U(N+1, T\to 0, \chi) = -\lim_{T\to 0} N\chi \left(\frac{e^{\chi/kT} - e^{-\chi/kT}}{e^{\chi/kT} + e^{-\chi/kT}}\right) = -N\chi$$

10.5 The Ising Model

Further

$$C_V = \left(\frac{\partial U}{\partial T}\right)_{N,V} = 4Nk\left(\frac{\chi}{kT}\right)^2 \frac{1}{(e^{\chi/kT} + e^{-\chi/kT})^2} \quad \text{(10.5-13)}$$

These results can be analyzed in terms of cooperative and anticooperative behavior. In particular, suppose that the parameter χ is positive, so that the $\uparrow\uparrow$ is a higher energy state than the $\uparrow\downarrow$ state. Therefore, the lowest energy state is that of alternating spins, and this is the most likely state of the system at low temperatures. We consider this to be anticooperativity in that a lattice point being in one state makes it more likely that its neighbors will be in the opposite state. However, if χ is negative in value (that is, attractive), $\uparrow\uparrow$ is a lower energy state than the $\uparrow\downarrow$ state, therefore, the lowest energy state is that of parallel spins, and is cooperative behavior in that a lattice point being in one state makes it more likely that its neighbors will be in the same state at low temperatures. Note that with either cooperative or anticooperative behavior, the energy of the system at low temperatures is $-N\chi$, even though in one case the state is aligned parallel spins and in the other it is alternating antiparallel spins.

The analysis above is for the simple one-dimensional Ising model. Conceptually, the model is easily extended to two and three dimensions, though these extensions are mathematically much more difficult. In fact, Lars Onsager received the Nobel Prize in Chemistry in 1968 for his solution of the two-dimensional Ising model;[18] there is no known solution for the three-dimensional Ising model. (As an aside, it is interesting to note that Onsager was the Gibbs Professor of Theoretical Chemistry at Yale University.)

Though the one-dimensional Ising model is a great simplification of real systems, it has been used to obtain insights into some real phenomena. The underlying assumption of such models is that only nearest-neighbor lattice interactions are involved. For example, in this way the Ising model is related the helix-coil conformations in polymers. A simple model for this is that a monomer in a chain can be either in the helix (H) conformation or the coil (C) conformation, and that the interaction energy of each monomer is only the result of interactions with adjacent monomers. We provide a simple generalization of the model here using the notation that ε_{HH} is the H-H interaction energy, ε_{CC} is the C-C interaction energy, and $\varepsilon_{CH} = \varepsilon_{HC}$ is the H-C or C-H interaction energy.

Consider the two-monomer chain where we do not know in advance whether the first element in the chain is in the helix or coil conformation. The possible conformations are HH, HC, CH, and CC. Therefore, the partition function is

$$\begin{aligned}Q(2, V, T) &= e^{-\varepsilon_{HH}/kT} + e^{-\varepsilon_{HC}/kT} + e^{-\varepsilon_{CH}/kT} + e^{-\varepsilon_{CC}/kT}\\ &= e^{-\varepsilon_{HH}/kT} + 2e^{-\varepsilon_{HC}/kT} + e^{-\varepsilon_{CC}/kT}\\ &= (e^{-\varepsilon_{HH}/kT} + e^{-\varepsilon_{HC}/kT}) + (e^{-\varepsilon_{HC}/kT} + e^{-\varepsilon_{CC}/kT})\end{aligned} \quad \text{(10.5-14)}$$

and the probabilities p of finding a two monomer chain in each of the HH, CH, and CC conformations are

$$p(HH) = \frac{e^{-\varepsilon_{HH}/kT}}{e^{-\varepsilon_{HH}/kT} + 2e^{-\varepsilon_{HC}/kT} + e^{-\varepsilon_{CC}/kT}};$$

---

[18] L. Onsager, *Phys. Rev.* **65**, 117 (1944).

182 Chapter 10: Simple Lattice Models for Fluids

$$p(CC) = \frac{e^{-\varepsilon_{CC}/kT}}{e^{-\varepsilon_{HH}/kT} + 2e^{-\varepsilon_{HC}/kT} + e^{-\varepsilon_{CC}/kT}};$$

$$\text{and } p(CH) = \frac{2e^{-\varepsilon_{CH}/kT}}{e^{-\varepsilon_{HH}/kT} + 2e^{-\varepsilon_{HC}/kT} + e^{-\varepsilon_{CC}/kT}} \quad (10.5\text{-}15)$$

where the factor of 2 arises because the CH and HC conformations are identical, differing only in the conformation of the starting monomer.

By the same reasoning, the conformations for a three-monomer chain, the possible conformations are HHH, HHC (and CHH), HCH, HCC (and CCH), CHC, and CCC, and the partition function can be shown to be

$$Q(3, V, T) = (e^{-\varepsilon_{HH}/kT} + e^{-\varepsilon_{HC}/kT})^2 + (e^{-\varepsilon_{HC}/kT} + e^{-\varepsilon_{CC}/kT})^2 \quad (10.5\text{-}16)$$

From this, expressions can be obtained for the likelihood of finding a three-monomer chain in each of the conformations listed above. A similar analysis can be done, for longer polymer chains (Problem 10.6).

The Zimm-Bragg model for proteins is somewhat more complicated in that it contains more detail. In that model, it is assumed that it is easier to start a polymerization chain from a coiled amino acid sequence than from a helix sequence, and that it is easier to add a coiled sequence to a coiled sequence than it is to add a helix sequence. Only a very simplified version of the model will be considered here, in which $\varepsilon_{CC} = \varepsilon_{HC} = \varepsilon_{HH} \neq \varepsilon_{CH}$—that is, there is no energy difference in adding a coil to a coil, a coil to a helix, or a helix to a helix; however, there is a different (and greater) energy required to add a helix to a coil. Also, in this model it is assumed that it is more difficult to start a chain from a helix than from a coil, and the energy difference is $\varepsilon_{CH} - \varepsilon_{CC}$. (Note that this is also required by symmetry, so that the CCH and HCC trimers are equally likely; and the same is true for the CHH and HHC trimers.)

The partition function for this system is

$$Q(3, V, T) = q_{CCC} + q_{CCH} + q_{CHC} + q_{CHH} + q_{HCC} + q_{HCH} + q_{HHC} + q_{HHH}$$

$$= e^{-2\varepsilon_{CC}/kT}[1 + e^{-(\varepsilon_{CH}-\varepsilon_{CC})/kT} + e^{-(\varepsilon_{CH}-\varepsilon_{CC})/kT} + e^{-(\varepsilon_{CH}-\varepsilon_{CC})/kT}$$

$$+ e^{-(\varepsilon_{CH}-\varepsilon_{CC})/kT}(1 + e^{-(\varepsilon_{CH}-\varepsilon_{CC})/kT} + 1 + 1)]$$

$$= e^{-2\varepsilon_{CC}/kT}[1 + 3s + 3s + s^2] = e^{-2\varepsilon_{CC}/kT}[1 + 6s + s^2] \quad (10.5\text{-}17)$$

where $s = e^{-(\varepsilon_{CH}-\varepsilon_{CC})/kT}$ is the Boltzmann factor of the energy penalty on forming a CH bond rather than CC bond, and also for initiating the chain from a helix rather than a coil. The probability of occurrence of the various conformations are then

$$p(CCC) = \frac{e^{-2\varepsilon_{CC}/kT}}{e^{-2\varepsilon_{CC}/kT}[1 + 3s + 3s + s^2]} = \frac{1}{[1 + 6s + s^2]}$$

$$p(CCH) = p(CHC) = p(CHH) = p(HCC) = p(HHC) = p(HHH)$$

$$= \frac{s}{[1 + 6s + s^2]}$$

$$\text{and } p(HCH) = \frac{s^2}{[1 + 6s + s^2]} \quad (10.5\text{-}18)$$

10.5 The Ising Model

These are shown in Fig. 10.5-1 as a function of the parameter s. Note that for amino acids, s is thought to be of the order of 10^{-3} to 10^{-4}, so that only coiled trimers would be expected with such a large energy penalty.

Another property that we can calculate is the expected helicity of the trimer—that is, the average number of helices in the chain from

$$\langle H \rangle = 0 \times p(\text{CCC}) + 1 \times [p(\text{CCH}) + p(\text{CHC}) + p(\text{HCC})] + 2 \times [p(\text{CHH}) + p(\text{HCH}) + p(\text{HHC})] + 3 \times p(\text{HHH})$$

$$= \frac{1 \times (s + s + s) + 2 \times (s + s^2 + s) + 3 \times s}{1 + 6s + s^2}$$

$$= \frac{10s + 3s^2}{1 + 6s + s^2} \tag{10.5-19}$$

The degree of helicity is shown in Fig. 10.5-2.

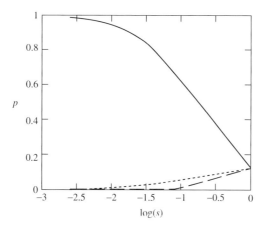

Figure 10.5-1 Probability of occurrences of various trimers as a function of the energy penalty s: CCC (solid line), CCH, CHC, CHH, HCC, HHC or HHH (dotted line) and HCH (dashed line).

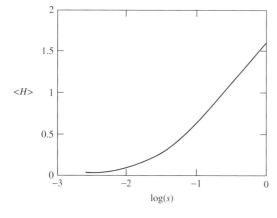

Figure 10.5-2 Average helicity of amino acid trimers as a function of the energy penalty function s.

Chapter 10: Simple Lattice Models for Fluids

CHAPTER 10 PROBLEMS

10.1 Draw the excess free energy versus composition diagram for the regular solution model (simple lattice model) for various values of χ/kT.

10.2 For the regular solution model, find the composition of the coexisting phases and draw the phase boundary as a function of χ/kT. Develop the equations to be solved for the compositions of the coexisting phases. Show that if χ has a very large value, the two phases are relatively insoluble in each other with the compositions of the coexisting phases being

$$x_1^1 = x_2^2 = e^{-\chi}$$

10.3 The spinodal curve for liquid-liquid equilibrium in a mixture described is the locus of points for which

$$\left(\frac{\partial \mu_i}{\partial x_i}\right)_{T,P} = 0$$

as compositions for which $\left(\frac{\partial \mu_i}{\partial x_i}\right)_{T,P} < 0$ are not physically possible, as the chemical potential of a substance must increase as its concentration increases (just as for a pure gas the situation in which $\left(\frac{\partial P}{\partial V}\right)_T > 0$ is unphysical). Develop an expression for the spinodal composition for the regular solution (simple lattice) model.

10.4 Obtain expressions for the third virial coefficients for the (a) van der Waals, (b) Redlich-Kwong, and (c) Peng-Robinson equations of state.

10.5 Develop expressions for the probabilities of occurrence of the HHH, HHC (and CHH), HCH, HCC (and CCH), CHC, and CCC conformations of the three-monomer chain 1-D lattice.

10.6 Develop expressions for the partition function and the probability of occurrence of each of the possible conformations of the four-monomer chain 1-D lattice.

10.7 Develop a simplified Zimm-Bragg model for a four-monomer chain, and compute probabilities of the different possible chains and the average helicity as a function of the energy penalty function s.

10.8 Derive Eq. 10.4-13.

10.9 Show that for p polymer chains (each of c monomer units on a lattice of N sites), the number of possible conformations is

$$\left(\frac{z}{N}\right)^p \left(\frac{z-1}{N}\right)^{p(c-2)} \frac{N!}{(N - p \times c)!}$$

10.10 Obtain the average energy for a five-site one-dimensional Ising model by enumerating all the states, and compare the result with that obtained from Eq. 10.5-12.

10.11 Write an expression for the chemical potentials of the solvent and polymer species in a mixture described by the Flory-Huggins model.

10.12 Write the equations that would be used to predict liquid-liquid equilibrium using the Flory-Huggins model.

10.13 The spinodal curve for liquid-liquid equilibrium in a mixture described by the Flory-Huggins model is the locus of points for which

$$\left(\frac{\partial \mu_i}{\partial \phi_i}\right)_{T,P} = 0$$

as compositions for which $\left(\frac{\partial \mu_i}{\partial \phi_i}\right)_{T,P} < 0$ are not physically possible, since the chemical potential of a substance must increase as its concentration (or volume fraction) increases. Develop an equation for the spinodal curve for the Flory-Huggins model.

10.14 An alternative to lattice models are cell models. In the simplest cell model, the total volume of the system V is divided into N equal cells of volume $\Delta = V/N$, and each cell is occupied by a single molecule. Each atom is restricted to remain in a single cell, all cells are occupied, and the energy of an atom in a cell is $-\varepsilon$. What is:
the canonical partition function for this system?
the Helmholtz energy of the system?
the energy of the system?
the heat capacity of the system?
the chemical potential of an atom in this system?
the vapor pressure of the system at a low enough pressure that the vapor can be considered to be an ideal gas?

Chapter 11

Interacting Molecules in a Dense Fluid. Configurational Distribution Functions

In this chapter, we are interested in the statistical mechanical description of a dense fluid, such as a liquid. In many ways, this is the most difficult class of systems to treat. A liquid is a dense fluid, and at high densities, the virial equation is very slow to converge, so that many high-order virial coefficients would be needed. However, it is not possible to evaluate analytically the very complex multidimensional integrals involved, so the virial equation of state cannot be used to describe liquids. In the past, statistical mechanical lattice and cell models have been used to describe liquids, and such models treat a liquid as a crystalline structure using sophisticated refinements of the simple models presented in the previous chapter. Such models are not very accurate, and miss some of the essential features of liquid behavior—such as that the molecules do not remain at fixed positions. However, there is a completely different, more theoretically based method used to describe liquids, the discussion of which begins in this chapter and continues through to Chapter 14.

INSTRUCTIONAL OBJECTIVES FOR CHAPTER 11

The goals for this chapter are for the student to:

- Understand the concept of a radial distribution function
- Understand the relationship between the radial distribution function and thermodynamic properties
- Have an introduction to the methods by which the radial distribution function is obtained

11.1 REDUCED SPATIAL PROBABILITY DENSITY FUNCTIONS

As background, it is useful to start by considering at the configuration integral for a system of N identical atoms

$$Z(N, V, T) = \int \ldots \int e^{-u/kT} d\underline{r}_1 d\underline{r}_2 \ldots d\underline{r}_N \qquad (11.1\text{-}1)$$

186 Chapter 11: Interacting Molecules in a Dense Fluid. Configurational Distribution Functions

Note that here, as in the evaluation of the virial coefficients, we can choose any convenient origin of the coordinate system—for example, the location of molecule 1 (or any other single molecule)—so that the configuration integral can be written as

$$Z(N, V, T) = \int \left[\int \cdots \int e^{-u/kT} d\underline{r}_{12} d\underline{r}_{13} \cdots d\underline{r}_{1N} \right] d\underline{r}_1$$

$$= V \int \cdots \int e^{-u/kT} d\underline{r}_{12} d\underline{r}_{13} \cdots d\underline{r}_{1N} \qquad (11.1\text{-}2)$$

where, for example, $\underline{r}_{12}$ is the vector between the origin of the coordinate system (here the location of molecule 1) and molecule 2, etc. However, since we can choose the origin of the coordinate system only once, further simplification of the integral by choice of the coordinate system is not possible. So instead of trying to directly evaluate the integral in Eq. 11.1-2, we will proceed differently.

Consider for the moment a collection of N identical (but distinguishable) atoms or molecules in a volume V at temperature T. We are now interested in obtaining an expression for the probability of finding molecules in specific locations near each other. However, since position is a continuous variable, there are an infinite number of positions, so the probability of finding a molecule at any point is essentially zero. Instead, as is usually the case with the statistics of continuous variables, we will use a probability distribution or probability density function, and consider the probability that a molecule is located in a finite, but differential, volume element $d\underline{r}$ about a specific location $\underline{r}$. In fact, we will initially consider a volume element $d\underline{r}$ that is so small that it can contain at most a single molecule.

We start by considering the probability that molecule 1 is in a small volume element $d\underline{r}_1$ around the location (or position vector) $\underline{r}_1$. This probability is just $d\underline{r}_1/V$ since, given no other information, the molecule is equally likely to be any place in the volume; so the probability that the molecule is in a specific small volume element is just equal to the fraction of the total system volume that the volume element occupies. However, since the molecules are indistinguishable, we really should consider the likelihood that any of the N molecules is in the volume element $d\underline{r}_1$ independent of their identity. That is,

Likelihood that any of the N molecules is in volume element $d\underline{r}_1 = \dfrac{N}{V} d\underline{r}_1 = \rho d\underline{r}_1.$

$$(11.1\text{-}3)$$

This is being referred to as a likelihood rather than a probability for semantic reasons. By definition, a probability has a value between 0 and unity—as does $d\underline{r}_1/V$, which approaches 0 if $d\underline{r}_1$ becomes infinitely small, and is unity when the volume element is equal to the system volume. However, the likelihood function goes from 0 to N over this range, so it cannot strictly be considered a probability function; more correctly, it is the probable number of molecules.

Now consider the probability that molecule 1 is in a volume element $d\underline{r}_1$ around $\underline{r}_1$, and simultaneously that molecule 2 is in volume element $d\underline{r}_2$ around $\underline{r}_2$, molecule 3 is in volume element $d\underline{r}_3$ around $\underline{r}_3$, ..., etc., regardless of their translational motions. This is more complicated, since there can be an energy of interaction between the molecules in the different volume elements, which would influence this probability. In fact, this probability is equal to the Boltzmann factor of the interaction energy in this configuration, normalized by the configuration integral (the sum of probabilities

11.1 Reduced Spatial Probability Density Functions

of all configurations), and consequently is given by[1]

$$\frac{e^{-u(\underline{r}_1,\underline{r}_2,\ldots\underline{r}_N)/kT} d\underline{r}_1 \ldots d\underline{r}_N}{\int \ldots \int e^{-u(\underline{r}_1,\underline{r}_2,\ldots\underline{r}_N)/kT} d\underline{r}_1 \ldots d\underline{r}_N} = \frac{e^{-u(\underline{r}_1,\underline{r}_2,\ldots\underline{r}_N)/kT} d\underline{r}_1 \ldots d\underline{r}_N}{Z(N,V,T)} \quad (11.1\text{-}4)$$

where Z is the N particle configuration integral discussed earlier. (Note that for simplicity in the following equations, we will frequently write $u(\underline{r}_1,\underline{r}_2,\ldots\underline{r}_N)$ as simply u and $Z(N,V,T)$ as Z_N.)

Of greater interest are reduced-probability functions involving fewer numbers of molecules. For example, the probability that molecule 1 is in the volume element $d\underline{r}_1$ about $\underline{r}_1$, and that molecule 2 is in the volume element $d\underline{r}_2$ about $\underline{r}_2,\ldots$, and that molecule n is in a volume element $d\underline{r}_n$ about $\underline{r}_n$ regardless of the locations of molecules $n+1, n+2,\ldots, N$ is

$$\frac{d\underline{r}_1 d\underline{r}_2 \ldots d\underline{r}_n \int \ldots \int e^{-u/kT} d\underline{r}_{n+1} d\underline{r}_{n+2} \ldots d\underline{r}_N}{\int \int e^{-u/kT} d\underline{r}_1 \ldots d\underline{r}_N}$$

$$= \frac{d\underline{r}_1 d\underline{r}_2 \ldots d\underline{r}_n \int \ldots \int e^{-u/kT} d\underline{r}_{n+1} d\underline{r}_{n+2} \ldots d\underline{r}_N}{Z_N} \quad (11.1\text{-}5)$$

However, we must again remember that identical molecules are indistinguishable. Therefore, instead of inquiring about the probability that a specific molecule is in a volume element $d\underline{r}$, now we can only ask about the likelihood that any one of the N indistinguishable molecules could be in that volume element. The likelihood that (simultaneously) one of the N molecules is in the volume element $d\underline{r}_1$ about $\underline{r}_1$, one of the N-1 remaining molecules is in the volume element $d\underline{r}_2$ about $\underline{r}_2,\ldots$, and that one of the remaining $N-n+1$ molecules is in the volume element $d\underline{r}_n$ about $\underline{r}_n$ regardless of the positions of the other $N-n$ molecules is

$$N(N-1)\ldots(N-n+1)\frac{d\underline{r}_1 d\underline{r}_2 \ldots d\underline{r}_n \int \ldots \int e^{-u/kT} d\underline{r}_{n+1} d\underline{r}_{n+2} \ldots d\underline{r}_N}{Z(N,V,T)}$$

$$= \frac{N!}{(N-n)!} \frac{d\underline{r}_1 d\underline{r}_2 \ldots d\underline{r}_n \int \ldots \int e^{-u/kT} d\underline{r}_{n+1} d\underline{r}_{n+2} \ldots d\underline{r}_N}{Z(N,V,T)}$$

$$\equiv \rho^n g^{(n)}(\underline{r}_1,\ldots,\underline{r}_n; T, \rho) d\underline{r}_1 d\underline{r}_2 \ldots d\underline{r}_n \quad (11.1\text{-}6)$$

The factors before the integral arises as follows. There are N choices for the first molecule identified by the index 1, N-1 remaining choices for the molecule designated by the index 2 (since one of the identical molecules has already been chosen), N-2 choices for the third molecule, etc. Also, in the last part of the equation above, we have defined an n-body correlation function as

$$g^{(n)}(\underline{r}_1,\ldots,\underline{r}_n; T, \rho) = \frac{N!}{(N-n)!} \left(\frac{V}{N}\right)^n \frac{\int \ldots \int e^{-u/kT} d\underline{r}_{n+1} d\underline{r}_{n+2} \ldots d\underline{r}_N}{\int \ldots \int e^{-u/kT} d\underline{r}_1 d\underline{r}_2 \ldots d\underline{r}_N}$$

$$(11.1\text{-}7)$$

---

[1] For each spatial vector $\underline{r}$, the integral is over the total system volume V. For simplicity of notation, the range of integration is not shown, except where it is needed for clarity.

Chapter 11: Interacting Molecules in a Dense Fluid. Configurational Distribution Functions

In most cases we will be interested in small values of n—that is, $n = 2$ or 3—so that the functions of interest are

$$g^{(2)}(\underline{r}_1, \underline{r}_2; T, \rho) = N(N-1)\frac{V^2}{N^2}\frac{\int\cdots\int e^{-u/kT}d\underline{r}_3\ldots d\underline{r}_N}{\int\cdots\int e^{-u/kT}d\underline{r}_1\ldots d\underline{r}_N} \cong \frac{V^2\int\cdots\int e^{-u/kT}d\underline{r}_3\ldots d\underline{r}_N}{\int\cdots\int e^{-u/kT}d\underline{r}_1\ldots d\underline{r}_N}$$

(11.1-8a)

$$g^{(3)}(\underline{r}_1, \underline{r}_2, \underline{r}_3; T, \rho) \cong \frac{V^3\int\cdots\int e^{-u/kT}d\underline{r}_4\ldots d\underline{r}_N}{\int\cdots\int e^{-u/kT}d\underline{r}_1\ldots d\underline{r}_N} \cong \frac{V^3\int\cdots\int e^{-u/kT}d\underline{r}_4\ldots d\underline{r}_N}{Z(N,V,T)}$$

(11.1-9a)

or more commonly written as[2]

$$\rho^2 g^{(2)}(\underline{r}_1, \underline{r}_2; T, \rho) = \frac{N(N-1)\int\cdots\int e^{-u(\underline{r}_1,\underline{r}_2,\underline{r}_3,\underline{r}_4,\ldots)/kT}d\underline{r}_3\ldots d\underline{r}_N}{Z_N}$$

$$\cong \frac{N^2\int\cdots\int e^{-u(\underline{r}_1,\underline{r}_2,\underline{r}_3,\underline{r}_4,\ldots)/kT}d\underline{r}_3\ldots d\underline{r}_N}{Z_N}$$

(11.1-8b)

and

$$\rho^3 g^{(3)}(\underline{r}_1, \underline{r}_2, \underline{r}_3; T, \rho) = \frac{N(N-1)(N-2)\int\cdots\int e^{-u(\underline{r}_1,\underline{r}_2,\underline{r}_3,\underline{r}_4,\ldots)/kT}d\underline{r}_4\ldots d\underline{r}_N}{Z_N}$$

$$\cong \frac{N^3\int\cdots\int e^{-u(\underline{r}_1,\underline{r}_2,\underline{r}_3,\underline{r}_4,\ldots)/kT}d\underline{r}_4\ldots d\underline{r}_N}{Z_N}$$

(11.1-9b)

The physical interpretation of the first of these correlation functions is as follows. The likelihood that a molecule is located in a volume element $d\underline{r}_1$ about the position vector $\underline{r}_1$ (which for simplicity can be taken to be the origin of a coordinate system) and simultaneously that a second molecule is in the volume element $d\underline{r}_2$ about $\underline{r}_2$ is, from the discussion above:

$$\frac{N!}{(N-2)!}\frac{d\underline{r}_1 d\underline{r}_2\int\cdots\int e^{-u/kT}d\underline{r}_3 d\underline{r}_4\ldots d\underline{r}_N}{Z_N} \approx N^2\frac{d\underline{r}_1 d\underline{r}_2\int\cdots\int e^{-u/kT}d\underline{r}_3 d\underline{r}_4\ldots d\underline{r}_N}{Z_N}$$

$$= \frac{N^2}{V^2}d\underline{r}_1 d\underline{r}_2\frac{V^2\int\cdots\int e^{-u/kT}d\underline{r}_3 d\underline{r}_4\ldots d\underline{r}_N}{Z_N} = \rho^2 g^{(2)}(\underline{r}_1, \underline{r}_2; \rho, T)d\underline{r}_1 d\underline{r}_2$$

(11.1-10)

---

[2] Here again, for simplicity of notation, we have used $Z_N = Z(N, V, T)$.

11.1 Reduced Spatial Probability Density Functions

As is clear from its definition above, the two-particle or pair correlation function $g^{(2)}(\underline{r}_1, \underline{r}_2; \rho, T)$ is a function of the position vectors $\underline{r}_1$ and $\underline{r}_2$, and also a function of the molecular density ρ and temperature T. For simplicity of notation, we will usually write $g^{(2)}(\underline{r}_1, \underline{r}_2; \rho, T)$ simply as $g^{(2)}(\underline{r}_1, \underline{r}_2)$. The function $g^{(2)}(\underline{r}_1, \underline{r}_2; \rho, T)$ is commonly referred to as the *radial distribution function*. We will use both terms interchangeably. Among the properties of the pair correlation function is that its value is unity if there are no intermolecular interactions, and also when the distance between the position vectors $\underline{r}_1$ and $\underline{r}_2$ is large on a molecular scale so that each molecule no longer feels the presence of the other.

Note that if there were no interactions among the molecules—that is, $u = 0$—then $Z_N = V^N$, and the integral in the numerator becomes equal to V^{N-2}. In this case, the likelihood of a molecule being in the volume element $d\underline{r}_1$ and simultaneously a second molecule being in the volume element $d\underline{r}_2$ is just $\rho^2 d\underline{r}_1 d\underline{r}_2$ and $g^{(2)}(\underline{r}_1, \underline{r}_2; \rho, T) = 1$. This result is only valid if there is no energy of interaction and therefore no correlation between the molecules in the volume elements $d\underline{r}_1$ and $d\underline{r}_2$, and is a simple extension of the discussion at the beginning of this section. That is, $\rho^2 d\underline{r}_1 d\underline{r}_2 = (\rho d\underline{r})^2$ is what is obtained if the molecules are uniformly distributed, so the number of molecules in any volume element is just the average density times the size of the volume element. However, in general, there is a connection between the two volume elements, since the molecules they contain can interact. That is, if the two volume elements are sufficiently close on a molecular scale, the presence of a molecule in one of the volume elements influences the likelihood of a molecule being in the second volume element. At very low density, this correlation is given by the Boltzmann factor of the interaction energy. However, at high density the correlation is more complicated, and the likelihood of both volume elements containing molecules is given by Eq. 11.1-10 above.

Next, we consider a somewhat different question: what is the probable number of molecules in volume element $d\underline{r}_2$, given that there is a molecule in $d\underline{r}_1$? This is computed as the likelihood of molecules being simultaneously in $d\underline{r}_1$ and $d\underline{r}_2$, divided by the likelihood of a molecule being in the volume element $d\underline{r}_1$, and is given by

Probable number of molecules in $d\underline{r}_2$ given that there is a molecule at $d\underline{r}_1$

$$= \frac{\text{probable number of molecules simultaneously in } d\underline{r}_1 \text{ and } d\underline{r}_2}{\text{probable number of molecules in } d\underline{r}_1}$$

$$= \frac{\dfrac{N!}{(N-2)!} \dfrac{d\underline{r}_1 d\underline{r}_2 \int \cdots \int e^{-u/kT} d\underline{r}_3 d\underline{r}_4 \ldots d\underline{r}_N}{Z_N}}{\dfrac{N!}{(N-1)!} \dfrac{d\underline{r}_1 \int \cdots \int e^{-u/kT} d\underline{r}_2 d\underline{r}_3 d\underline{r}_4 \ldots d\underline{r}_N}{Z_N}} \approx \frac{N^2 \dfrac{d\underline{r}_1 d\underline{r}_2 \int \cdots \int e^{-u/kT} d\underline{r}_3 d\underline{r}_4 \ldots d\underline{r}_N}{Z_N}}{N \dfrac{d\underline{r}_1 \int \cdots \int e^{-u/kT} d\underline{r}_2 d\underline{r}_3 d\underline{r}_4 \ldots d\underline{r}_N}{Z_N}}$$

$$= \frac{N \dfrac{d\underline{r}_1 d\underline{r}_2 \int \cdots \int e^{-u/kT} d\underline{r}_3 d\underline{r}_4 \ldots d\underline{r}_N}{Z_N}}{\dfrac{d\underline{r}_1}{V}} = \frac{NV}{V^2} d\underline{r}_2 \frac{V^2 \int \cdots \int e^{-u/kT} d\underline{r}_3 d\underline{r}_4 \ldots d\underline{r}_N}{Z_N}$$

$$= \rho g^{(2)}(\underline{r}_1, \underline{r}_2; \rho, T) d\underline{r}_2 \qquad (11.1\text{-}11)$$

190 Chapter 11: Interacting Molecules in a Dense Fluid. Configurational Distribution Functions

since by a change in coordinates from $d\underline{r}_2 d\underline{r}_3 \ldots$ to relative coordinates $d\underline{r}_{12} d\underline{r}_{13} \ldots$ it is easily shown that

$$\frac{\int \ldots \int e^{-u/kT} d\underline{r}_2 d\underline{r}_3 d\underline{r}_4 \ldots d\underline{r}_N}{Z_N} = \frac{1}{V}$$

11.2 THERMODYNAMIC PROPERTIES FROM THE PAIR CORRELATION FUNCTION

The importance of the two-body correlation function or radial distribution function becomes evident in the computation of the thermodynamic properties of dense fluids. For example, the average value of the interaction energy (also called the *configurational energy*) of an assembly of molecules is

$$<U^C> = \frac{\int \ldots \int u(\underline{r}_1, \underline{r}_2, \ldots \underline{r}_N) e^{-u(\underline{r}_1, \underline{r}_2, \ldots \underline{r}_N)/kT} d\underline{r}_1 d\underline{r}_2 \ldots d\underline{r}_N}{\int \ldots \int e^{-u(\underline{r}_1, \underline{r}_2, \ldots \underline{r}_N)/kT} d\underline{r}_1 d\underline{r}_2 \ldots d\underline{r}_N}$$

$$= \frac{\int \ldots \int u(\underline{r}_1, \underline{r}_2, \ldots \underline{r}_N) e^{-u(\underline{r}_1, \underline{r}_2, \ldots \underline{r}_N)/kT} d\underline{r}_1 d\underline{r}_2 \ldots d\underline{r}_N}{Z_N} \quad \textbf{(11.2-1)}$$

Now assuming, as before (and this is a strong assumption) that the potential is pairwise additive—that is, that

$$u(\underline{r}_1, \underline{r}_2, \ldots \underline{r}_N) = \sum_i \sum_{\substack{j \\ i<j}} u(r_{ij}) \quad \textbf{(11.2-2)}$$

then

$$<U^C> = \frac{\int \ldots \int u(\underline{r}_1, \underline{r}_2, \ldots \underline{r}_N) e^{-u(\underline{r}_1, \underline{r}_2, \ldots \underline{r}_N)/kT} d\underline{r}_1 d\underline{r}_2 \ldots d\underline{r}_N}{\int \ldots \int e^{-u(\underline{r}_1, \underline{r}_2, \ldots \underline{r}_N)/kT} d\underline{r}_1 d\underline{r}_2 \ldots d\underline{r}_N}$$

$$= \frac{\int \ldots \int \sum \sum u(r_{ij}) e^{-u/kT} d\underline{r}_1 d\underline{r}_2 \ldots d\underline{r}_N}{\int \ldots \int e^{-u/kT} d\underline{r}_1 d\underline{r}_2 \ldots d\underline{r}_N}$$

$$= \frac{N(N-1)}{2} \int \int u(r_{12}) \left[\frac{\int \ldots \int e^{-u/kT} d\underline{r}_3 \ldots d\underline{r}_N}{\int \ldots \int e^{-u/kT} d\underline{r}_1 \ldots d\underline{r}_N} \right] d\underline{r}_1 d\underline{r}_2 \quad \textbf{(11.2-3)}$$

and

$$<U^C> = \frac{N(N-1)}{2} \int\int u(r_{12}) \left[\frac{(N-2)!}{N!} \rho^2 g^{(2)}(\underline{r}_1, \underline{r}_2; \rho, T) \right] d\underline{r}_1 d\underline{r}_2$$

$$= \frac{\rho^2}{2} \int\int u(r_{12}) g^{(2)}(r_{12}; \rho, T) d\underline{r}_1 d\underline{r}_2 = \frac{\rho^2}{2} \cdot V \int u(r_{12}) g^{(2)}(r_{12}; \rho, T) d\underline{r}_{12}$$

$$= \frac{\rho^2 \cdot V}{2} \cdot 4\pi \int u(r_{12}) g^{(2)}(r_{12}; \rho, T) r_{12}^2 dr_{12} = 2\pi N\rho \int u(r) g(r; \rho, T) r^2 dr$$

$$\textbf{(11.2-4)}$$

11.2 Thermodynamic Properties from the Pair Correlation Function

In writing this last equation we have recognized that for spherical molecules, the radial distribution function depends not on the two position vectors $\underline{r}_1$ and $\underline{r}_2$, but only the scalar distance between them, r_{12}, which for simplicity of notation we have designated in the integral as r. The physical meaning of Eq. 11.2-4 is as follows. Given that there is a molecule at the origin of the coordinate system, its energy of interaction with a single molecule a distance r away is $u(r)$. The number of molecules in a spherical shell of thickness dr (shown in Fig. 11.2-1) at a distance r away is $4\pi\rho r^2 g(r)\,dr$. To obtain the total interaction energy of that one molecule with all others, we need to integrate over all values of r from 0 to ∞. Finally, to obtain the total energy of the system, we need to recognize that there are N molecules in the system, so the previous result has to be multiplied by the factor N. However, there is one final complication: in the argument above we would have overcounted by a factor of 2 since, for example, the single interaction between molecules 1 and 2 would be counted twice, once when 1 was the central molecule at the origin of the coordinate system, and again when molecule 2 was the central molecule. Consequently, we have to divide the result by a factor of 2. This simple analysis explains the origin of each of the terms in Eq. 11.2-4.

To obtain a relationship between the pressure (or equation of state) and the radial distribution function is a bit more complicated. The starting point is

$$P = kT \left(\frac{\partial \ln Q}{\partial V}\right)_{N,T} = kT \left(\frac{\partial \ln Z}{\partial V}\right)_{N,T} \quad (11.2\text{-}5)$$

with

$$Z(N, V, T) = \int_V \cdots \int_V e^{-u(\underline{r}_1,\ldots,\underline{r}_N)/kT}\, d\underline{r}_1 \ldots d\underline{r}_N \quad (11.2\text{-}6a)$$

which in rectangular coordinates is

$$Z(N, V, T) = \int_0^{V^{1/3}} \cdots \int_0^{V^{1/3}} e^{-u(x_1,y_1,z_1,\ldots,x_N,y_N,z_N)/kT}\, dx_1 dy_1 dz_1 dx_2 dy_2 dz_2 \ldots dx_N dy_N dz_N \quad (11.2\text{-}6b)$$

The difficulty in evaluating the derivative $(\partial \ln Z/\partial V)_{N,T}$ is that the volume V appears in limits of the configuration integral Z, rather than in the integrand. To remove this difficulty, we define a set of dimensionless variables $x_i^* = x_i/V^{1/3}$,

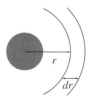

Figure 11.2-1 Volume of Shell $= 4\pi r^2 dr$.

Chapter 11: Interacting Molecules in a Dense Fluid. Configurational Distribution Functions

$y_i^* = y_i/V^{1/3}$, and $z_i^* = z_i/V^{1/3}$, so that the integral becomes

$$Z(N, V, T) = V^N \int_0^1 \cdots \int_0^1 e^{-u(x_1,\ldots,z_N)/kT} dx_1^* dy_1^* \ldots dz_N^* \qquad (11.2\text{-}7)$$

To proceed, we again use the pairwise additivity assumption for the interaction potential

$$u(\underline{r}_1, \ldots, \underline{r}_N) = \sum\sum_{1\le i\ i<j\le N} u(\underline{r}_i, \underline{r}_j) = \sum\sum_{1\le i\ i<j\le N} u(r_{ij}) \qquad (11.2\text{-}8)$$

with

$$r_{ij} = \left[(x_i - x_j)^2 + (y_i - y_j)^2 + (z_i - z_j)^2\right]^{1/2} = \left[\sum_{\ell=x,y,z}(\ell_i - \ell_j)^2\right]^{1/2}$$

$$= V^{1/3}\left[\sum_{\ell^*=x^*,y^*,z^*}(\ell_i^* - \ell_j^*)^2\right]^{1/2}$$

Also

$$\frac{d(r_{ij})}{dV} = \frac{d}{dV}\left\{V^{1/3}\left[\sum(\ell_i^* - \ell_j^*)^2\right]^{1/2}\right\} = \frac{1}{3V^{2/3}}\left[\sum(\ell_i^* - \ell_j^*)^2\right]^{1/2} = \frac{r_{ij}}{3V}$$

so that

$$\left[\frac{\partial Z(N, V, T)}{\partial V}\right]_{N,T} = \frac{\partial}{\partial V}\left\{V^N \int_0^1 \cdots \int_0^1 e^{-u(r_1,\ldots r_N)/kT} dx_1^* \ldots dz_N^*\right\}$$

$$= NV^{N-1}\int_0^1 \cdots \int_0^1 e^{-u(r_1,\ldots r_N)/kT} dx_1^* \ldots dz_N^*$$

$$+ V^N \int_0^1 \cdots \int_0^1 \left(\frac{-1}{kT}\right)\frac{du(\underline{r}_1,\ldots \underline{r}_N)}{dV} e^{-u(\underline{r}_1,\ldots \underline{r}_N)/kT} dx_1^* \ldots dz_N^*$$

and

$$\left[\frac{\partial Z(N, V, T)}{\partial V}\right]_{N,T} = \frac{N}{V}\left\{V^N \int_0^1 \cdots \int_0^1 e^{-u(r_1..r_N)/kT} dx_1^* \ldots dz_N^*\right\}$$

$$- \frac{V^N}{kT} \int_0^1 \cdots \int_0^1 \frac{d}{dV}\sum\sum u(r_{ij}) e^{-u/kT} dx_1^* \ldots dz_N^* =$$

11.2 Thermodynamic Properties from the Pair Correlation Function

$$= \frac{N}{V} Z - \frac{V^N}{kT} \frac{(N)(N-1)}{2} \int_0^1 \cdots \int_0^1 \frac{du(r_{12})}{dV} e^{-u/kT} dx_1^* \ldots dz_N^*$$

$$= \frac{N}{V} Z - \frac{V^N}{kT} \frac{(N)(N-1)}{2} \int_0^1 \cdots \int_0^1 \frac{du(r_{12})}{dr_{12}} \frac{dr_{12}}{dV} e^{-u/kT} dx_1^* \ldots dz_N^*$$

$$= \frac{N}{V} Z - \frac{V^N}{kT} \frac{(N)(N-1)}{2} \int_0^1 \cdots \int_0^1 \frac{du(r_{12})}{dr_{12}} \frac{r_{12}}{3V} e^{-u/kT} dx_1^* \ldots dz_N^* \qquad (11.2\text{-}9)$$

Consequently

$$\frac{1}{Z}\left(\frac{\partial Z}{\partial V}\right)_{N,T} = \left(\frac{\partial \ln Z}{\partial V}\right)_{N,T} = \rho - \frac{N(N-1)}{6kTV} \frac{\int_V \cdots \int_V \frac{du(r_{12})}{dr_{12}} r_{12} e^{-u/kT} d\underline{r}_1 \ldots d\underline{r}_N}{Z}$$

$$\left(\frac{\partial \ln Z}{\partial V}\right)_{N,T} = \rho - \frac{N^2}{6kTV} \iint_V \frac{du(r_{12})}{dr_{12}} r_{12} \frac{\left[\int_V \cdots \int_V e^{-u/kT} d\underline{r}_3 \ldots d\underline{r}_N\right]}{Z} d\underline{r}_1 d\underline{r}_2$$

$$= \rho - \frac{N^2}{6kTV^3} \iint_V \frac{du(r_{12})}{dr_{12}} r_{12} g^{(2)}(r_{12}) d\underline{r}_1 d\underline{r}_2 \qquad (11.2\text{-}10)$$

so that

$$P = kT\left(\frac{\partial \ln Z}{\partial V}\right)_{N,T} = \rho kT - \frac{\rho^2}{6V} \iint_V \frac{du(r_{12})}{dr_{12}} r_{12} g(r_{12}) d\underline{r}_1 d\underline{r}_{12}$$

$$= \rho kT - \frac{\rho^2}{6} \int_V \frac{du(r_{12})}{dr_{12}} r_{12} g(r_{12}) d\underline{r}_{12}$$

$$= \rho kT - \frac{\rho^2}{6} 4\pi \int_0^\infty \frac{du(r_{12})}{dr_{12}} r_{12} g(r_{12}) r_{12}^2 dr_{12}$$

$$P = \rho kT - \frac{2\pi}{3}\rho^2 \int_0^\infty \frac{du(r)}{dr} g(r) r^3 dr \qquad (11.2\text{-}11)$$

This is the desired relation between the radial distribution function, the two-body interaction potential and the pressure. It is this relation, referred to as the *pressure equation*, which will lead to the volumetric equation of state. An alternative relation between the radial distribution function and the pressure (or volumetric equation of state) will be developed later in this chapter.

11.3 THE PAIR CORRELATION FUNCTION (RADIAL DISTRIBUTION FUNCTION) AT LOW DENSITY

From Eq. 11.1-8, the pair correlation function is defined as

$$g^{(2)}(\underline{r}_1, \underline{r}_2; \rho, T) = \frac{V^2 \int \cdots \int e^{-u(\underline{r}_1, \underline{r}_2, \ldots \underline{r}_N)/kT} d\underline{r}_3 d\underline{r}_4 \ldots d\underline{r}_N}{Z_N} \quad \text{(11.1-8a)}$$

with

$$Z_N = Z(N, V, T) = \int \cdots \int e^{-u(\underline{r}_1, \underline{r}_2, \ldots \underline{r}_N)/kT} d\underline{r}_1 d\underline{r}_2 \ldots d\underline{r}_N$$

To begin the evaluation of the configuration integral for a low-density fluid, we will follow the same procedure used in Chapter 7 for deriving the expression for the second virial coefficient—that is, we will assume the interaction energy is pairwise additive

$$u(\underline{r}_1, \ldots, \underline{r}_N) = \sum_i \sum_{\substack{j \\ 1 \le i < j \le N}} u(r_{ij}) \quad \text{(7.3-3)}$$

so that

$$e^{-u(\underline{r}_1, \ldots \underline{r}_N)/kT} = \prod_i \prod_{\substack{j \\ 1 \le i \le j \le N}} e^{-u(r_{ij})/kT} \quad \text{(7.3-4)}$$

and

$$Z(N, V, T) = \int_V \cdots \int_V \prod_i \prod_{\substack{j \\ 1 \le i \le j \le N}} e^{-u(r_{ij})/kT} d\underline{r}_1 \ldots d\underline{r}_N \quad \text{(7.3-5)}$$

Again using the Mayer cluster function $f_{ij}(r_{ij}) = e^{-u(r_{ij})/kT} - 1$, we obtain

$$Z(N, V, T) = \int_V \cdots \int_V \prod_i \prod_{\substack{j \\ 1 \le i \le j \le N}} (1 + f_{ij}) d\underline{r}_1 \ldots d\underline{r}_N \quad \text{(7.4-3)}$$

which was expanded into a sum of products

$$Z(N, V, T) = \int_V \cdots \int_V \left(1 + \sum_i \sum_{\substack{j \\ j > i}} f_{ij} + \sum_i \sum_{\substack{j \\ j > i}} \sum_{i'} \sum_{\substack{j' \\ j' > i'}} f_{ij} f_{i'j'} + \cdots \right) d\underline{r}_1 \ldots d\underline{r}_N \quad \text{(7.4-4)}$$

Now, keeping only the first two terms in the expansion, and following the procedure in Chapter 7, we have

$$Z(N, V, T) = V^N + V^N \left(\frac{N^2}{2}\right)\left(\frac{\beta_1}{V}\right) = V^N \left[1 + \frac{N\rho\beta_1}{2}\right] \quad \text{(11.3-1)}$$

11.3 The Pair Correlation Function (Radial Distribution Function) at Low Density

where

$$\beta_1 = \frac{1}{V} \int \int f(\underline{r}_i, \underline{r}_j) d\underline{r}_i d\underline{r}_j = \int_V f(r) d\underline{r} = 4\pi \int_0^\infty f(r) r^2 dr \quad (11.3\text{-}2)$$

The evaluation of the numerator is slightly more complicated by the fact that the integral does not include the position vectors for molecules 1 and 2. Consequently, the Mayer function expansion is

$$\int \cdots \int e^{-u(\underline{r}_1,\underline{r}_2,\ldots\underline{r}_N)/kT} d\underline{r}_3 d\underline{r}_4 \ldots d\underline{r}_N$$

$$= \int_V \cdots \int_V \left(1 + f_{12} + \sum_{j>3} f_{1j} + \sum_{j>3} f_{2j} + \sum_{i>3} \sum_{\substack{j \\ j>i}} f_{ij} + \cdots \right) d\underline{r}_3 \ldots d\underline{r}_N$$

$$= (1 + f_{12}) \int_V \cdots \int_V d\underline{r}_3 \ldots d\underline{r}_N$$

$$+ \int_V \cdots \int_V \left((N-2)f_{13} + (N-2)f_{23} + \frac{(N-3)(N-4)}{2} f_{34} + \cdots \right) d\underline{r}_3 \ldots d\underline{r}_N$$

$$= (1 + f_{12})V^{N-2} + 2(N-2)V^{N-3} \int_V f_{13} d\underline{r}_3$$

$$+ \frac{(N-3)(N-4)}{2} V^{N-4} \int_V \int_V f_{34} d\underline{r}_3 d\underline{r}_4 + \cdots$$

$$\approx V^{N-2} \left[1 + f_{12} + 2\frac{N}{V} \int_V f(r) d\underline{r} + \frac{N^2}{2V} \int_V f(r) d\underline{r} + \cdots \right]$$

$$= V^{N-2} \left[1 + f_{12} + 2\rho\beta_1 + \frac{N\rho\beta_1}{2} + \cdots \right] \quad (11.3\text{-}3)$$

Now, combining Eqs. 11.3.-1 and 11.3-3

$$g^{(2)}(\underline{r}_1, \underline{r}_2; \rho, T) = \frac{V^2 \int \cdots \int e^{-u(\underline{r}_1,\underline{r}_2,\ldots\underline{r}_N)/kT} d\underline{r}_3 d\underline{r}_4 \ldots d\underline{r}_N}{Z_N}$$

$$= V^2 \frac{V^{N-2} \left[1 + f_{12} + 2\rho\beta_1 + \frac{N\rho\beta_1}{2} + \cdots \right]}{V^N \left[1 + \frac{N\rho\beta_1}{2} \right]} \quad (11.3\text{-}4a)$$

and keeping only the leading (density-independent) terms

$$g^{(2)}(\underline{r}_1, \underline{r}_2; \rho=0, T) = 1 + f_{12} = 1 + (e^{-u(\underline{r}_1,\underline{r}_2)/kT} - 1) = e^{-u(\underline{r}_1,\underline{r}_2)/kT} \quad (11.3\text{-}4b)$$

Or, as it usually as written for spherical particles

$$g(r; \rho = 0, T) = e^{-u(r)/kT} \qquad (11.3\text{-}5)$$

where r is the distance from a molecule at the origin of the coordinate system.

Equation 11.3-5 is interesting, because we see that when the intermolecular separation distance r is very small, so that the interaction potential is infinite (for example, if r were less than σ of the hard-sphere potential), the radial distribution function, $g(r)$ is 0. That is, there cannot be a second molecule so close to the first that the molecules would overlap. At very large separation distances—that is, $r \gg \sigma$—the interaction potential $u(r)$ is 0, and $g(r) = 1$. This corresponds to there being no spatial correlation between the molecules. That is, at large separations, the number of molecules in a volume element far removed from a central molecule is just equal to the average density times the size of the volume element. Finally, since the radial distribution function at low density is equal to the Boltzmann factor of the interaction energy, the ranges of these two functions are the same. (Here, by range we mean the extent of the intermolecular separation distance over which there is a spatial correlation among the molecules, so that the value of the radial distribution function is different from unity, indicating no correlation and a completely random distribution of molecules.) That the ranges of intermolecular potential and the radial distribution function are the same at low density is illustrated in Figs. 11.3-1a and 11.3-1b for the Lennard-Jones 12-6 potential. In these figures, the energy u^* is u/ε and r^* is r/σ.

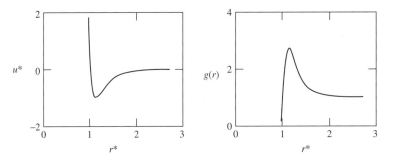

Figure 11.3-1 (a) Lennard-Jones 12-6 potential u^* as a function of r^* and (b) the low density radial distribution function for this potential as a function of r^*.

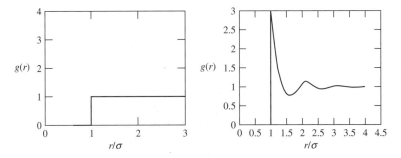

Figure 11.3-2 The (a) low density and (b) higher density radial distribution for the hard-sphere potential as a function of r/σ.

Note that the region around the peak in the radial distribution function corresponds to the first coordination shell of the molecule. For the hard-sphere potential, the radial distribution function is even simpler. Since the interaction energy for the hard-sphere potential is infinite for $r < \sigma$, and zero for $r \geqslant \sigma$, $g(r)$ is zero for $r < \sigma$ and unity for $r \geqslant \sigma$. This is shown in Fig. 11.3-2a. The radial distribution function is more complicated at higher density as shown in Fig. 11.3-2b.

11.4 METHODS OF DETERMINATION OF THE PAIR CORRELATION FUNCTION AT HIGH DENSITY

At high density, the radial distribution function is more complicated, extending several molecular diameters. since in a dense fluid there are correlations over larger intermolecular separation distances. This is a result of there being both a direct correlation, between molecules 1 and 2, and a collection of indirect correlations, such as the correlation between the locations of molecules 1 and 3 combined with the correlation between molecules 3 and 2, and the correlations between molecules 1 and 3, 3 and 4, and 4 and 1, etc. Consequently, the radial distribution function in a dense fluid is of longer range and has several peaks (corresponding to coordination shells) and valleys (resulting from a steric hindrance to molecules just outside each coordination shell from the molecules in that shell). In fact, we will use the idea of direct and indirect correlations in Chapter 12 to derive one of the important statistical mechanical equations, the Ornstein-Zernike equation, used in the determination of radial distribution functions in dense fluids.

Obtaining the radial distribution function in a liquid or dense fluid is very much more complicated than analysis for the dilute gas. There are a number of very different methods that are used. The first is to obtain the radial distribution from laboratory scattering experiments on the fluid of interest. Typically, x-ray or neutron beams are used for the scattering, since the wavelengths of these beams are comparable to molecular dimensions. The intensity of scattered radiation by the fluid is measured as a function of the scattering angle; and by a mathematical analysis (Fourier transformation), the radial distribution is obtained as discussed in Section 11.6. This direct method does not require the assumption of pairwise additivity and is exact for atomic fluids for which there is only a single radial distribution function. However, it is more difficult in the study multiatom molecular fluids that are generally of interest, since in such cases there are a number atom-atom correlation functions. This is shown in Fig. 11.4-1 for HCl, where there are three different intermolecular correlation functions: the hydrogen-hydrogen, chlorine-chlorine, and hydrogen-chlorine correlation functions. These three correlation functions cannot be obtained from a single x-ray or neutron scattering measurement. Note that each correlation function is the result of averages over many pairs of molecules in different configurations. Also, for the case here the two hydrogen-chlorine separation distances shown in this configuration, both contribute to the single hydrogen-chlorine pair correlation function. For molecules with more atoms, the number of different correlation functions increases.

A second method of determining the radial distribution function is by use of statistical mechanical theory, and there are several ways to proceed. One method has been to use the assumption of pairwise additivity of the potential and the graph theory of clusters, as was done in the development of expressions for the virial coefficients from the canonical ensemble, to develop integral equations for the radial distribution function. The graph theory development is extremely complicated, so approximations are made leading to models with names such as the Percus-Yevick, hypernetted chain,

198 Chapter 11: Interacting Molecules in a Dense Fluid. Configurational Distribution Functions

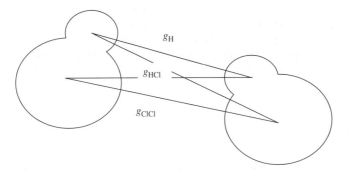

Figure 11.4-1 Atom-atom correlation functions for the two-center HCl molecule.

and mean spherical approximations, as examples. Some of these will be discussed in the Chapter 12. Such integral equations are usually solved numerically to obtain the radial distribution function at various temperatures and densities. However, after numerical evaluation, one only has a table of numbers and not analytic expressions for the radial distribution function as a function of separation distance, temperature, and density for the model pairwise-additive potential obtained from an approximate theory, and not an analytic expression.

A third method of obtaining the radial distribution function is by molecular-level computer simulation, which is discussed in some detail in Chapter 13. A brief introduction, for the sake of continuity, is given here. In this method, by computer programming, molecules are described by model potentials (usually, but not always, pairwise additive) and are considered to be in a box that exists in the memory of a computer. Many different configurations of the collection of molecules are considered, and averages over these configurations are used to obtain the radial distribution and thermodynamic properties of these model molecules. There are two general simulation methods in use; they will be described here in the next few sentences, in the simplest terms, and in more detail in Chapter 13. The first method involves considering many configurations of the system, essentially randomly generated, and using a Boltzmann weighted-averaging process. (The method actually used is slightly different because, for computational efficiency as explained in Chapter 13, a chain of configurations is generated with a likelihood proportional to the Boltzmann factor of the energy, and then a linear average is used to obtain the radial distribution function and the properties.) Such a procedure is called Monte Carlo simulation, in reference to the random nature of roulette or other games of pure chance.

The other general computer simulation method is molecular-dynamics. In this procedure, an initial configuration and distribution of velocities for the molecules is chosen, and then Newtonian mechanics is used to follow the evolution of the system as a function of time (as discussed in Chapter 13). Once the system has equilibrated—as evidenced by fluctuations being random rather than systematic in the various calculated properties, such as energy and pressure—average values of both thermodynamic and dynamic (that is, transport) properties and the radial distribution function can be obtained.

The efficient implementation of these simulation methods is much more intricate and sophisticated than the simple descriptions given above. If implemented properly, and if the simulations are run long enough, one obtains essentially exact property values for a fluid whose molecules interact with the model potential used. This is very valuable for testing theories and obtaining insight into the behavior of fluids. However, it must be remembered that at each state point, one obtains only numerical values of the properties, and not an analytical expression such as an equation of state or an explicit equation for the radial distribution function. However, one may try to fit the set of numerical values obtained with an equation. One of the important advantages of these simulation methods is that, by clever computational algorithms, they can be used for very complicated molecular fluids, including polymers.

11.5 FLUCTUATIONS IN THE NUMBER OF PARTICLES AND THE COMPRESSIBILITY EQUATION

For later reference, it is useful to have information about the fluctuations in the number of particles in a system described by the grand canonical ensemble. We can get this information as follows. The average number of particles in the system is computed from

$$\overline{N} = \sum_{N,E} N P(N, E, V) = \frac{\sum_{N,E} N e^{N\mu/kT} e^{-E_i(N,V)/kT}}{\Xi(V, T, \mu)}$$

$$= \frac{kT}{\Xi(V, T, \mu)} \left(\frac{\partial \Xi(V, T, \mu)}{\partial \mu}\right)_{V,T} = kT \left(\frac{\partial \ln \Xi(V, T, \mu)}{\partial \mu}\right)_{V,T} \quad (11.5\text{-}1)$$

and

$$\overline{N^2} = \sum_{N,E} N^2 P(N, E, V) = \frac{\sum_{N,E} N^2 e^{N\mu/kT} e^{-E_i(N,V)/kT}}{\Xi(V, T, \mu)}$$

$$= \frac{(kT)^2}{\Xi(V, T, \mu)} \left(\frac{\partial^2 \Xi(V, T, \mu)}{\partial \mu^2}\right)_{V,T}$$

$$= (kT)^2 \left(\frac{\partial^2 \ln \Xi(V, T, \mu)}{\partial \mu^2}\right)_{V,T} + (kT)^2 \left(\frac{\partial \ln \Xi(V, T, \mu)}{\partial \mu}\right)_{V,T}^2 \quad (11.5\text{-}2)$$

Now, assuming that the fluctuations in the number of particles is a result of a Gaussian distribution, to obtain the standard deviation we look at

$$\overline{N^2} - \overline{N}^2 = (kT)^2 \left(\frac{\partial^2 \ln \Xi(V, T, \mu)}{\partial \mu^2}\right)_{V,T} \quad (11.5\text{-}3)$$

However, we have already shown that

$$\frac{PV}{kT} = \ln \Xi \quad (6.2\text{-}12)$$

so that

$$\overline{N^2} - \overline{N}^2 = kTV \left(\frac{\partial^2 P}{\partial \mu^2}\right)_{V,T} \quad (11.5\text{-}4)$$

From classical thermodynamics, we have

$$dG = d(\mu\overline{N}) = \mu d\overline{N} + \overline{N} d\mu = V dP - S dT + \mu d\overline{N} \quad (11.5\text{-}5)$$

therefore

$$\overline{N} d\mu = V dP - S dT, \text{ so that } \left(\frac{\partial P}{\partial \mu}\right)_{V,T} = \frac{\overline{N}}{V} \text{ and } \left(\frac{\partial^2 P}{\partial \mu^2}\right)_{V,T} = \frac{1}{V}\left(\frac{\partial \overline{N}}{\partial \mu}\right)_{V,T} \quad (11.5\text{-}6)$$

In classical thermodynamics, the properties are considered to be average properties; so when comparing classical and statistical thermodynamics, the number of particles in classical thermodynamics is interpreted to be the average number of particles in statistical thermodynamics. Also

$$\overline{N} d\mu = V dP - S dT \text{ so that } \left(\frac{\partial \mu}{\partial \overline{N}}\right)_{V,T} = \frac{V}{\overline{N}}\left(\frac{\partial P}{\partial \overline{N}}\right)_{V,T}$$

$$= \frac{1}{\overline{N}}\left(\frac{\partial P}{\partial \frac{\overline{N}}{V}}\right)_{V,T} = \frac{1}{\overline{N}}\left(\frac{\partial P}{\partial \rho}\right)_{V,T} \quad (11.5\text{-}7)$$

and the isothermal compressibility κ is defined as

$$\kappa = -\frac{1}{V}\left(\frac{\partial V}{\partial P}\right)_T = \frac{1}{\rho}\left(\frac{\partial \rho}{\partial P}\right)_T \quad (11.5\text{-}8)$$

Therefore

$$\overline{N^2} - \overline{N}^2 = kT\left(\frac{\partial \overline{N}}{\partial \mu}\right)_{V,T} = \overline{N}kT\left(\frac{\partial \rho}{\partial P}\right)_T = \overline{N}kT\rho\kappa \quad (11.5\text{-}9)$$

and

$$\frac{\overline{N^2} - \overline{N}^2}{\overline{N}} = kT\left(\frac{\partial \rho}{\partial P}\right)_T = \rho kT\kappa \quad (11.5\text{-}10)$$

This equation provides a relationship between the likelihood of fluctuations in the number of particles in the grand canonical ensemble and the compressibility of the fluid, a derivative property. It is much like the equation relating the energy fluctuations in the canonical ensemble to the constant volume heat capacity, another derivative property as shown in Chapter 3.

To proceed, we note that in the canonical ensemble, the definition of the radial distribution function is

$$\rho^2 g^{(2)}(\underline{r}_1, \underline{r}_2; \rho, T) = N(N-1)\frac{\int \ldots \int e^{-u/kT} d\underline{r}_3 d\underline{r}_4 \ldots d\underline{r}_N}{Z_N} \quad (11.1\text{-}8b)$$

The analogous expression in the grand canonical ensemble is

$$\rho^2 g^{(2)}(\underline{r}_1, \underline{r}_2; \rho, T) = \overline{N(N-1)}\frac{\int \ldots \int e^{-u/kT} d\underline{r}_3 d\underline{r}_4 \ldots d\underline{r}_N}{Z_N}$$

$$= (\overline{N^2} - \overline{N})\frac{\int \ldots \int e^{-u/kT} d\underline{r}_3 d\underline{r}_4 \ldots d\underline{r}_N}{Z_N} \quad (11.5\text{-}11)$$

11.5 Fluctuations in the Number of Particles and the Compressibility Equation

Therefore

$$\iint \rho^2 g^{(2)}(\underline{r}_1, \underline{r}_2; \rho, T)\, d\underline{r}_1 d\underline{r}_2 = (\overline{N^2} - \overline{N}) \frac{\int \cdots \int e^{-u/kT} d\underline{r}_3 d\underline{r}_4 \ldots d\underline{r}_N d\underline{r}_1 d\underline{r}_2}{Z_N}$$

$$= \overline{N^2} - \overline{N} \qquad (11.5\text{-}12)$$

and

$$\iint \rho^2\, d\underline{r}_1 d\underline{r}_2 = \rho^2 \iint d\underline{r}_1 d\underline{r}_2 = \rho^2 V^2 = (\overline{N})^2 \qquad (11.5\text{-}13)$$

Combining these last two equations gives

$$\rho^2 \iint \left[g^{(2)}(\underline{r}_1, \underline{r}_2; \rho, T) - 1 \right] d\underline{r}_1 d\underline{r}_2 = \overline{N^2} - \overline{N} - (\overline{N})^2 \qquad (11.5\text{-}14)$$

and

$$\frac{\rho^2}{\overline{N}} \iint \left[g^{(2)}(\underline{r}_1, \underline{r}_2; \rho, T) - 1 \right] d\underline{r}_1 d\underline{r}_2 = \frac{\overline{N^2} - (\overline{N})^2}{\overline{N}} - 1 = kT \left(\frac{\partial \rho}{\partial P} \right)_T - 1$$

$$(11.5\text{-}15)$$

that by a change of coordinates and integration can be written as

$$kT \left(\frac{\partial \rho}{\partial P} \right)_T - 1 = \frac{\rho}{V} \iint \left[g^{(2)}(\underline{r}_1, \underline{r}_2; \rho, T) - 1 \right] d\underline{r}_1 d\underline{r}_2$$

$$= \frac{\rho}{V} \iint \left[g^{(2)}(\underline{r}_{12}; \rho, T) - 1 \right] d\underline{r}_{12} d\underline{r}_1$$

$$= \rho \int \left[g^{(2)}(\underline{r}_{12}; \rho, T) - 1 \right] d\underline{r}_{12} = 4\pi\rho \int_0^\infty [g^{(2)}(r; \rho, T) - 1] r^2\, dr$$

$$(11.5\text{-}16)$$

or simply

$$kT \left(\frac{\partial \rho}{\partial P} \right)_T - 1 = 4\pi\rho \int_0^\infty \left[g^{(2)}(r; \rho, T) - 1 \right] r^2\, dr \qquad (11.5\text{-}17)$$

This last equation is referred to as the *compressibility equation*. Its use is that it provides an alternate relationship between the radial distribution function and the equation of state to the pressure equation (Eq. 11.2-11). In particular, if we have the radial distribution function as a function of separation distance between the molecules, temperature and density (for all densities down to zero density where we know that the equation of state is that of an ideal gas), then we can integrate the compressibility equation to obtain the equation of state.

We now have two relations between the pressure (or volumetric equation of state), and the radial distribution function, the pressure equation (Eq. 11.2-11), and the compressibility equation (Eq. 11.5-17). These two relationships are completely independent of each other, and of quite different character. To use the pressure equation, the radial distribution function is needed as a function of molecular separation distance, but at only the density and temperature of interest; while to compute the pressure using the compressibility equation, the radial distribution function is needed for all densities from zero density to the density of interest. Consequently, the pressure and compressibility equations are complimentary.

There is no method that provides an exact expression for the radial distribution function; only approximate expressions are obtained using the methods discussed in the following chapters. However, one way to test the accuracy of the results obtained from these approximate methods is to use the radial distribution functions obtained in both the pressure and compressibility equations. It is then presumed that an approximate radial distribution function of greater accuracy will lead to pressures obtained from these two independent equations that are in closer agreement with each other.

11.6 DETERMINATION OF THE RADIAL DISTRIBUTION FUNCTION OF FLUIDS USING COHERENT X-RAY OR NEUTRON DIFFRACTION

The radial distribution function of an atomic liquid can be experimentally determined by doing a diffraction experiment using radiation with a wavelength smaller than or comparable to the size of a molecule. Commonly, x-ray or neutron beams are used, as their wavelengths are of the same order as atomic dimensions (see Table 11.6-1). In x-ray scattering, the electrons of an atom or molecule do the scattering, while it is the nucleus of the atom that is the scattering center of neutrons. In x-ray diffraction, an x-ray generator (which is readily available in crystallographic and polymer laboratories) is needed, while a nuclear reactor for a neutron beam source is needed for neutron scattering. There are a number of details of both experiments that are quite sophisticated, such as scattered radiation detection, filtering, etc., which will not be dealt with here. Instead, we consider only the briefest outline of the experiments, and how such experiments are interpreted to determine the radial distribution function of a monatomic fluid.

Table 11.6-1 Radiation Source and Wavelength in meters and Angstroms

Radiation will scatter from particles of approximately the same size as their wavelength. Therefore, x-rays and neutrons, which have wavelengths in the size range of atoms, can be used to probe atomic structure and the radial correlation function.

Radiation Source	λ (m)	λ Å
100 kV electrons (electron microscope)		0.037 Å
X-rays	10^{-9} to 10^{-12} m	0.01–10 Å
Thermal neutrons (ambient temperature)	10^{-11} to 10^{-12} m	1–10 Å
Visible light	10^{-6} to 10^{-7} m	1000–10,000 Å
UV	10^{-7} to 10^{-8} m	100–1000 Å
IR	10^{-4} to 10^{-5} m	10^5–10^6 Å
Microwaves	10^{-1} to 10^{-3} m	
Radiowaves	10^5 to 10^{-2} m	

$$1 \text{ Å} = 10^{-10} \text{ m} = 0.1 \text{ nm} = 10^{-4} \text{ } \mu\text{m}$$

11.6 The Radial Distribution Function of Fluids using Coherent X-ray or Neutron Diffraction

The essence of the diffraction experiment is that a liquid is subjected to a monochromatic (fixed wavelength) beam that has been collimated in such a way that all rays are parallel and in phase. The resulting scattered radiation is then measured as a function of the scattering angle. The experiment may be done in either the transmission mode or in the reflection mode, as shown in Fig. 11.6-1. The instrument geometry is chosen such that each ray is likely to be scattered only once between the radiation source and the radiation detector (or counter); and after passing through a collimator and monochromator, only parallel beams of the same wavelength as the incident beam enter the detector. (As the wavelength (and energy) of the incident and scattered radiation are the same, this is referred to as coherent scattering.)

In the discussion here, we will use $\underline{S}_o$ to denote the unit vector in the direction of the incident radiation, $\underline{S}$ for the vector in the direction of the scattered radiation, and the vector $\underline{s}$ to be

$$\underline{s} = 2\pi \frac{(\underline{S}_o - \underline{S})}{\lambda} \tag{11.6-1}$$

where λ is the wavelength of the radiation. The magnitude of this vector is

$$s = 4\pi \frac{\sin \theta}{\lambda} \tag{11.6-2}$$

where θ is one-half the scattering angle (the angle between $\underline{S}$ and $\underline{S}_o$). Also, we will use $f(s)$ to represent the atomic scattering factor or the amplitude of scattered radiation from one atom at s (i.e., at the angle θ of radiation of wavelength λ). In neutron scattering, the scattering site is the atomic nucleus, and the scattering factor is a constant, i.e., $f(s) = f$, which is proportional to the neutron scattering cross-section. Neutron scattering factors (or cross sections) have been determined from experiments, and values can be looked up in tables.[3] These scattering factors cannot be calculated from theory and can vary widely among different atoms. For example, the scattering cross section is about 10 times larger for hydrogen than deuterium, is quite small for some metals, and can even be negative. In x-ray diffraction the electrons are the scattering sites; there can be one (in hydrogen) or many of them (in larger atoms), and the electrons are in constant motion. In this case, from basic x-ray scattering theory (which we will not discuss here), the scattering cross section is related to the Fourier transform of the electron density around the atom

$$f(s) = \alpha \int \rho_e(\underline{r}) e^{i \underline{s} \cdot \underline{r}} d\underline{r} \tag{11.6-3}$$

where $\rho_e(\underline{r})$ is the electron density at the vectorial position $\underline{r}$ in a coordinate system centered at the origin of the atom, and α is a constant of proportionality (which is canceled out in the normalization we discuss later). The electron density around an atom is generally computed from quantum mechanics, so that x-ray scattering

---

[3]Neutron scattering lengths and cross sections of the elements and their isotopes can be found in *Neutron News*, Vol. 3, No. 3, 1992, pp. 29–37.

204 Chapter 11: Interacting Molecules in a Dense Fluid. Configurational Distribution Functions

function $f(s)$ can be calculated from theory and is tabulated for most atoms of interest, and consequently is a known function.[4]

Experimentally, what is determined in a scattering experiment is the intensity of scattered radiation at each scattering angle. If there were no interference (constructive or destructive) between scattered rays, the intensity (which is the square of the amplitude) of scattered radiation would be

$$I(s) = Nf^2(s) \tag{11.6-4}$$

where N is the number of scattering centers or molecules. However, there is interference between the different scattered rays (or neutrons); since while the rays were initially in phase, due to their different path lengths, the rays reaching the detector are no longer in phase. For two scatters, this is indicated in Fig. 11.6-1. The difference in path length of the two scattered rays is the distance $x_1 - x_2$ given by

$$x_2 = \underline{S}_o \cdot \underline{r}_{21} \quad \text{and} \quad x_1 = \underline{S} \cdot \underline{r}_{21} \quad \text{then} \quad (x_2 - x_1) = (\underline{S}_o - \underline{S}) \cdot \underline{r}_{21}$$

the phase difference $\Delta \phi$ between the two rays is then

$$\Delta \varphi = \frac{2\pi}{\lambda}(\underline{S} - \underline{S}_o) \cdot \underline{r}_{21} = \underline{s} \cdot \underline{r}_{21} \tag{11.6-5}$$

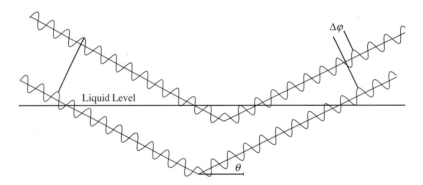

Figure 11.6-1 Schematic diagram of a reflection scattering experiment. The incident beams (entering from the left) and the scattered beams (exiting on the right) pass through a monochromator so they are of the same fixed wavelength, and then they pass through a collimator so that the incident and scattered beams are at the same angle with respect to the sample. The incident beams are initially in phase and the scattered beams are out of phase by the angle $\Delta \phi$. The complete scattering spectrum is obtained by moving the beam source and detector in tandem throughout all values of θ from 0 to 90° that are accessible with the equipment. Note that here the scattering angle is 2θ.

---

[4]X-ray scattering factors can be found in C. T. Chantler, K. Olsen, R. A. Dragoset, J. Chang, A. R. Kishore, S. A. Kotochigova, and D. S. Zucker, (2005), *X-Ray Form Factor, Attenuation and Scattering Tables* (version 2.1). National Institute of Standards and Technology, Gaithersburg, MD, available at http://physics.nist.gov/ffast (2008, April 11).

11.6 The Radial Distribution Function of Fluids using Coherent X-ray or Neutron Diffraction

and the total amplitude of scattered radiation $A(s)$ at the angle corresponding to s from the two atoms 1 and 2 is

$$A(s) = f(s) + f(s)e^{i\underline{s}\cdot\underline{r}_{21}} \tag{11.6-6}$$

where the first term is the contribution to the scattering amplitude from molecule 1 and the second term is the out-of-phase contribution of atom 2. Now, generalizing this result for scattering from N identical atoms

$$A(s) = f(s) \sum_{m=1}^{N} e^{i\underline{s}\cdot\underline{r}_{m1}} \tag{11.6-7}$$

Note the similarity between this equation for scattering from multiple atoms and Eq. 11.6-3 for scattering from multiple electrons within an atom.

The intensity of scattered radiation at the angle corresponding to s, $A(s)$ is the square of the amplitude

$$\begin{aligned} I(s) &= |A(s)|^2 = A^*(s) \cdot A(s) \\ &= \left[\sum_{n=1}^{N} f(s)e^{-i\underline{s}\cdot\underline{r}_{n1}}\right] \times \left[\sum_{m=1}^{N} f(s)e^{i\underline{s}\cdot\underline{r}_{m1}}\right] = \sum_{n=1}^{N}\sum_{m=1}^{N} f^2(s)e^{i\underline{s}\cdot\underline{r}_{mn}} \\ &= Nf^2(s) + f^2(s) \sum_{\substack{n=1 \\ n\neq m}}^{N}\sum_{m=1}^{N} e^{i\underline{s}\cdot\underline{r}_{mn}} \end{aligned} \tag{11.6-8}$$

where we have separated the $n = m$ and the $n \neq m$ terms in the sum, and used that $r_{m1} - r_{n1} = r_{mn}$.

In a liquid, the scattering sites are not fixed, but move during the course of a scattering experiment (which can take minutes or hours, depending on the intensity of the incident and scattered radiation). From the definition of the radial distribution function, and using the ergodic hypothesis that allows us to replace a time average with an average over states, we have

$$\left\langle \sum_{\substack{n=1 \\ n\neq m}}^{N}\sum_{m=1}^{N} e^{i\underline{s}\cdot\underline{r}_{mn}} \right\rangle = \left\langle \int\int e^{i\underline{s}\cdot\underline{r}_{mn}} d\underline{r}_n d\underline{r}_m \right\rangle = \frac{N^2}{V}\int e^{i\underline{s}\cdot\underline{r}} g(r) d\underline{r} = N\rho \int e^{i\underline{s}\cdot\underline{r}} g(r) d\underline{r}$$

$$\tag{11.6-9}$$

where the brackets $\langle\rangle$ denote an average over time, and the integrals are over all positions of a pair of scattering atoms (and therefore is the average over states). Therefore

$$I(s) = Nf^2(s) + N\rho f^2(s) \int e^{i\underline{s}\cdot\underline{r}} g(r) d\underline{r} \tag{11.6-10}$$

The factor N, the number of scatters in the target region of the liquid that is subjected to the beam, is not known. What is done is to normalize the measured large scattering angle portion of the intensity of scattered radiation to the scattering from the single atoms without interference. We will explain later why this can be done. In this way, a new intensity of scattered radiation, $I'(s)$ is as follows:

$$\lim_{s \to \infty} I(s) = NI'(s) = Nf^2(s) \quad \text{so that} \quad I'(s) = \frac{I(s)}{N} = f^2(s) \qquad (11.6\text{-}11)$$

We then rewrite Eq. 11.6-10 as

$$I'(s) = f^2(s) + \rho f^2(s) \int e^{i\underline{s}\cdot\underline{r}}[g(r) - 1] \, d\underline{r} + \rho f^2(s) \int e^{i\underline{s}\cdot\underline{r}} \, d\underline{r} \qquad (11.6\text{-}12)$$

Next we choose a coordinate system in which the vector $\underline{s}$ is along the z axis, so that $\underline{s}\cdot\underline{r} = sr\cos\theta$, and polar-spherical coordinates are be used for integration over the vector $\underline{r}$ leading to

$$I'(s) = f^2(s) + \rho f^2(s) \int e^{isr\cos\theta}[g(r) - 1]r^2 \sin\theta \, d\theta \, d\phi \, dr$$

$$+ \rho f^2(s) \int e^{isr\cos\theta} r^2 \sin\theta \, d\theta \, d\phi \, dr$$

$$= f^2(s) + 2\pi\rho f^2(s) \int e^{isr\cos\theta}[g(r) - 1]r^2 \sin\theta \, d\theta \, dr$$

$$+ 2\pi\rho f^2(s) \int e^{isr\cos\theta} r^2 \sin\theta \, d\theta \, dr \qquad (11.6\text{-}13)$$

To proceed further, we note that by two changes of the variable of integration

$$\int_0^\pi e^{isr\cos\theta} \sin\theta \, d\theta = \int_{-1}^{+1} e^{isrx} dx = \frac{1}{ikr} \int_{-isr}^{+isr} \exp(y) \, dy = \frac{e^{isr} - e^{-isr}}{isr} = \frac{\sin(sr)}{sr}$$

$$(11.6\text{-}14)$$

and so we obtain

$$I'(s) = f^2(s) + 2\pi\rho f^2(s) \int_0^\infty [g(r) - 1]\frac{\sin(sr)}{sr} r^2 dr + 2\pi\rho f^2(s) \int_0^\infty \frac{\sin(sr)}{sr} r^2 dr$$

$$(11.6\text{-}15)$$

The interpretation of the terms on the right-hand side of this equation is as follows. The first term is the intensity of scattered radiation if there were no interference from adjacent atoms. The second term is the result of the scattering interference between nearby atoms. The last term is proportional to the Fourier transform of the Kronecker delta function, and thus only makes a contribution to the intensity of scattered radiation at $s = 0$—that is, when $\theta = 0$—which occurs when the beam generator is pointing directly into the detector. Experimentally, it is not possible to measure the intensity of radiation at very small scattering angles (i.e., near $s = 0$), so this term is neglected.

11.6 The Radial Distribution Function of Fluids using Coherent X-ray or Neutron Diffraction

Therefore, the final diffraction equation is

$$I'(s) = f^2(s) + 2\pi\rho f^2(s) \int_0^\infty [g(r) - 1]\frac{\sin(sr)}{sr} r^2\, dr \qquad (11.6\text{-}16a)$$

or, defining the total structure function $H(s)$

$$\frac{I'(s) - f^2(s)}{f^2(s)} = H(s) = 2\pi\rho \int_0^\infty [g(r) - 1]\frac{\sin(sr)}{sr} r^2\, dr \qquad (11.6\text{-}16b)$$

Note that in this equation the radial distribution function $g(r)$ appears within an integral. However, this integral is of the Fourier type, so that the function obtained from experiment, $H(s)$, is in fact the Fourier transform of what is referred to as the total correlation function $h(r) = g(r) - 1$. Thus

$$h(r) = g(r) - 1 = \frac{1}{2\pi^2 \rho r} \int_0^\infty H(s) \sin(sr) s\, ds \qquad (11.6\text{-}17)$$

Therefore, the radial distribution function $g(r)$—or, equivalently, the total correlation function $h(r)$—is directly obtainable from an x-ray or neutron scattering experiment. However, to obtain a value of the radial distribution function at each value of the interatomic separation r, information is needed on the structure function $H(s)$ at all values of s to evaluate the integral.

Finally, note that the structure function $H(s)$ at zero angle ($s = 0$) cannot be measured due to the interference in the detector from unscattered radiation. However, it can be obtained in another way. To see this, we note that

$$\lim_{s \to 0} \frac{\sin(sr)}{sr} = \lim_{s \to 0} \frac{1}{sr}\left[sr - \frac{(sr)^3}{3!} + \frac{(sr)^5}{5!} - \cdots\right] = 1$$

so that

$$H(s = 0) = 2\pi\rho \int_0^\infty [g(r) - 1] r^2\, dr \qquad (11.6\text{-}18)$$

Now, we have previously shown (Eq. 11.5-7) that by starting with the grand canonical ensemble and using fluctuation theory, the following result was obtained:

$$kT\left(\frac{\partial \rho}{\partial P}\right)_T - 1 = 4\pi\rho \int_0^\infty [g(r) - 1] r^2\, dr \qquad (11.5\text{-}7)$$

where k is the Boltzmann constant. Consequently,

$$H(s = 0) = \frac{1}{2}\left[kT\left(\frac{\partial \rho}{\partial P}\right)_{N,T} - 1\right] \qquad (11.6\text{-}19)$$

208 Chapter 11: Interacting Molecules in a Dense Fluid. Configurational Distribution Functions

So even though we cannot measure the scattering function at zero angle ($s = 0$, because of the unscattered radiation), we can nonetheless obtain the value of $H(s = 0)$ from the measured value of the isothermal compressibility.

An example of the measured scattering intensity $I(s)$ for liquid argon is shown in Fig. 11.6-2, and the structure function $H(s)$ derived from that data is shown in Fig. 11.6-3. Then the radial distribution $g(r)$ computed from that data using Eq. 11.6-17 is shown in Fig. 11.6-4.[5] There we see an excluded region in which the center of a second molecule is not present, due to the repulsive forces, followed by a relatively strong peak in the radial distribution function near $r = 4$ Å (0.4 nm) corresponding to the first coordination shell of an atom. Next is a region of lower-than-average density of centers of mass ($g<1$) as a result of partial exclusion of additional molecules, resulting from the molecules in the first coordination shell. Beyond that we see increasingly weaker peaks and valleys in the radial distribution function due to the continuing but imperfect order in the liquid, with the radial distribution function reaching an asymptotic value of 1 at a separation distance of several atomic diameters.

It should be noted that the oscillations of the radial distribution function at separation distances before the first peak sometimes found from Fourier transforming scattering data are experimental artifacts. They arise as follows. The scattering equations contain an exponential of the imaginary part of the dot product of the scattering

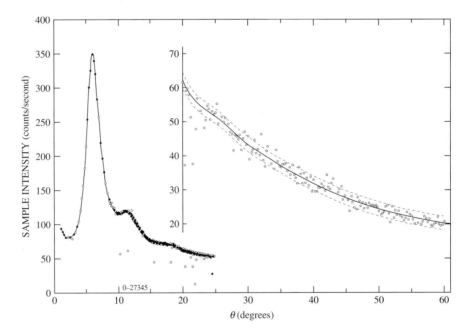

Figure 11.6-2 Intensity of scattered x-ray radiation from liquid argon at $T = -125°C$ and density of 0.982 g/cc normalized to the scattering of an argon atom at a large scattering angle. (As the intensity is low at large scattering angle, the 90 percent confidence interval is shown by the dotted lines). Reprinted with permission from P. J. Mikolaj and C. J. Pings, *J. Chem. Phys.* Vol. 46, 1404 (1967). Copyright 1967, American Institute of Physics.

---

[5]P. J. Mikolaj and C. J. Pings, *J. Chem. Phys.* **46**, 1401, 1967.

11.6 The Radial Distribution Function of Fluids using Coherent X-ray or Neutron Diffraction

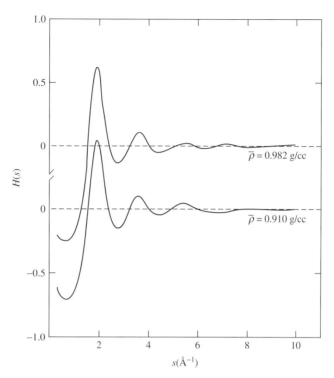

Figure 11.6-3 The structure function $H(s)$ derived from the x-ray diffraction data from liquid argon of Fig. 11.6-2. Reprinted with permission from P. J. Mikolaj and C. J. Pings, J. Chem. Phys. Vol. 46, 1405 (1967). Copyright 1967, American Institute of Physics.

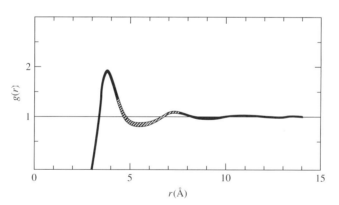

Figure 11.6-4 Argon atom-argon atom radial distribution function computed from the measured x-ray diffraction scattering intensity (with hatched region showing likely uncertainty. Reprinted with permission from P. J. Mikolaj and C. J. Pings, J. Chem. Phys. Vol. 46, 1408 (1967). Copyright 1967, American Institute of Physics.

vector s and the separation distance r between the two scattering sites. Since this exponential function decays to 0 for large values of its argument, this means that scattering from atoms close together (small r) make a significant contribution at large scattering angles (large s values), while scattering from atoms far apart make a significant contribution at small scattering angles (small s values). Also, note from Fig. 11.6-2 that the intensity of scattered radiation decreases significantly at large scattering angles, so the signal-to-noise ratio is decreasing. Finally, because the beam source and detector cannot be in the same place, it is not possible to measure the intensity of scattered radiation at $2\theta = 180°$—or even very close to that. Consequently, because of the combination of decreased accuracy at large scattering angles

210 Chapter 11: Interacting Molecules in a Dense Fluid. Configurational Distribution Functions

(large values of s), and the truncation of data below $2\theta = 180°$, oscillatory behavior in $g(r)$ may be found at very small interatomic separations (before the first peak); but fortunately, this artifact decays rapidly with increasing r, and so does not contaminate the results near and after the first peak.

As is evident from the discussion above, there is a reciprocal relationship between s-space (that is, scattering angle) and physical r-space. That is, scattering at large s values determines the radial distribution function at small r values, and scattering at small s values most affect the radial distribution function at large separations. However, there is not a simple one-to-one relationship, since the radial distribution function at each value of r is determined by an integral of the structure factor over all values of s.

The direct inversion scattering (x-ray or neutron) data to obtain the radial distribution function as used by Mikolaj and Pings is not often used. One reason is the accuracy and range (in s-space) of the data that can be obtained. Because unscattered radiation would overwhelm scattered radiation at small scattering angles (small values of s) and the intensity of scattered radiation becomes very low and therefore noisy (i.e., the signal-to-noise ratio becomes low) at large scattering angles, as shown in Fig. 11.6-2, values of the scattering function $H(s) = (I'(s) - f^2(s))/f^2(s)$ are available over only a limited range of angle (or s), and tend to be of not high accuracy at large s. Consequently, another procedure is used to obtain the radial distribution function from scattering data, which is the reverse of what has so far been described. What is done in this case is to assume a model for the pairwise interactions between the atoms and then use integral equation theory (Chapter 12), computer simulation (Chapter 13), or perturbation theory (Chapter 14) to compute the radial distribution function for this model potential. The radial distribution function so obtained is used in Eq. 11.6-16b to obtain $H(s)$. If the measured and calculated scattering functions agree in the range of s in which the measured $H(s)$ is thought to be accurate, the calculated radial distribution is taken to be that of the fluid (though there may be a question of uniqueness because different interaction models or model parameters may lead to similar predictions for $H(s)$). If the measured and calculated scattering functions do not agree, the interaction model is adjusted by changing the potential parameters or the form of the potential, and the procedure is repeated until reasonable agreement is obtained.

11.7 DETERMINATION OF THE RADIAL DISTRIBUTION FUNCTIONS OF MOLECULAR LIQUIDS

Using x-ray or neutron diffraction to determine the radial distribution function (or, more generally, the structure) of molecular liquids is much more complicated than is the case for atomic liquids that has been described in the previous section. The main problem is that, as shown in Section 11.4, in multiatom molecules there are a number of site-site distribution functions, while only a single scattering function can be obtained from a single measurement. The best that can be obtained is an averaged correlation function between centers of mass. Such information is frequently useful.

Even if a scattering experiment is done on a homonuclear diatomic molecule—for example, bromine—and the single atom scattering function $f(s)$ is computed for a bromine atom in a bromine molecule (which will have a different electron density than an isolated bromine atom), the scattering function $I(s)$ can be deconvoluted using Eq. 11.6-17 to get something like an atom-atom correlation function. However, the real atom-atom pair correlation function will not be obtained. The problem is that the actual atom-atom pair correlation function is not a spherically symmetric function, since there is steric interference along bond that prevents the presence of nonbonded atoms that does not occur in other directions. Nonetheless, such correlation function information can be useful.

For a molecule containing two different atoms—for example, HCl or even water, H_2O—there are three different atom-atom correlation functions. In such cases it is possible to get information on the atom-atom correlation functions (subject to the same caveat as above) by doing three different scattering measurements. For example, for water, because of the large differences in the neutron scattering cross sections of hydrogen and deuterium, and with the assumption that the site-site correlation functions for H_2O and D_2O are identical, neutron scattering measurements have been done for H_2O, D_2O, and a $64H_2O:36D_2O$ mixture.[6] From such data, approximate g_{HH}, g_{OO} and g_{HO} correlation functions can be obtained. An alternative would be to do two of the three neutron scattering measurements mentioned above and one x-ray scattering measurement. From such information one can ascertain, for example, the extent of hydrogen binding and the hydrogen-bond distance.

Generally, the way x-ray or neutron diffraction data are used to understand molecular liquids is the reverse of the procedure so far discussed. What is done, as described in the previous section, is to make an interaction potential model (or force field) for the molecule of interest, use computer simulation or some form of molecular theory to compute all the site-site distribution functions, and then use the multisite version of Eq. 11.6-16b to compute the structure factor $H(s)$ corresponding to the force field used. (Presenting the multiatom form of this equation is beyond the scope of this text.) If the computed and measured structure factors are largely in agreement, then one has some confidence that the force field is reasonably correct, and the site-site correlation functions are meaningful. However, because it is the integrals of several site-site correlation functions that lead to the single measured structure factor, there is the question of uniqueness—that is, it is possible that different force fields that lead to different site-site distribution functions would still result in essentially similar structure factors. However, it is expected that the site-site distribution functions from these different models would be quite similar.

11.8 DETERMINATION OF THE COORDINATION NUMBER FROM THE RADIAL DISTRIBUTION FUNCTION

The physical picture of the structure of an atomic liquid is as follows. As a result of steric hindrance, the atoms cannot overlap. That is, there is a center-to-center separation distance below which there can be not neighboring molecules. For the hard-sphere fluid, this center-to-center distance is equal to twice the radius of the hard-sphere. The radial distribution function is zero in this region, as shown in Fig. 11.6-3 for small values of r. Just beyond this excluded volume separation distance, there will be a first coordination shell of atoms with the result that the density of atoms will be greater than the bulk density, so that $g(r)$ will be greater than unity. For the hard-sphere fluid, $g(r)$ rises as a step function at the hard-sphere diameter and shows a decreasing oscillating behavior around a value of 1 at greater distances (See Figs. 11.3-2b and 11.6-4). For real atoms that interact with softer potentials (such as the Lennard-Jones 12-6 potential), there is a rapid but smooth increase in

---

[6] J. G. Powles, J. C. Dore, and D. I. Page, *Molec. Phys.*, **24**, 1025 (1972).

$g(r)$ at distances near the molecular diameter as shown in Fig. 11.6-4. In this case $g(r)$ reaches a maximum close to the minimum in the interaction potential well, and also exhibits oscillatory thereafter.

The separation distances around the maximum in the radial distribution function corresponds to the location of the first coordination shell around the atom and atom concentrations greater than the average density. As a result of the presence of the atoms in this first coordination shell, there is a steric hindrance that blocks out other molecules, which results in a less-than-average atomic density (that is, $g(r)$ less than 1) beyond the first coordination shell until the formation of a second coordination shell in the vicinity of twice the atomic diameter. However, as the atomic packing in the first coordination shell is not perfect (as seen from the peak in $g(r)$ not being a delta function), the packing beyond the first coordination shell becomes increasingly more disordered in successive coordination shells, which results in a dampening of the further maxima and minima in the radial distribution function with increasing distance. Depending on the density, successive coordination shells (as evidenced by maxima in $g(r)$) may be barely visible beyond 3 to 5 atomic diameters.

The region of the first coordination shell is generally taken to be from the separation distance at which the radial distribution function first increases from 0 to the first minimum in $g(r)$ designated as $r_{min,1}$, and the number of molecules in this first coordination shell is

$$N_1 = 4\pi\rho \int_0^{r_{min,1}} g(r) r^2 \, dr$$

(Note that the integral should not start at $r = 0$, but rather at the value of r where $g(r)$ is first nonzero. However, since there is no contribution to the integral when $g(r)$ is equal to 0, it is easiest to start the integration at $r = 0$.) The coordination number is useful in several ways. First, as we saw in Chapter 10 (Lattice Models), the coordination number is a parameter in molecular modeling. Second, this parameter can be an indicator of the type of crystal structure most similar to that of the disordered liquid. As reference points, the close-packed face centered cubic lattice has a coordination number of 12—that is, 12 nearest neighbors. The body centered cubic lattice has 8, the simple cubic lattice has 6, and the very open diamond structure has a coordination number of 4. The coordination number for liquids at ambient conditions is commonly taken to be 10.

Coordination shells beyond the first can also be defined in terms of successive minima in the radial distribution function. That is the range of the first coordination shell is from 0 to $r_{min,1}$, the range of the second is from $r_{min,1}$ to $r_{min,2}$, etc. where $r_{min,2}$ is the location of the second minimum is $g(r)$, and so on. The number of atoms in the second coordination shell N_2 would be obtained from

$$N_2 = 4\pi\rho \int_{r_{min,1}}^{r_{min,2}} g(r) r^2 \, dr$$

and similarly for the higher coordination shells.

Another way that the radial distribution function can be used is to identify types of molecular bonding. This is done as follows. Atoms have been assigned values of their "van der Waals radii," which is a radius that is used if the atom is modeled as a hard-sphere. Some van der Waals radii are: Hydrogen 1.2 Å, Carbon 1.7 Å, Nitrogen 1.55 Å, Oxygen 1.52 Å, and Fluorine 1.35 Å

Species that are chemically bonded will have bond distances shorter than the sum of their van der Waals radii. For example, in hydrogen fluoride, the H-F bond distance is only 0.92 Å (compared to the sum of the van der Waal radii of 2.55 Å), and the typical C-H bond distance is 1.09 Å. In contrast, the separation distances between nonbonded atoms in liquids is usually equal to or greater than the sum of the atomic radii of the atoms. In addition, there can be associative, nonbonded interactions such as hydrogen bonds. In such cases, the atom-atom separation distances are greater than the chemical bond distance, but somewhat less than the sum of the atomic van der Waals radii. For example, the separation distance of a hydrogen-bonding O-H pair is typically in the range of 2.3 to 2.5 Å, which is less than the sum of their atomic radii of 2.72 Å. Consequently, the location of the first peak in the radial distribution function, and also its magnitude, gives an indication of the type and strength of atom-atom interactions.

11.9 DETERMINATION OF THE RADIAL DISTRIBUTION FUNCTION OF COLLOIDS AND PROTEINS

The diffraction techniques described here have also been used to determine the centers-of-mass correlation function or radial distribution function for large, approximately spherical molecules. In such cases neutron diffraction is generally used, since the scattering length for such a molecule can be obtained from the scattering cross sections of the individual atoms knowing the structure of the molecule. (However, obtaining the x-ray scattering factor would require a quantum mechanical calculation to determine the electron density of the molecule, a very difficult calculation for a molecule with a very large number of atoms.) Since colloids and proteins are composed of many molecules, the centers-of-mass separations are large compared to the size of an atom. Based on the discussion above about the reciprocity of s-space and r-space, this means that scattering intensities need to be measured at low values of s—that is, at very small scattering angles. The technique used is referred to as *small angle neutron scattering* (SANS) using specialized equipment.

For colloids and polymers, the molecular assemblies are so large and with so many atoms that the atom-atom (or site-site) correlations cannot be used. However, if the colloid or polymer can be considered to be approximately spherical (or globular) and the neutron scattering cross section for the macromolecule has been computed, then neutron scattering data can be used to obtain the center-to-center radial distribution. The procedure would be the same as has been described earlier, in that a center-to-center interaction potential would be suggested and integral equation theory, computer simulation, or perturbation theory would be used to compute the radial distribution function from which the scattering function would be calculated and compared with that obtained from neutron scattering. The interaction potential would then be adjusted until agreement with the scattering function was obtained, in which case the radial distribution function is accepted. In the case of globular proteins or colloids, simple potential models such as a simple hard-sphere model may be sufficient; but with a diameter characteristic of the macromolecule[7] a sticky hard-sphere model such as discussed in Section 8.3 can be used. We discuss more sophisticated models in the context of the potential of mean force in the Chapter 12.

---

[7] P. N. Pusey and W. van Megen, *Nature*, **320**, 340 (1986).

CHAPTER 11 PROBLEMS

11.1 The thermodynamic properties calculated from the radial distribution function can be very sensitive to the intermolecular potential function used in the calculation. You will establish this by direct calculation. Below are the values of Mikolaj and Pings (*J. Chem. Phys.* **46**, 1401 and 1412 (1967)) for the radial distribution function at argon as determined from x-ray scattering.

Calculate the pressure and configurational contribution to the internal energy of argon at the indicated conditions assuming a Lennard-Jones 12-6 potential for argon with the following parameters:

a. $\sigma = 3.405$ Å (Pings)
$\varepsilon/k = 120$ K

b. $\sigma = 3.499$ Å (2nd Virial)
$\varepsilon/k = 118.13$ K

c. $\sigma = 3.434$ Å (gas viscosity)
$\varepsilon/k = 120.72$

Note that the configurational contribution to the internal energy, U^C, and the compressibility factor, $(P/\rho kT)_c$, can be obtained from the radial distribution function by the relations

$$U^C = 2\pi N\rho \int_0^\infty u(r)g(r)r^2\,dr \quad \text{and}$$

$$\left(\frac{P}{\rho kT}\right)^C = -\frac{2\pi\rho}{3kT}\int_0^\infty \frac{du(r)}{dr}g(r)r^3\,dr$$

Values of $[g(r)-1]$ for argon in the critical region at $T = -130°C$ and $\rho = 0.910$ g/cc.

r(Å)	g(r)-1	r(Å)	g(r)-1	r(Å)	g(r)-1
3.0	−1.0276	7.0	0.0479	12.0	0.0008
3.1	−0.8508	7.1	0.0604	12.2	−0.0004
3.2	−0.5663	7.2	0.0699	12.4	−0.0013
3.3	−0.2156	7.3	0.0751	12.6	−0.0020
3.4	0.1455	7.4	0.0768	12.8	−0.0019
3.5	0.4639	7.5	0.0758	13.0	−0.0019
3.6	0.7002	7.6	0.0718	13.2	−0.0015
3.7	0.8429	7.7	0.0650	13.4	−0.0010
3.8	0.8975	7.8	0.0562	13.6	−0.0003
3.9	0.8787	7.9	0.0465	13.8	0.0003
4.0	0.8064	8.0	0.0362	14.0	0.0007
4.1	0.6985	8.1	0.0252	14.2	0.0012
4.2	0.5654	8.2	0.0147	14.4	0.0014
4.3	0.4375	8.3	0.0038	14.6	0.0016
4.4	0.3145	8.4	−0.0045	14.8	0.0017
4.5	0.1990	8.5	−0.0108	15.0	0.0017
4.6	0.1041	8.6	−0.0169	15.2	0.0018
4.7	0.0236	8.7	−0.0215	15.4	0.0015
4.8	−0.0277	8.8	−0.0240	15.6	0.0015
4.9	−0.0700	8.9	−0.0253	15.8	0.0014
5.0	−0.0963	9.0	−0.0258	16.0	0.0012
5.1	−0.1182	9.1	−0.0256	16.2	0.0011
5.2	−0.1319	9.2	−0.0243	16.4	0.0011
5.3	−0.1445	9.3	−0.0225	16.6	0.0009
5.4	−0.1511	9.4	−0.0200	16.8	0.0008
5.5	−0.1533	9.5	−0.0170	17.0	0.0007
5.6	−0.1498	9.6	−0.0135	17.2	0.0007
5.7	−0.1446	9.7	−0.0099	17.4	0.0007
5.8	−0.1362	9.8	−0.0069	17.6	0.0005
5.9	−0.1241	9.9	−0.0027	17.8	0.0005
6.0	−0.1127	10.0	0.0003	18.0	0.0004
6.1	−0.0966	10.2	0.0059	18.2	0.0004
6.2	−0.0830	10.4	0.0102	18.4	0.0003
6.3	−0.0679	10.6	0.0132	18.6	0.0003
6.4	−0.0524	10.8	0.0127	18.8	0.0003
6.5	−0.0329	10.0	0.0117	19.0	0.0003
6.6	−0.0180	10.2	0.0092	19.2	0.0003
6.7	0.0006	10.4	0.0071	19.4	0.0003
6.8	0.0163	10.6	0.0045	19.6	0.0003
6.9	0.0317	10.8	0.0023	19.8	0.0003
				20.0	0.0003

11.2 Show that the internal energy of a mixture is

$$U(\rho, \underline{x}, T) = \frac{3}{2}NkT + 2\pi N\rho \sum_i \sum_j x_i x_j \int_0^\infty u_{ij}(r)g_{ij}(r)r^2\,dr$$

11.3 One very simple approximation for the radial distribution functions in mixtures that has been made in the past is the one-fluid model for which

$$g_{ii}(r) = g_{ij}(r) = g_{jj}(r)$$

Show that with this approximation the internal energy of the mixture can be written as

$$U(\rho, \underline{x}, T) = \frac{3}{2}NkT + 2\pi N\rho \int_0^\infty \bar{u}(r, \underline{x})g(r)r^2 dr$$

where $\bar{u}(r, \underline{x}) = \sum_i \sum_j x_i x_j u(r, \underline{x})$ is an effective mixture potential. Obtain an expression for this effective mixture potential in terms of $u_{ii}(r), u_{ij}(r)$ and $u_{jj}(r)$. Further, if each of these intermolecular potentials are of the Lennard-Jones 12-6 form, how are the parameters of $u(r)$ related to those in $u_{ii}(r), u_{ij}(r)$ and $u_{jj}(r)$? (Note that this model is also referred to as a one-fluid model.)

11.4 Draw the three low-density radial distribution functions for a binary mixture of Lennard-Jones 12-6 fluids for which $\varepsilon_{11} = 2\varepsilon_{22}$ and $\sigma_{11} = \sigma_{22}$, assuming the Lorentz-Berthelot combining rules of Eqs. 8.2-4 and 8.2-5 apply with $k_{12} = 0$.

11.5 The average of any property $\theta(N, V, T, \underline{r}_1, \underline{r}_2, \ldots, \underline{r}_n)$ in the canonical ensemble, represented by $\bar{\theta}(N, V, T)$, is computed from

$$\bar{\theta}(N, V, T) = \frac{\frac{q^N}{N!\Lambda^{3N}} \int \cdots \int \theta(N, V, T, \underline{r}_1, \underline{r}_2, \ldots, \underline{r}_N) e^{-\frac{u(\underline{r}_1, \underline{r}_2, \ldots, \underline{r}_N)}{kT}} d\underline{r}_1 d\underline{r}_2 \ldots d\underline{r}_N}{\frac{q^N}{N!\Lambda^{3N}} \int \cdots \int e^{-\frac{u(\underline{r}_1, \underline{r}_2, \ldots, \underline{r}_N)}{kT}} d\underline{r}_1 d\underline{r}_2 \ldots d\underline{r}_N}$$

$$= \frac{\int \cdots \int \theta(N, V, T, \underline{r}_1, \underline{r}_2, \ldots, \underline{r}_N) e^{-\frac{u(\underline{r}_1, \underline{r}_2, \ldots, \underline{r}_N)}{kT}} d\underline{r}_1 d\underline{r}_2 \ldots d\underline{r}_N}{\int \cdots \int e^{-\frac{u(\underline{r}_1, \underline{r}_2, \ldots, \underline{r}_N)}{kT}} d\underline{r}_1 d\underline{r}_2 \ldots d\underline{r}_N} \equiv \langle \theta(N, V, T, \underline{r}_1, \underline{r}_2, \ldots \underline{r}_N) \rangle_N$$

where the notation $\langle \rangle$ has been used to indicate that the property in the system of interaction molecules is to be averaged using the configuration integral.

a. Explain why the chemical potential of a species can be computed using

$$\mu(N, V, T) = A(N+1, V, T) - A(N, V, T)$$

b. Show that

$$\mu(N, V, T) = A(N+1, V, T) - A(N, V, T) = -kT \ln \frac{Q(N+1, V, T)}{Q(N, V, T)}$$

$$= -kT \ln \left[\frac{q}{(N+1)\Lambda^3} \frac{\int \cdots \int e^{-\frac{u(\underline{r}_1, \underline{r}_2, \ldots, \underline{r}_{N+1})}{kT}} d\underline{r}_1 d\underline{r}_2 \ldots d\underline{r}_{N+1}}{\int \cdots \int e^{-\frac{u(\underline{r}_1, \underline{r}_2, \ldots, \underline{r}_N)}{kT}} d\underline{r}_1 d\underline{r}_2 \ldots d\underline{r}_N} \right]$$

c. Finally, show that

$$\mu(N, V, T) = -kT \ln \left[\frac{q}{(N+1)\Lambda^3} \int \left\langle e^{-\left[\frac{u(\underline{r}_1, \underline{r}_2, \ldots, \underline{r}_{N+1})}{kT} - \frac{u(\underline{r}_1, \underline{r}_2, \ldots, \underline{r}_N)}{kT}\right]} \right\rangle_N d\underline{r}_{N+1} \right]$$

Chapter 12

Integral Equation Theories for the Radial Distribution Function

In this chapter we consider the method of determining radial distribution functions based on the use of integral equations. There are two very different starting points for deriving integral equations for the determination of the radial distribution function. We first consider one method (Section 12.1) and then a completely different, complementary method (Section 12.3).

INSTRUCTIONAL OBJECTIVES FOR CHAPTER 12

The goals for this chapter are for the student to:

- Understand the basis for integral equation methods used to obtain the radial distribution function
- Understand the basis for the Ornstein-Zernike integral equation
- Understand the origin of the Percus-Yevick equation for hard-spheres

12.1 THE YVON-BORN-GREEN (YBG) EQUATION

The method we consider first starts from the definition of the radial distribution function, Eq. 11.1-8a

$$g^{(2)}(\underline{r}_1, \underline{r}_2; T, \rho) = \frac{V^2 \int \cdots \int e^{-u(\underline{r}_1, \underline{r}_2, \ldots, \underline{r}_N)/kT} d\underline{r}_3 \ldots d\underline{r}_N}{\int \cdots \int e^{-u(\underline{r}_1, \underline{r}_2, \ldots, \underline{r}_N)/kT} d\underline{r}_1 \ldots d\underline{r}_N} \quad (11.1.8a)$$

Now, taking the derivative of the radial distribution function with respect to the vector $\underline{r}_2$, we have

$$\frac{\partial g^{(2)}(\underline{r}_1, \underline{r}_2; T, \rho)}{\partial \underline{r}_2} = \frac{V^2 \int \cdots \int \frac{\partial}{\partial \underline{r}_2} e^{-u(\underline{r}_1, \underline{r}_2, \ldots, \underline{r}_N)/kT} d\underline{r}_3 \ldots d\underline{r}_N}{\int \cdots \int e^{-u(\underline{r}_1, \underline{r}_2, \ldots, \underline{r}_N)/kT} d\underline{r}_1 \ldots d\underline{r}_N}$$

$$= \frac{-\frac{V^2}{kT} \int \cdots \int e^{-u(\underline{r}_1, \underline{r}_2, \ldots, \underline{r}_N)/kT} \frac{\partial u(\underline{r}_1, \underline{r}_2, \ldots, \underline{r}_N)}{\partial \underline{r}_2} d\underline{r}_3 \ldots d\underline{r}_N}{\int \cdots \int e^{-u(\underline{r}_1, \underline{r}_2, \ldots, \underline{r}_N)/kT} d\underline{r}_1 \ldots d\underline{r}_N} \quad (12.1\text{-}1)$$

12.1 The Yvon-Born-Green (YBG) Equation

(Note that since the integration over all position vectors has been carried out in the denominator, it is a function of only N, V, and T, and not $\underline{r}_2$.) Now to proceed, we will assume pairwise additivity of the interaction potential, that is

$$u(\underline{r}_1,\ldots,\underline{r}_N) = \sum_i \sum_{\substack{j \\ 1 \leq i < j \leq N}} u(r_{ij}) = u(r_{12}) + u(r_{13}) + u(r_{14}) + \cdots + u(r_{N-1,N})$$

(7.3-3)

Therefore

$$\frac{\partial u(\underline{r}_1,\ldots,\underline{r}_N)}{\partial \underline{r}_2} = \frac{\partial}{\partial \underline{r}_2}\left(\sum_i \sum_{\substack{j \\ 1 \leq i < j \leq N}} u(r_{ij})\right)$$

$$= \frac{\partial u(r_{12})}{\partial \underline{r}_2} + \frac{\partial u(r_{23})}{\partial \underline{r}_2} + \frac{\partial u(r_{24})}{\partial \underline{r}_2} + \cdots + \frac{\partial u(r_{2N})}{\partial \underline{r}_2} \quad (12.1\text{-}2)$$

and

$$\frac{\partial g^{(2)}(\underline{r}_1,\underline{r}_2;T,\rho)}{\partial \underline{r}_2} = \frac{-\dfrac{V^2}{kT}\int\cdots\int e^{-u(\underline{r}_1,\underline{r}_2,\ldots,\underline{r}_N)/kT}\dfrac{\partial u(\underline{r}_1,\underline{r}_2,\ldots,\underline{r}_N)}{\partial \underline{r}_2}\,d\underline{r}_3\ldots d\underline{r}_N}{\int\cdots\int e^{-u(\underline{r}_1,\underline{r}_2,\ldots,\underline{r}_N)/kT}\,d\underline{r}_1\ldots d\underline{r}_N}$$

$$= \frac{-\dfrac{V^2}{kT}\int\cdots\int e^{-u(\underline{r}_1,\underline{r}_2,\ldots,\underline{r}_N)/kT}\left[\dfrac{\partial u(r_{12})}{\partial \underline{r}_2}+\dfrac{\partial u(r_{23})}{\partial \underline{r}_2}+\dfrac{\partial u(r_{24})}{\partial \underline{r}_2}+\cdots+\dfrac{\partial u(r_{2N})}{\partial \underline{r}_2}\right]d\underline{r}_3\ldots d\underline{r}_N}{\int\cdots\int e^{-u(\underline{r}_1,\underline{r}_2,\ldots,\underline{r}_N)/kT}\,d\underline{r}_1\ldots d\underline{r}_N}$$

$$= -\frac{\dfrac{V^2}{kT}\dfrac{\partial u(r_{12})}{\partial \underline{r}_2}\int\cdots\int e^{-u(\underline{r}_1,\underline{r}_2,\ldots,\underline{r}_N)/kT}\,d\underline{r}_3\ldots d\underline{r}_N}{\int\cdots\int e^{-u(\underline{r}_1,\underline{r}_2,\ldots,\underline{r}_N)/kT}\,d\underline{r}_1\ldots d\underline{r}_N}$$

$$-\frac{\dfrac{V^2}{kT}\int\cdots\int e^{-u(\underline{r}_1,\underline{r}_2,\ldots,\underline{r}_N)/kT}\left[\dfrac{\partial u(r_{23})}{\partial \underline{r}_2}+\dfrac{\partial u(r_{24})}{\partial \underline{r}_2}+\cdots+\dfrac{\partial u(r_{2N})}{\partial \underline{r}_2}\right]d\underline{r}_3\ldots d\underline{r}_N}{\int\cdots\int e^{-u(\underline{r}_1,\underline{r}_2,\ldots,\underline{r}_N)/kT}\,d\underline{r}_1\ldots d\underline{r}_N}$$

$$= -\frac{1}{kT}\frac{\partial u(r_{12})}{\partial \underline{r}_2}g^{(2)}(\underline{r}_1,\underline{r}_2;T,\rho) - \frac{1}{kT}\frac{V^2(N-2)\int\cdots\int e^{-u(\underline{r}_1,\underline{r}_2,\ldots,\underline{r}_N)/kT}\left[\dfrac{\partial u(r_{23})}{\partial \underline{r}_2}\right]d\underline{r}_3\ldots d\underline{r}_N}{\int\cdots\int e^{-u(\underline{r}_1,\underline{r}_2,\ldots,\underline{r}_N)/kT}\,d\underline{r}_1\ldots d\underline{r}_N}$$

(12.1-3)

or
$$\frac{\partial g^{(2)}(\underline{r}_1, \underline{r}_2; T, \rho)}{\partial \underline{r}_2}$$
$$= -\frac{1}{kT} \frac{\partial u(r_{12})}{\partial \underline{r}_2} g^{(2)}(\underline{r}_1, \underline{r}_2; T, \rho) \qquad (12.1\text{-}4)$$
$$- \frac{V^2(N-2)}{kT} \int \frac{\partial u(r_{23})}{\partial \underline{r}_2} \left[\frac{\int \cdots \int e^{-u(\underline{r}_1, \underline{r}_2, \ldots, \underline{r}_N)/kT} d\underline{r}_4 \ldots d\underline{r}_N}{\int \cdots \int e^{-u(\underline{r}_1, \underline{r}_2, \ldots, \underline{r}_N)/kT} d\underline{r}_1 \ldots d\underline{r}_N} \right] d\underline{r}_3$$

However, from Eq. 11.1-9a
$$g^{(3)}(\underline{r}_1, \underline{r}_2, \underline{r}_3; T, \rho) = \frac{V^3 \int \cdots \int e^{-u/kT} d\underline{r}_4 \ldots d\underline{r}_N}{\int \cdots \int e^{-u/kT} d\underline{r}_1 \ldots d\underline{r}_N} \qquad (11.1\text{-}9a)$$

so that
$$\frac{\partial g^{(2)}(\underline{r}_1, \underline{r}_2; T, \rho)}{\partial \underline{r}_2}$$
$$= -\frac{1}{kT} \frac{\partial u(r_{12})}{\partial \underline{r}_2} g^{(2)}(\underline{r}_1, \underline{r}_2; T, \rho) - \frac{(N-2)}{VkT} \int \frac{\partial u(r_{23})}{\partial \underline{r}_2} g^{(3)}(\underline{r}_1, \underline{r}_2, \underline{r}_3; T, \rho) \, d\underline{r}_3$$
$$= -\frac{1}{kT} \frac{\partial u(r_{12})}{\partial \underline{r}_2} g^{(2)}(\underline{r}_1, \underline{r}_2; T, \rho) - \frac{\rho}{kT} \int \frac{\partial u(r_{23})}{\partial \underline{r}_2} g^{(3)}(\underline{r}_1, \underline{r}_2, \underline{r}_3; T, \rho) \, d\underline{r}_3$$
$$(12.1\text{-}5)$$

where we have neglected the difference between N and $N-2$. This last equation can be rewritten as
$$kT \frac{\partial \ln g^{(2)}(\underline{r}_1, \underline{r}_2; T, \rho)}{\partial \underline{r}_2} = -\frac{\partial u(r_{12})}{\partial \underline{r}_2} - \rho \int \frac{\partial u(r_{23})}{\partial \underline{r}_2} \frac{g^{(3)}(\underline{r}_1, \underline{r}_2, \underline{r}_3; T, \rho)}{g^{(2)}(\underline{r}_1, \underline{r}_2; T, \rho)} \, d\underline{r}_3 \qquad (12.1\text{-}6)$$

This is known as the equation of Yvon-Born-Green, each of whom independently derived this integro-differential equation.

The Yvon-Born-Green (YBG) equation relates the unknown two-body correlation function to the even less known three-body correlation function. So to proceed, one needs some information about the three-body correlation function. However, before we consider this, notice that in the limit of zero density ($\rho \to 0$), this equation reduces to
$$kT \frac{\partial \ln g^{(2)}(\underline{r}_1, \underline{r}_2; T, \rho)}{\partial \underline{r}_2} = -\frac{\partial u(r_{12})}{\partial \underline{r}_2} \qquad \text{as } \rho \to 0 \qquad (12.1\text{-}7)$$

which on integration gives us
$$g^{(2)}(\underline{r}_1, \underline{r}_2; T, \rho) = C e^{-u(r_{12})/kT} = e^{-u(r_{12})/kT} \qquad \text{as } \rho \to 0 \qquad (12.1\text{-}8)$$

where the constant of integration C has been set to unity by invoking the condition that $g^{(2)}(\underline{r}_1, \underline{r}_2; T, \rho) = 1$, as the molecular separation distance becomes very large so that $u(r_{12}) = 0$. Note that previously we had obtained Eq. 12.1-8 using the Mayer cluster expansion (see Section 11.3 and Eq. 11.3-5).

12.2 THE KIRKWOOD SUPERPOSITION APPROXIMATION

At higher densities we cannot solve Eq. 12.1-6 for the two-particle radial distribution function $g^{(2)}(\underline{r}_1, \underline{r}_2; T, \rho)$, because this equation also contains the three-particle distribution function $g^{(3)}(\underline{r}_1, \underline{r}_2, \underline{r}_3; T, \rho)$. The earliest assumption or closure made was the superposition approximation of Kirkwood that

$$g^{(3)}(\underline{r}_1, \underline{r}_2, \underline{r}_3; T, \rho) = g^{(2)}(\underline{r}_1, \underline{r}_2; T, \rho) g^{(2)}(\underline{r}_1, \underline{r}_3; T, \rho) g^{(2)}(\underline{r}_2, \underline{r}_3; T, \rho) \quad \textbf{(12.2-1)}$$

or that the three-particle distribution is the product of all two-particle distribution functions. Note that in the limit of zero density, this assumption is consistent with the assumption of pairwise additivity of the potential. This is seen as follows:

$$\ln g^{(3)}(\underline{r}_1, \underline{r}_2, \underline{r}_3; T, \rho = 0) = \ln g^{(2)}(\underline{r}_1, \underline{r}_2; T, \rho = 0) + \ln g^{(2)}(\underline{r}_1, \underline{r}_3; T, \rho = 0)$$

$$+ \ln g^{(2)}(\underline{r}_2, \underline{r}_3; T, \rho = 0) - \frac{u(\underline{r}_1, \underline{r}_2, \underline{r}_3)}{kT}$$

$$= -\frac{u(\underline{r}_1, \underline{r}_2)}{kT} - \frac{u(\underline{r}_1, \underline{r}_3)}{kT} - \frac{u(\underline{r}_2, \underline{r}_3)}{kT}$$

or

$$u(\underline{r}_1, \underline{r}_2, \underline{r}_3) = u(\underline{r}_1, \underline{r}_2) + u(\underline{r}_1, \underline{r}_3) + u(\underline{r}_2, \underline{r}_3) \quad \textbf{(12.2-2)}$$

which is the pairwise additivity assumption.

Now, using the Kirkwood superposition approximation in the YBG equation, we get

$$kT \frac{\partial \ln g^{(2)}(\underline{r}_1, \underline{r}_2; T, \rho)}{\partial \underline{r}_2} = -\frac{\partial u(r_{12})}{\partial \underline{r}_2} - \rho \int \frac{\partial u(r_{23})}{\partial \underline{r}_2} \frac{g^{(3)}(\underline{r}_1, \underline{r}_2, \underline{r}_3; T, \rho)}{g^{(2)}(\underline{r}_1, \underline{r}_2; T, \rho)} d\underline{r}_3 =$$

$$= -\frac{\partial u(r_{12})}{\partial \underline{r}_2} - \rho \int \frac{\partial u(r_{23})}{\partial \underline{r}_2} \frac{g^{(2)}(\underline{r}_1, \underline{r}_2; T, \rho) g^{(2)}(\underline{r}_1, \underline{r}_3; T, \rho) g^{(2)}(\underline{r}_2, \underline{r}_3; T, \rho)}{g^{(2)}(\underline{r}_1, \underline{r}_2; T, \rho)} d\underline{r}_3$$

$$= -\frac{\partial u(r_{12})}{\partial \underline{r}_2} - \rho \int \frac{\partial u(r_{23})}{\partial \underline{r}_2} g^{(2)}(\underline{r}_1, \underline{r}_3; T, \rho) g^{(2)}(\underline{r}_2, \underline{r}_3; T, \rho) d\underline{r}_3$$

or

$$kT \frac{\partial \ln g^{(2)}(\underline{r}_1, \underline{r}_2; T, \rho)}{\partial \underline{r}_2}$$

$$= -\frac{\partial u(r_{12})}{\partial \underline{r}_2} - \rho \int \frac{\partial u(r_{23})}{\partial \underline{r}_2} g^{(2)}(\underline{r}_1, \underline{r}_3; T, \rho) g^{(2)}(\underline{r}_2, \underline{r}_3; T, \rho) d\underline{r}_3 \quad \textbf{(12.2-3)}$$

This is an integro-differential equation for the two-particle distribution function $g^{(2)}(\underline{r}_1, \underline{r}_2; T, \rho)$ that can, in principle, be solved (but only with considerable difficulty). In fact, the results obtained from solving this equation for any potential function are not in good agreement with computer simulation results for the same potential.

Consider the procedure used to arrive at Eq. 12.2-3. We started with the definition of $g^{(2)}(\underline{r}_1, \underline{r}_2; T, \rho)$ and took its derivative with respect to the molecular separation distance, which resulted in an equation (Eq. 12.1-6) that contained $g^{(3)}(\underline{r}_1, \underline{r}_2, \underline{r}_3; T, \rho)$. To be able to solve this equation, we then made an assumption about $g^{(3)}(\underline{r}_1, \underline{r}_2, \underline{r}_3; T, \rho)$—that is, we made an assumption (Eq. 12.2-1) about the three-particle distribution function in order to be able to solve for the two-particle distribution function. We can then imagine that a way to improve upon this procedure is to develop a hierarchy of such equations. The next level would be to start with the definition of $g^{(3)}(\underline{r}_1, \underline{r}_2, \underline{r}_3; T, \rho)$ and take a derivative with respect to a position vector, which, following the procedure that led to Eq. 12.1-6, would now give an integro-differential equation in containing $g^{(4)}(\underline{r}_1, \underline{r}_2, \underline{r}_3, \underline{r}_4; T, \rho)$. We could now make an assumption about this quantity (for example, like the Kirkwood superposition approximation, that it was a product of all three-particle distribution functions) and solve the resulting equation for $g^{(3)}(\underline{r}_1, \underline{r}_2, \underline{r}_3; T, \rho)$, which would be an improvement over Eq. 12.2-1. The resulting expression for $g^{(3)}(\underline{r}_1, \underline{r}_2, \underline{r}_3; T, \rho)$ would then be used in Eq. 12.1-6 to obtain an improved estimate for $g^{(2)}(\underline{r}_1, \underline{r}_2; T, \rho)$. If this was found to be unsatisfactory, one could then go to the next level of deriving an integro-differential equation for $g^{(4)}(\underline{r}_1, \underline{r}_2, \underline{r}_3, \underline{r}_4; T, \rho)$ in terms of $g^{(5)}(\underline{r}_1, \underline{r}_2, \underline{r}_3, \underline{r}_4, \underline{r}_5; T, \rho)$, solving the resulting equation for $g^{(4)}(\underline{r}_1, \underline{r}_2, \underline{r}_3, \underline{r}_4; T, \rho)$, using the result in the equation for $g^{(3)}(\underline{r}_1, \underline{r}_2, \underline{r}_3; T, \rho)$, solving it, and then using that result in Eq. 12.1-6 to solve for $g^{(2)}(\underline{r}_1, \underline{r}_2; T, \rho)$. However, the method outlined here is so tedious that it is rarely used.

The key idea of this hierarchy of equations is that if one wants to obtain a good lower-order (i.e., two-particle) distribution, one should not make the superposition approximation about that distribution function, but rather about some higher-order particle distribution such as the three-particle, four-particle, or higher-order distribution functions. The expectation, then, is that the higher the order of the distribution function about which the superposition approximation is made, the more accurate the resulting two-particle distribution function will be.

12.3 THE ORNSTEIN-ZERNIKE EQUATION

The second integral equation method we consider for determining the radial distribution function has a very different basis, and makes use of a concept similar to that discussed previously in trying to understand the behavior of the radial distribution function. In particular, the idea is that there is a direct correlation between molecules 1 and 2 (though it is not simply the low-density result of a Boltzmann factor in the interaction energy), and then an indirect effect as a result of the correlations between intervening single molecules (1-3-2) and chains of molecules (1-3-4-2, 1-3-·····-2, etc.) between molecules 1 and 2. We describe this by introducing the following two functions:

$$\text{the total correlation function } h(\underline{r}_1, \underline{r}_2) = g^{(2)}(\underline{r}_1, \underline{r}_2) - 1$$
$$\text{and the direct correlation function } c(\underline{r}_1, \underline{r}_2). \quad (12.3\text{-}1)$$

The expectation is that the range of the direct correlation function $c(\underline{r}_1, \underline{r}_2)$ will be approximately that of the intermolecular potential.

The simplest (and least accurate) approximation is that the total correlation and the direct correlation are equal—that is, $h(\underline{r}_1, \underline{r}_2) = c(\underline{r}_1, \underline{r}_2)$. The next level of approximation is that the total correlation is equal to the sum of the direct correlation between

12.3 The Ornstein-Zernike Equation

molecules 1 and 2 and the effect on the total correlation function due to the indirect correlations from all third molecules, wherever they are located (consequently, one will have to integrate over their position), that is

$$h(\underline{r}_1, \underline{r}_2) = c(\underline{r}_1, \underline{r}_2) + \frac{N}{V} \int c(\underline{r}_1, \underline{r}_3) c(\underline{r}_3, \underline{r}_2) \, d\underline{r}_3$$

$$= c(\underline{r}_1, \underline{r}_2) + \rho \int c(\underline{r}_1, \underline{r}_3) c(\underline{r}_3, \underline{r}_2) \, d\underline{r}_3 \quad (12.3\text{-}2)$$

where the factor of N arises from the fact that we have N (actually N-2) choices for the third molecule, and V is the normalization factor, since the integration is carried out over the total volume. Continuing to higher levels of accuracy, we can consider additional indirect correlations from increasingly greater numbers of intermediate molecules (1-3-4-2, 1-3-4-5-2, etc.) and obtain

$$\begin{aligned}
h(\underline{r}_1, \underline{r}_2) = & \, c(\underline{r}_1, \underline{r}_2) + \rho \int c(\underline{r}_1, \underline{r}_3) c(\underline{r}_3, \underline{r}_2) \, d\underline{r}_3 \\
& + \rho^2 \iint c(\underline{r}_1, \underline{r}_3) c(\underline{r}_3, \underline{r}_4) c(\underline{r}_4, \underline{r}_2) \, d\underline{r}_3 \, d\underline{r}_4 \\
& + \rho^3 \iiint c(\underline{r}_1, \underline{r}_3) c(\underline{r}_3, \underline{r}_4) c(\underline{r}_4, \underline{r}_5) c(\underline{r}_5, \underline{r}_2) \, d\underline{r}_3 \, d\underline{r}_4 \, d\underline{r}_5 + \cdots
\end{aligned}$$
$$(12.3\text{-}3)$$

It is easy to show that this last equation is equivalent to

$$h(\underline{r}_1, \underline{r}_2) = c(\underline{r}_1, \underline{r}_2) + \rho \int c(\underline{r}_1, \underline{r}_3) h(\underline{r}_3, \underline{r}_2) \, d\underline{r}_3 \quad (12.3\text{-}4)$$

which is the Ornstein-Zernike equation.[1] The way that one proves the equivalence between Eqs. 12.3-3 and 12.3-4 is by repeatedly substituting Eq. 12.3-3 into itself. For example, after one substitution we have

$$\begin{aligned}
h(\underline{r}_1, \underline{r}_2) = & \, c(\underline{r}_1, \underline{r}_2) + \rho \int c(\underline{r}_1, \underline{r}_3) c(\underline{r}_3, \underline{r}_2) \, d\underline{r}_3 \\
& + \rho^2 \iint c(\underline{r}_1, \underline{r}_3) c(\underline{r}_3, \underline{r}_4) h(\underline{r}_4, \underline{r}_2) \, d\underline{r}_3 \, d\underline{r}_4
\end{aligned} \quad (12.3\text{-}5)$$

Repeating this process leads to Eq. 12.3-4.

Much like the discussion of graphs with connecting points and vertices used in deriving the virial equation of state from the canonical ensemble, we can use graphs to represent the various integrals comprising the Ornstein-Zernike equation. For example,

the first term $c(\underline{r}_1, \underline{r}_2)$ will be represented by ○—○

the next term $\int c(\underline{r}_1, \underline{r}_3) c(\underline{r}_3, \underline{r}_2) d\underline{r}_3$ is ○—●—○

and $\iint c(\underline{r}_1, \underline{r}_3) c(\underline{r}_3, \underline{r}_4) c(\underline{r}_4, \underline{r}_2) d\underline{r}_3 d\underline{r}_4$ is ○—●—●—○

---

[1] L. S. Ornstein and F. Zernike, *Proc. Sect. Sci. K. ned. Akad. Wet.* **17**, 793 (1914).

222   Chapter 12: Integral Equation Theories for the Radial Distribution Function

Higher order terms will result in graphs such as

and so on. In each of these cases, an unfilled circle represents the position vector $\underline{r}_1$ or $\underline{r}_2$ that is not integrated over, and a filled circle represents a position vector in an integration.

12.4 CLOSURES FOR THE ORNSTEIN-ZERNIKE EQUATION

The Ornstein-Zernike equation is a single equation in two unknowns, $h(\underline{r}_1, \underline{r}_2)$ and $c(\underline{r}_1, \underline{r}_2)$. Therefore, in order to use this equation to obtain a solution for the radial distribution function, $g(\underline{r}_1, \underline{r}_2)$—or, equivalently, $h(\underline{r}_1, \underline{r}_2)$—we need to either specify the function $c(\underline{r}_1, \underline{r}_2)$ or provide a relationship between $h(\underline{r}_1, \underline{r}_2)$ and $c(\underline{r}_1, \underline{r}_2)$. There are various approximations or closures that have been used. The simplest is the so-called mean spherical approximation:[2]

$$\begin{array}{ll} h(\underline{r}_1, \underline{r}_2) = -1 \quad \text{or} \quad g(\underline{r}_1, \underline{r}_2) = 0 & \text{for } r_{12} < d \\ c(\underline{r}_1, \underline{r}_2) = -u(\underline{r}_1, \underline{r}_2)/kT & \text{for } r_{12} > d \end{array} \quad (12.4\text{-}1)$$

In this equation, d is a characteristic diameter—for example, the diameter of the hard-sphere molecule or some other distance measure for soft interaction potentials.

Another commonly used closure is the Percus-Yevick (or PY) approximation, obtained as follows. Defining a new function, $y(r) = g(r)e^{u(r)/kT}$ results in

$$c(r) = g(r) - y(r) = e^{-u(r)/kT} y(r) - y(r) = f(r) y(r) \quad (12.4\text{-}2)$$

and when used in the Ornstein-Zernike equation gives

$$y(r_{12}) = 1 + \rho \int f(r_{13}) y(r_{13}) h(r_{23}) \, d\underline{r}_3$$

or the following equation containing only $g(r)$

$$g(r_{12})e^{u(r_{12})/kT} = 1 + \rho \int (e^{u(r_{13})/kT} - 1) g(r_{13}) e^{u(r_{13})/kT} (g(r_{23}) - 1) \, d\underline{r}_3 \quad (12.4\text{-}3)$$

This is an integral equation for $g(r)$, known as the Percus-Yevick equation can be solved (with some difficulty) for the radial distribution function for a specified interaction potential. Table 12.4-4 lists the results for the radial distribution function of the hard-sphere fluid obtained in the Percus-Yevick approximation as a function of $r^* = r/\sigma$ using the MATLAB® program PYHS available on the website www.wiley.com/college/sandler.

A different approximation is the hypernetted chain (HNC)[3] closure

$$c(r) = f(r) y(r) + y(r) - 1 - \ln y(r) \quad (12.4\text{-}4)$$

---

[2] J. L. Lebowitz and J. K. Percus, *Phys. Rev.* **144**, 251 (1966).

[3] J. M. J. van Leeuwen, J. Groeneveld, and J. de Boer, *Physica* **25**, 792 (1959).

12.4 Closures for the Ornstein-Zernike Equation

leading to the HNC equation

$$\ln y(r_{12}) = \rho \int \left[h(r_{13}) - \ln g(r_{13}) - \frac{u(r_{13})}{kT} \right] [g(r_{23}) - 1] \, d\underline{r}_3$$

$$= \rho \int \left[h(r_{13}) - \ln g(r_{13}) - \frac{u(r_{13})}{kT} \right] h(r_{23}) \, d\underline{r}_3 \qquad (12.4\text{-}5)$$

that can also be solved numerically (again, with difficulty) for the radial distribution function for a specified interaction potential.

Though the closures of the Ornstein-Zernike equation have been introduced in an apparently arbitrary manner here, they can in fact be derived by use of complicated graph theoretic methods[4] by summing the contributions of graphs such as those shown in Section 12.3, in which one approximates the sum of this infinite collection of graphs by including some of the terms and neglecting others. Such a discussion is well beyond the scope of this book. Each of the equations in this section is most commonly solved numerically rather than analytically.

It is useful to note that in terms of the notation in this section, we can rewrite the compressibility equation of Eq. 11.5-17

$$kT \left(\frac{\partial \rho}{\partial P} \right)_T - 1 = 4\pi \rho \int_0^\infty [g^{(2)}(r; \rho, T) - 1] r^2 \, dr \qquad (11.5\text{-}17)$$

as

$$kT \left(\frac{\partial \rho}{\partial P} \right)_T - 1 = 4\pi \rho \int_0^\infty h(r; \rho, T) r^2 \, dr \qquad (12.4\text{-}6)$$

A method to solve the Ornstein-Zernike equation is by use of a three-dimensional Fourier transform. A three-dimensional Fourier transform pair is defined as follows:

$$\hat{h}(s) = \rho \int h(r) e^{-i\underline{s}\cdot\underline{r}} \, d\underline{r} \quad \text{and} \quad h(r) = \frac{1}{(2\pi)^3 \rho} \int \hat{h}(s) e^{i\underline{s}\cdot\underline{r}} \, d\underline{s} \qquad (12.4\text{-}7)$$

where $\underline{r}$ and $\underline{s}$ are vectors—or, more simply, in polar-spherical coordinates as

$$\hat{h}(s) = 4\pi \rho \int h(r) \left(\frac{\sin(sr)}{sr} \right) r^2 \, dr \quad \text{and} \quad h(r) = \frac{1}{2\pi^2 \rho} \int \hat{h}(s) \left(\frac{\sin(sr)}{sr} \right) s^2 \, ds$$

$$(12.4\text{-}8)$$

With this notation, the Ornstein-Zernike equation

$$h(\underline{r}_1, \underline{r}_2) = c(\underline{r}_1, \underline{r}_2) + \rho \int c(\underline{r}_1, \underline{r}_3) h(\underline{r}_3, \underline{r}_2) \, d\underline{r}_3 \qquad (12.3\text{-}4)$$

---

[4] Historically, the integral equation closures discussed here have been obtained from well-defined approximations using graph theory (see Chapter 7 and the previous section) by neglecting some graphs and retaining others. A nice description of this is given in the book *Liquids and Liquid Mixtures, 3rd ed.*, by J. S. Rowlinson and F. L. Swinton, Butterworths & Co., London, 1982, Sections 7.3 and 7.4.

can be rewritten as

$$h(r_{12}) = c(r_{12}) + \rho \int c(r_{13}) h(r_{23}) \, d\underline{r}_3 \tag{12.4-9}$$

Multiplying this equation by $e^{i\underline{s}\cdot\underline{r}_{12}}$ and integrating gives us

$$\int h(r_{12}) e^{-i\underline{s}\cdot\underline{r}_{12}} \, d\underline{r}_{12} = \int c(r_{12}) e^{-i\underline{s}\cdot\underline{r}_{12}} \, d\underline{r}_{12} + \rho \iint c(r_{13}) h(r_{23}) e^{-i\underline{s}\cdot\underline{r}_{12}} \, d\underline{r}_3 \, d\underline{r}_{12}$$

$$= \int c(r_{12}) e^{-i\underline{s}\cdot\underline{r}_{12}} \, d\underline{r}_{12} + \rho \iint c(r_{13}) h(r_{23}) e^{-i\underline{s}\cdot\underline{r}_{13}} e^{i\underline{s}\cdot\underline{r}_{23}} \, d\underline{r}_{13} \, d\underline{r}_{12} \tag{12.4-10}$$

Here (as in the discussion of multiple integrals for the virial equation of state) we are free to choose the origin of the coordinate system at any convenient location. Finally, rewriting this equation using the carat (^) symbol for a Fourier transform as in Eq. 12.4-7, we have

$$\hat{h}(s) = \hat{c}(s) + \rho \hat{c}(s) \hat{h}(s) \quad \text{or} \quad \hat{c}(s) = \frac{\hat{h}(s)}{1 + \rho \hat{h}(s)} \quad \text{and} \quad \hat{h}(s) = \frac{\hat{c}(s)}{1 - \rho \hat{c}(s)} \tag{12.4-11}$$

So by making an assumption about either $c(r)$ or $h(r)$, then taking the Fourier transform, using Eq. 12.4-11, and finally taking the inverse transform, the radial distribution function is obtained. In fact, different numerical methods are typically used.

The website for this book contains the MATLAB® program PYHS for calculating the Percus-Yevick approximation for the radial distribution function of the hard-sphere fluid as a function of either the packing fraction $\eta = \pi N \sigma^3 / 6V = \pi \rho \sigma^3 / 6$ or reduced density $\rho \sigma^3$. Note that with the definition of packing fraction given above, the fluid-solid transition for hard-spheres occurs at $\eta_f = 0.494$ ($\rho_f = 0.943$), and the most dense phase, the face-centered close-packed solid, occurs at $\eta_{fcc} = 0.7405$ ($\rho_{fcc} = 1.414$). Figure 12.4-1 contains some example results of the radial distribution function obtained with this program at different packing fractions η. Also note the increasing scale of the $g(r)$ with increasing density (packing fraction η). As an example, Table 12.4-1 contains values the temperature-independent hard-sphere radial distribution function for one high density case computed with this same program.

There is an important point with regard to the temperature dependence of the radial distribution function of the hard-sphere fluid that should be mentioned here. In general, the two-particle or radial distribution function $g^{(2)}(r; T, \rho)$ is a function of temperature and density, as well as radial separation r_{12}. However, the radial distribution function for the hard-sphere fluid, while dependent on density, is independent of temperature. The reason for this is that it is obtained from integrals involving the Boltzmann factors in the interaction energy $e^{-u(r)/kT}$ where $u(r) = \infty$ and $e^{-u(r)/kT} = 0$ for $r < \sigma$, and $u(r) = 0$ and $e^{-u(r)/kT} = 1$ for $r > \sigma$. So there is no temperature dependence remaining in the integrals defining the radial distribution function. It is for the same reason that the second virial coefficient for the hard-sphere fluid was found to be a constant independent of temperature.

12.4 Closures for the Ornstein-Zernike Equation

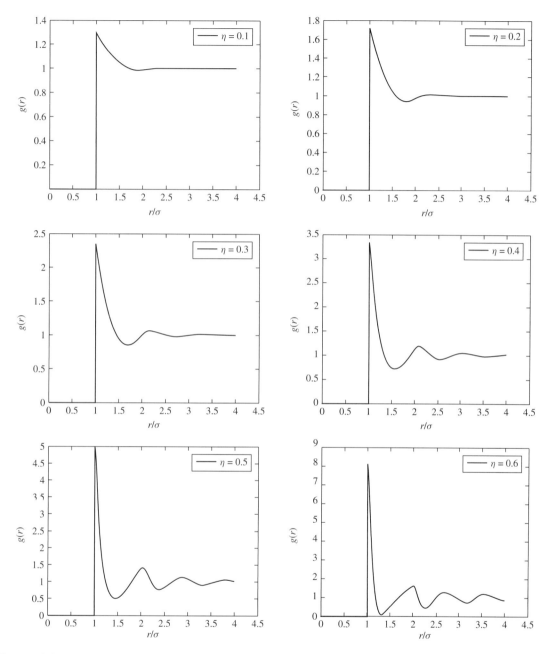

Figure 12.4-1 The Percus-Yevick solution for the radial distribution function of the hard-sphere fluid as a function of interatomic separation at various values of the packing fraction $\eta = \pi N\sigma^3/6V = \pi\rho\sigma^3/6$. Note that the vertical scales are increased at the higher packing fractions, and that the radial distribution function becomes more like that of a solid with sharper, better-defined peaks. Also, the number of molecules in the first coordination shell of the hard-sphere fluid in this approximation is 5.1, 8.7, 11.0, 12.0, 12.3, and 12.8 at $\eta = 0.1, 0.2, 0.3, 0.4, 0.5$, and 0.6, respectively. However, as $\eta_f = 0.494$ is the freezing density of the hard-sphere fluid, the liquid state Percus-Yevick results at higher densities should not be expected to be correct.

Table 12.4-1 Pair correlation function of the hard-sphere fluid in the Percus-Yevick approximation at $\rho\sigma^3 = 0.9$ calculated using the MATLAB® program PYHS

r/σ	$g(r/\sigma)$	r/σ	$g(r/\sigma)$	r/σ	$g(r/\sigma)$
1.0000	4.4194	1.3500	0.7291	1.7000	0.7930
1.0100	4.2118	1.3600	0.7039	1.7100	0.8086
1.0200	4.0111	1.3700	0.6814	1.7200	0.8245
1.0300	3.8172	1.3800	0.6615	1.7300	0.8406
1.0400	3.6303	1.3900	0.6439	1.7400	0.8568
1.0500	3.4503	1.4000	0.6287	1.7500	0.8732
1.0600	3.2771	1.4100	0.6157	1.7600	0.8897
1.0700	3.1108	1.4200	0.6047	1.7700	0.9064
1.0800	2.9512	1.4300	0.5957	1.7800	0.9232
1.0900	2.7983	1.4400	0.5885	1.7900	0.9401
1.1000	2.6519	1.4500	0.5831	1.8000	0.9571
1.1100	2.5121	1.4600	0.5794	1.8100	0.9741
1.1200	2.3785	1.4700	0.5772	1.8200	0.9912
1.1300	2.2513	1.4800	0.5765	1.8300	1.0084
1.1400	2.1301	1.4900	0.5772	1.8400	1.0256
1.1500	2.0149	1.5000	0.5792	1.8500	1.0429
1.1600	1.9055	1.5100	0.5824	1.8600	1.0602
1.1700	1.8018	1.5200	0.5868	1.8700	1.0776
1.1800	1.7036	1.5300	0.5923	1.8800	1.0950
1.1900	1.6108	1.5400	0.5987	1.8900	1.1124
1.2000	1.5232	1.5500	0.6061	1.9000	1.1298
1.2100	1.4407	1.5600	0.6143	1.9100	1.1473
1.2200	1.3631	1.5700	0.6234	1.9200	1.1648
1.2300	1.2902	1.5800	0.6332	1.9300	1.1823
1.2400	1.2220	1.5900	0.6438	1.9400	1.1999
1.2500	1.1581	1.6000	0.6550	1.9500	1.2175
1.2600	1.0985	1.6100	0.6667	1.9600	1.2351
1.2700	1.0430	1.6200	0.6791	1.9700	1.2527
1.2800	0.9915	1.6300	0.6919	1.9800	1.2704
1.2900	0.9438	1.6400	0.7052	1.9900	1.2882
1.3000	0.8998	1.6500	0.7190	2.0000	1.3059
1.3100	0.8592	1.6600	0.7331		
1.3200	0.8220	1.6700	0.7476		
1.3300	0.7880	1.6800	0.7625		
1.3400	0.7571	1.6900	0.7776		

12.5 THE PERCUS-YEVICK HARD-SPHERE EQUATION OF STATE

One result from the use of the PY approximation in the Ornstein-Zernike equation is that an analytic equation is obtained for the equation of state for the hard-sphere fluid;[5,6] in fact, the two equations given below:

$$\left(\frac{P}{\rho kT}\right)_P = \frac{1 + 2\eta + 3\eta^2}{(1-\eta)^2} \quad \text{and} \quad \left(\frac{P}{\rho kT}\right)_C = \frac{1 + \eta + \eta^2}{(1-\eta)^3} \quad (12.5\text{-}1)$$

where $\eta = \pi N\sigma^3/6V = \pi\rho\sigma^3/6$. The first equation above with the subscript P arises from using the radial distribution function obtained in the pressure equation (Eq. 11.2-11), and the second equation is a result of using the same distribution function in the compressibility equation (Eq. 11.5-17). Note that if the PY approximation were exact, one should obtain the same equation of state regardless of whether the pressure or compressibility equation was used. Therefore, comparison of the results obtained from both equations can be used as a consistency test of the solution.

It is interesting to note that the best description of the hard-sphere fluid is obtained from a weighted average of the two results above:

$$\left(\frac{P}{\rho kT}\right) = \frac{1}{3}\left(\frac{P}{\rho kT}\right)_P + \frac{2}{3}\left(\frac{P}{\rho kT}\right)_C = \frac{1 + \eta + \eta^2 - \eta^3}{(1-\eta)^3} \quad (12.5\text{-}2)$$

This is the Carnahan-Starling equation.[7] It is one of the most used results arising from statistical mechanics. One application is that it has been used as the basis for developing equations of state. For example, if a molecule is considered to consist of a central hard core plus other interactions (attractive dispersion forces, electrostatic forces, etc.), the Carnahan-Starling term can be used to represent the hard-core part, and the effects of other terms added as a Taylor series expansion. This is referred to as perturbation theory and is discussed in Chapter 14. Such an analysis has been widely used for systems of scientific and engineering interest, such as large globular molecules, proteins, and colloids.

For later reference, there are other results that follow from the Carnahan-Starling equation. The first is the radial distribution for hard-spheres at contact. To derive this, we start from Eq. 11.2-11

$$P = \rho kT - \frac{2\pi}{3}\rho^2 \int_0^\infty \frac{du(r)}{dr} g(r) r^3 dr \quad (11.2\text{-}11)$$

and note that the derivative $du(r)/dr$ is not well behaved at $r = \sigma$ since $u(r)$ goes from infinity at r incrementally less than σ, and to 0 for r incrementally greater than σ. The Boltzmann factor in the intermolecular potential, $e^{-u(r)/kT}$, is better behaved in that its value goes to 0 at r incrementally less than σ, to 1 for r incrementally greater than σ, and its derivative is related to the delta function. Therefore, to evaluate

---

[5] E. Thiele, *J. Chem. Phys.* **39**, 474 (1963).
[6] M. S. Wertheim, *J. Math. Phys.* **5**, 643 (1964).
[7] N. F. Carnahan and K. E. Starling, *J. Chem. Phys.* **51**, 635 (1969).

228 Chapter 12: Integral Equation Theories for the Radial Distribution Function

the derivative of the hard-sphere potential we do the following:

$$\frac{d}{dr}e^{-u(r)/kT} = -\frac{1}{kT}e^{-u(r)/kT}\frac{du(r)}{dr}$$

or $\quad \dfrac{du(r)}{dr} = -kT e^{u(r)/kT}\dfrac{d}{dr}e^{-u(r)/kT} = -kT e^{u(r)/kT}\delta(r-\sigma)$

where $\delta(x)$ is the delta function whose value is 1 at $x = 0$, and is 0 elsewhere. Therefore

$$P = \rho kT + \frac{2\pi kT}{3}\rho^2 \int_0^\infty e^{-u(r)/kT} g^{hs}(r)\delta(r-\sigma)r^3\,dr$$

$$= \rho kT + \frac{2\pi kT}{3}\rho^2 g^{hs}(r=\sigma^+)\sigma^3 = \rho kT + \frac{2\pi kT}{3}\rho^2 g^{hs}(\sigma^+)\sigma^3$$

where σ^+ indicates a value of σ incrementally greater than the hard-sphere diameter (and σ^- does not appear since the Boltzmann factor is 0 there). Then using

$$\frac{P}{\rho kT} = 1 + \frac{2\pi}{3}\rho g^{hs}(r=\sigma^+)\sigma^3 = \frac{1+\eta+\eta^2-\eta^3}{(1-\eta)^3}$$

it is simple algebra to show that

$$g^{hs}(\sigma^+) = \frac{2-\eta}{2(1-\eta)^3} \tag{12.5-3}$$

(Note that $g^{hs}(\sigma^-) = 0$.)

Other interesting results from the Carnahan-Starling equation of state are that the Helmholtz energy and pressure above that of an ideal gas at the same density are

$$\frac{A(N,V,T) - A^{IG}(N,V,T)}{NkT} = \frac{3-2\eta}{(1-\eta)^2}$$

and $\quad \dfrac{(P(N,V,T) - P^{IG}(N,V,T))V}{NkT} = \dfrac{4\eta - 2\eta^2}{(1-\eta)^3}$

(12.5-4)

12.6 THE RADIAL DISTRIBUTION FUNCTIONS AND THERMODYNAMIC PROPERTIES OF MIXTURES

While it is difficult to obtain the radial distribution for a pure fluid, the problem is even more difficult for a mixture; the best way to get such information is from computer simulation, as discussed in the next chapter. Before simulation, a collection of approximations were made in integral equation theory. The starting point is that by using the pairwise additivity assumption, the internal energy of a mixture of monatomic molecules can be shown to be (Problem 11.2)

$$U(N,V,T) = \frac{3}{2}NkT + 2\pi N\rho \sum_i \sum_j x_i x_j \int_0^\infty u_{ij}(r_{ij})g_{ij}(r_{ij};\rho,T)r_{ij}^2\,dr_{ij} \tag{12.6-1}$$

12.6 The Radial Distribution Functions and Thermodynamic Properties of Mixtures

A very simple assumption is that all the radial distribution functions have the same dependence on intermolecular separation, that is

$$g_{11}(r) = g_{22}(r) = g_{12}(r) = \cdots = g(r) \tag{12.6-2}$$

Furthermore, by writing

$$u(r) = \sum_i \sum_j x_i x_j u_{ij}(r) \tag{12.6-3}$$

Equation 12.6-1 becomes

$$U(N, V, T) = \frac{3}{2} NkT + 2\pi N\rho \int_0^\infty u(r) g(r) r^2 \, dr \tag{12.6-4}$$

This is referred to as the random mixture or one-fluid model and can be used with any intermolecular potential. Note that if the Lennard-Jones potential is used, we have

$$\varepsilon \left[\left(\frac{\sigma}{r}\right)^{12} - \left(\frac{\sigma}{r}\right)^6 \right] = \sum_i \sum_j x_i x_j \varepsilon_{ij} \left[\left(\frac{\sigma_{ij}}{r}\right)^{12} - \left(\frac{\sigma_{ij}}{r}\right)^6 \right] \tag{12.6-5}$$

which has the solution

$$\varepsilon = \frac{\left(\sum_i \sum_j x_i x_j \varepsilon_{ij} \sigma_{ij}^6\right)^2}{\sum_i \sum_j x_i x_j \varepsilon_{ij} \sigma_{ij}^{12}} \quad \text{and} \quad \sigma = \left(\frac{\sum_i \sum_j x_i x_j \varepsilon_{ij} \sigma_{ij}^{12}}{\sum_i \sum_j x_i x_j \varepsilon_{ij} \sigma_{ij}^6}\right)^{\frac{1}{6}} \tag{12.6-6}$$

In the case of the one-fluid model, the mixture is treated as a pure component with an effective potential (and potential parameters) that changes with changing composition.

A somewhat better approximation is to assume that the like molecule and unlike molecule potentials have the same form (for example, the Lennard-Jones 12-6 potential, although not restricted here to that form):

$$u_{ij}(r) = \varepsilon_{ij} F\left(\frac{r}{\sigma_{ij}}\right) \tag{12.6-7}$$

and further assuming a universal form for the radial distribution function in reduced variables r/σ

$$g_{11}\left(\frac{r}{\sigma_{11}}\right) = g_{22}\left(\frac{r}{\sigma_{22}}\right) = g_{12}\left(\frac{r}{\sigma_{12}}\right) = g\left(\frac{r}{\sigma}\right) \tag{12.6-8}$$

This leads to

$$U(N, V, T) = \frac{3}{2} NkT + 2\pi N\rho \sum_i \sum_j x_i x_j \varepsilon_{ij} \sigma_{ij}^3 \int_0^\infty F(r) g(r; \rho, T) r^2 \, dr$$

$$= \frac{3}{2} NkT + 2\pi N\rho \varepsilon \sigma^3 \int_0^\infty F(r) g(r; \rho, T) r^2 \, dr \tag{12.6-9}$$

230 Chapter 12: Integral Equation Theories for the Radial Distribution Function

with
$$\varepsilon\sigma^3 = \sum_i \sum_j x_i x_j \varepsilon_{ij} \sigma_{ij}^3 \qquad (12.6\text{-}10)$$

Equations 12.6-7 to 12.6-10 are referred to as the *van der Waals 1 theory*.

12.7 THE POTENTIAL OF MEAN FORCE

We have shown that at low density, the radial distribution function is
$$g(r_{12}, T, \rho \to 0) = e^{-u(r_{12})/kT} \qquad (12.7\text{-}1)$$

A new function, the potential of mean force, $w(r_{12}, T, \rho)$, is defined by its relationship to the radial distribution function at all densities, temperatures, and intermolecular separation distances by
$$g(r_{12}, T, \rho) = e^{-w(r_{12}, T, \rho)/kT} \qquad (12.7\text{-}2)$$

That is, when $w(r_{12}, T, \rho)$ is substituted in Eq. 12.7-2, it reproduces the radial distribution function between two particles at the temperature and density of the system. It can be interpreted as follows. If $u(r_{12})$ is the interaction potential between two particles in a vacuum, $w(r_{12}, T, \rho)$ is the effective potential between these two particles in a fluid, where their interaction is affected by the presence of intermediate molecules and by all indirect interactions. That is, in addition to the interaction between atom 1 and atom 2, atom 1 interacts with atom 3 that also interacts with atom 2, etc. As a result of these indirect interactions, the range of the potential of mean force $w(r_{12}, T, \rho)$ is considerably longer than the (direct) interaction potential $u(r_{12})$. Also, unlike the true intermolecular potential, the potential of mean force depends on temperature and density (and also the concentration of other species if the atoms of interest are in a mixture).

To obtain a more formal expression for the potential of mean force, we note that the force between two atoms in a vacuum as they are moved further apart (or closer together) is
$$\underline{F}_{12} = -\frac{du(\underline{r}_{12})}{d\underline{r}_{12}} \qquad (12.7\text{-}3)$$

In a fluid, the force between atoms 1 and 2 is affected by the presence of all other atoms. What we want to compute is the force between atom 1 at $\underline{r}_1$ and atom 2 at $\underline{r}_2$ in the fluid obtained by averaging over the locations of all the other atoms. That is

$$\begin{aligned}
-\langle \underline{F}_{12}\rangle &= -\frac{dw(\underline{r}_1, \underline{r}_2)}{d\underline{r}_{12}} = -\left\langle \frac{du(\underline{r}_1, \underline{r}_2, \underline{r}_3 \ldots \underline{r}_N)}{d\underline{r}_{12}}\right\rangle_{\underline{r}_1, \underline{r}_2} \\
&= \frac{-\int \cdots \int e^{-u/kT} \dfrac{du(\underline{r}_1, \underline{r}_2, \underline{r}_3 \ldots \underline{r}_N)}{d\underline{r}_{12}} d\underline{r}_3\, d\underline{r}_4 \ldots d\underline{r}_N}{\int \cdots \int e^{-u/kT} d\underline{r}_3 d\underline{r}_4 \ldots d\underline{r}_N} \\
&= \frac{kT \dfrac{d}{d\underline{r}_{12}} \int \cdots \int e^{-u/kT} d\underline{r}_3 d\underline{r}_4 \ldots d\underline{r}_{1N}}{\int \cdots \int e^{-u/kT} d\underline{r}_3\, d\underline{r}_4 \ldots d\underline{r}_N} \\
&= kT \frac{d}{d\underline{r}_{12}} \ln \int \cdots \int e^{-u/kT} d\underline{r}_3\, d\underline{r}_4 \ldots d\underline{r}_N
\end{aligned} \qquad (12.7\text{-}4)$$

12.7 The Potential of Mean Force

Separately we note that since neither the configuration integral $Z(N, V, T)$ nor N and V are a function of $\underline{r}_{12}$, we have

$$\frac{d}{d\underline{r}_{12}}\left(\frac{V^2}{N(N-1)Z(N,V,T)}\right) = \frac{d}{d\underline{r}_{12}} \ln\left(\frac{V^2}{N(N-1)Z(N,V,T)}\right) = 0 \quad \text{(12.7-5)}$$

Using this relation in Eq. 11.10-4 we

$$-\langle \underline{F}_{12} \rangle = -\frac{dw(\underline{r}_1, \underline{r}_2)}{d\underline{r}_{12}} = kT \frac{d}{d\underline{r}_{12}} \ln \left[\frac{V^2 \int \cdots \int e^{-u/kT} d\underline{r}_3\, d\underline{r}_4 \ldots d\underline{r}_N}{N(N-1)Z(N,V,T)} \right]$$

$$= kT \frac{d}{d\underline{r}_{12}} \ln g(\underline{r}_1, \underline{r}_2) \quad \text{(12.7-6)}$$

On integration this gives us

$$-w(\underline{r}_1, \underline{r}_2) = kT \ln g(\underline{r}_1, \underline{r}_2) + c \quad \text{or} \quad g(\underline{r}_1, \underline{r}_2) = e^{-w(\underline{r}_1, \underline{r}_2)/kT} e^{-c/kT} \quad \text{(12.7-7)}$$

where c is the constant of integration. When atoms 1 and 2 are at infinite separation—that is, when $|\underline{r}_1 - \underline{r}_2| \to \infty$—we know that $g(\underline{r}_1, \underline{r}_2) = 1$ so that $w(\underline{r}_1, \underline{r}_2) = 0$. Consequently, the integration constant c must be zero, and therefore $g(\underline{r}_1, \underline{r}_2) = e^{-w(\underline{r}_1, \underline{r}_2)/kT}$ or $g(r) = e^{-w(r)/kT}$, which is Eq. 12.7-2.

Alternatively, it is easily shown (Problem 12.9) that

$$\frac{\partial w(r; \rho, T)}{\partial \underline{r}_1} = -\frac{\int_V \cdots \int_V e^{-u(\underline{r}_1, \underline{r}_2 \ldots \underline{r}_N)/kT} \frac{\partial u(\underline{r}_1, \underline{r}_2, \ldots \underline{r}_N)}{\partial \underline{r}_1} d\underline{r}_2\, d\underline{r}_3 \ldots d\underline{r}_N}{\int_V \cdots \int_V e^{-u(\underline{r}_1, \underline{r}_2 \ldots \underline{r}_N)/kT} d\underline{r}_1\, d\underline{r}_2 \ldots d\underline{r}_N}$$

(12.7-8)

where $-\partial u(\underline{r}_1, \underline{r}_2, \ldots \underline{r}_N)/\partial \underline{r}_1$ the force on atom 1 at vectorial position $\underline{r}_1$ averaged over the locations of all other atoms.

An interesting interpretation for the potential of mean force $w(r_{12}, T, \rho)$ is that since it is the integral of force over distance, it is also the work done to bring a pair of particles in a dense fluid from infinite separation to a separation distance of r. Since this work is done at fixed N, V, and T, we then have $w(r)$ is the Helmholtz energy change for this process.

As already mentioned, because of the indirect interactions (that is, the interactions between atoms 1 and 2 through all possible intermediate atoms), the potential of mean force, or PMF as it is frequently referred to, is of much longer range than the two-body potential between two isolated atoms, and in fact does not resemble it. This is evident in Fig. 12.7-1 that is the potential of mean force (divided by kT) computed from the Percus-Yevick solution for the hard-sphere potential. Note that while for hard-spheres the two-body potential is infinite for $r/\sigma < 1$, and zero for $r/\sigma \geqslant 1$, the potential of mean force for this interaction is also infinite for $r/\sigma < 1$, but then is nonzero for $r/\sigma \geqslant 1$, and the range over which it is nonzero increases with density. Also, while the hard-sphere potential is either zero or infinity and independent of temperature, the potential of mean force depends on temperature (which is why $w(r)/kT$ is plotted in the figures). Both the extended range of the potential of mean force (beyond that of the underlying true two-body potential) and its temperature dependence are the result of solvation forces—that is, the effect of the two particles of interest being in a fluid rather than isolated in a vacuum.

It is especially interesting to note that in Fig. 12.7-1, there are regions in which the potential of mean force between the atoms is attractive (negative values of the potential of mean force), even though the hard-sphere potential has no attractive region. The simplest way to understand how this occurs is from a kinetic argument. Consider two atoms sufficiently close to each other than no other atoms can get between them, as shown in Fig. 12.7-2. On each collision of the atoms of interest with surrounding atoms in the fluid, there is a force as indicated by the arrows. However, when two atoms are sufficiently close together, there is a region between them shielded from collisions with other atoms (indicated by a lack of arrows directed toward the center of each atom). Consequently, because of this imbalance of possible collisions, there is a net force of each atom in the pair of interest toward the other as a result of the force imbalance—that is, there is an apparent attraction. This is what is shown in the PMF in Fig. 12.7-1. At high densities, a similar argument can be made to explain the weaker regions of attraction at large atomic separations resulting from the higher coordination shells.

What should be clear from this discussion is that it is not simple to develop an accurate model for the potential of mean force, so that developing the PMF is not a "cheap" way of obtaining the radial distribution function. Indeed, in the discussion above, the potential of mean force was obtained from known values of the radial distribution and not the reverse.

A problem of some interest is to try to understand the behavior of polymers, colloids or proteins in solution. Predicting the radial distribution function of these macromolecules (with so many atoms) from rigorous theory is not possible, though some progress can be made using computer simulation that is discussed in Chapter 13. Instead, a common procedure for obtaining an approximate radial distribution function for such molecules is to use physical insight to make a model for the potential of mean force with adjustable parameters and then fit the parameters to some available experimental data, for example, osmotic pressure data (as discussed in the following section) or precipitation data. The underlying idea is that since the proteins or colloids are so large compared to the small solvent molecules (and the ions from salt that are generally in such solutions), the solvent can be considered to be a continuum (rather than a collection of individual atoms or molecules), so that the interactions between the proteins or colloids can be described by a potential of mean force in the solution. For example, to model the precipitation of globular proteins in aqueous solution using a polymer to induce precipitation, the following potential of mean force model has been used:[8]

$$w(r, T, \rho) = w_{\text{hs}}(r) + w_{\text{att}}(r) + w_{\text{elect}}(r, T, \rho) \qquad (12.7\text{-}9)$$

where $w_{\text{hs}}(r)$ is the hard-sphere interaction (resulting in an excluded volume between the molecules). This effective hard-sphere diameter can be obtained from information on the size of protein or colloid. The next term is the weak van der Waals interactions between the molecules in the presence of the solvent, which is usually attractive. This term could be obtained by summing all the simultaneous atom-atom interactions in

---

[8] F. W. Tavares and S. I. Sandler, *AIChE Journal* **43**, 218 (1997).

12.7 The Potential of Mean Force 233

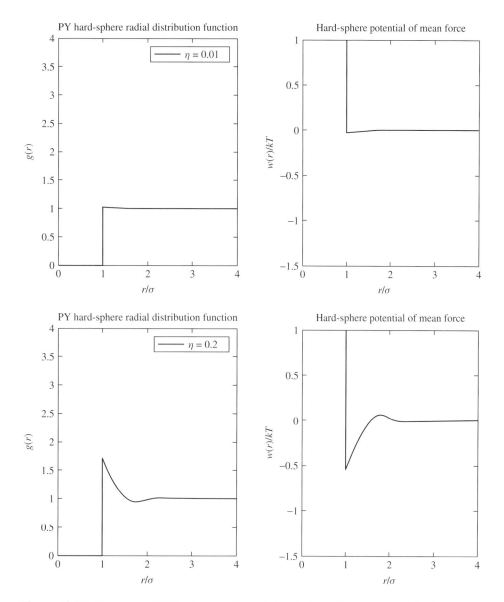

Figure 12.7-1 The radial distribution function $g(r)$ and dimensionless potential of mean force $w(r)/kT$ for the hard-sphere fluid at various densities calculated with the MATLAB® program PYHSPMF.

the molecules, in which each might be represented, for example, by the Lennard-Jones 12-6 or similar interactions, or even simpler (since the short-range repulsion is represented by the hard-sphere term) by

$$w_{\text{vdW}}(r) = -\frac{C}{r^n} \quad (12.7\text{-}10)$$

where C is a parameter specific to each system and n, the exponent of the intermolecular separation distance, depends on the types of interactions and the number of interacting sites on the globular protein, colloid, or polymer. For single neutral

234 Chapter 12: Integral Equation Theories for the Radial Distribution Function

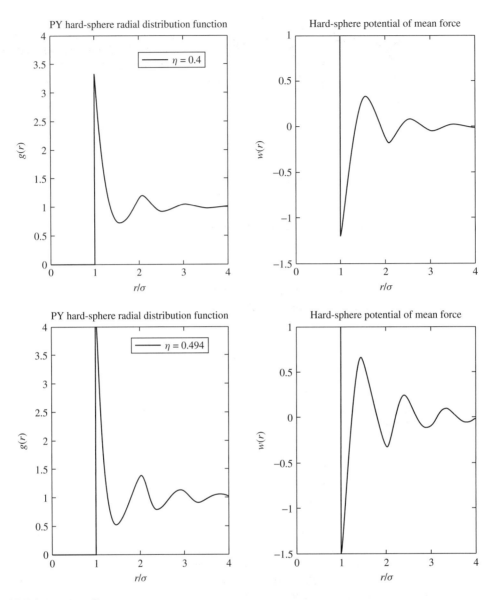

Figure 12.7-1 (*continued*)

atoms, n is frequently taken to be 6, as in the Lennard-Jones 12-6 potential. In the colloid and protein literature, this attractive term is frequently written as

$$w_{\text{vdW}}(r) = -\frac{H}{36(r/\sigma)^6} \qquad (12.7\text{-}11)$$

and H is referred to as the Hamaker constant.

The electrostatic terms in the potential of mean force could also include charge-charge, charge-dipole, and charge-induced dipole interactions if the particles are charged (as is frequently the case for macromolecules unless the pH of the solution is at the isoelectric point of the macromolecule—that is, the pH at which the

12.7 The Potential of Mean Force

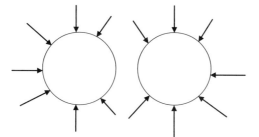

Figure 12.7-2 Forces on two hard-spheres as a result of collisions with other hard-spheres. Note that there is a region between the two spheres that cannot be penetrated by other spheres (the region with a lack of arrows). The resulting force imbalance pushing the spheres together behaves as an attraction.

molecule has no net charge), and dipole-dipole, dipole-induced dipole, quadrupole-quadrupole, and other interactions. However, these are also affected by the presence of the solvent and the counter-ions in solution. Consequently, the charge-charge interaction to the potential of mean force would not be the simple Coulomb interaction

$$u(r) = \frac{q_1 q_2}{r} \qquad (12.7\text{-}12)$$

where q_1 and q_2 are the charges on the atoms; but, as derived in Chapter 15 (for identical proteins or colloids of like charge q), the shielded Coulomb potential is used

$$w_{\text{elect}}(r) = \frac{q^2 e^{\kappa(\sigma-r)}}{\varepsilon(1+\kappa\sigma)r} \qquad (15.2\text{-}39)$$

in which the Debye length κ is given by $\kappa^2 = \frac{4\pi}{\varepsilon kT}\sum \rho_i q_i^2$ where ρ_i is the density of charged species i and ε is the dielectric constant of the solvent.

In dimensionless form, the overall potential of mean force is

$$\frac{w(r)}{kT} = \begin{cases} \infty & \text{for } r^* = r/\sigma < 1 \\ -\frac{H^*}{36(r^*)^6} + \frac{A^* e^{\kappa^*(1-r^*)}}{(1+\kappa^*)r^*} & \text{for } r^* = r/\sigma \geq 1 \end{cases} \qquad (12.7\text{-}13)$$

where $r^* = r/\sigma$, $H^* = H/kT$, $\kappa^* = \kappa\sigma$ and $A^* = q^2/\varepsilon\sigma kT$

This potential of mean force, for some representitive values of the parameters, is shown in Figs. 12.7-3.

If a polymer is added to the solution (usually to precipitate the protein), an attractive force arises as a result of the exclusion of the polymer from the region between two macromolecules when they are very close (see Fig. 12.7-2). This effect is included by addition an osmotic contribution $w_{\text{osm}}(r, T, \rho)$ to the potential of Eq. 12.7-9. This is usually described by the Asakura and Oosawa[9] potential:

$$w_{\text{osm}}(r,T,\rho) = \begin{cases} -\frac{4\pi \left(\frac{\sigma+d_P}{2\sigma}\right)^3 \rho_3^* kT}{3\left(\frac{d_P}{\sigma}\right)^3}\left(1 - \frac{3r}{4\left(\frac{\sigma+d_P}{2\sigma}\right)} + \frac{r^3}{16\left(\frac{\sigma+d_P}{2\sigma}\right)^3}\right) & \sigma < r < \sigma + d_P \\ 0 & r > \sigma + d_P \end{cases}$$

$$(12.7\text{-}14)$$

where σ is the diameter of the macromolecule, and d_P is the diameter of the polymer.

---

[9] S. Asakura and F. Oosawa, *J. Chem. Phys.* **22**, 1255 (1958), and *J. Poly. Sci.* **33**, 183 (1958).

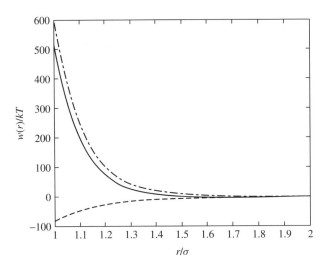

Figure 12.7-3a The dimensionless potential of mean force $w(r)/kT$ for a colloid or protein for $H^* = 10$, $A^* = 10$ and $\kappa^* = 8$. The solid line is total potential, the dashed line is the Hamaker contribution, and the dash-dot line is the electrostatic (shielded Coulomb) contribution. Though hard to see on this graph, the potential is attractive for $r/\sigma > 1.5$, and has a minimum of $\varepsilon/kT = -2.14$ at $r/\sigma = 1.68$.

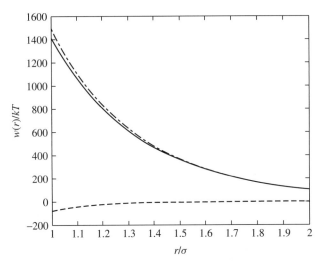

Figure 12.7-3b The dimensionless potential of mean force $w(r)/kT$ for a colloid or protein for $H^* = 10$, $A^* = 10$ and $\kappa^* = 2$. The solid line is total potential, the dashed line is the Hamaker contribution, and the dash-dot line is the electrostatic (shielded Coulomb) contribution. For this set of parameters, the potential is always repulsive.

A more sophisticated model for the potential of mean force for colloids and proteins that takes into better account the phenomena mentioned above—including, more explicitly, the effects of the intervening solvent, pH, and charge—is due to Derjagun and Landau,[10] and to Verwey and Overbeck,[11] and is referred to as the DLVO model. The interested reader can find information about this model in the book by Israelachvili.[12]

---

[10] B. V. Derjaguin and L. Landau, Acta Physiochim. *URSS* **14**, 633 (1941).

[11] E. J. W. Verwey and J. Th. Overbeek, *The Theory of Stability of Lyophobic Colloids*, Elsevier, Amsterdam, 1948.

[12] See, for example, J. Israelachvili, *Intermolecular & Surface Forces*, 2nd ed., Academic Press, London, 1991.

12.8 Osmotic Pressure and the Potential of Mean Force for Protein and Colloidal Solutions

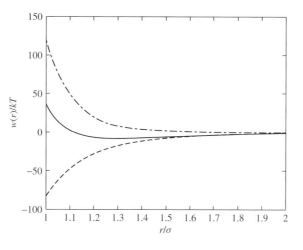

Figure 12.7-3c The dimensionless potential of mean force $w(r)/kT$ for a colloid or protein for $H^* = 10$, $A^* = 2$ and $\kappa^* = 8$. The solid line is total potential, the dashed line is Hamaker contribution, and the dash-dot line is the electrostatic (shielded Coulomb) contribution. Though hard to see on this graph, the potential is attractive for $r/\sigma > 1.1$, and has a minimum of $\varepsilon/kT = -8.91$ at $r/\sigma = 1.27$.

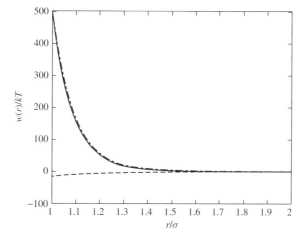

Figure 12.7-3d The dimensionless potential of mean force $w(r)/kT$ for a colloid or protein for $H^* = 2$, $A^* = 10$ and $\kappa^* = 10$. The solid line is total potential, the dashed line is the Hamaker contribution, and the dash-dot line is the electrostatic (shielded Coulomb) contribution. Though hard to see on this graph, the potential is attractive for $r/\sigma > 1.56$, and has a minimum of $\varepsilon/kT = -0.42$ at $r/\sigma = 1.72$.

12.8 OSMOTIC PRESSURE AND THE POTENTIAL OF MEAN FORCE FOR PROTEIN AND COLLOIDAL SOLUTIONS

The derivation of the virial equation of state is based on a dilute concentration of atoms (or molecules) in a vapor phase, so that the space between the atoms is a vacuum. Another type of the virial equation is based on a hybrid continuum/atomistic model in which the species of atoms or molecules of interest in a solution are treated as if they were immersed in a continuum solvent. That is, the solvent occupying the space between the molecules of interest is not treated as a collection of atoms, but rather as a continuous fluid in which the molecules are floating. This solvent, before the solute is added, is characterized by its temperature T, pressure P, dielectric constant (if ions are involved), and chemical potential μ.

As solute is added, the properties of the solution will change. Now consider the addition of the solute in a way that the chemical potential of the solvent is unchanged. As solute is added at constant temperature and chemical potential of the solvent, the equilibrium pressure above the solution, which we will refer to as P_{sol}, will have to change (increase) to keep the solvent chemical potential constant to compensate for the dilution of the solvent by the solute. If the concentration of added solute is not very high, this equilibrium pressure can be written in virial form as

$$P_{sol} - P = \Pi = \rho_s kT[1 + B_2(T,\mu)\rho_s + B_3(T,\mu)\rho_s^2 + \cdots\cdots] \quad (12.8\text{-}1)$$

In this equation, ρ_s is the density of solute; the pressure difference between the solution P_{sol} and that of the pure solvent is referred to as the *osmotic pressure* and is designated by the symbol Π, and $B_2(T,\mu)$ and $B_3(T,\mu)$, etc. are referred to as the *osmotic virial coefficients*. It is important to remember that this expansion is applicable at constant solvent chemical potential so that the pressure change as a result of a solute addition is measured after the solution has adjusted to maintain a constant chemical potential of the solvent. This situation is referred to as the McMillan-Mayer standard state, as they were the ones that derived this equation from statistical mechanics.

An experimental realization of this process is two cells separated by a membrane that is permeable to solvent, but not to the solute. At initial equilibrium, the temperature, pressure (and therefore the height of the solvent), and the chemical potential of the solvent is the same in both cells. Upon addition of the solute on one side of the membrane (which dilutes the solvent in that cell), some solvent will migrate across the membrane increasing the volume and the hydrostatic pressure in that cell until the chemical potential of the solute-containing solvent on one side of the membrane equals that of the pure solvent on the other side. When this occurs, there will be a difference in pressure (as manifested by a difference in height of the solutions) between the solutions separated by the membrane. This difference in pressure is the osmotic pressure referred to above. This is illustrated in Fig. 12.8-1.

A derivation of Eq. 12.8-1 following the procedure used to derive the virial equation, but now considering the solvent as a background continuum (instead of a vacuum) and replacing the interaction potential with the potential of mean force leads to the following expression of the osmotic second virial coefficient:

$$B_2(T,\mu) = 2\pi \int_0^\infty (1 - e^{-w(r,\rho,T)/kT}) r^2 \, dr \quad (12.8\text{-}2)$$

While this expression for the osmotic second virial coefficient looks like Eq. 7.5-4

$$B_2(T) = 2\pi \int_0^\infty (1 - e^{-u(r)/kT}) r^2 \, dr \quad (7.5\text{-}4)$$

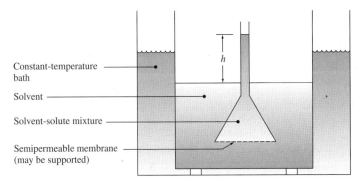

Figure 12.8-1 A Schematic Diagram of a Simple Osmometer.

for the second virial coefficient of a gas, there is a subtle but extremely important difference between the two. To compute a value for the second virial coefficient of a gas, one needs the true interaction potential $u(r)$ between two atoms interacting in a vacuum. However, to compute the osmotic virial coefficient, one needs the potential of mean force $w(r, \rho, T)$ between two solute molecules separated by a continuum of the solvent molecules. The potential of mean-force models, discussed in the previous section (which are functions of temperature and concentration), can be used for these calculations. Also, as the charge on proteins and other biomolecules have a strong dependence on the pH of the solution, that is an important variable as well.

It is difficult (or impossible) to know the potential of mean force exactly for every type of molecule in every solvent and in solvents with additional added components (salts, polymers, etc.), and so additional terms to those above may be added. It is useful to note that a measured value of the osmotic second virial coefficient can provide the information as to whether the Boltzmann average of the potential of mean force, as occurs in the osmotic second virial coefficient, results in a net attractive force (negative value of B_2) or net repulsive force (positive value of B_2). If there is a net repulsion between proteins in solution, it is not likely that they will crystallize out of solution. On the other hand, if there is a strong net attraction, the proteins will rapidly agglomerate and form an amorphous precipitate. But if there is a slight attraction, as evidenced by a slightly negative osmotic second virial coefficient, the protein molecules in solution will slowly come together and have time to reorient so as to form crystals. Consequently, information on the osmotic second virial coefficient has been especially useful in rapidly identifying solution conditions leading to protein crystallization.[13,14]

Having mentioned all these complications in dealing with macromolecules, I will close this chapter with evidence that in some cases a very simple potential of mean force can be used. Pusey and van Megen[15] have shown that colloidal spheres polymethylmethacrylate (PMMA) of 610 nm, with a 10 to 20 nm stabilizing layer of poly-12-hydroxysteric acid, in a mixed solvent of decalin and carbon disulfide (to match the refractive index of the colloid) behaved very much like a simple hard-sphere fluid. The phase diagram of this colloid system showed transitions from a fluid phase to a crystal phase and then to glass at the volume fractions expected for the hard-sphere fluid (in a vacuum) using slightly rescaled colloid diameter.

CHAPTER 12 PROBLEMS

—**12.1** The Carnahan-Starling recipe for the compressibility factor of a rigid sphere fluid is

$$Z = \frac{PV}{NkT} = \frac{1 + \eta + \eta^2 - \eta^3}{(1-\eta)^3}$$

where $\eta = \frac{\pi \sigma^3}{6} \frac{N}{V}$

Derive from this expression the following

a. $\alpha = \frac{1}{V}\left(\frac{dV}{dT}\right)_P$

= coefficient of thermal expansion

b. $\beta = -\frac{1}{V}\left(\frac{dV}{dP}\right)_T$

= coefficient of isothermal compressibility

12.2 Derive Eq. 12.5-4 for the excess Helmholtz energy over that of an ideal gas at the same temperature and

---

[13] A. George, W. W. Wilson, *Acta Cryst.* D**50**, 361 (1994).

[14] P. M. Tessier, S. D. Vandrey, B. W. Berger, R. Pazhianur, S. I. Sandler, and A. M. Lenhoff, *Acta Cryst.* D**58**, 15361 (2002).

[15] P. N. Pusey and W. van Megen, *Nature*, **320**, 340 (1986).

density for a fluid obeying the Carnahan-Starling equation.

12.3 Below are experimental values for α and β (as defined in Problem 12.1) for liquid argon and nitrogen. Use these values to compute an effective hard-sphere diameter for these liquids.

Liquid	T	α (K^{-1})	β (cm^2/dyne)
Ar	84.0	4.45×10^{-3}	21.0×10^{-11}
N$_2$	70.0	5.34×10^{-3}	27.0×10^{-11}

12.4 The Carnahan-Starling equation for a mixture of hard-spheres is

$$P = \frac{6kT}{\pi} \left[\frac{\eta_0}{1-\eta_3} + \frac{3\eta_1\eta_2}{(1-\eta_3)^2} + \frac{\eta_2^3(3-\eta_3)}{(1-\eta_3)^3} \right]$$

where

$$\eta_k = \frac{\pi}{6} \sum_i \rho x_i \sigma_i^k$$

Derive the second and third virial coefficients for a hard-sphere mixture from this equation and determine their mole fraction dependence. Is the mole fraction dependence of the second virial coefficient quadratic and cubic for the third virial coefficient as expected from Chapter 7?

12.5 Develop an equation of state for the triangular-well fluid (see Problem 8.13) using the Carnahan-Starling expression for the hard-sphere compressibility factor, assuming that the low-density representation of the radial distribution function $g(r, \rho, T) = e^{-u(r)/kT}$ can be used at all densities.

12.6 Obtain the radial distribution function for the hard-sphere fluid in the Percus-Yevick approximation at the following packing fractions: $\eta = 0.1, 0.2, 0.3, 0.4,$ and 0.45.

12.7 Calculate the osmotic second virial coefficient as a function of reduced temperature for the potential shown in:
a. Figure 12.7-3a
b. Figure 12.7-3b
c. Figure 12.7-3c
d. Figure 12.7-3d

12.8 Derive equations for the excess internal and Gibbs energies, entropy, and chemical potential over that of an ideal gas at the same temperature and density for a fluid that obeys the Carnahan-Starling equation.

12.9 Derive Eq. 12.7-8.

12.10 Graphically compare the Percus-Yevick hard-sphere compressibility factors $P/\rho kT$ from the compressibility and pressure equations with that obtained from the Carnahan-Starling equation.

12.11 Develop the equation for the chemical potential for hard-spheres from the Carnahan-Starling equation. (Extra credit: do the same using the Percus-Yevick equations of state obtained from the compressibility and pressure equations.)

Chapter 13

Determination of the Radial Distribution Function and Fluid Properties by Computer Simulation

At present, the most common and accurate method of obtaining the radial distribution function (and some thermodynamic properties) for a model fluid of interacting molecules is by direct computer simulation. The two general methods used are known as Monte Carlo and molecular-dynamics computer simulation, and these will be very briefly described in this chapter.

One can trace the evolution of the calculation of radial distribution functions to the development of high-speed computers. After the development of integral equations, it was possible—with great effort at that time—to obtain numerical solutions generally based on the assumption of pairwise additivity of the intermolecular potential. The results so obtained for a chosen pair potential were approximate as a result of the assumptions made in deriving the integral equations used. An alternative that developed following the enormous improvements in the speed of computers was to use molecular-level simulation to obtain the radial distribution function and thermodynamic properties for a given potential model directly by taking averages over many possible configurations of the molecules, rather than the numerical solution of an approximate integral equation.

With the ever-increasing speed and memory of computers, molecular-level simulation is now the "gold standard" method of obtaining the properties of a model fluid once the force field (intermolecular potential) is specified. Such methods were first used in the 1950s for relatively small numbers of simple molecules (for example, hard-spheres), and with improvements in computational power, they have progressed to larger numbers of frequently very complicated molecules including polymers and proteins. In simplest terms, molecular simulation methods consider a collection of virtual molecules that are present in a box that exists only in computer memory, many different configurations of the molecules considered, and then average values of the radial distribution function and the macroscopic properties (for example, internal energy and pressure) are computed.

The number of molecules that can be placed in such a box has increased tremendously over the years from less than a hundred to thousands and tens of thousands and, on supercomputers and parallel computers, to even hundreds of thousands now. However, such numbers are still small compared to Avogadro's number; so one characteristic of computer simulation is that there are fluctuations in the properties (you should remember from earlier chapters that fluctuations generally decrease as $N^{-0.5}$, where N here is roughly the product of the number of molecules in the simulation and the number of configurations considered for those molecules.)

INSTRUCTIONAL OBJECTIVES FOR CHAPTER 13

The goals of this chapter are to provide the student with

- An introduction to Monte Carlo molecular computer simulation
- An introduction to molecular-dynamics computer simulation

13.1 INTRODUCTION TO MOLECULAR LEVEL COMPUTER SIMULATION[1]

Molecular simulation methods can be broadly classified into two groups: molecular dynamics (MD) and Monte Carlo (MC). However, in each of these categories, specific techniques have been developed to deal with certain situations, such as the calculation of phase behavior, dealing with very dense fluids, or dealing with long chain molecules, to name but a few.

In a molecular-dynamics simulation, an initial state of the system is specified, including the system volume, the number of molecules, and the location and velocity of each molecule. The trajectory of each molecule is then followed by numerically integrating Newton's laws of motion over very small time steps, which includes solving for the change in trajectory as each molecule moves in the force field of the other molecules in the system, including molecule-molecule collisions. As discussed below, average values of properties over the time of the simulation are obtained. Also, dynamic properties and transport properties (for example, diffusion coefficients) can be computed. The first molecular-dynamics simulations were done by Alder and Wainwright for a fluid of hard-spheres.[2]

In a Monte Carlo simulation, an initial state including the positions of every molecule and, frequently, temperature and density or volume (but possibly instead other macroscopic properties such as pressure) are specified. Then a long sequence of positional changes of the molecules is made, usually one molecule at a time, and instantaneous properties are calculated, which are then averaged over many configurations of the system. In this way an average over states is obtained. If the same force field and macroscopic state parameters (for example, N, T, and V) are used, then by the ergodic hypothesis, results from molecular-dynamics and Monte Carlo

---

[1] Detailed discussions of molecular computer simulation methods can be found in texts devoted to this subject, such as *Understanding Molecular Simulation: From Algorithms to Applications* by D. Frenkel and B. Smit, Academic Press, 1996; *Computer Simulation of Liquids* by M. P. Allen and D. J. Tildesley, Oxford Univ. Press, 1987; *Molecular Modelling: Principles and Applications, 2nd ed.*, A. R. Leach, Prentice Hall, 2001.
[2] B. J. Alder and T. E. Wainwright, *J. Chem. Phys.* **27**, 1208 (1957).

simulations should be in agreement. The earliest successful Monte Carlo simulation of fluids is due to Metropolis et al.[3]

While the molecular-dynamics and Monte Carlo methods are different, they do share several features in common. First is that for a new simulation, the initial state (molecular positions in Monte Carlo, and in addition velocities in molecular-dynamics) may be arbitrarily chosen. Therefore, to allow equilibration of the system from the initial state, many of the initial cycles in the simulation (integration time steps in molecular-dynamics and positional changes in Monte Carlo) should be ignored before starting to collect data to obtain average values of the system properties.

Second, many simulations—especially those run on PCs or other small computers—because of speed and memory limitations, are done using numbers of molecules that are small compared to Avogadro's number. In fact, typically only thousands or tens of thousands molecules are used. Such a small number of molecules would correspond to a very small aerosol drop rather than a bulk liquid. In such a droplet, the ratio of the number of molecules at or near the surface to the number of molecules in the bulk is very much greater than in a large amount of liquid, resulting in incorrect simulation results for a bulk fluid. Also, there would be a surface tension contribution to the energy of the liquid (due to the high curvature of the droplet surface) that is unimportant in a bulk liquid. So that the MD or MC simulation represents a bulk fluid rather than a small droplet, both methods use periodic boundary conditions in which a cubic (or other regularly shaped) simulation box is considered to be surrounded by its periodic images. Thus, a molecule near an edge of the central simulation box interacts with molecules in its closest periodic image, or if a move takes a molecule outside the central box, its complementary image molecule enters the central box. This is shown in two dimensions in Fig. 13.1-1. This procedure is referred to as *the use of periodic boundary conditions*.

As a consequence of using periodic boundary conditions, when calculating the interaction between molecules, a cutoff distance of one-half the box length is used; otherwise the interaction energy between, for example, molecules 1 and 2 would also include all the periodic images of molecule 2, introducing a false periodicity into the liquid structure. To compensate for this cutoff distance, it is common to add a "long-range correction" to the calculated values. For example, the total interaction energy—that is, the configurational energy—for a system of N molecules would be computed (for a monatomic fluid) from

$$U^C = \sum_i \sum_{\substack{i > j \\ r_{ij} < L/2}} u(r_{ij}) + 2\pi \frac{N^2}{L^3} \int_{L/2}^{\infty} u(r) r^2 \, dr \qquad (13.1\text{-}1)$$

where L is the length of a cubic box and $L/2$ is the cutoff distance. The integral in this equation should also contain the radial distribution function (see Eq. 11.2-4). However, with present computing power, the box is usually of sufficient size that the radial distribution function at r near and greater than $L/2$ has only very small amplitude oscillations around a value of unity, so that setting $g(r \geq L/2) = 1$ introduces only a very small error and results in Eq. 13.1-1.

---

[3]N. Metropolis, A. W. Rosenbluth, M. N. Rosenbluth, A. H. Teller and E. Teller, *J. Chem. Phys*. 21, 1087 (1953).

244 Chapter 13: Determination of the Radial Distribution Function

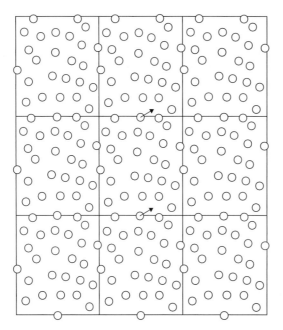

Figure 13.1-1 Illustration of periodic boundary conditions. The central box is the simulation box, and the surrounding boxes are the periodic images. The movement of one atom out of the central box and the insertion of its periodic image is illustrated.

The discussion above deals only with the simplest cubic periodic boundary conditions. For specialized applications, or for efficiency, in some simulations orthorhombic, hexagonal prism, rhombic dodecahedral, and other geometric shapes have been used instead of a simple cube for the central simulation box. A significant complication occurs in the simulation of fluids containing charged species regardless of which of the simple periodic boundary conditions described above is used. We can see how the problem arises by noting that using the Coulomb potential, which decays as $1/r$, results in a divergent integral for the long-range correction of Eq. 13.1-1. This problem is generally treated in simulation by what is referred to as the *Ewald sum method*,[4] which is nicely described by Frenkel and Smit.[5] The underlying physics is that an ion cloud of a net opposite charge forms around each ion, shielding the central charge and thereby reducing its effective range, and this must be accounted for. This is discussed in Chapter 15 in the context of the derivation of Debye-Hückel theory.

While the discussion above was concerned with the common features of molecular-dynamics and Monte Carlo simulations, there are some very important differences. One, already mentioned, is that since molecular-dynamics follows the dynamics of the system, a representative trajectory in time is eventually obtained, and transport as well as equilibrium properties can be computed. However, for molecular-dynamics results to be accurate, the time step must be chosen to be less than the mean time between collisions. Consequently, in dense fluids (and also with long chain molecules), the time steps are very short; and long computer run times are necessary to obtain accurate results. Even with long computer runs, it may not be possible in simulation to obtain some molecular configurations; for example, a transition in a dense fluid that may require the coordinated movement of several molecules for another molecule

---

[4]P. P. Ewald, *Ann. Phys.* **64**, 253 (1921).
[5]See Chapter 12 of D. Frenkel and B. Smit, *Understanding Molecular Simulation, 2nd ed.*, Academic Press, 2001.

to pass. That is, at high density, some or all molecules may be trapped in certain configurations for the duration of the simulation. This problem is less severe in Monte Carlo simulations, since, as we shall see, a molecule can be moved from one location to another without regard to the intermediate states. Further, with the use of some biasing methods, which we will discuss briefly in the following section, Monte Carlo simulations of dense liquids and long chain molecules are possible, that would be much more difficult—or impossible—using molecular-dynamics simulations.

Molecular computer simulation is sometimes referred to as an *in silico* (within silicon or a computer chip) experiment. Much like a real experiment, what one obtains is a measured (or, in this case, computed) value of properties at a given state point. That is, if the interatomic potential, temperature, and density (number of atoms and volume) are fixed, from simulation one can obtain values of the pressure, internal energy, and radial distribution function as a function of interatomic separation. However, one does not obtain equations describing the properties. Therefore, one only obtains information about a specific state point, but does not obtain an equation of state (unless one does many calculations and then correlates the data with some empirical relation). This is quite different from previous methods, such as the graph theory expansion we did of the partition function in Chapters 7 and 8 that led to a specific equation of state, the virial equation, and values of the second virial coefficient in terms of the interatomic potential.

13.2 THERMODYNAMIC PROPERTIES FROM MOLECULAR SIMULATION

In the following sections we will provide a brief description of the details of molecular simulation. Whether it be Monte Carlo or molecular-dynamics simulation, the common feature is that one considers a great many (typically millions) different arrangements of the molecules in the simulation box, and then computes the thermodynamic properties as averages over these different arrangements. Monte Carlo and molecular-dynamics differ in how these arrangements ate generated.

Assuming pairwise additivity, interaction energy contribution to the internal energy at a fixed temperature T and density ρ in any one configuration or arrangement of the molecules can be computed using

$$U^C(T, \rho) = \sum_i \sum_{\substack{i > j \\ r_{ij} < L/2}} u(r_{ij}) + 2\pi \frac{N^2}{L^3} \int_{L/2}^{\infty} u(r) r^2 \, dr$$

$$= \sum_i \sum_{\substack{i > j \\ r_{ij} < L/2}} u(r_{ij}) + 2\pi N \rho \int_{L/2}^{\infty} u(r) r^2 \, dr \qquad (13.2\text{-}1)$$

where the sum is over all molecules i and j in the simulation box. (Note that in this calculation and elsewhere, when the cutoff distance around a molecule in the simulation box extends into one or more periodic images of the simulation box, interactions with the molecules in those periodic images are used.) Also, the last term on the right-hand side of this equation is not a function of the arrangement of the molecules and needs to be computed only once for each density.

Typically, what is done in a simulation is to compute the configurational energy at many fixed intervals during the course of the simulation (after the initial equilibration), and keep a running-average value. When a reasonable steady average value is obtained, with only small fluctuations, the simulation is ended and the average value is recorded. This calculation would then be repeated at other densities and temperatures to obtain some insight into how the configurational energy changes with these variables. Remember that since fluctuations about the mean value depend on $1/\sqrt{M}$, where M is the sample size—here, the product of the number of molecules and the number of samples in the averaging process—and since this product is finite (and much less than Avogadro's number), the value of the average configurational energy will fluctuate about a mean value during the course of the simulation.

The radial or pair distribution function $g(r)$ is defined by Eq. 11.1-8a. As we have seen, its interpretation is that the number of molecules in a three-dimensional volume element $d\underline{r}$ at a distance r from a central molecule in a fluid of density ρ is $\rho g(r) d\underline{r}$, while the number of molecules in that volume element in a completely random fluid (or ideal gas) at the same density would be $\rho \, d\underline{r}$. Therefore, to compute $g(r)$ in a simulation of spherical molecules, the following procedure is used. A series of thin shells is constructed around each molecule. Typically, each shell is of thickness δr, so that the volume of a shell is $\frac{4}{3}\pi[(r+\delta r)^3 - r^3]$, where δr is sufficiently small. Using i to represent the ith shell at a distance r_i from a central molecule and n_i to be the number of molecules in the ith shell, what is done is to obtain mean value of n_i by averaging over each molecule in the simulation box over the course of the simulation. Note that in this calculation, all molecules within each of the shells are counted, even though some of these molecules are in the periodic images of the simulation box. The pair correlation function, attributed to a position halfway between r_i and $r_i + \delta r$, is then computed from

$$g\left(r_i + \frac{1}{2}\delta r, \rho, T\right) = \frac{n_i}{\frac{4\pi}{3}\rho\left[(r_i + \delta r)^3 - r_i^3\right]} \tag{13.2-2}$$

This calculation is then repeated for all other shells in the simulation of one state point, and then separately later at other densities and temperatures.

Once the pair correlation function has been computed, Eq. 11.2-4

$$U^C(T, \rho) = 2\pi N\rho \int_0^\infty u(r)g(r; \rho, T)r^2 \, dr \tag{11.2-4}$$

can also be used to compute the configuration contribution to the internal energy. There is, however, an important difference between using Eq. 13.2-1 and Eq. 11.2-4. That is, Eq. 13.2-1 is used during the course of the simulation to obtain the configurational energy in each configuration of the molecules, which is then used to obtain the average value and to monitor the progress of the simulation. Equation 11.2-4, however, is best used after the simulation is completed and a pair correlation function has been obtained.

The pressure can be computed from Eq. 11.2-11:

$$P = \rho k T - \frac{2\pi}{3}\rho^2 \int_0^\infty \frac{du(r)}{dr} g(r; \rho, T) r^3 \, dr \tag{10.2-11}$$

13.2 Thermodynamic Properties from Molecular Simulation 247

However, this equation, like Eq. 11.2-4 above, is best used at the end of the simulation once the pair correlation is known, rather than during the simulation to compute averages or monitor the convergence of the simulation. For these purposes—that is, to compute the pressure "on-the-fly" during a simulation—a different method of computing pressure based on the virial theorem (not the virial equation of state) is used. This is described below.

From Newtonian classical mechanics, we have that the force on a particle i in the x-direction $F_{i,x}$ is related to its acceleration in that direction as follows:

$$m \frac{dv_{i,x}}{dt} = m \frac{d^2 r_{i,x}}{dt^2} = F_{i,x} \tag{13.2-3}$$

Now, multiplying this equation by $r_x/2$ and rearranging gives

$$\frac{1}{2} m r_{i,x} \frac{d^2 r_x}{dt^2} = \frac{1}{4} \frac{d^2}{dt^2}(m r_{i,x}^2) - \frac{1}{2} m \left(\frac{dr_{i,x}}{dt}\right)^2 = \frac{1}{4} \frac{d^2}{dt^2}(m r_{i,x}^2) - \frac{1}{2} m v_{i,x}^2 = \frac{1}{2} r_{i,x} F_{i,x} \tag{13.2-4}$$

Next, we integrate this equation over the time period from some initial time $t = 0$ after which the system has equilibrated to some later time τ, and divide by τ to obtain average values

$$\frac{1}{4\tau} \int_0^\tau \frac{d^2}{dt^2}(m r_{i,x}^2)\, dt - \frac{1}{2\tau} m \int_0^\tau v_{i,x}^2\, dt = \frac{1}{4\tau}\left[\frac{d}{dt}(m r_{i,x}^2)\bigg|_\tau - \frac{d}{dt}(m r_{i,x}^2)\bigg|_{t=0}\right] - \frac{m}{2}\langle v_{i,x}^2 \rangle$$

$$= \frac{1}{2\tau} \int_0^\tau r_{i,x} F_{i,x}\, dt = \frac{1}{2}\langle r_{i,x} F_{i,x} \rangle \tag{13.2-5}$$

where the brackets $\langle\ \rangle$ denote the average over the time interval. Now summing over all N molecules, and since we are only considering a time interval after which the system has equilibrated, even though there are fluctuations, there is no net change with time; that is, the time derivative terms in the equation above are zero, and we obtain

$$\sum_{i=1}^N \frac{m}{2}\langle v_{i,x}^2 \rangle = -\frac{1}{2}\sum_{i=1}^N \langle r_{i,x} F_{i,x} \rangle \tag{13.2-6}$$

Next, we generalize this expression to all three coordinate directions and obtain

$$\sum_{i=1}^N \frac{m}{2}\langle v_{i,x}^2 + v_{i,y}^2 + v_{i,u}^2 \rangle = \frac{3}{2} NkT = -\frac{1}{2}\sum_{i=1}^N \langle \mathbf{r}_i \cdot \mathbf{F}_i \rangle \quad \text{or} \quad \sum_{i=1}^N \langle \mathbf{r}_i \cdot \mathbf{F}_i \rangle = -3NkT \tag{13.2-7}$$

where $\mathbf{r}$ is the position vector and $\mathbf{F}$ is the force vector, and we have used that the average kinetic energy of the molecules is $3NkT/2$. There are two contributions to the so-called virial $\sum_{i=1}^N \langle \mathbf{r}_i \cdot \mathbf{F}_i \rangle$. The first comes from the interactions between the gas molecules and the wall, which are independent of the interactions between the molecules (that is, the force needed to keep even ideal gas molecules in a box), and the second arises from interactions among the molecules. Since we know that

Chapter 13: Determination of the Radial Distribution Function

$PV = NkT$ when the interactions between the molecules can be neglected, we have

$$\sum_{i=1}^{N} \langle \mathbf{r}_i \cdot \mathbf{F}_i \rangle_{\text{only wall interactions}} = -3NkT = -3PV \qquad (13.2\text{-}8a)$$

and

$$\sum_{i=1}^{N} \langle \mathbf{r}_i \cdot \mathbf{F}_i \rangle_{\text{intermolecular interactions}} + \sum_{i=1}^{N} \langle \mathbf{r}_i \cdot \mathbf{F}_i \rangle_{\text{only wall interaction}}$$

$$= \sum_{i=1}^{N} \langle \mathbf{r}_i \cdot \mathbf{F}_i \rangle_{\text{intermolecular interactions}} - 3PV = -3NkT \qquad (13.2\text{-}8b)$$

Then

$$P = \frac{NkT}{V} + \frac{1}{3V} \sum_{i=1}^{N} \langle \mathbf{r}_i \cdot \mathbf{F}_i \rangle_{\text{intermolecular interactions}} = \frac{NkT}{V} - \frac{1}{3V} \sum_{i=1}^{N} \sum_{j>i}^{N} \left\langle \mathbf{r}_i \cdot \left(\frac{du(r_{ij})}{d\mathbf{r}_i} \right) \right\rangle$$

$$= \frac{NkT}{V} - \frac{1}{3V} \sum_{i=1}^{N} \sum_{j>i}^{N} \left\langle r_{i,x} \left(\frac{du(r_{ij})}{dr_{i,x}} \right) + r_{i,y} \left(\frac{du(r_{ij})}{dr_{i,y}} \right) + r_{i,z} \left(\frac{du(r_{ij})}{dr_{i,z}} \right) \right\rangle$$

$$(13.2\text{-}9)$$

Now, using that $r_{ij} = \sqrt{(r_{i,x} - r_{j,x})^2 + (r_{i,y} - r_{j,y})^2 + (r_{i,z} - r_{j,z})^2}$, the equation above can be written as

$$P = \frac{NkT}{V} - \frac{1}{3V} \sum_{i=1}^{N} \sum_{j>i}^{N} \left\langle r_{ij} \left(\frac{du(r_{ij})}{dr_{ij}} \right) \right\rangle \qquad (13.2\text{-}10)$$

It is Eq. 13.2-10 that is used to calculate the pressure in any configuration of the molecules in the simulation, and is then averaged over many configurations to obtain the pressure. In this way the pressure can be obtained during the simulation. Indeed, during the course of a simulation, it is common to monitor the interaction energy using Eq. 13.2-1 and the pressure using Eq. 13.2-10 to determine whether equilibrium has been obtained in the simulation. Also, it is easy to show (see Problem 13.4) that Eq. 11.2-11 and Eq. 13.2-10 are equivalent. As will be discussed below, there is an interesting type of conceptual symmetry that occurs.

In Monte Carlo simulation, only the energies need to be computed when considering transitions between configurations, so there is an additional small computational penalty to also compute the forces on the molecules to calculate the pressure. In molecular-dynamics simulations, it is only the forces that need to be computed when considering transitions between configurations, so there is an additional small computational penalty to also compute the configurational energy.

In order to compute the entropy from simulation, we would have to use

$$S(N, V, E) = k \ln \Omega(N, V, E) \qquad (13.2\text{-}11)$$

if the microcanonical ensemble were used, where $\Omega(N, V, E)$ is the number of configurations available to the system at fixed volume, number of molecules, and total energy. Alternatively, using the canonical ensemble, the entropy is computed from

$$S(N, V, T) = k \ln Q(N, V, T) + kT \left(\frac{\partial \ln Q(N, V, T)}{\partial T}\right)_{N,V}$$

$$\text{where} \quad Q(N, V, T) = \sum_{\text{states } i} e^{-E_i(N,V)/kT} \tag{13.2-12}$$

Even for thousands of molecules, the number of different configurations is so large (only one molecule moved a very small distance is a new configuration) that neither $\Omega(N, V, E)$ nor $Q(N, V, T)$ can be computed. Consequently, the entropy (and therefore also the Gibbs and Helmholtz energies) cannot be computed directly in the molecular simulations that are discussed in this chapter. The properties that can be computed are the configurational energy U, the pressure, and the radial distribution function. All of these can be obtained from averages over a large number of configurations, but do not require the impossible task of considering all possible configurations in order to evaluate the partition functions $\Omega(N, V, E)$ and $Q(N, V, T)$.

13.3 MONTE CARLO SIMULATION

In its very simplest form, a Monte Carlo simulation could be done by starting with an empty (virtual) box, and then using a random number generator (or, more precisely, a pseudo-random number generator, since computers are deterministic not random) to generate a position for each molecule in the box. This would be repeated for each molecule until the required density is obtained. The simulated pressure, configurational energy, radial distribution function, and so forth would then obtained by averaging the results from many individual box fillings, each weighted with the normalized Boltzmann factor

$$\frac{e^{-U^C/kT}}{\sum_{\substack{\text{all} \\ \text{configurations}}} e^{-U^C/kT}} \tag{13.3-1}$$

The problem with this very simple Monte Carlo approach is that by randomly inserting molecules into a box, it is likely that one or more pairs of molecules will overlap, and the probability of this happening increases rapidly with density. Since even a single pair of overlapping molecules has infinite energy, the Boltzmann factor for such a configuration is zero. Consequently, at moderate and high density, few (if any) randomly generated configurations will contribute to the average. There is also the conceptual problem that to evaluate the denominator in the above equation, one has to sum over all possible states of the system in order to normalize the probabilities—an impossible task.

Therefore, more efficient Monte Carlo simulation methods have been developed. Each of these methods is based on some form of importance sampling. The basic idea is to start with an acceptable configuration (i.e., one that does not have infinite energy, usually obtained by placing molecules on a regular lattice) and, from this configuration, develop other configurations in a way that is biased to result in configurations of lower energy. The bias in choosing successive configurations is then accounted for in the averaging process. The most common procedure, and the only one we will discuss here, is due to Metropolis et al.[3]

The Metropolis[3] algorithm is based on generating a Markov chain of states. The two characteristics of Markov chains are that there are a finite (or countable) set of states of the system, and that the probability of transition from one state to another depends only on the properties of each state and not on other states—in particular, not on states that the system may previously have occupied. The Markov chain is characterized by a set of states $(1, 2, 3, \ldots, n)$ and a $n \times n$ matrix of transition probabilities between each of the states.

To illustrate the properties of a Markov chain, we consider the following simple example. Suppose there is a system that consists of three states and the probability that the system is in each of these three states is p_1, p_2, and p_3, respectively, which we denote by the vector $\mathbf{P} = (p_1, p_2, p_3)$. Clearly $p_1 + p_2 + p_3 = 1$. Next, a set of probabilities are formulated for the transition from any state to any other state; for example, $T_{1 \to 2}$ is the transition probability from state 1 to state 2, $T_{1 \to 3}$ from state 1 to state 3, etc. The list of all transition probabilities is usually presented in matrix form. As an example here, we will use the following transition probability matrix.

$$\begin{bmatrix} T_{1 \to 1} & T_{1 \to 2} & T_{1 \to 3} \\ T_{2 \to 1} & T_{2 \to 2} & T_{2 \to 3} \\ T_{3 \to 1} & T_{3 \to 2} & T_{3 \to 3} \end{bmatrix} = \begin{bmatrix} 0.5 & 0.2 & 0.3 \\ 0.4 & 0.4 & 0.2 \\ 0.4 & 0.3 & 0.3 \end{bmatrix}$$

Notice that the transition matrix is not necessarily symmetric—that is, the probability of going from state 1 to state 2 does not have to be the same as going from state 2 to state 1. Also note that the sum of the transition probabilities along any row is unity, and that each element along the diagonal of the matrix is the probability that the system remains in its current state.

Next, we consider how the probability distribution among the three states of the system (p_1, p_2, p_3) changes whenever a transition occurs. The second column in the Table 13.3-1 shows how the probability distribution among the three possible states changes on 10 successive transitions if we start from the system in state 1—that is, the initial probability distribution among the states is (1 0 0). In column 3 of the table, the calculation is repeated starting with the different initial probability distribution of (0 1 0). The last column shows the results of repeating the calculation starting with an initial probability distribution of (0 0 1).

The somewhat surprising observation from this table is that the after a sufficient number of transitions, the probability distribution among the possible states of the system does not depend on the initial state; therefore, it can only depend on the transition matrix between the states. This is an important observation for Monte Carlo simulation, since it indicates that after many transitions, when the probability

Table 13.3-1 Change in State Probabilities as a Function of Initial State and Number of Transitions

Initial state	(1.000 0.000 0.000)	(0.000 1.000 0.000)	(0.000 0.000 1.000)
Transition 1	(0.500 0.200 0.300)	(0.400 0.400 0.200)	(0.400 0.300 0.300)
Transition 2	(0.450 0.270 0.280)	(0.440 0.300 0.260)	(0.440 0.290 0.271)
Transition 3	(0.445 0.282 0.273)	(0.444 0.286 0.270)	(0.444 0.285 0.272)
Transition 4	(0.445 0.284 0.272)	(0.444 0.284 0.271)	(0.444 0.284 0.272)
....			
Transition 10	(0.444 0.284 0.272)	(0.444 0.284 0.272)	(0.444 0.284 0.272)

distribution between states is no longer changing (which is taken to be the equilibrium state), this probability distribution does not depend on the initial state of the system, but only on the transition matrix. Also, this result means that in order to evaluate the probability distribution of the states of the system, we do not have to evaluate the summation in the denominator of Eq. 13.3-1; we only need information about the transition matrix.

With this background, the Metropolis algorithm of Monte Carlo molecular simulation in the canonical ensemble (fixed N, V, and T) can now be presented. In the canonical ensemble, the likelihood of occurrence of any state is proportional to the Boltzmann factor in its energy. Therefore, the transition probability for a change from state m to state n is proportional to $e^{-(E_n - E_m)/kT}$. The procedure then is to start with the collection of atoms in some arbitrarily chosen state without overlap. A change in the state is made, usually by moving a randomly chosen atom. This is accomplished by generating a random integer number in the range from 1 to N, where N is the number of atoms in the system. The atom chosen in this way at location (x_o, y_o, z_o) is moved in the x, y, and z directions. The extent of the movement is determined by generating three additional random numbers between -1 and 1, which we designate as α_x, α_y, and α_z. The new possible location of the atom is $(x_o + \alpha_x \times \delta_m, y_o + \alpha_y \times \delta_m, z_o + \alpha_z \times \delta_m)$, where δ_m is the maximum allowed distance of a proposed move in each coordinate direction. The value of δ_m is usually adjusted during the early stages of a simulation so that approximately half of the proposed moves are accepted according to the acceptance criterion discussed below. This acceptance ratio has been found to be a suitable compromise between moves that are so small that the simulation is very slow to converge and moves that are so large that few moves are accepted.

To properly sample the system, microscopic reversibility must be satisfied; that is, the transition probability of generating a state m from state n must be equal to the transition probability of generating state n from state m. This is ensured by the completely random method of generating possible moves discussed above. However, after a possible move is generated, importance sampling is used to decide whether the move is accepted; if not, the initial state is retained and counted again in the averaging. This is done with importance sampling as follows. Using the configurational energy of the old and new states, we calculate

$$\mathcal{P} = e^{-(E_n - E_m)/kT} \tag{13.3-2}$$

If $\mathcal{P}$ is greater than 1 (that is $E_n < E_m$), the move is accepted. If $\mathcal{P}$ is less than 1 as a result of $E_n > E_m$, then an additional random number R between 0 and 1 is generated. If R is smaller than $\mathcal{P}$, the move is accepted; if not, it is rejected. That is, the conditions for the acceptance or rejection of a move are

$\mathcal{P}$ greater than 1, move is accepted

$\mathcal{P}$ less than 1, move accepted if $\mathcal{P} > R$

$\mathcal{P}$ less than 1, move rejected if $\mathcal{P} < R$ (13.3-3)

The final condition that must be met is that a sufficiently large number of states of the system must be sampled (that is, the simulation must be long enough) to be representative of all the possible states of interest of the system. Such a simulation is said to be *ergodic*. The properties of each of the states in the simulation (after discarding the early states that reflect the initial configuration and are far from equilibrium) are then averaged without any weighting. Note that this is the essential difference from the simple Monte Carlo simulation, in which the states are generated randomly and then properties averaged with a Boltzmann factor weighting. Here, the states are generated with a Boltzmann factor weighting and then the properties are linearly averaged.

The above is a brief description of the simple Monte Carlo NVT simulation for an atomic system. Monte Carlo simulation techniques have evolved much beyond this stage, and the reader is referred to books on this subject for details.[1] The obvious improvements are to polyatomic systems, in which moves also consist of rotations of the whole molecule as well as around bonds; to different biasing methods to allow the study of chain molecules and polymers; to mixtures in which moves can include molecule identity swaps; and to the use of other ensembles such as the grand canonical ensemble in which $VT\mu$ are fixed (which is especially useful for the simulation of adsorption and osmotic equilibrium). Monte Carlo simulation has also been used for the NPT ensemble. The reason this is of interest is that most experimental measurements are made at fixed temperature and pressure, rather than temperature and density. Also, excess thermodynamic properties on mixing are determined at fixed temperature and pressure. In this case, a possible Monte Carlo move can be either a particle displacement or a volume change of the simulation box. The Markov chain generation acceptance criteria are different than in the NVT ensemble (and the simulation is somewhat slower), because all particle locations must be scaled with each volume change and the long-range correction changes.

Another very useful method is the so-called *Gibbs ensemble simulation*,[6] which involves two simulation boxes (of different densities for pure fluids, and also different compositions for mixtures) and allows for the calculation of vapor-liquid and other phase equilibria. In this simulation, the total number of molecules (to be distributed between the two boxes) and the total volume (to be divided between the two boxes) are fixed, and temperatures in both simulation boxes are identical, satisfying one of the conditions for phase equilibrium. The simulation then includes three types of moves for a pure fluid and a fourth type of move for mixtures. First are the particle movements within each box to ensure equilibrium within each box. Next are volume changes of both boxes (at fixed total volume) to ensure that the pressure in both simulation boxes will be equal, a second condition for phase equilibrium. The third type of move is the transfer of a molecule from one box to the other to ensure equality of chemical potentials (the third condition for phase equilibrium). For mixtures, an identity swap move (i.e., interchange a species 1 molecule with a species 2 molecule) is the fourth type of move, and it is needed to ensure the equality of chemical potentials for all species in both boxes.

After equilibration, the common pressure in both simulation boxes, the different molecular densities in each simulation box, and (if a mixture) the different compositions in each box are computed. In this way, the vapor pressure of a pure fluid and densities of the coexisting phases can be computed as a function of temperature. In the case of a mixture, vapor-liquid or other phase boundaries can also be computed.

The website www.wiley.com/college/sandler contains simple Monte Carlo simulation programs in MATLAB®[7] for the square-well and Lennard-Jones 12-6 fluids. (The square-well program can be used for the hard-sphere fluid by setting the R_{sw} parameter in the square-well fluid equal to 1 on input.) These programs, MC_sqwell

---

[6] A. Z. Panagiotopoulos, *Molecular Physics* **61**, 813 (1987).

[7] MATLAB® is a registered trademark of The MathWorks, Inc.

and MC_LJ, can be found in the folder MATLAB®. The readme.txt file provides information on their use.

13.4 MOLECULAR-DYNAMICS SIMULATION

Molecular-dynamics simulation is very different from Monte Carlo simulation in that, although the initial placement of atoms in the simulation box and their initial velocities (speed and direction) may be arbitrarily chosen, beyond that point the simulation is completely deterministic. The procedure is that, for each atom i at a particular position, the forces acting on it in the x, y, and z directions (that is $F_{i,x}$, $F_{i,y}$, and $F_{i,z}$) are obtained by summing the forces resulting from its interactions with all other atoms. If the interaction is pairwise additive, the forces are

$$F_{i,x} = \sum_{\substack{j=1 \\ j \neq i}}^{N} F_{ij,x} = -\sum_{\substack{j=1 \\ j \neq i}}^{N} \frac{du(r_{ij})}{dx}, \quad F_{i,y} = \sum_{\substack{j=1 \\ j \neq i}}^{N} F_{ij,y} = -\sum_{\substack{j=1 \\ j \neq i}}^{N} \frac{du(r_{ij})}{dy} \quad \text{and}$$

$$F_{i,z} = \sum_{\substack{j=1 \\ j \neq i}}^{N} F_{ij,z} = -\sum_{\substack{j=1 \\ j \neq i}}^{N} \frac{du(r_{ij})}{dz} \tag{13.4-1}$$

where $F_{ij,x}$ is the force on atom i in the x direction as a result of atom j that is at a separation distance r_{ij}, and the sum is over all other atoms. There is the similar interpretation for the other coordinate directions. Newton's laws of motion for each atom i in each of the coordinate directions are

$$m\frac{d^2 x_i}{dt^2} = F_{i,x} = \sum_{\substack{j=1 \\ j \neq i}}^{N} F_{ij,x} = -\sum_{\substack{j=1 \\ j \neq i}}^{N} \frac{du(r_{ij})}{dx}, \quad m\frac{d^2 y_i}{dt^2} = F_{i,y} = \sum_{\substack{j=1 \\ j \neq i}}^{N} F_{ij,y} = -\sum_{\substack{j=1 \\ j \neq i}}^{N} \frac{du(r_{ij})}{dy}$$

$$\text{and} \quad m\frac{d^2 z_i}{dt^2} = F_{i,z} = \sum_{\substack{j=1 \\ j \neq i}}^{N} F_{ij,z} = -\sum_{\substack{j=1 \\ j \neq i}}^{N} \frac{du(r_{ij})}{dz} \tag{13.4-2}$$

The procedure, then, is to numerically integrate the equations of motion over such small time intervals or time steps Δt that there is not a significant change in the velocity of any molecule during this time interval. Also, since many time steps are needed (first for the system to evolve from the initial state to a near equilibrium state and then continuing for many more time steps to compute average properties), very accurate numerical integration procedures have to be used; otherwise, small numerical errors will accumulate during the simulation. Numerical methods such as finite difference and the more accurate predictor-corrector and other integration methods have been used.

Choosing a time step for the integrations is a balance between steps that are too small, so that the simulation to obtain accurate equilibrium averages will take too long; and steps that are too large and lead to errors due to numerical integration—and because of instabilities that arise if, as a result of an interaction (or collision), an atom undergoes a large velocity change or even velocity reversal within the time step. Time steps of the order of femtoseconds (one-quadrillionth of a second) are typical.

254 Chapter 13: Determination of the Radial Distribution Function

One problem of the molecular-dynamics method as described above is that we are usually interested in the values of properties at a specified temperature and density (in the canonical ensemble), while the molecular-dynamics method so far described is adiabatic—that is, of constant total energy. Here, by total energy, we mean the sum of kinetic and potential (or interaction) energy. Since the total energy is fixed in the simple molecular-dynamics calculation, as particles interact their interaction energy may increase at the expense of their kinetic energy, or vice versa. Also, since the temperature of the system is

$$\frac{3}{2}NkT = \frac{1}{2}m\sum_{i=1}^{N} v_i^2 \qquad (13.4\text{-}3)$$

the problem that arises in what has so far been described is that since there is a constant interchange between kinetic and interaction energies, the temperature of the system (which only depends on kinetic energy) is not fixed, but varies during the course of the simulation.

There are several ways that the adiabatic simulation method described above can be modified to be a constant-temperature molecular-dynamics simulation, all of which involve a continual change in the kinetic energy of the atoms. The simplest method is that during the simulation, the temperature T of the atoms is computed using Eq. 13.4-3 and compared with a desired set temperature T_s. Then a new velocity of each atom i in each coordinate direction j (denoted by $v_{i,j}^{\text{old}}$) is scaled to obtain a new velocity (denoted by $v_{i,j}^{\text{new}}$) as follows:

$$v_{i,j}^{\text{new}} = v_{i,j}^{\text{old}} \sqrt{\frac{T_s}{T}} \qquad (13.4\text{-}4)$$

A less abrupt way of changing the velocities is by use of a virtual thermostat to mimic the way energy is interchanged between a real system and thermostatic bath. Starting from the idea that the rate of temperature change of the system (by heat input from a thermostat) should be proportional to the difference between the system temperature T and the bath temperature T_b, we have

$$\frac{dT}{dt} = \frac{\Delta T}{\Delta t} = \tau(T_b - T) \qquad (13.4\text{-}5)$$

In this equation, Δt is the time step used in the numerical integration of Eq. 13.4-2, and τ is a parameter coupling the system and the bath. (For a physical system, τ would be the ratio of the total heat transfer coefficient (product of heat transfer coefficient and area) to the total heat capacity of the system (product of mass and constant volume heat capacity)). If τ is small in value, the coupling is weak and the system temperature changes slowly. If τ is large, there is a tight coupling and the system temperature changes rapidly. With this model,[8] the velocity scaling is

$$v_{i,j}^{\text{new}} = v_{i,j}^{\text{old}} \sqrt{1 + \tau \Delta t \left(\frac{T_B}{T_s} - 1\right)} \qquad (13.4\text{-}6)$$

---

[8]H. J. C. Berendsen, J. P. M. Postma, W. F. van Gunsteren, A. Di Nola, and J. R. Haak, *J. Chem. Phys.* **81**, 3684 (1984).

The value of τ is adjusted to give good simulation results, which will also depend on the time step Δt used in the integration. Empirically, a value of $\tau \Delta t \approx 0.0025$ has been found to give reasonable results. Other, more sophisticated thermostats for molecular dynamics are also used. However, none of these isothermal ensemble molecular-dynamics simulations result in a true constant temperature system, because the temperature will fluctuate during the simulation.

It should be noted that the adiabatic molecular-dynamics simulation is at a fixed number of atoms, volume, and energy (NVE) and therefore corresponds to a simulation in the microcanonical ensemble. The isothermal molecular-dynamics simulation is at a fixed number of atoms, volume, and temperature (NVT) and corresponds to a simulation in the canonical ensemble (though, since there are temperature fluctuations, in principle it is not a true canonical ensemble simulation).

The website www.wiley.com/college/sandler contains a simple molecular-dynamics simulation MATLAB®[7] program for the Lennard-Jones 12-6 fluid. These programs, MD_LJ and MD_LJ2, can be found in the MATLAB® programs folder. The readme.txt file provides information on its use. There is also the program MD_LJ2 that does an isothermal molecular dynamics for the LJ 12-6 fluid. Finally the LJ_MD_MC program does a Monte Carlo simulation followed by an isothermal molecular-dynamics simulation for the Lennard-Jones 12-6 fluid at the same state conditions. This is useful for comparing the properties computed from both types of simulations. Note that exactly the same results should be obtained for an infinite number of moves in a Monte Carlo simulation and an infinite number of time steps in a molecular-dynamics simulation. This is the ergodic hypothesis of Chapter 1. However, because of the stochastic nature of simulations (the random numbers generated in a Monte Carlo simulation and initial assignments of positions and velocities in a molecular-dynamics simulation), the results obtained from the two simulation techniques will not be identical for finite simulations, though they will converge as the simulation lengths increase.

An excellent two-dimensional illustration of adiabatic and isothermal molecular-dynamics simulations, prepared by Professor David Kofke and Dr. Andrew Schultz, can be downloaded from http://www.etomica.org/wiki/LennardJones:Simulator and run on your personal computer. In this simulation, the user can set the state conditions and see the fluctuations in the pressure, temperature, potential and kinetic energies, and the radial distribution as a function of interatomic separation as the system evolves to equilibrium from an initial lattice configuration. The user can adjust the temperature and atom number density in the isothermal simulation, and the energy and number density in the adiabatic simulation. (To choose the energy in the adiabatic simulation, first choose the isothermal simulation, set a reduced temperature that will fix the initial energy in the system, and then choose "adiabatic simulation" and run.)

CHAPTER 13 PROBLEMS

13.1 Use the MATLAB® isothermal molecular-dynamics program MD_LJ2 to compute the radial distribution function, internal energy, and pressure for the Lennard-Jones 12-6 potential at one of following conditions:
  a. $T^* = kT/\varepsilon = 1.0$ and $\rho^* = \rho\sigma^3 = 0.4$
  b. $T^* = 0.9$ and $\rho^* = 0.3$
  c. $T^* = 1.5$ and $\rho^* = 0.45$

13.2 One of the results of the MATLAB® isothermal molecular-dynamics program MD_LJ2 is the velocity distribution function. Compute the Maxwell distribution of speeds (Eq. 3.9-5) at one of the conditions in Problem 13.1 and compare the results to those obtained from the molecular-dynamics program at one of the conditions in that problem.

13.3 From having done molecular simulations at a collection of state points (either by yourself our sharing simulation results with colleagues) for the Lennard-Jones 12-6 potential, comment on how the maximum in the first peak of the radial distribution function changes with T^* and ρ^*.

13.4 Show that Eqs. 11.2-11 and 13.2-10 are equivalent.

13.5 For a state point of your choice for the hard-sphere fluid, compare the results for the radial distribution function obtained from the Monte Carlo program MC_sqwell (setting $R_{sw} = 1$) with those obtained from the Percus-Yevick solution using the program Percus-Yevick HS.

13.6 Compare the values of radial distribution function for the hard-sphere fluid at contact given by Eq. 12.5-3 at several densities with the values obtained from Monte Carlo simulation using the MATLAB® program Monte Carlo program MC_sqwell (setting $R_{sw} = 1$).

13.7 At a fixed packing fraction, calculate and plot the radial distribution function for the square-well fluid using the MATLAB® Monte Carlo program MC_sqwell at a collection of reduced temperatures kT/ε for the well width $R_{sw} = 1.5$. Comment on the temperature dependence of the radial distribution function.

13.8 At a fixed packing fraction, calculate and plot the radial distribution function for the square-well fluid using the MATLAB® Monte Carlo program MC_sqwell at a fixed value of reduced temperature kT/ε for varying well widths. Comment on the well width dependence of the radial distribution program.

13.9 One model for globular proteins is as a square-well fluid with a very deep well ε and a very narrow well width R_{sw}. Keeping the intensity of the interaction approximately constant by choosing a fixed value of $(R_{sw}^3 - 1)\varepsilon$, determine how the radial distribution function, pressure, and compressibility change over $R_{sw} - \varepsilon$ space using the MATLAB® program MC_sqwell.

13.10 A test of the ergodic hypothesis (Section 1.4) would be to study a system using Monte Carlo simulation (average over possible states) and molecular dynamics (average over a long time interval) and see if the two results agree, at least to within the statistical fluctuations that arise from the simulations being for a small number of atoms (compared to Avogadro's number) and over a relatively small numbers of states (MC) and short time intervals (MD). Use the MATLAB® program LJ_MC_MD for the Lennard-Jones 12-6 fluid, and compare the results for the radial distribution function and the thermodynamic properties for each of the following cases. How does the extent of agreement change with the length of the simulation? Use one of the following conditions:
  a. $T^* = kT/\varepsilon = 1.0$ and $\rho^* = \rho\sigma^3 = 0.8$
  b. $T^* = 0.9$ and $\rho^* = 0.2$
  c. $T^* = 0.5$ and $\rho^* = 0.8$
  d. $T^* = 2.0$ and $\rho^* = 0.75$
  e. $T^* = 0.2$ and $\rho^* = 0.75$
  f. $T^* = 0.2$ and $\rho^* = 1.0$

13.11 From having done molecular simulations at a collection of state points (either by yourself our sharing simulation results with colleagues) for the square-well potential, comment on how the range of the radial distribution function (that is, the number of easily visible peaks and valleys) changes with T^* and ρ^*.

13.12 Since the MATLAB® isothermal molecular-dynamics program MD_LJ2 does a simulation at constant temperature, the kinetic energy of the system is unchanged in the simulation and depends only on the temperature T^* as shown in the graph produced in the simulation. However, as also shown in the graph, the average total energy (sum of the kinetic and potential energies) depends on both T^* and ρ^*. Comment on the behavior of the total energy on T^* and ρ^*, and why the total energy is negative at some densities.

Chapter 14

Perturbation Theory

As should be evident from the previous chapters, considerable effort is involved in obtaining the values of the thermodynamic properties and the radial distribution function for a chosen interaction potential at a single temperature and density, regardless of whether an integral equation method or computer simulation is used. Also, either of these methods only results in numerical values of these properties at the chosen temperature and density; they do not provide an analytical equation for use in calculations with other interaction potentials or at other state points.

One method of extending the usefulness of the thermodynamic properties and radial distributions functions that have been obtained for one interaction potential (which we call the reference potential) for use with a different potential is by using *perturbation theory*, wherein one does a Taylor series expansion of a thermodynamic property or the radial distribution function in the difference between the new potential of interest and the reference potential. An introduction to this method is the subject of this chapter.

INSTRUCTIONAL OBJECTIVES FOR CHAPTER 14

The goals for this chapter are for the student to:

- Understand perturbation theory using the hard-sphere as the reference potential
- Understand perturbation theory for other reference fluids
- See how perturbation theory is used to develop thermodynamic models of use in chemistry and chemical engineering

14.1 PERTURBATION THEORY FOR THE SQUARE-WELL POTENTIAL

As one can see from the previous sections, the calculation of the thermodynamic properties and radial distribution function for a liquid is a difficult task. One case where we do know the thermodynamic properties with reasonable accuracy is the hard-sphere fluid. (The thermodynamic properties on other fluids are generally known only from simulation and from fitting general simulation results with complicated polynomials in reduced temperature and reduced density.) Therefore, an obvious question is, can we use the information for the hard-sphere fluid to estimate the

258 Chapter 14: Perturbation Theory

properties of other fluids? This is what is done in perturbation theory,[1] which is based on a Taylor series expansion of a property about the values of that property for some other potential model or in some other state. As a reminder, a Taylor series expansion of some property $\Theta(x, \lambda)$, whose value is a function of the variables x and λ, can be expanded as a function of λ as follows:

$$\Theta(x, \lambda) = \Theta(x, \lambda = 0) + \lambda \left(\frac{\partial \Theta(x, \lambda)}{\partial \lambda} \right)_{\lambda=0,x} + \frac{\lambda^2}{2!} \left(\frac{\partial^2 \Theta(x, \lambda)}{\partial \lambda^2} \right)_{\lambda=0,x} + \frac{\lambda^3}{3!} \left(\frac{\partial^3 \Theta(x, \lambda)}{\partial \lambda^3} \right)_{\lambda=0,x} + \cdots$$

(14.1-1)

What is generally done in statistical mechanical perturbation theory is to split the interaction potential in a reference part ($\lambda = 0$ part) and a perturbation part and then do a Taylor series expansion in λ.

To be explicit, consider the hard-sphere potential

$$u_{\text{hs}}(r) = \begin{cases} \infty & \text{for} \quad r < \sigma \\ 0 & \text{for} \quad \sigma < r \end{cases}$$

(14.1-2a)

and, for example, the square-well potential

$$u_{\text{sw}}(r) = \begin{cases} \infty & \text{for} \quad r \leq \sigma \\ -\varepsilon & \text{for} \quad \sigma < r < R_{\text{sw}}\sigma \\ 0 & \text{for} \quad r \geq R_{\text{sw}}\sigma \end{cases}$$

(14.1-2b)

Note that the only difference between these two potentials is that in the region between σ and $R_{\text{sw}}\sigma$, the square-well potential is nonzero (in fact, $-\varepsilon$) in value. We can think of this attractive well as a perturbation to the hard-sphere potential. That is, the perturbation potential is

$$u_{\text{pert}}(r) = \begin{cases} 0 & \text{for} \quad r \leq \sigma \\ -\varepsilon & \text{for} \quad \sigma < r < R_{\text{sw}}\sigma \\ 0 & \text{for} \quad r \geq R_{\text{sw}}\sigma \end{cases}$$

(14.1-3)

so that $u(r, \lambda) = u_{\text{hs}}(r) + \lambda u_{\text{pert}}(r)$. In this way $u(r, \lambda = 1) = u_{\text{hs}}(r) + u_{\text{pert}}(r) = u_{\text{sw}}(r)$, while the hard-sphere potential is recovered for $\lambda = 0$. We now explore the possibility of obtaining the properties of the square-well fluid from the known properties of the hard-sphere fluid by doing an expansion around $\lambda = 0$.

To illustrate the ideas behind perturbation theory in its simplest form, consider first how perturbation theory could be used to determine the second virial coefficient for the square-well fluid from the known expression for the second virial coefficient of the hard-sphere fluid. We start by noting that

$$B_2^{\text{sw}}(T) = B_2(T, \lambda = 1) = B_2(T, \lambda = 0) + \lambda \left(\frac{\partial B_2(T, \lambda)}{\partial \lambda} \right)_{\lambda=0} + \frac{\lambda^2}{2!} \left(\frac{\partial^2 B_2(T, \lambda)}{\partial \lambda^2} \right)_{\lambda=0} + \cdots$$

$$= B_2^{\text{hs}} + \lambda \left(\frac{\partial B_2(T, \lambda)}{\partial \lambda} \right)_{\lambda=0} + \frac{\lambda^2}{2!} \left(\frac{\partial^2 B_2(T, \lambda)}{\partial \lambda^2} \right)_{\lambda=0} + \cdots$$

$$= \frac{2\pi\sigma^3}{3} + \lambda \left(\frac{\partial B_2(T, \lambda)}{\partial \lambda} \right)_{\lambda=0} + \frac{\lambda^2}{2!} \left(\frac{\partial^2 B_2(T, \lambda)}{\partial \lambda^2} \right)_{\lambda=0} + \cdots$$

(14.1-4)

since $B_2(T, \lambda = 0) = B_2^{\text{hs}} = 2\pi\sigma^3/3$.

---
[1]The idea of using perturbation theory, or a Taylor series expansion around a known simple potential, in statistical mechanics of dense fluids was first developed by R. Zwanzig, *J. Chem. Phys.* **22**, 1420 (1954).

14.1 Perturbation Theory for the Square-Well Potential

Now using

$$B_2(T) = 2\pi \int_0^\infty \left(1 - e^{-u(r)/kT}\right) r^2 \, dr$$

$$= 2\pi \int_0^\infty \left(1 - e^{-u_{\text{hs}}(r)/kT - \lambda u_{\text{pert}}(r)/kT}\right) r^2 \, dr$$

$$= 2\pi \int_0^\infty \left(1 - e^{-u_{\text{hs}}(r)/kT} e^{-\lambda u_{\text{pert}}(r)/kT}\right) r^2 \, dr \qquad \text{(14.1-5)}$$

we have

$$\left(\frac{\partial B_2(T,\lambda)}{\partial \lambda}\right)_T = 2\pi \frac{d}{d\lambda} \int_0^\infty \left(1 - e^{-u_{\text{hs}}(r)/kT} e^{-\lambda u_{\text{pert}}(r)/kT}\right) r^2 \, dr$$

$$= 2\pi \int_0^\infty \left(\frac{u_{\text{pert}}(r)}{kT} e^{-u_{\text{hs}}(r)/kT} e^{-\lambda u_{\text{pert}}(r)/kT}\right) r^2 \, dr$$

so that

$$\left(\frac{\partial B_2(T,\lambda)}{\partial \lambda}\right)_{T,\lambda=0} = \frac{2\pi}{kT} \int_0^\infty u_{\text{pert}}(r) e^{-u_{\text{hs}}(r)/kT} r^2 \, dr$$

$$= \frac{2\pi}{kT} \int_\sigma^{R_{\text{sw}}\sigma} -\varepsilon r^2 \, dr$$

$$= -\frac{2\pi\sigma^3}{3} \frac{\varepsilon}{kT} \left(R_{\text{sw}}^3 - 1\right) \qquad \text{(14.1-6a)}$$

since $e^{-u_{\text{hs}}(r)/kT} = 1$ for $r > \sigma$. Similarly

$$\left(\frac{\partial^2 B_2(T,\lambda)}{\partial \lambda^2}\right)_T = \frac{2\pi}{kT} \frac{d}{d\lambda} \int_0^\infty \left(u_{\text{pert}}(r) e^{-u_{\text{hs}}(r)/kT} e^{-\lambda u_{\text{pert}}(r)/kT}\right) r^2 \, dr$$

$$= -\frac{2\pi}{(kT)^2} \int_0^\infty \left((u_{\text{pert}}(r))^2 e^{-u_{\text{hs}}(r)/kT} e^{-\lambda u_{\text{pert}}(r)/kT}\right) r^2 \, dr$$

so that

$$\left(\frac{\partial^2 B_2(T,\lambda)}{\partial \lambda^2}\right)_{T,\lambda=0} = -\frac{2\pi}{(kT)^2} \int_0^\infty (u_{\text{pert}}(r))^2 e^{-u_{\text{hs}}(r)/kT} r^2 \, dr$$

$$= -\frac{2\pi\sigma^3}{3} \left(\frac{\varepsilon}{kT}\right)^2 \left(R_{\text{sw}}^3 - 1\right) \qquad \text{(14.1-6b)}$$

260 Chapter 14: Perturbation Theory

Therefore, to second order in the Taylor (perturbation) series expansion, for the full square-well potential ($\lambda = 1$) we have

$$B_2^{sw}(T) = \frac{2\pi}{3}\sigma^3 - \frac{2\pi}{3}\sigma^3 \left(R_{sw}^3 - 1\right) \left[\frac{\varepsilon}{kT} + \frac{1}{2}\left(\frac{\varepsilon}{kT}\right)^2\right] \quad \text{(14.1-7a)}$$

In reduced units

$$\frac{B_2(T)}{\frac{2\pi}{3}\sigma^3} = 1 - \left(R_{sw}^3 - 1\right)\left[\frac{\varepsilon}{kT} + \frac{1}{2}\left(\frac{\varepsilon}{kT}\right)^2\right]$$

$$= 1 - \left(R_{sw}^3 - 1\right)\left[\frac{1}{T^*} + \frac{1}{2(T^*)^2}\right] \quad \text{2}^{\text{nd}} \text{ order perturbation result}$$

$$\frac{B_2(T)}{\frac{2\pi}{3}\sigma^3} = 1 - \left(R_{sw}^3 - 1\right)\frac{\varepsilon}{kT} = 1 - \frac{\left(R_{sw}^3 - 1\right)}{T^*} \quad \text{1}^{\text{st}} \text{ order perturbation result}$$

(14.1-7b)

This is to be compared with the exact solution of Eq. 8.1-6:

$$B_2(T) = \frac{2\pi\sigma^3}{3}\left[1 + \left(1 - e^{\varepsilon/kT}\right)\left(R_{sw}^3 - 1\right)\right] \quad \text{or}$$

$$\frac{B_2(T)}{\frac{2\pi\sigma^3}{3}} = B_2^*(T) = 1 + \left(1 - e^{1/T^*}\right)\left(R_{sw}^3 - 1\right) \quad \text{(8.1-6)}$$

The agreement between the first- and second-order perturbation theory results and the exact result for the second virial coefficient is poor at low temperatures, as shown in Fig. 14.1-1; while at high reduced temperatures agreement is reasonably good as would be expected since, as the temperature increases, the influence of the well in the potential is less important. (Note that the λ expansion is equivalent to an expansion in $1/T$. Can you explain why?) Also, as would be expected, second-order perturbation theory is more accurate than the first-order theory.

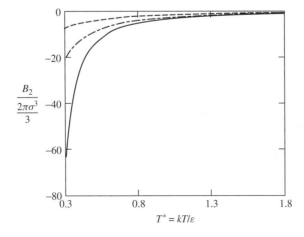

Figure 14.1-1 A comparison of the exact reduced second virial coefficient for the square-well fluid (solid line) as a function of reduced temperature $T^* = kT/\varepsilon$ with results of first-order (dotted line) and second-order (dash-dot line) perturbation theory.

14.1 Perturbation Theory for the Square-Well Potential

The example of perturbation theory above was a very simple one. The application to a dense fluid is much more involved. The starting point, then, in the canonical ensemble is the equation for the Helmholtz free energy:

$$\frac{A(N,V,T)}{kT} = -\ln Q(N,V,T) = -\ln\left\{\frac{\left(\frac{2\pi mkT}{h^2}\right)^{3N/2} q_{\text{int}}^N Z(N,V,T)}{N!}\right\}$$

$$= -\ln\left\{\frac{\left(\frac{2\pi mkT}{h^2}\right)^{3N/2} q_{\text{int}}^N}{N!}\right\} - \ln Z(N,V,T)$$

$$= -\ln\left\{\frac{\left(\frac{2\pi mkT}{h^2}\right)^{3N/2} q_{\text{int}}^N}{N!}\right\} - \ln \int_V \cdots \int_V e^{-u(\underline{r}_1,\ldots \underline{r}_N)/kT}\, d\underline{r}_1 \ldots d\underline{r}_N$$

(14.1-8)

To proceed, it is convenient to write the potential as

$$u_{\text{sw}}(r) = u_{\text{hs}}(r) + \lambda u_{\text{pert}}(r) \tag{14.1-9}$$

where λ is a parameter whose value goes from 0 (no perturbation resulting in the hard-sphere potential) to 1 (full perturbation leading to the square-well potential). Next, we do a Taylor series expansion of the Helmholtz free energy in the perturbation parameter λ:

$$\frac{A(N,V,T;\lambda)}{kT} = \frac{A(N,V,T;\lambda=0)}{kT} + \lambda\left(\frac{\partial\left(\frac{A(N,V,T;\lambda)}{kT}\right)}{\partial\lambda}\right)_{\lambda=0}$$

$$+ \frac{\lambda^2}{2!}\left(\frac{\partial^2\left(\frac{A(N,V,T;\lambda)}{kT}\right)}{\partial\lambda^2}\right)_{\lambda=0} + \cdots \tag{14.1-10}$$

The translational and internal parts of the partition function do not depend on the interaction potential (unless the molecules are for example, very long polymers, where molecular entanglement is affected by the interaction potential that could change the translational motions; however, we will not consider that complication here). Therefore, we have

$$\frac{A(N,V,T;\lambda)}{kT} = \frac{A(N,V,T;\lambda=0)}{kT} - \lambda\left(\frac{\partial(\ln Z(N,V,T;\lambda))}{\partial\lambda}\right)_{\lambda=0}$$

$$- \frac{\lambda^2}{2!}\left(\frac{\partial^2(\ln Z(N,V,T;\lambda))}{\partial\lambda^2}\right)_{\lambda=0} + \cdots \tag{14.1-11}$$

14.2 FIRST ORDER BARKER-HENDERSON PERTURBATION THEORY[2]

To proceed, we will assume pairwise additivity of the potential, which yields

$$u(\underline{r}_1, \underline{r}_2, \ldots, \underline{r}_N, \lambda) = \sum_i \sum_{j>i} u(r_{ij}, \lambda) = \sum_i \sum_{j>i} u_{\text{hs}}(r_{ij}) + \lambda \sum_i \sum_{j>i} u_{\text{pert}}(r_{ij})$$

(14.2-1)

and

$$\left(\frac{\partial (\ln Z(N, V, T; \lambda))}{\partial \lambda}\right)_{\lambda=0}$$

$$= \left(\frac{1}{Z(N, V, T; \lambda)} \frac{\partial \int \cdots \int e^{-u(\underline{r}_1, \underline{r}_2, \ldots, \underline{r}_N, \lambda)/kT} d\underline{r}_1 \ldots d\underline{r}_N}{\partial \lambda}\right)_{\lambda=0}$$

$$= \left(\frac{1}{Z(N, V, T; \lambda)} \frac{\partial \int \cdots \int e^{-\left[\sum_i \sum_{j>i} u_{\text{hs}}(r_{ij}) + \lambda \sum_i \sum_{j>i} u_{\text{pert}}(r_{ij})\right]/kT} d\underline{r}_1 \ldots d\underline{r}_N}{\partial \lambda}\right)_{\lambda=0}$$

$$= -\frac{1}{kT}\left(\frac{1}{Z(N, V, T; \lambda)} \int \cdots \int \sum_i \sum_{j>i} u_{\text{pert}}(r_{ij}) e^{-\left[\sum_i \sum_{j>i} u_{\text{hs}}(r_{ij}) + \lambda \sum_i \sum_{j>i} u_{\text{pert}}(r_{ij})\right]/kT} d\underline{r}_1 \ldots d\underline{r}_N\right)_{\lambda=0}$$

(14.2-2)

However

$$\left(\frac{\int \cdots \int \sum_i \sum_{j>i} u_{\text{pert}}(r_{ij}) e^{-\left[\sum_i \sum_{j>i} u_{\text{hs}}(r_{ij}) + \lambda \sum_i \sum_{j>i} u_{\text{pert}}(r_{ij})\right]/kT} d\underline{r}_1 \ldots d\underline{r}_N}{Z(N, V, T; \lambda)}\right)_{\lambda=0}$$

$$= \frac{\int \cdots \int \sum_i \sum_{j>i} u_{\text{pert}}(r_{ij}) e^{-\left[\sum_i \sum_{j>i} u_{\text{hs}}(r_{ij})\right]/kT} d\underline{r}_1 \ldots d\underline{r}_N}{Z(N, V, T; \lambda = 0)}$$

$$= \frac{N(N-1)}{2} \frac{\int \cdots \int u_{\text{pert}}(r_{12}) e^{-\left[\sum_i \sum_{j>i} u_{\text{hs}}(r_{ij})\right]/kT} d\underline{r}_1 \ldots d\underline{r}_N}{\int \cdots \int e^{-\left[\sum_i \sum_{j>i} u_{\text{hs}}(r_{ij})\right]/kT} d\underline{r}_1 \ldots d\underline{r}_N}$$

(14.2-3)

where, in obtaining this last equation, we have recognized that the derivative is to be evaluated at $\lambda = 0$, and that since all the molecules are identical, there are

---

[2] J. A. Barker and D. Henderson, *J. Chem. Phys*. **47**, 2856 (1967).

14.2 First Order Barker-Henderson Perturbation Theory

$N(N-1)/2$ identical contributions to the integral. So we need only to evaluate the integral for one pair of molecules and multiply the result by this factor. Now, remembering that the radial distribution function is defined as

$$g^{(2)}(\underline{r}_1, \underline{r}_2; T, \rho) = \frac{V^2 \int \cdots \int e^{-u(\underline{r}_1, \underline{r}_2, \ldots, \underline{r}_N)/kT} \, d\underline{r}_3 \ldots d\underline{r}_N}{\int \cdots \int e^{-u(\underline{r}_1, \underline{r}_2, \ldots, \underline{r}_N)/kT} \, d\underline{r}_1 \ldots d\underline{r}_N}$$

which for the pairwise additive, hard-sphere potential is[3]

$$g^{(2)}_{\text{hs}}(\underline{r}_1, \underline{r}_2; \rho) = \frac{V^2 \int \cdots \int e^{-\left[\sum_i \sum_{j>i} u_{\text{hs}}(r_{ij})\right]/kT} \, d\underline{r}_3 \ldots d\underline{r}_N}{\int \cdots \int e^{-\left[\sum_i \sum_{j>i} u_{\text{hs}}(r_{ij})\right]/kT} \, d\underline{r}_1 \ldots d\underline{r}_N} \quad (14.2\text{-}4)$$

we obtain

$$\left(\frac{\partial \ln Z(N, V, T; \lambda)}{\partial \lambda}\right)_{\lambda=0}$$

$$= -\frac{1}{kT} \frac{N(N-1)}{2V^2} \frac{V^2 \iint u_{\text{pert}}(r_{12}) \left[\int \cdots \int e^{-\sum_i \sum_{j>i} u_{\text{hs}}(r_{ij})/kT} \, d\underline{r}_3 \ldots d\underline{r}_N\right] d\underline{r}_1 \, d\underline{r}_2}{\int \cdots \int e^{-\sum_i \sum_{j>i} u_{\text{hs}}(r_{ij})/kT} \, d\underline{r}_1 \ldots d\underline{r}_N}$$

$$= -\frac{1}{kT} \frac{N(N-1)}{2V^2} \iint u_{\text{pert}}(r_{12}) g^{(2)}_{\text{hs}}(r_{12}; \rho) \, d\underline{r}_1 \, d\underline{r}_2$$

$$= -\frac{1}{kT} \frac{N(N-1)}{2V^2} \int d\underline{r}_1 \int u_{\text{pert}}(r_{12}) g^{(2)}_{\text{hs}}(r_{12}; \rho) \, d\underline{r}_{12} = -\frac{1}{kT} \frac{N(N-1)}{2V} \int u_{\text{pert}}(r_{12}) g^{(2)}_{\text{hs}}(r_{12}; \rho) \, d\underline{r}_{12}$$

$$= -\frac{1}{kT} \frac{N(N-1) 2\pi}{V} \int_0^\infty u_{\text{pert}}(r) g^{(2)}_{\text{hs}}(r; \rho) r^2 \, dr = -\frac{2\pi \rho N}{kT} u_{\text{pert}} \int_0^\infty (r) g^{(2)}_{\text{hs}}(r; \rho) r^2 \, dr \quad (14.2\text{-}5)$$

---

[3]There is an important point with regard to the temperature dependence of the radial distribution function for the hard-sphere fluid. In general, the two-particle or radial distribution function $g^{(2)}(r; T, \rho)$ is a function of temperature and density, as well as radial separation r_{12}. However, while the radial distribution function for the hard-sphere fluid is dependent of density, it is independent of temperature. The reason for this is, as discussed earlier, that it is obtained from integrals involving the Boltzmann factors in the interaction energy $e^{-u(r)/kT}$, where $u(r) = \infty$ and $e^{-u(r)/kT} = 0$ for $r < \sigma$, and $u(r) = 0$ and $e^{-u(r)/kT} = 1$ $r \geq \sigma$. So there is no temperature dependence in the integrals defining the radial distribution function for the hard-sphere fluid. This is to be compared with, for example, the case of the square-well fluid of Eq. 14.1-2, for which where $u(r) = \infty$ and $e^{-u(r)/kT} = 0$ for $r < \sigma$, $u(r) = -\varepsilon$, and $e^{-u(r)/kT} = e^{\varepsilon/kT}$ for $\sigma < r < R_{\text{sw}}\sigma$; and $u(r) = 0$ and $e^{-u(r)/kT} = 1$ for $r > R_{\text{sw}}\sigma$. This temperature dependence of the Boltzmann factor results in a temperature-dependent radial distribution function. Consequently, the square-well potential—and, in fact, any potential that has a region in which the interaction is finite in value (rather than infinite or zero) over some range of intermolecular separations—will have a radial distribution function that is dependent on temperature as well as density.

Here, we have replaced the variable of integration r_{12} simply with r and, as usual, ignored the difference between N and $N-1$.

Therefore, to first order in the perturbation expansion, we have

$$\frac{A(N, V, T; \lambda)}{kT} = \frac{A(N, V, T; \lambda = 0)}{kT} + \lambda \frac{2\pi \rho N}{kT} \int u_{\text{pert}}(r) g_{\text{hs}}^{(2)}(r; \rho) r^2 \, dr$$

or, for $\lambda = 1$

$$A(N, V, T; \lambda = 1) = A(N, V, T; \lambda = 0) + 2\pi \rho N \int u_{\text{pert}}(r) g_{\text{hs}}^{(2)}(r; \rho) r^2 \, dr \quad (14.2\text{-}6)$$

and specifically for the perturbation potential given by Eq. (14.1-3)

$$A_{\text{sw}}(N, V, T) = A_{\text{hs}}(N, V, T) - 2\pi \rho N \varepsilon \int_{\sigma}^{R_{\text{sw}}\sigma} g_{\text{hs}}^{(2)}(r; \rho) r^2 \, dr \quad (14.2\text{-}7\text{a})$$

or

$$\frac{A_{\text{sw}}(N, V, T)}{NkT} = \frac{A_{\text{hs}}(N, V, T)}{NkT} - 2\pi \rho \frac{\varepsilon}{kT} \int_{\sigma}^{R_{\text{sw}}\sigma} g_{\text{hs}}^{(2)}(r; \rho) r^2 \, dr \quad (14.2\text{-}7\text{b})$$

In these last equations, we have used that $A(N, V, T; \lambda = 1) = A_{\text{sw}}(N, V, T)$, and that $A(N, V, T; \lambda = 0) = A_{\text{hs}}(N, V, T)$.

Now using that

$$P = -\left(\frac{\partial A}{\partial V}\right)_{N,T} = \frac{\rho^2}{N}\left(\frac{\partial A}{\partial \rho}\right)_T \quad (14.2\text{-}8)$$

we obtain the following equation of state to first-order term in the perturbation expansion around the hard-sphere fluid

$$P_{\text{sw}}(\rho, T) = P_{\text{hs}}(\rho, T) - \frac{\rho^2}{N} \frac{\partial}{\partial \rho} \left(2\pi N \rho \varepsilon \int_{\sigma}^{R_{\text{sw}}\sigma} g_{\text{hs}}^{(2)}(r, \rho) r^2 \, dr \right)_T$$

$$= P_{\text{hs}}(\rho, T) - 2\pi \rho^2 \varepsilon \int_{\sigma}^{R_{\text{sw}}\sigma} g_{\text{hs}}^{(2)}(r, \rho) r^2 \, dr - 2\pi \varepsilon \rho^3 \left(\frac{\partial}{\partial \rho} \int_{\sigma}^{R_{\text{sw}}\sigma} g_{\text{hs}}^{(2)}(r, \rho) r^2 \, dr \right)_T$$

$$(14.2\text{-}9)$$

The expressions obtained so far are sometimes referred to as *mean-field approximations*. The name arises from the fact that, to the order so far considered, the radial distribution function is unchanged from that in the reference potential (here the hard-sphere fluid), and the perturbation contribution is obtained as an average of the perturbation potential over the structure (that is, the radial distribution function) of the reference fluid.

Though the results above are explicit for the square-well fluid, they are easily generalized to any other potential that can be represented as the sum of a hard-sphere potential $u_{hs}(r)$ plus a perturbation potential $u_{pert}(r)$ of any form. The results in this case are

$$\frac{A(N,V,T)}{NkT} = \frac{A_{hs}(N,V,T)}{NkT} + \frac{2\pi\rho}{kT}\int_\sigma^\infty g_{hs}^{(2)}(r;\rho) u_{pert}(r) r^2 \, dr \quad \text{and} \quad (14.2\text{-}10)$$

$$P(\rho,T) = P_{hs}(\rho,T) + 2\pi\rho^2 \int_\sigma^\infty g_{hs}^{(2)}(r,\rho) u_{pert}(r) r^2 \, dr$$

$$+ 2\pi\varepsilon\rho^3 \left(\frac{\partial}{\partial\rho} \int_\sigma^{R_{sw}\sigma} g_{hs}^{(2)}(r,\rho) u_{pert}(r) r^2 \, dr \right)_T \quad (14.2\text{-}11)$$

Finally, the obvious extension to perturbation theory would be to the radial distribution. However, this becomes quite complicated involving higher-order distribution functions (which can be simplified somewhat by using the Kirkwood superposition approximation, Section 12.2). See Problem 14.9.

14.3 SECOND-ORDER PERTURBATION THEORY

To improve the accuracy of perturbation expansion, one should consider higher order terms in the series. The next term in the series comes from the following equation:

$$\left(\frac{\partial^2 \ln Z(N,V,T;\lambda)}{\partial \lambda^2}\right)_{\lambda=0}$$

$$= \frac{\partial}{\partial \lambda}\bigg|_{\lambda=0} \left(\frac{\partial \ln Z(N,V,T;\lambda)}{\partial \lambda}\right)_{\lambda=0}$$

$$= -\frac{1}{kT}\frac{\partial}{\partial \lambda}\bigg|_{\lambda=0}\left(\frac{1}{Z(N,V,T;\lambda)}\int\cdots\int \sum_i\sum_{j>i} u_{\text{pert}}(r_{ij}) e^{-\left[\sum_i\sum_{j>i} u_{\text{hs}}(r_{ij}) + \lambda \sum_i\sum_{j>i} u_{\text{pert}}(r_{ij})\right]/kT} dr_1\ldots dr_N\right)_{\lambda=0}$$

$$= \frac{1}{(kT)^2}\left(\frac{1}{Z(N,V,T;\lambda)}\int\cdots\int \sum_i\sum_{j>i} u_{\text{pert}}(r_{ij}) \sum_k\sum_{l>k} u_{\text{pert}}(r_{kl}) e^{-\left[\sum_i\sum_{j>i} u_{\text{hs}}(r_{ij}) + \lambda \sum_i\sum_{j>i} u_{\text{pert}}(r_{ij})\right]/kT} dr_1\ldots dr_N\right)_{\lambda=0}$$

$$+ \frac{1}{kT}\left(\frac{1}{Z(N,V,T;\lambda)^2}\int\cdots\int \sum_i\sum_{j>i} u_{\text{pert}}(r_{ij}) e^{-\left[\sum_i\sum_{j>i} u_{\text{hs}}(r_{ij}) + \lambda \sum_i\sum_{j>i} u_{\text{pert}}(r_{ij})\right]/kT} dr_1\ldots dr_N \left(\frac{\partial Z(N,V,T;\lambda)}{\partial \lambda}\right)\right)_{\lambda=0}$$

$$= \frac{1}{(kT)^2}\left(\frac{1}{Z(N,V,T;\lambda)}\int\cdots\int \sum_i\sum_{j>i} u_{\text{pert}}(r_{ij}) \sum_k\sum_{l>k} u_{\text{pert}}(r_{kl}) e^{-\left[\sum_i\sum_{j>i} u_{\text{hs}}(r_{ij}) + \lambda \sum_i\sum_{j>i} u_{\text{pert}}(r_{ij})\right]/kT} dr_1\ldots dr_N\right)_{\lambda=0}$$

$$+ \frac{1}{kT}\left(\frac{1}{Z(N,V,T;\lambda)}\int\cdots\int \sum_i\sum_{j>i} u_{\text{pert}}(r_{ij}) e^{-\left[\sum_i\sum_{j>i} u_{\text{hs}}(r_{ij}) + \lambda \sum_i\sum_{j>i} u_{\text{pert}}(r_{ij})\right]/kT} dr_1\ldots dr_N \left(\frac{\partial \ln Z(N,V,T;\lambda)}{\partial \lambda}\right)\right)_{\lambda=0}$$

(14.3-1)

Continuing, we have

$$\left(\frac{\partial^2 \ln Z(N,V,T;\lambda)}{\partial \lambda^2}\right)_{\lambda=0}$$

$$= \frac{1}{(kT)^2}\left(\frac{1}{Z(N,V,T;\lambda)}\int\cdots\int \sum_i \sum_{j>i} u_{\text{pert}}(r_{ij}) \sum_k \sum_{l>k} u_{\text{pert}}(r_{kl}) e^{-\left[\sum_i \sum_{j>i} u_{\text{hs}}(r_{ij}) + \lambda \sum_i \sum_{j>i} u_{\text{pert}}(r_{ij})\right]/kT} dr_1 \ldots dr_N\right)_{\lambda=0}$$

$$+ \frac{1}{kT}\left(\frac{1}{Z(N,V,T;\lambda)}\int\cdots\int \sum_i \sum_{j>i} u_{\text{pert}}(r_{ij}) e^{-\left[\sum_i \sum_{j>i} u_{\text{hs}}(r_{ij}) + \lambda \sum_i \sum_{j>i} u_{\text{pert}}(r_{ij})\right]/kT} dr_1 \ldots dr_N \left(\frac{\partial \ln Z(N,V,T;\lambda)}{\partial \lambda}\right)\right)_{\lambda=0}$$

$$= \frac{\int\cdots\int\left(\frac{N(N-1)}{2}\frac{(N-2)(N-3)}{2} u_{\text{pert}}(r_{12}) u_{\text{pert}}(r_{34}) + \frac{N(N-1)}{2}\frac{(N-2)2}{2} u_{\text{pert}}(r_{12}) u_{\text{pert}}(r_{13}) + \frac{N(N-1)}{2}\frac{2\times 1}{2}(u_{\text{pert}}(r_{12}))^2\right) e^{-\left[\sum_i \sum_{j>i} u_{\text{hs}}(r_{ij})\right]/kT} dr_1 \ldots dr_N}{Z(N,V,T;\lambda=0)(kT)^2}$$

$$+ \frac{1}{kT}\left(\frac{1}{Z(N,V,T;\lambda=0)}\int\cdots\int \frac{N(N-1)}{2} u_{\text{pert}}(r_{12}) e^{-\left[\sum_i \sum_{j>i} u_{\text{hs}}(r_{ij})\right]/kT} dr_1 \ldots dr_N \left(\frac{\partial \ln Z(N,V,T;\lambda)}{\partial \lambda}\right)_{\lambda=0}\right) \tag{14.3-2}$$

268 Chapter 14: Perturbation Theory

and

$$\left(\frac{\partial^2 \ln Z(N, V, T; \lambda)}{\partial \lambda^2}\right)_{\lambda=0}$$

$$= \frac{N^4}{8V^4(kT)^2} \int \cdots \int u_{\text{pert}}(r_{12}) u_{\text{pert}}(r_{34}) g_{\text{hs}}^{(4)}(\underline{r}_1, \underline{r}_2, \underline{r}_3, \underline{r}_4; \rho) \, d\underline{r}_1 \ldots d\underline{r}_4$$

$$+ \frac{N^3}{4V^3(kT)^2} \int \cdots \int u_{\text{pert}}(r_{12}) u_{\text{pert}}(r_{13}) g_{\text{hs}}^{(3)}(\underline{r}_1, \underline{r}_2, \underline{r}_3; \rho) \, d\underline{r}_1 \, d\underline{r}_2 \, d\underline{r}_3$$

$$+ \frac{N^2}{4V^2(kT)^2} \iint (u_{\text{pert}}(r_{12}))^2 g_{\text{hs}}^{(2)}(\underline{r}_1, \underline{r}_2; \rho) \, d\underline{r}_1 \, d\underline{r}_2$$

$$+ \frac{N^2}{2V^2 kT} \iint u_{\text{pert}}(r_{12}) g_{\text{hs}}^{(2)}(\underline{r}_1, \underline{r}_2; \rho) \, d\underline{r}_1 \, d\underline{r}_2 \left(\frac{\partial \ln Z(N, V, T; \lambda)}{\partial \lambda}\right)_{\lambda=0}$$

(14.3-3)

However

$$\frac{N^2}{2V^2 kT} \iint u_{\text{pert}}(r_{12}) g_{\text{hs}}^{(2)}(\underline{r}_1, \underline{r}_2; \rho) \, d\underline{r}_1 \, d\underline{r}_2 \left(\frac{\partial \ln Z(N, V, T; \lambda)}{\partial \lambda}\right)_{\lambda=0}$$

$$= \frac{2\pi N^2}{VkT} \iint u_{\text{pert}}(r) g_{\text{hs}}^{(2)}(r; \rho) r^2 \, dr \left(\frac{\partial \ln Z(N, V, T; \lambda)}{\partial \lambda}\right)_{\lambda=0}$$

$$= \frac{2\pi \rho N}{kT} \int u_{\text{pert}}(r) g_{\text{hs}}^{(2)}(r; \rho) r^2 \, dr \left(\frac{\partial \ln Z(N, V, T; \lambda)}{\partial \lambda}\right)_{\lambda=0}$$

$$= \frac{2\pi \rho N}{kT} \int u_{\text{pert}}(r) g_{\text{hs}}^{(2)}(r; \rho) r^2 \, dr \left(-\frac{2\pi \rho N}{kT} \int u_{\text{pert}}(r) g_{\text{hs}}^{(2)}(r; \rho) r^2 \, dr\right)$$

$$= -\left(\frac{2\pi \rho N}{kT} \int u_{\text{pert}}(r) g_{\text{hs}}^{(2)}(r; \rho) r^2 \, dr\right)^2 \quad (14.3\text{-}4)$$

Therefore

$$\left(\frac{\partial^2 \ln Z(N, V, T; \lambda)}{\partial \lambda^2}\right)_{\lambda=0}$$

$$= \frac{\rho^4}{8(kT)^2} \int \cdots \int u_{\text{pert}}(r_{12}) u_{\text{pert}}(r_{34}) g_{\text{hs}}^{(4)}(\underline{r}_1, \underline{r}_2, \underline{r}_3, \underline{r}_4; \rho) \, d\underline{r}_1 \ldots d\underline{r}_4$$

$$+ \frac{\rho^3}{4(kT)^2} \int \cdots \int u_{\text{pert}}(r_{12}) u_{\text{pert}}(r_{13}) g_{\text{hs}}^{(3)}(\underline{r}_1, \underline{r}_2, \underline{r}_3; \rho) \, d\underline{r}_1 \, d\underline{r}_2 \, d\underline{r}_3$$

$$+ \frac{\rho^2}{4(kT)^2} \iint (u_{\text{pert}}(r_{12}))^2 g_{\text{hs}}^{(2)}(\underline{r}_1, \underline{r}_2; \rho) \, d\underline{r}_1 \, d\underline{r}_2$$

$$- \left(\frac{2\pi \rho N}{kT} \int u_{\text{pert}}(r) g_{\text{hs}}^{(2)}(r; \rho) r^2 \, dr\right)^2 \quad (14.3\text{-}5)$$

and finally

$$A(N, V, T; \lambda = 1)$$
$$= A_{\text{hs}}(N, V, T) + 2\pi\rho N \int u_{\text{pert}}(r) g_{\text{hs}}^{(2)}(r; \rho) r^2 \, dr$$

$$+ \frac{1}{2} \begin{pmatrix} \frac{\rho^4}{8(kT)^2} \int \cdots \int u_{\text{pert}}(r_{12}) u_{\text{pert}}(r_{34}) g_{\text{hs}}^{(4)}(\underline{r}_1, \underline{r}_2, \underline{r}_3, \underline{r}_4; \rho) \, d\underline{r}_1 \ldots d\underline{r}_4 \\ + \frac{\rho^3}{4(kT)^2} \int \cdots \int u_{\text{pert}}(r_{12}) u_{\text{pert}}(r_{13}) g_{\text{hs}}^{(3)}(\underline{r}_1, \underline{r}_2, \underline{r}_3; \rho) \, d\underline{r}_1 \, d\underline{r}_2 \, d\underline{r}_3 \\ + \frac{\rho^2}{4(kT)^2} \iint (u_{\text{pert}}(r_{12}))^2 g_{\text{hs}}^{(2)}(\underline{r}_1, \underline{r}_2; \rho) \, d\underline{r}_1 \, d\underline{r}_2 \\ - \left(\frac{2\pi\rho N}{kT} \int u_{\text{pert}}(r) g_{\text{hs}}^{(2)}(r; \rho) r^2 \, dr \right)^2 \end{pmatrix}$$

(14.3-6)

The important thing to notice in this equation is that, while the first-order perturbation term involves the two-body correlation function for the hard-sphere fluid, the second order perturbation term also involves the three-body and four-body correlation functions. While generally we have some information about the two-body correlation function from either theory or computer simulation, there is very little information on the higher-order correlation functions. This greatly limits the use of perturbation theory, though there are some approximations that have been proposed. However, a discussion of those is beyond the scope of this introduction to perturbation theory; and in that which follows, the discussion will largely be limited to first-order perturbation theory. Nonetheless, for completeness, we mention one such approximate second-order result, the local compressibility model.[4]

$$\frac{A(N, V, T)}{NkT} = \frac{A_{\text{hs}}(N, V, T)}{NkT} + \frac{2\pi\rho}{kT} \int_\sigma^\infty u_{\text{pert}}(r) g_{\text{hs}}^{(2)}(r_{12}; \rho) r^2 \, dr$$
$$- \frac{\pi\rho}{(kT)^2} \int_\sigma^\infty g_{\text{hs}}^{(2)}(r_{12}; \rho) \left[u_{\text{pert}}(r) \right]^2 \left[\frac{\partial}{\partial P} \left(\rho g_{\text{hs}}^{(2)}(r_{12}; \rho) \right) \right]_T r^2 \, dr$$

(14.3-7)

14.4 PERTURBATION THEORY USING OTHER REFERENCE POTENTIALS

While the idea behind perturbation theory using the hard-sphere fluid as the reference fluid is conceptually very appealing, complications arise when one considers all types of interaction potentials that might be of interest. One complication is that the hard-sphere interaction potential is not a realistic representation of the interactions between real molecules (atoms). An interaction potential with a softer repulsion (that is, different from a hard wall) would be a better reference potential. However, the radial distribution function for such a potential is a then a function of temperature as well as density and radial separation distance. This makes perturbation theory more

---
[4]W. R. Smith, D. Henderson and J. A. Barker, *J. Chem. Phys.* **55**, 4027 (1971).

Chapter 14: Perturbation Theory

difficult to use, though we consider in this section how perturbation theory can be used for soft (i.e., Lennard-Jones-like) interaction potentials.

A more general way to proceed is to use some other model potential for which the thermodynamic properties and radial distribution function are known—for example, from previous Monte Carlo or molecular dynamics computer simulations—as the reference fluid and the difference between the potential of interest and the reference potential as the perturbation. That is, the generalization of Eq. 14.1-9 would be

$$u(r, \lambda) = u_{\text{ref}}(r) + \lambda u_{\text{pert}}(r) \quad \text{where} \quad u_{\text{pert}}(r) = u(r, \lambda = 1) - u_{\text{ref}}(r) \quad (14.4\text{-}1)$$

Similarly, $g_{\text{hs}}^{(2)}(r, \rho)$ in the perturbation analysis would be replaced by $g_{\text{ref}}^{(2)}(r, \rho, T)$ throughout. In what follows, we discuss how such a perturbation analysis is done.

Consider, for example, the Lennard-Jones 12-6 potential:

$$u(r) = 4\varepsilon \left[\left(\frac{\sigma}{r}\right)^{12} - \left(\frac{\sigma}{r}\right)^{6} \right]$$

This potential can be considered to be a combination of attractive and repulsive parts. One way of making this separation is based on energy, in which case the potential is said to be repulsive when the interaction energy is positive (that is for $r \leq \sigma$) and attractive when the interaction energy is negative (when $r > \sigma$). However, this division is not unique. Another possibility is to define the attractive part of the potential as that part in which the force between the molecules is attractive, and the repulsive part of the potential as where the force is negative (repelling the molecules). In this way, the repulsive and attractive parts of the potential are defined in terms of the derivative of the interaction energy. That is, when $du(r)/dr \leq 0$, which occurs when $r \leq 2^{1/6}\sigma$, the interaction is considered to be repulsive; and it is attractive when $du(r)/dr > 0$, which occurs when $r > 2^{1/6}\sigma$. We will refer to the separation of attractive and repulsive potentials used earlier based on energy as the Barker-Henderson (BH) division, and the modified version described here based on force as the Weeks-Chandler-Anderson[5] (WCA) division.

With the Barker-Henderson division, the repulsive and attractive parts of the L-J 12-6 potential are

$$u_{\text{rep}}(r) = 4\varepsilon \left[\left(\frac{\sigma}{r}\right)^{12} - \left(\frac{\sigma}{r}\right)^{6} \right] \text{ for } r \leq \sigma \quad \text{and} \quad u_{\text{rep}}(r) = 0 \text{ for } r > \sigma$$

$$u_{\text{att}}(r) = 0 \text{ for } r \leq \sigma \quad \text{and} \quad u_{\text{att}}(r) = 4\varepsilon \left[\left(\frac{\sigma}{r}\right)^{12} - \left(\frac{\sigma}{r}\right)^{6} \right] \text{ for } r > \sigma \quad (14.4\text{-}2)$$

as shown in Fig. 14.4-1.

In using the Weeks-Chandler-Anderson division of the Lennard-Jones potential in their perturbation theory, it has been found advantageous to write the potential as

$$u_{\text{rep}}(r) = 4\varepsilon \left[\left(\frac{\sigma}{r}\right)^{12} - \left(\frac{\sigma}{r}\right)^{6} \right] + \varepsilon \text{ for } r \leq 2^{1/6}\sigma \quad \text{and} \quad u_{\text{rep}}(r) = 0 \text{ for } r > 2^{1/6}\sigma$$

(14.4-3)

---

[5] J. D. Weeks, D. Chandler, and H. C. Andersen, *J. Chem. Phys.* **54**, 5237 (1971), and later papers by the same authors.

14.4 Perturbation Theory Using Other Reference Potentials

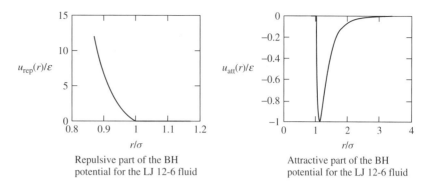

Figure 14.4-1 The Barker-Henderson division of the LJ 12-6 potential in the repulsive and attractive parts. Note very different scales for the repulsive and attractive potentials.

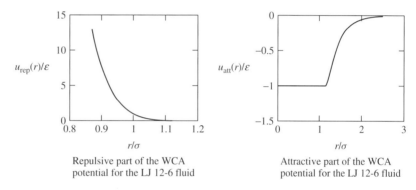

Figure 14.4-2 The Weeks-Chandler-Anderson separation of the LJ 12-6 potential in the repulsive and attractive parts. Note very different scales for the repulsive and attractive potentials.

and

$$u_{\text{att}}(r) = -\varepsilon \text{ for } r \leq 2^{1/6}\sigma \quad \text{and} \quad u_{\text{att}}(r) = 4\varepsilon\left[\left(\frac{\sigma}{r}\right)^{12} - \left(\frac{\sigma}{r}\right)^{6}\right] \text{ for } r > 2^{1/6}\sigma$$

as shown in Fig. 14.4-2.

The use of either of these potentials in perturbation theory introduces the complication that the repulsive part of the potential is not the simple hard-sphere potential for which the radial distribution function is only a function of density and not temperature. Here, as a result of the softness of the repulsive part of the reference potential, the radial distribution function is now a function of both temperature and density. So, in principle, it would be necessary to obtain the radial distribution functions (by simulation or some other method) for the repulsive parts of the potentials in both the BH and WCA divisions at each intermolecular separation distance, density, and temperature. This is avoided in practice by choosing an effective hard-sphere diameter that depends on temperature, in the case of Barker-Henderson perturbation theory, and temperature and density in the case of Weeks-Chandler-Anderson perturbation theory. In Barker-Henderson theory, the effective hard-sphere diameter is given by

$$d(T) = \int_0^\sigma \left[1 - e^{-u(r)/kT}\right] dr \qquad (14.4\text{-}4)$$

272 Chapter 14: Perturbation Theory

which arises from setting the first perturbation term to 0 in a more complicated form of perturbation theory based on a double Taylor series (or perturbation series) expansion in the depth of the attractive well, and the inverse of the steepness of the repulsive part of the potential (which would be 0 for the hard-sphere potential). The Weeks-Chandler-Anderson choice of diameter is a little easier to explain because it is chosen to result in the same compressibility for both the reference and hard-sphere fluids, at each density and temperature. From Eq. 11.5-17, the compressibility is

$$kT \left(\frac{\partial \rho}{\partial P}\right)_T - 1 = 4\pi\rho \int_0^\infty [g^{(2)}(r; \rho, T) - 1] r^2 \, dr \qquad (11.5\text{-}17)$$

which from Eq. 12.4-2 using $y(r) = g^{(2)}(r) e^{u(r)/kT}$ can be rewritten as

$$kT \left(\frac{\partial \rho}{\partial P}\right)_T - 1 = 4\pi\rho \int_0^\infty [y(r; \rho, T) e^{-u(r)/kT} - 1] r^2 \, dr \qquad (14.4\text{-}5)$$

Equating the compressibilities of the hard-sphere and repulsive potentials results in the following implicit equation for the effective hard-sphere diameter:

$$\int_0^\infty [y_{\text{hs}}(r; \rho, T, d) e^{-u_{\text{hs}}(r; d)/kT} - 1] r^2 \, dr = \int_0^\infty [y_{\text{hs}}(r; \rho, T, d) e^{-u_{\text{rep}}(r)/kT} - 1] r^2 \, dr$$

$$(14.4\text{-}6)$$

which needs to be solved by iteration at each temperature and density.

A more complicated situation is the case of nonspherical, multiatom molecules. In this case, the application of perturbation theory is much more difficult, as there is no universal reference potential that can be used. Also, the application of perturbation theory to nonspherical molecules that results in a change of shape from a sphere, and therefore a change in the packing fraction or free volume (that is, the volume unoccupied by molecules), can be very slow to converge, and several terms in the perturbation series are required. Consequently, we will not consider perturbation theory for nonspherical molecules here.

14.5 ENGINEERING APPLICATIONS OF PERTURBATION THEORY

Statistical mechanical perturbation theory can be used as the basis for deriving volumetric equations of state of a different form than the virial equation of state considered earlier. For example, referring to Eq. 14.2-9

$$P_{\text{sw}}(\rho, T) = P_{\text{hs}}(\rho, T) - 2\pi \rho^2 \varepsilon \int_\sigma^{R_{\text{sw}}\sigma} g_{\text{hs}}^{(2)}(r; \rho) r^2 \, dr$$

$$- \pi \frac{\varepsilon}{kT} \rho^3 \left(\frac{\partial}{\partial \rho} \int_\sigma^{R_{\text{sw}}\sigma} g_{\text{hs}}^{(2)}(r; \rho) r^2 \, dr\right)_T \qquad (14.2\text{-}9)$$

14.5 Engineering Applications of Perturbation Theory

if the assumption is now made that the radial distribution function is independent of density—that is, that $g_{hs}^{(2)}(r;\rho) \approx g_{hs}^{(2)}(r)$—then the equation above reduces to

$$P_{sw}(\rho, T) = P_{hs}(\rho, T) - 2\pi\rho^2 \varepsilon C; \quad \text{where} \quad C = \int_{\sigma}^{R_{sw}\sigma} g_{hs}^{(2)}(r) r^2 \, dr = \text{constant}$$

(14.5-1)

Different equations of state can be developed from Eq. 14.5-1, depending on the assumptions that are made. For example, if we use the simple van der Waals excluded volume approximation

$$Z_{hs}(\rho) = \frac{P\underline{V}}{RT} = \frac{1}{1 - \frac{N\beta}{V}}$$

(14.5-2)

and use the notation that $a = 2\pi n^2 \varepsilon C$, $\beta = 2\pi\sigma^3/3$, and $b = N_{Av}\beta$ where $n = N/N_{Av}$ yields

$$P = \frac{RT}{\underline{V} - b} - \frac{a}{\underline{V}^2}$$

(14.5-3)

where N is the number of molecules, N_{Av} is Avogadro's number, β is the excluded volume per molecule (that is, the volume not available to other molecules due to the presence of a molecule), R is the gas constant, and $\underline{V}$ is the molar volume. Equation 14.5-3 is the famous van der Waals equation of state, and was the first equation to describe the vapor phase, the liquid phase, and the vapor-liquid transition.

The van der Waals equation was the first member in the family of volumetric equations of state referred to as *cubic equations*, as discussed in Section 10.2. Hundreds of other cubic equations have been proposed, and several are commonly used in engineering design and analysis. Two of the most common, as already mentioned in Section 10.2, are the Soave-Redlich-Kwong equation

$$P = \frac{RT}{\underline{V} - b} - \frac{a(T)}{\underline{V}(\underline{V} - b)}$$

(14.5-4)

and the Peng-Robinson equation

$$P = \frac{RT}{\underline{V} - b} - \frac{a(T)}{\underline{V}(\underline{V} + b) + b(\underline{V} - b)}$$

(14.5-5)

Both of these equations are cubic in the compressibility factor Z (see Section 10.2), both retain the simple van der Waals excluded volume term of Eq. 14.5-2, and can be thought of as resulting from different approximations being made for the first integral term in Eq. 14.2-9.

The van der Waals excluded volume expression for hard-spheres in not very accurate, as will be shown in Chapter 16. Indeed, we know from Eq. 12.5-2 that a better approximation (written in terms of molar properties) is

$$Z_{hs} = \left(\frac{P\underline{V}}{RT}\right) = \frac{1 + \eta + \eta^2 - \eta^3}{(1 - \eta)^3}$$

(14.5-6)

where $\eta = \pi N\sigma^3/6V = \pi\rho\sigma^3/6$. Using this expression in Eq. 14.2-9 and elsewhere, making the same approximations that led to the van der Waals equation, leads to an extended van der Waals equation:

$$P = \frac{RT}{\underline{V}} \frac{1 + \eta + \eta^2 - \eta^3}{(1-\eta)^3} - \frac{a}{\underline{V}^2} \qquad (14.5\text{-}7)$$

This equation is one of the family of volumetric equations of state that is referred to as the perturbed hard-sphere equations. Other members of this family use more complicated and accurate expressions for the last term in Eq. 14.5-7. Note, however, because of the more complicated volume (or η) dependence of the first term on the right-hand side, this family of equations is no longer of the cubic form.

These types of equations have been extended still further by replacing the first term on the right-hand side that is specific to hard spherical molecules with expressions derived from statistical mechanics for chain molecules, giving rise to the family of perturbed hard-chain equations of state. Among the other related equations are the perturbed soft-sphere and soft-chain equations of state that have been developed by replacing Eq. 14.5-7 with expressions derived from statistical mechanics for soft-spheres (for example, Lennard-Jones spheres) and soft chains, respectively (see Chapter 16).

Protein, Colloidal and Polymer Solutions

An interesting (but approximate) use of the single-site model to represent more complicated molecules, especially in solution, is their application to some proteins, colloids, and polymers that fold into conformations that are almost spherical or globular. To get an understanding of such solutions that is qualitatively correct, it is common to represent these approximately globular molecules as spheres (and when in electrolyte solutions, which is frequently the case, as charged spheres). The description of such solutions frequently treats the solvent (usually water) as a homogeneous medium containing the globular molecules, and the interactions between these spheres in solution is represented by a potential of mean force, which may be the models considered in earlier chapters (i.e., square well, Lennard-Jones 12-6) with appropriately chosen parameters, or the more refined models of Section 12.7.

CHAPTER 14 PROBLEMS

14.1 Compute the vapor-liquid equilibrium phase behavior for the square-well fluid using first-order perturbation theory around the hard-sphere fluid and the Percus-Yevick hard-sphere radial distribution function obtained from the MATLAB® program PYHS for a range of reduced temperatures for a parameters set of your choice.

14.2 The van der Waals equation of state is

$$P = \frac{RT}{\underline{V} - b} - \frac{a}{\underline{V}^2}$$

where a and b are constants. The first term on the right is a simple approximation for the hard-sphere pressure. By retaining only the first-order term in the perturbation expansion for the configuration integral, and by assuming that $u_{\text{pert}}(r)$ is weak and very long range, so that $g_{\text{ref}}(r)$ can be taken to be approximately equal to unity over all of the range of r values, show that the a parameter in the van der Waals equation is a constant, independent of temperature and density. Also, derive a relation between a and $u_{\text{pert}}(r)$.

14.3 Using the results for the pair correlation function for hard-spheres in the Percus-Yevick approximation (obtained from the MATLAB® program PYHS), compute the pressure, the compressibility factor, and the Helmholtz energy of the square-well fluid using

first-order perturbation theory for one of the following conditions:
   a. $T^* = kT/\varepsilon = 1.5$; $R_{sw} = 1.5$ and $\rho\sigma^3 = 0.2$;
   b. $T^* = kT/\varepsilon = 1.0$; $R_{sw} = 1.3$ and $\rho\sigma^3 = 0.3$; or
   c. $T^* = kT/\varepsilon = 2.5$; $R_{sw} = 1.7$ and $\rho\sigma^3 = 0.4$.

14.4 Compare the results for the pressure computed in Problem 14.3 with the results of the MATLAB® program MC_LJ or MD_LJ for the same conditions.

14.5 Develop a plot of the effective hard-sphere diameter in Barker-Henderson perturbation theory as a function of reduced temperature for the Lennard-Jones 12-6 potential.

14.6 Show that the first-order perturbation expansion about the hard-sphere result for the second virial coefficient for a hard-sphere molecule containing a dipole, Eq. 8.3-4, diverges.

14.7 The triangular-well potential is

$$u(r) = \begin{cases} \infty & r \leq \sigma \\ -\varepsilon \dfrac{R_{tw}\sigma - r}{R_{tw}\sigma - \sigma} & \sigma < r < R_{tw}\sigma \\ 0 & r \geq R_{tw}\sigma \end{cases}$$

For this potential, using first-order perturbation theory,
   a. Develop an expression for its second virial coefficient.
   b. Find the internal energy and entropy deviations from ideal gas behavior, for this potential at sufficiently low density and high temperature that only the first-order perturbation term for the second virial coefficient is needed in the equation of state.
   c. Find its chemical potential at the same conditions as in part b.

14.8 Compare the calculated results from first-order perturbation theory for the triangular-well potential of the previous problem for various values of $T^* = kT/\varepsilon$ and $R_{tw} = 1.5$ with the exact values obtained by numerical integration.

14.9 Perturbation theory for the radial distribution function was not considered in this chapter because of its complexity. To see this, show that

$$\left.\frac{\partial}{\partial\lambda}\right|_{N,V,T,\lambda=0} \int \cdots \int e^{-\left[\sum_i\sum_{j>i} u_{hs}(r_{ij}) + \lambda\sum_i\sum_{j>i} u_{pert}(r_{ij})\right]/kT} dr_3 \ldots dr_N$$

$$= \frac{Z_N}{kTV^2}\left[u_{pert}(r_{12})g_{hs}^{(2)}(r_{12})\right.$$

$$+ \rho\int u_{pert}(r_{13})g_{hs}^{(3)}(\underline{r}_1,\underline{r}_2,\underline{r}_3)\,d\underline{r}_3$$

$$\left. + \frac{\rho^2}{2}\int u_{pert}(r_{34})g_{hs}^{(4)}(\underline{r}_1,\underline{r}_2,\underline{r}_3,\underline{r}_4)\,d\underline{r}_3\,d\underline{r}_4\right]$$

Chapter 15

A Theory of Dilute Electrolyte Solutions and Ionized Gases

In the preceding chapters, we have considered several statistical thermodynamic methods of describing the properties of real gases and liquids. Now we will develop another theory that is applicable to dilute electrolyte solutions and ionized gases. A question that should occur to the reader is, why aren't the methods that have already been developed applicable to systems with charged particles? We will answer this first, and then develop the Debye-Hückel theory for fluids containing charged particles.

INSTRUCTIONAL OBJECTIVES FOR CHAPTER 15

The goals for this chapter are for the student to:

- Understand why the methods of the previous chapters used for neutral molecules are not applicable to charged particles
- Understand the basis for the Debye-Hückel theory for fluids containing charged particles
- Understand why mean ionic activity coefficients are used
- Understand the assumptions inherent in the Debye-Hückel limiting law
- Understand how the Debye-Hückel theory is extended for solutions of moderate and high ionic strengths for engineering use

15.1 SOLUTIONS CONTAINING IONS (AND ELECTRONS)

Electrolyte solutions and ionized gases (called plasmas) are different from the mixtures we have considered because the interactions between charged particles are of very long range—that is, they extend over many molecular diameters. To see this, consider the interaction potential that is typically used for charged particles i and j (each of which might be either an ion or an electron), which consist of

15.1 Solutions Containing Ions (and Electrons)

a hard impenetrable core and a charge-charge interaction described by a Coulomb potential:[1]

$$u_{ij}(r) = \begin{cases} \infty; & r < a \\ \dfrac{q_i q_j}{\varepsilon r}; & r \geq a \end{cases} \qquad (15.1\text{-}1)$$

Here, q_i is the charge on a particle of species i (which may be $+$ or $-$ and has units of the number of electrons) and ε is the dielectric constant (also called the static permeability) of the medium within which the ions are present, and is the product of the temperature-dependent dielectric constant (or the relative static permeability of the medium), ε_r, and the static permeability of a vacuum ε_0, that is, $\varepsilon = \varepsilon_r \varepsilon_0$, where $\varepsilon_0 = 8.854 \times 10^{-12}$ C^2/(J m). In an ionized gas there is only a vacuum between the particles, so $\varepsilon_r = 1$. It is approximately this value for air, while for water and aqueous solutions, ε_r is the dielectric constant of water, equal to 78.2 at 25°C. In this model, the charged particles act as hard-spheres at very close range (center-of-mass separation r less than a), and interact with a Coulomb force law beyond the hard-sphere radius. (When ions are present in a liquid solution, we will assume that this same intermolecular potential function is valid at small separation distances. This approximation is equivalent to assuming that the neutral solvent molecules in the system behave like a continuum, even for very small values of r—that is, even for ion-ion separation distances that are so small that only one or two solvent molecules can get between the two interacting ions. This approximation is reasonable, especially for very dilute solutions in which large ion-ion separation distances predominate.)

The Coulomb potential is very long range (i.e., it decays very slowly as only $1/r$) compared to the much shorter-range potentials we considered earlier, such as the Lennard-Jones 12-6 potential that decays as $1/r^6$. So we can infer that long-range interactions will be important in determining the solution behavior of charged particles. To see the problems caused by such a potential, consider charged particles in a gas. As we have seen, one method of correcting for nonideal behavior in a gas is to use a virial expansion. Using the interaction potential function of Eq. 15.1-1, we can try to compute a value for the second virial coefficient as follows:

$$B(T) = 2\pi \int_0^\infty \left[1 - e^{-u(r)/kT}\right] r^2 \, dr$$

$$= 2\pi \int_0^a \left[1 - e^{-\infty/kT}\right] r^2 \, dr + 2\pi \int_a^\infty \left[1 - e^{-z_i z_j e^2/\varepsilon rkT}\right] r^2 \, dr \qquad (15.1\text{-}2)$$

This second integral is infinite in value. To see this, we expand the exponential term in a Taylor series

$$B(T) = \frac{2\pi a^3}{3} + 2\pi \int_a^\infty \left[1 - \left\{1 - \frac{z_i z_j e^2}{\varepsilon rkT} + \cdots\right\}\right] r^2 \, dr$$

$$\approx \frac{2\pi a^3}{3} + \frac{z_i z_j e^2}{\varepsilon kT} \int_a^\infty r \, dr \qquad (15.1\text{-}3)$$

---

[1] The charge q on an electron, denoted by e, is 1.609×10^{-19} C where C designates the charge unit Coulomb.

278 Chapter 15: A Theory of Dilute Electrolyte Solutions and Ionized Gases

The integral on the right diverges, so the value of the second virial coefficient is infinite and the virial equation cannot be used when the Coulomb potential describes the interactions between particles. Also, the virial expansion itself diverges. This is because, as we saw in Chapter 7, the virial expansion arises from considering the interactions first between pairs of molecules (β_1 integrals), then triplets of molecules (β_2 integrals), etc., with increasing density; however, as a result of the long-range character of the Coulomb interaction, many ions interact simultaneously even at very low density. This is because the interaction strength decreases as $1/r$, but the number of charged particles in a spherical shell around a central molecule increases as r^2. Therefore, as one goes further away from a central particle, the number of charged particles increases more rapidly than their interaction with the central particle decays, so that the total interaction energy for the central particle increases without bound. What this simple argument neglects is that as the distance between the two interacting particles increases, other charged particles will be present between them (shielding the charges of the two particles of interest), so that the effective interaction between the charged particles decays more rapidly than $1/r$ for any nonzero density of charged particles. It is this shielding that is described by the Debye-Hückel theory.

A more direct proof of the divergence of the virial integral (without the need of a series expansion) is as follows. Consider the integral

$$I = \int_a^\infty \left[1 - e^{-C/r}\right] r^2 \, dr \quad \text{where} \quad C = z_i z_j e^2 / \varepsilon kT \quad (15.1\text{-}4)$$

Now, integrating by parts using $u = 1 - e^{-C/r}$, $v = \frac{1}{3}r^3$, $du = -e^{-C/r} \left(\frac{C}{r^2}\right) dr$, and $dv = r^2 \, dr$, we have

$$I = \int_a^\infty \left[1 - e^{-C/r}\right] r^2 \, dr = \frac{1}{3} r^3 \left(1 - e^{-C/r}\right) \Big|_{r=a}^{r=\infty} - \frac{1}{3} C \int_a^\infty e^{-C/r} r \, dr$$

$$= \frac{1}{3} a^3 \left(1 - e^{-C/a}\right) + \frac{C}{3} \int_a^\infty e^{-C/r} r \, dr$$

and using $y = 1/r$, and $dr = r^2 \, dy = dy/y^2$, we obtain

$$I = \frac{1}{3} a^3 \left(1 - e^{-C/a}\right) - \frac{C}{3} \underbrace{\int_{1/a}^{0} e^{-Cy} y^{-3} \, dy}_{\text{divergent integral}}$$

The discussion above dealt with ionized gases, in which the charged particles are ions and electrons. However, a similar situation arises in electrolyte solutions, in which all the charged particles are ions. In this chapter we use a method to describe systems of charged particles based on the radial distribution function (actually the potential of mean force) obtained from electrostatics. This approach uses a combination of molecular and continuum models in that the ions are treated atomistically, while the solvent is considered to be a continuum characterized only by its macroscopic properties (here its dielectric constant).

The goal of the analysis here is to develop expressions for the activity coefficients of ions in solution. To begin, it is useful to consider the different types of activity

15.1 Solutions Containing Ions (and Electrons)

coefficients that might be of use. So far, we have only considered the most common activity coefficients (for example, in the UNIQUAC and Flory-Huggins models) based on the reference state for each species in the mixture being its pure component state. In this case

$$\mu_i = \mu_i^\circ + RT \ln x_i \gamma_i \quad \text{with} \quad RT \ln \gamma_i = \left(\frac{\partial G^{\text{ex}}}{\partial N_i}\right)_{T,P,N_{j \neq i}} \quad (15.1\text{-}5a)$$

with μ_i° being the pure component chemical potential (molar Gibbs energy) in the same state of aggregation (liquid or solid) of the species as in the mixture, and at the same temperature and pressure. With this definition, γ_i is unity in the pure component limit and departs from that value as the species is diluted. However, this definition of the activity coefficient is not useful for a salt, which exists as dissolved ions in solution and as a solid as a pure component. An alternate way to proceed is to use a reference state based on the dissolved species at infinite dilution. This is the Henry's law reference state, and the activity coefficient is defined by

$$\mu_i = \mu_i^\circ + RT \ln x_i \gamma_i^* \quad (15.1\text{-}5b)$$

where μ_i° is now the chemical potential of the dissolved species at infinite dilution in the solvent. With this definition, the redefined activity coefficient $\gamma_i^* \to 1$ as $x_i \to 0$, and departs from that value as the solution becomes more concentrated in the solute. But even this redefinition of the activity coefficient is not useful in the case for electrolyte solutions, in which the degree of ionization (and consequently the mole fractions of the anions, cations, and solvent) may not be known *a priori*.

This has resulted in other versions of Henry's law; the one we use here, based on molality (number of moles of solute per 1000 g of solvent), is the following:

$$\mu_i = \mu_i^\circ + RT \ln \left(\frac{M_i \gamma_i^\square}{M_i = 1}\right) \quad (15.1\text{-}1c)$$

where M_i is the molality of species i, μ_i° is the chemical potential of the species in a hypothetical ideal 1 molal solution (obtained by extrapolating extremely dilute solution behavior), and, with this definition, $\gamma_i^\square \to 1$ as $M_i \to 0$. For solutions of neutral molecules in most solvents, the Henry's law activity coefficient (whether it is based on mole fraction, molality, or molarity) is generally close to unity in value in very dilute solutions. However, for electrolytes in water, the deviations from ideal solution behavior are very large in very dilute solutions. For example, in a 1:1 electrolyte (such as NaCl), $\gamma_i^\square \approx 0.8$ in a 0.01 molal aqueous solution, which corresponds to a mole fraction of solute of only about 4×10^{-4}. Therefore, deviations from ideal solution behavior are much more important in ionic solutions than for molecular solutes at the same concentration.

The discussion that follows is applicable both to ionized gases and liquid electrolyte solutions, the difference being the value of the dielectric constant is $\varepsilon = 1$ for ionized gases and is a function of the solvent and temperature in liquid solutions. For simplicity of discussion, the term *ions* will be used below and electrolyte solutions will be considered explicitly. However, the discussion is also applicable to an ionized gas by setting $\varepsilon = 1$ and recognizing that one of the species of ions is really an electron.

Before proceeding, there is one characteristic of ions that should be remembered. Since ions originate from a compound with no net charge, there is a charge balance restriction on the number of ions of each type that can be present. To be specific, consider a single binary electrolyte

$$A_{\nu_+}^{q+}B_{\nu_-}^{q-} \to \nu_+ A^{q+} + \nu_- B^{q-} \qquad (15.1\text{-}6)$$

that fully dissociates to give ν_+ cations of type A with charge q_+, and ν_- anions of type B with charge q_-. If N is the initial number of undissociated molecules $A_{\nu_+}B_{\nu_-}$ in a volume V, and ρ is their number density, then after ionization there are N_+ cations and N_- anions. The following stoichiometric conditions apply to maintain electrical neutrality:

$$\nu_+ q_+ + \nu_- q_- = 0; \quad N_+ q_+ + N_- q_- = 0; \quad \rho_+ q_+ + \rho_- q_- = 0;$$
$$\text{where} \quad N_+ = \nu_+ N \quad \text{and} \quad N_- = \nu_- N \qquad (15.1\text{-}7)$$

15.2 DEBYE-HÜCKEL THEORY

The most successful method of describing the behavior of very dilute electrolytes and ionized gases has been by using the model of Debye and Hückel. The main idea of this model is to treat the solution of ions in a solvent as if the ions were in a continuum of the solvent. That is, we take a microscopic view as far as the ions are concerned, but treat the neutral solvent molecules as a continuum, characterized by only a dielectric constant. A limitation of this approach is that the microscopic nature of the solvent will be important when the separation distance between ions is small. Consequently, we expect this model based on a continuum solvent to be applicable to dilute solutions in which there are many (i.e., a continuum) of solvent molecules between widely spaced charged particles, but not to concentrated electrolyte solutions. We shall see that the results obtained will be useful for the calculation of deviations from ideal solution behavior only in the limit of very low electrolyte concentrations, and not for the properties of concentrated electrolyte solutions. Since the model that is developed will be correct in the limit of infinite dilution of the charged particles, it is referred to as a limiting law. However, as will be shown, with judicious empirical adjustment, the model can be used to correlate measured activity coefficients for more concentrated solutions.

The basic idea of the model is that deviations from ideal solution behavior are determined by considering only charged particle interactions, which are predominantly electrostatic. Consequently, the equations of electrostatics are used to develop the theory; so it is worth reviewing the basic equations. In a medium of dielectric constant ε, the electrostatic potential $\phi(\underline{r})$ at a point $\underline{r}$ due to a set of point charges q_i at points given by the vectors $\underline{r}'_i$ is

$$\phi(\underline{r}) = \sum_i \frac{q_i}{\varepsilon |\underline{r} - \underline{r}'_i|} \qquad (15.2\text{-}1)$$

If, instead of a set of point charges, there is a continuous charge distribution $\eta(\underline{r}')$, then the expression for the potential at $\underline{r}$ is

$$\phi(\underline{r}) = \int_V \frac{\eta(\underline{r}')\, d\underline{r}'}{\varepsilon |\underline{r} - \underline{r}'|} \qquad (15.2\text{-}2)$$

15.2 Debye-Hückel Theory

Taking the Laplacian derivative of this equation gives us

$$\nabla^2 \phi(\underline{r}) = \nabla^2 \int_V \frac{\eta(\underline{r}')\, d\underline{r}'}{\varepsilon |\underline{r} - \underline{r}'|} \qquad (15.2\text{-}3)$$

However, it is known from electrostatic potential theory that

$$\nabla^2 \int \frac{\eta(\underline{r}')\, d\underline{r}'}{\varepsilon |\underline{r} - \underline{r}'|} = \frac{-4\pi}{\varepsilon} \eta(\underline{r}) \qquad (15.2\text{-}4)$$

so that the Poisson equation is obtained

$$\nabla^2 \phi(\underline{r}) = -\frac{4\pi}{\varepsilon} \eta(\underline{r}) \qquad (15.2\text{-}5)$$

which is a fundamental equation of electrostatics. Equation 15.2-3 is the solution of Eq. 15.2-5. So, given any charge distribution $\eta(\underline{r})$, one can calculate the resulting potential field by solving Eq. 15.2-5 for $\phi(r)$ subject to the boundary conditions.

In the Debye-Hückel theory, we are concerned with the charge distribution and electrostatic potential around any ion due to the electrostatic atmosphere created by the other ions. In particular, we are interested in the charge distribution that results from averaging over all possible configurations of the ions. Also, we will sometimes be interested in the electrostatic potential resulting from all the ions, at other times from only a single ion, or from all but a single ion. To do this we need to solve the Poisson equation for various charge distributions. We can do this taking advantage of the superposition principle for this (and other) linear differential equations. Let $\eta_1(\underline{r})$ be a charge distribution that gives rise to a potential $\phi_1(\underline{r})$, which satisfies the Poisson equation

$$\nabla^2 \phi_1(\underline{r}) = -\frac{4\pi}{\varepsilon} \eta_1(\underline{r}) \qquad (15.2\text{-}6)$$

and let $\eta_2(\underline{r})$ be a second charge distribution which gives rise to $\phi_2(\underline{r})$, another electrostatic potential. Clearly, then, for a charge distribution that is the sum $\eta_1(\underline{r}) + \eta_2(\underline{r})$, the potential is the sum of potentials $\phi_1(\underline{r}) + \phi_2(\underline{r})$ and also satisfies the Poisson equation

$$\nabla^2 (\phi_1 + \phi_2) = -\frac{4\pi}{\varepsilon} (\eta_1 + \eta_2) \qquad (15.2\text{-}7)$$

This is the superposition principle of electrostatics.

A convenient starting point for Debye-Hückel theory is the observation that the change in Helmholtz energy of a system of ions above that for the same system if the ions were uncharged can be computed from the difference in the configuration integrals of both systems (since the ideal gas parts of the partition function are unchanged). That is

$$A(N, V, T, q) - A(N, V, T, q = 0) =$$
$$- kT \left[\ln \int \cdots \int e^{-u(\underline{r}_1, \ldots \underline{r}_N, q)/kT} \, d\underline{r}_1 \ldots d\underline{r}_N \right.$$
$$\left. - \ln \int \cdots \int e^{-u(\underline{r}_1, \ldots \underline{r}_N, q=0)/kT} \, d\underline{r}_1 \ldots d\underline{r}_N \right] \qquad (15.2\text{-}8a)$$

282 Chapter 15: A Theory of Dilute Electrolyte Solutions and Ionized Gases

or

$$e^{-\frac{(A(N,V,T,q) - A(N,V,T,q=0))}{kT}} = \frac{\int \cdots \int e^{-u(\underline{r}_1,\ldots\underline{r}_N,q)/kT}\, d\underline{r}_1 \ldots d\underline{r}_N}{\int \cdots \int e^{-u(\underline{r}_1,\ldots\underline{r}_N,q=0)/kT}\, d\underline{r}_1 \ldots d\underline{r}_N} \quad (15.2\text{-}8b)$$

where

$$u(\underline{r}_1,\ldots \underline{r}_N, q) = u(\underline{r}_1,\ldots \underline{r}_N, q=0) + \frac{1}{2}\sum_i \sum_{\substack{j \\ i\neq j}} \frac{q_i q_j}{\varepsilon |\underline{r}_i - \underline{r}_j|} \quad (15.2\text{-}9)$$

Now, consider a specific ion 1 located at the position vector $\underline{r}_1$ that we will take as the origin of the coordinate system (i.e., $\underline{r}_1 = (0, 0, 0)$, and all other position vectors will be measured with respect to this origin). The electrostatic potential acting on this ion (which is designated as ψ_1) from all the other ions in the system is

$$\psi_1 = \sum_{\substack{i \\ i \neq 1}} \frac{q_i}{\varepsilon |\underline{r}_i|} \quad (15.2\text{-}10)$$

Note that ψ_1 is a function of the positions of all the other ions in the system. The total electrostatic potential at the position vector $\underline{r}$ is given by Eq. 15.2-1; and the total electrostatic potential energy of the system in which ion 1 is located at the origin, obtained by summing over all the ions, is

$$\phi(\underline{r}) = u(\underline{r}_1,\ldots,\underline{r}_N, q) - u(\underline{r}_1,\ldots,\underline{r}_N, q=0) = \frac{1}{2}\sum_j q_j \psi_1 \quad (15.2\text{-}11)$$

In fact, what we are interested in is the average value of ψ_1, designated as $\overline{\psi_1}$, computed as a canonical average

$$\overline{\psi_1} = \frac{\int \cdots \int \psi_1 e^{-u(\underline{r}_1,\ldots\underline{r}_N,q)/kT}\, d\underline{r}_1 \ldots d\underline{r}_N}{\int \cdots \int e^{-u(\underline{r}_1,\ldots\underline{r}_N,q)/kT}\, d\underline{r}_1 \ldots d\underline{r}_N} \quad (15.2\text{-}12)$$

which, since we have integrated over all positions, is a function only of temperature and density of ions. Note also, for later reference, that

$$\left(\frac{\partial A}{\partial q_j}\right)_{N,V,T} = \frac{\partial}{\partial q_j}\bigg|_{N,V,T}\left[\begin{array}{l} A(N, V, T, q=0) \\ -kT\left[\ln \int \cdots \int e^{-u(\underline{r}_1,\ldots\underline{r}_N,q)/kT}\, d\underline{r}_1 \ldots d\underline{r}_N \right. \\ \left. -\ln \int \cdots \int e^{-u(\underline{r}_1,\ldots\underline{r}_N,q=0)/kT}\, d\underline{r}_1 \ldots d\underline{r}_N \right] \end{array}\right]$$

$$= -kT \frac{\partial}{\partial q_j}\bigg|_{N,V,T} \ln \int \cdots \int e^{-u(\underline{r}_1,\ldots\underline{r}_N,q)/kT}\, d\underline{r}_1 \ldots d\underline{r}_N$$

$$= \frac{\int \cdots \int \frac{du(\underline{r}_1,\ldots\underline{r}_N,q)}{dq_j} e^{-u(\underline{r}_1,\ldots\underline{r}_N,q)/kT}\, d\underline{r}_1 \ldots d\underline{r}_N}{\int \cdots \int e^{-u(\underline{r}_1,\ldots\underline{r}_N,q)/kT}\, d\underline{r}_1 \ldots d\underline{r}_N} \quad (15.2\text{-}13)$$

15.2 Debye-Hückel Theory

However

$$\frac{du(\underline{r}_1,\ldots \underline{r}_N,q)}{dq_j} = \frac{d}{dq_j}\left[u(\underline{r}_1,\ldots \underline{r}_N,q=0) + \frac{1}{2}\sum_i \sum_{\substack{j \\ i \neq j}} \frac{q_i q_j}{\varepsilon|\underline{r}_i - \underline{r}_j|}\right]$$

$$= \sum_{\substack{i \\ i \neq j}} \frac{q_i}{\varepsilon|\underline{r}_i - \underline{r}_j|} \quad \textbf{(15.2-14)}$$

so that

$$\left(\frac{\partial A}{\partial q_j}\right)_{N,V,T} = \frac{\int \cdots \int \frac{du(\underline{r}_1,\ldots \underline{r}_N,q)}{dq_j} e^{-u(\underline{r}_1,\ldots \underline{r}_N,q)/kT}\, d\underline{r}_1 \ldots d\underline{r}_N}{\int \cdots \int e^{-u(\underline{r}_1,\ldots \underline{r}_N,q)/kT}\, d\underline{r}_1 \ldots d\underline{r}_N}$$

$$= \frac{\int \cdots \int \sum_{\substack{i \\ i \neq j}} \frac{q_i}{\varepsilon|\underline{r}_i - \underline{r}_j|} e^{-u(\underline{r}_1,\ldots \underline{r}_N,q)/kT}\, d\underline{r}_1 \ldots d\underline{r}_N}{\int \cdots \int e^{-u(\underline{r}_1,\ldots \underline{r}_N,q)/kT}\, d\underline{r}_1 \ldots d\underline{r}_N} = \overline{\psi_j} \quad \textbf{(15.2-15)}$$

Therefore, if we can compute the average electrostatic potential $\overline{\psi_j}$ acting on ion j from all the other ions in the system as a function of temperature and ion density, we will be able to compute the change in free energy of the system as the charges are turned on.

Defining $\phi(\underline{r})$ to be the total electrostatic potential at $\underline{r}$, given that ion 1 is at the origin, its average value is

$$\overline{\phi(\underline{r})} = \frac{\int \cdots \int \phi(\underline{r}) e^{-u(\underline{r}_1,\ldots \underline{r}_N,q)/kT}\, d\underline{r}_2 \ldots d\underline{r}_N}{\int \cdots \int e^{-u(\underline{r}_1,\ldots \underline{r}_N,q)/kT}\, d\underline{r}_2 \ldots d\underline{r}_N} \quad \textbf{(15.2-16)}$$

Next, taking the Laplacian derivative with respect to $\underline{r}$ and using the Poisson equation, we obtain

$$\nabla^2 \overline{\phi(\underline{r})} = \frac{\int \cdots \int \nabla^2 \phi(\underline{r}) e^{-u(\underline{r}_1,\ldots \underline{r}_N,q)/kT}\, d\underline{r}_2 \ldots d\underline{r}_N}{\int \cdots \int e^{-u(\underline{r}_1,\ldots \underline{r}_N,q)/kT}\, d\underline{r}_2 \ldots d\underline{r}_N}$$

$$= -\frac{4\pi}{\varepsilon} \frac{\int \cdots \int \eta(\underline{r}) e^{-u(\underline{r}_1,\ldots \underline{r}_N,q)/kT}\, d\underline{r}_2 \ldots d\underline{r}_N}{\int \cdots \int e^{-u(\underline{r}_1,\ldots \underline{r}_N,q)/kT}\, d\underline{r}_2 \ldots d\underline{r}_N} = -\frac{4\pi}{\varepsilon}\overline{\eta(r)} \quad \textbf{(15.2-17)}$$

where $\overline{\eta(r)}$ is the average charge density at $\underline{r}$, given that ion 1 is at the origin. However, this average charge density at $\underline{r}$ given that ion 1 is at the origin is related

284 Chapter 15: A Theory of Dilute Electrolyte Solutions and Ionized Gases

to the radial distribution (or pair correlation) function $g_{i1}^{(2)}(\underline{r})$ as follows:

$$\overline{\eta(\underline{r})} = \sum_{i=1}^{n} \rho_i q_i g_{i1}^{(2)}(\underline{r}) = \sum_{i=1}^{n} \rho_i q_i e^{-w_{i1}(\underline{r})/kT} \qquad (15.2\text{-}18)$$

where ρ_i is the density of ionic species i, q_i is its charge and w_{i1} is the potential of mean force defined in Sec. 12.7. Therefore

$$\nabla^2 \overline{\phi(\underline{r})} = -\frac{4\pi}{\varepsilon} \sum_{i=1}^{n} \overline{\eta(\underline{r})} = -\frac{4\pi}{\varepsilon} \sum_{i=1}^{n} \rho_i q_i g_{i1}^{(2)}(\underline{r}) = -\frac{4\pi}{\varepsilon} \sum_{i=1}^{n} \rho_i q_i e^{-w_{i1}(\underline{r})/kT}$$

$$(15.2\text{-}19\text{a})$$

Our model here is of an ion that is a sphere; therefore, using spherical coordinates and taking advantage of the spherical symmetry, the equation above becomes

$$\frac{1}{r^2} \frac{d}{dr} \left(r^2 \frac{d}{dr} \overline{\phi(r)} \right) = -\frac{4\pi}{\varepsilon} \sum_{i=1}^{n} \rho_i q_i g_{i1}^{(2)}(r) = -\frac{4\pi}{\varepsilon} \sum_{i=1}^{n} \rho_i q_i e^{-w_{i1}(r)/kT} \qquad (15.2\text{-}9\text{b})$$

There are two regions to consider in the solution of this equation. The first region, by the choice of the interaction potential of Eq. 15.1-1, has a hard, impenetrable core, so that no ion pairs can have a center-to-center distance in the region $r < a$. Therefore, the charge density and the radial distribution function are both zero in the region between spheres of radius $r = a/2$ (the edge of the central atom) and $r = a$:

$$g_{i1}^{(2)}(r) = 0 \quad \text{for } \frac{a}{2} \leq r < a, \qquad (15.2\text{-}20\text{a})$$

This is a result of the assumption of a hard-sphere central core, and that the charge on an ion resides only at its center. Therefore

$$\frac{1}{r^2} \frac{d}{dr} \left(r^2 \frac{d\overline{\phi(r)}}{dr} \right) = 0 \quad \frac{a}{2} < r < a \qquad (15.2\text{-}20\text{b})$$

which has the solution

$$\overline{\phi(r)} = C_1 + \frac{C_2}{r} \quad \frac{a}{2} < r < a \qquad (15.2\text{-}21)$$

where C_1 and C_2 are the constants of integration. (The second region is separation distances for which $r > a$ where the radial distribution function and charge density are not zero, and the Poisson equation for this region must be solved separately.) For notational simplicity, an overbar will not be used in what follows, but remember that in all cases we are considering average values of the electrostatic potential.

To evaluate these constants, we use Gauss' theorem—a general theorem of electrostatics and a basic relation of potential theory—which, in words, states that the surface integral of the normal component of the electric-displacement vector over any closed surface is 4π times the total charge enclosed within the surface. For a sphere, the electric displacement vector is $D(r) = -\varepsilon \frac{d\phi}{dr}$. For the case here, a sphere of radius $a/2$ enclosing ion 1 of charge q_1, we have

$$D\left(r = \frac{a}{2}\right) = -\varepsilon \left. \frac{d\phi}{dr} \right|_{r=\frac{a}{2}} = \frac{\varepsilon C_2}{\left(\frac{a}{2}\right)^2} \qquad (15.2\text{-}22)$$

15.2 Debye–Hückel Theory

and the surface area is $4\pi \left(\frac{a}{2}\right)^2$, so Gauss' theorem requires that

$$4\pi q_1 = 4\pi \left(\frac{a}{2}\right)^2 \frac{\varepsilon C_2}{\left(\frac{a}{2}\right)^2} \quad \text{or} \quad q_1 = \varepsilon C_2 \quad \text{and} \quad C_2 = \frac{q_1}{\varepsilon} \qquad (15.2\text{-}23)$$

so that

$$\phi(r) = C_1 + \frac{q_1}{\varepsilon r} \quad \frac{a}{2} \leq r \leq a \qquad (15.2\text{-}24)$$

Here, the term $q_1/\varepsilon r$ is the electrostatic potential due to the central ion, and the constant C_1 is the potential due to the charge distribution external to the sphere of radius r.

For the second region, $r > a$, we have to solve the equation

$$\frac{1}{r^2} \frac{d}{dr}\left(r^2 \frac{d\phi(r)}{dr}\right) = -\frac{4\pi}{\varepsilon} \sum_{i=1}^{n} \rho_i q_i g_{i1}^{(2)}(r) = -\frac{4\pi}{\varepsilon} \sum_{i=1}^{n} \rho_i q_i e^{-w_{i1}(r)/kT} \qquad (15.2\text{-}25)$$

An important assumption in Debye–Hückel theory is that the potential of mean force in this region is a result of the interaction of the ion 1 with all the ions, so that

$$w_{i1}(\underline{r}) = q_i \phi(\underline{r}) \qquad (15.2\text{-}26)$$

and

$$\frac{1}{r^2} \frac{d}{dr}\left(r^2 \frac{d\phi(r)}{dr}\right) = -\frac{4\pi}{\varepsilon} \sum_{i=1}^{n} \rho_i q_i e^{-q_i \phi(r)/kT} \qquad (15.2\text{-}27)$$

This nonlinear differential equation for the electrostatic potential is referred to as the Poisson–Boltzmann equation, and it is a basic equation of Debye–Hückel theory. Unfortunately, this equation cannot be solved analytically. However, assuming that the exponent is small, the exponential term can be expanded in a Taylor series.

$$e^{-q_i \phi(r)/kT} \approx 1 - \frac{q_i \phi(r)}{kT} + \frac{1}{2}\left(\frac{q_i \phi(r)}{kT}\right)^2 + \cdots \qquad (15.2\text{-}28)$$

By retaining only the first term in the series, the right hand side of Eq. 15.2-27 becomes

$$-\frac{4\pi}{\varepsilon} \sum_i \rho_i q_i e^{-q_i \phi(r)/kT} \approx -\frac{4\pi}{\varepsilon} \sum_i \rho_i q_i + \frac{4\pi}{\varepsilon kT} \sum \rho_i q_i^2 \phi(r) \qquad (15.2\text{-}29)$$

but since $\sum_i \rho_i q_i = 0$ by electrical neutrality, the approximate, linearized Poisson–Boltzmann equation is obtained

$$\frac{1}{r^2} \frac{d}{dr}\left(r^2 \frac{d\phi(r)}{dr}\right) = \frac{4\pi}{\varepsilon kT} \left(\sum \rho_i q_i^2\right) \phi(r) \qquad (15.2\text{-}30)$$

286 Chapter 15: A Theory of Dilute Electrolyte Solutions and Ionized Gases

It is useful to introduce the following variable in obtaining a solution to the linearized Poisson-Boltzmann equation:

$$\kappa^2 = \frac{4\pi}{\varepsilon kT}\left(\sum \rho_i q_i^2\right) \tag{15.2-31}$$

With this substitution, the linearized Poisson-Boltzmann equation is

$$\frac{1}{r^2}\frac{d}{dr}\left(r^2 \frac{d\phi(r)}{dr}\right) = \kappa^2 \phi(r) \tag{15.2-32}$$

which has as its general solution

$$\phi(r) = \frac{C_3 e^{-\kappa r}}{r} + \frac{C_4 e^{\kappa r}}{r} \tag{15.2-33}$$

At an infinite distance from the origin (the location of ion 1), the electrostatic potential must remain finite (and, in fact, vanish) which implies $C_4 = 0$. Therefore

$$\phi(r) = \frac{C_3 e^{-\kappa r}}{r} \quad r > a \tag{15.2-34}$$

which is the electrostatic potential field in the region $r > a$. For the region $a/2 \leq r \leq a$, we previously found that

$$\phi(r) = \frac{q_i}{\varepsilon r} + C_1 \quad \frac{a}{2} \leq r \leq a \tag{15.2-24}$$

Now, requiring that at $r = a$, the intersection of the two regions, that $\phi(r)$, and its derivative $d\phi/dr$ should be continuous, we obtain

$$C_3 \frac{e^{-\kappa a}}{a} = \frac{q_1}{a\varepsilon} + C_1 \tag{15.2-35a}$$

and

$$C_3 \left(\frac{e^{-\kappa a}}{a^2} + \frac{\kappa e^{-\kappa a}}{a}\right) = \frac{q_i}{\varepsilon a^2} \tag{15.2-35b}$$

from which

$$C_1 = -\frac{q_i \kappa}{\varepsilon(1+\kappa a)} \quad \text{and} \quad C_3 = \frac{q_i e^{\kappa a}}{\varepsilon(1+\kappa a)}; \tag{15.2-36}$$

Therefore,

$$\phi(r) = \frac{q_1}{\varepsilon r} - \frac{q_1 \kappa}{\varepsilon(1+\kappa a)} \quad \text{for} \quad \frac{a}{2} \leq r < a$$

$$\phi(r) = \frac{q_1 e^{\kappa(a-r)}}{\varepsilon(1+\kappa a)}\frac{1}{r} \quad \text{for} \quad r > a \tag{15.2-37}$$

15.2 Debye-Hückel Theory

The electrostatic potential due to all the ions other than ion 1 is

$$\psi_1 = \phi(r) - \frac{q_1}{\varepsilon r}$$

or

$$\psi_1 = \phi(r) - \frac{q_1}{\varepsilon r} = \frac{q_1}{\varepsilon r} - \frac{q_1 \kappa}{\varepsilon(1+\kappa a)} - \frac{q_1}{\varepsilon r} = -\frac{q_1 \kappa}{\varepsilon(1+\kappa a)} \quad \text{for} \quad \frac{a}{2} \leq r < a \quad \text{(15.2-38a)}$$

Also, and more generally, for any ion i chosen to be at the origin

$$\psi_i = \phi(r) - \frac{q_i}{\varepsilon r} = \frac{q_i}{\varepsilon r} - \frac{q_i \kappa}{\varepsilon(1+\kappa a)} - \frac{q_i}{\varepsilon r} = -\frac{q_i \kappa}{\varepsilon(1+\kappa a)} \quad \text{for} \quad \frac{a}{2} \leq r < a \quad \text{(15.2-38b)}$$

We see from this expression that ψ_i (previously designated as $\overline{\psi_i}$) is a function of temperature and density of ions through the parameter κ.

The electrostatic interaction energy of the central ion i with another ion j in the mixture, u_{ij}, is

$$u_{ij}(r_{ij}) = q_j \phi_i(r_{ij}) = \frac{q_i q_j}{\varepsilon r_{ij}} - \frac{q_i q_j \kappa}{\varepsilon(1+\kappa a)} \quad \text{for} \quad \frac{a}{2} \leq r_{ij} < a$$

$$= \frac{q_i q_j e^{\kappa(a-r)}}{\varepsilon(1+\kappa a)} \frac{1}{r_{ij}} \quad \text{for} \quad r_{ij} > a \quad \text{(15.2-39)}$$

The form of this interaction potential is interesting in that for very small ion-ion separation distances ($r < a$), the electrostatic interaction between two ions is proportional to $q_i q_j / r$, which is the Coulomb potential; while at large separation distances the electrostatic interaction energy is proportional to $q_i q_j e^{-\kappa r_{ij}}/r_{ij}$, which is referred to as the *shielded Coulomb potential*. Thus the interaction energy at large separations is the Coulomb potential damped by an exponential term that depends on κ, which is related to the concentration of ions, and is due to the presence of other ions between the two of interest. That is, at large separation distances, the presence of an ion cloud (of opposite charge) around a central ion shields that ion, and reduces its interactions with other ions.

Since we have $\nabla^2 \phi = -4\pi \eta/\varepsilon$ [from Eq. 15.2-5] and $\nabla^2 \phi = \kappa^2 \phi$ [from Eq. 15.2-32] we can obtain the charge density directly from the electrostatic potential as

$$\eta(r) = -\frac{\varepsilon \kappa^2 \phi(r)}{4\pi} \quad \text{(15.2-40)}$$

where η is the average total charge density at r around an ion of type i at the origin. Also, we can compute the charge density at r around the chosen central ion of type i due only to ions of type j, η_{ij} from

$$\eta_{ij}(r) = q_j \rho_j \underbrace{g_{ij}^{(2)}(r)}_{\text{radial distribution function}} = q_j \rho_j e^{-q_i \phi_i(r)/kT} \approx q_j \rho_j \left[1 - \frac{q_i \phi_i(r)}{kT}\right] \quad \text{(15.2-41)}$$

Chapter 15: A Theory of Dilute Electrolyte Solutions and Ionized Gases

and the total charge density is

$$\eta(r) = \eta_{ji}(r) + \eta_{ii}(r) = q_j \rho_j g_{ji}(r) + q_i \rho_i g_{ii}(r)$$

$$= q_j \rho_j e^{-q_j \phi(r)/kT} + q_i \rho_i e^{-q_i \phi(r)/kT}$$

$$\approx q_j \rho_j \left[1 - \frac{q_j \phi(r)}{kT}\right] + q_i \rho_i \left[1 - \frac{q_i \phi(r)}{kT}\right]$$

$$= q_j \rho_j - \frac{\rho_j q_j^2 \phi(r)}{kT} + q_i \rho_i - \frac{\rho_i q_i^2 \phi(r)}{kT} = -\frac{\phi(r)}{kT}(\rho_i q_i^2 + \rho_j q_j^2)$$

$$= -\frac{\varepsilon \kappa^2 \phi(r)}{4\pi} \quad (15.2\text{-}42)$$

which is identical to Eq. 15.2-40 above.

The total charge in a spherical shell between r and $r + dr$ around the central ion i is

$$\eta(r) 4\pi r^2 \, dr = -\frac{\varepsilon \kappa^2}{4\pi} \phi(r) 4\pi r^2 \, dr = -\frac{\varepsilon \kappa^2}{4\pi} \frac{q_i e^{\kappa(a-r)}}{\varepsilon r (1 + \kappa a)} 4\pi r^2 \, dr$$

$$= -\frac{q_i \kappa^2 e^{\kappa(a-r)}}{(1 + \kappa a)} r \, dr \quad (15.2\text{-}43)$$

To obtain the distance r_{max} surrounding any ion at which the charge density is a maximum, we set

$$\frac{d}{dr}\left(-\frac{q_i \kappa^2 e^{\kappa(a-r)}}{(1 + \kappa a)} r\right) = 0 = -\frac{q_i \kappa^2 e^{\kappa(a-r)}}{(1 + \kappa a)} \frac{d}{dr}(re^{-\kappa r})$$

or

$$\frac{d}{dr}(re^{-\kappa r}) = 0 = e^{-\kappa r} - r\kappa e^{-\kappa r} = e^{-\kappa r}(1 - r\kappa) \quad (15.2\text{-}44)$$

to obtain

$$r_{max} = 1/\kappa \quad (15.2\text{-}45)$$

Consequently, $(1/\kappa)$, which has units of length and is called *Debye length*[2], is the distance from an ion at which the charge density is a maximum. So a measure of the thickness of the ion atmosphere (or cloud) around an ion is

$$\frac{1}{\kappa} = \text{Debye length} = \left(\frac{\varepsilon kT}{4\pi} \frac{1}{\sum_\alpha \rho_\alpha q_\alpha^2}\right)^{1/2} \quad (15.2\text{-}46)$$

---

[2]The value of the Debye length at 25°C with water as the solute is approximately, in units of Angstroms (10^{-8} cm or 0.1 nanometers), $1/\kappa = 3.04/\sqrt{I}$ where $I = \frac{1}{2}\sum_i M_i z_i^2$ where z_i is the charge on the ion, M_i is its molarity, and I is the ionic strength in moles/liter. More generally, $1/\kappa = \sqrt{\frac{\varepsilon_0 \varepsilon_r kT}{2 N_{Av} e^2 I}}$ where N_{Av} is Avogadro's number, e is the charge on an electron, and here I is the ionic strength in moles/m^3.

15.2 Debye-Hückel Theory

The total charge outside the central ion of charge q_i is

$$\int_a^\infty \eta(r) 4\pi r^2 \, dr = -\int_a^\infty \frac{q_i \kappa^2 e^{-\kappa(a-r)}}{1+\kappa a} r \, dr = \underbrace{-q_i}_{\text{just opposite of charge of central ion}} \quad (15.2\text{-}47)$$

Thus, our approximate solution is self-consistent, at least to the extent that the total charge in the region beyond any ion is the negative of its charge.

We can now proceed to the calculation of the thermodynamic properties. The starting point is

$$\left(\frac{\partial A}{\partial q_j}\right)_{N,V,T} = \psi_j \quad (15.2\text{-}15)$$

with

$$\psi_j = -\frac{q_j \kappa}{\varepsilon(1+\kappa a)} \quad (15.2\text{-}38\text{b})$$

so that

$$\left(\frac{\partial A}{\partial q_j}\right)_{N,V,T} = -\frac{q_j \kappa}{\varepsilon(1+\kappa a)} \quad (15.2\text{-}48)$$

Therefore

$$dA = -\sum_j \frac{q_j \kappa}{\varepsilon(1+\kappa a)} dq_j \quad (15.2\text{-}49)$$

Note that the Debye parameter κ depends on the charge of the ions, but it is the charge of the ions that is the variable of integration. The simplest way to do the integration is to introduce a charging parameter λ so that the charge on any ion is λq, with λ ranging from 0 (no charge on the ions) to 1 (ions fully charged). Using this parameter, we have

$$dA = -\sum_j \frac{\lambda q_j^2 \kappa(\lambda)}{\varepsilon(1+a\kappa(\lambda))} d\lambda \quad (15.2\text{-}50)$$

where

$$\kappa(\lambda)^2 = \frac{4\pi}{\varepsilon kT}\sum \rho_i (\lambda q_i)^2 = \lambda^2 \frac{4\pi}{\varepsilon kT}\sum \rho_i q_i^2 = \lambda^2 \kappa(\lambda = 1)^2$$

and

$$\kappa(\lambda) = \lambda \kappa(\lambda = 1) = \lambda \kappa \quad (15.2\text{-}51)$$

where $\kappa = \frac{4\pi}{\varepsilon kT}\sum \rho_i q_i^2$ is of fixed value, since it is λ and not q_i that will now be the variable of integration. (For simplicity of notation, κ has been used for $\kappa(\lambda = 1)$. Therefore

$$dA = -\sum_j \frac{\lambda^2 q_j^2 \kappa}{\varepsilon(1+a\lambda\kappa)} d\lambda$$

and

$$A(\lambda = 1) - A(\lambda = 0) = A^{\text{el}} = -\sum_j \frac{q_j^2 \kappa}{\varepsilon} \int_0^1 \frac{\lambda^2}{(1 + a\lambda\kappa)} d\lambda \qquad (15.2\text{-}52)$$

where A^{el} has been used to indicate the electrostatic contribution to the Helmholtz energy. Completing the integration gives

$$\frac{A^{\text{el}}}{kT} = -\frac{V}{4\pi a^3}\left[\ln(1 + \kappa a) - \kappa a + \frac{\kappa^2 a^2}{2}\right] = -\frac{V\kappa^3}{12\pi}\tau(\kappa a)$$

where $\tau(\kappa a) = \dfrac{3}{(\kappa a)^3}\left[\ln(1 + \kappa a) - \kappa a + \dfrac{\kappa^2 a^2}{2}\right] \qquad (15.2\text{-}53)$

(Note that as $\kappa a \to 0$, $\tau(\kappa a) \to 1 - \frac{3}{4}\kappa a + \cdots$)

The chemical potential (or partial molar Gibbs energy) of an ion is then computed as follows:

$$\left(\frac{\partial A^{\text{el}}}{\partial N_i}\right)_{V,T,N_{j \neq i}} = -\frac{\partial}{\partial N_i}\left(\frac{VkT}{4\pi a^3}\left[\ln(1 + \kappa a) - \kappa a + \frac{\kappa^2 a^2}{2}\right]\right)_{V,T,N_{j \neq i}}$$

$$= -\frac{VkT}{4\pi a^3}\left(\frac{\partial[\ln(1 + \kappa a)]}{\partial N_i} - a\frac{\partial \kappa}{\partial N_i} + \frac{a^2}{2}\frac{\partial \kappa^2}{\partial N_i}\right)_{V,T,N_{j \neq i}}$$

$$= -\frac{VkT}{4\pi a^3}\left(\frac{a}{1 + \kappa a}\frac{\partial \kappa}{\partial N_i} - a\frac{\partial \kappa}{\partial N_i} + a^2\kappa\frac{\partial \kappa}{\partial N_i}\right) \qquad (15.2\text{-}54)$$

Now, since

$$\kappa = \left(\frac{4\pi}{\varepsilon kT}\sum\left(\frac{N_i}{V}\right)q_i^2\right)^{\frac{1}{2}}$$

then

$$\left(\frac{\partial \kappa}{\partial N_i}\right)_{V,T,N_{j \neq i}} = \frac{\frac{1}{2}\frac{4\pi}{\varepsilon kT}\frac{q_i^2}{V}}{\left(\frac{4\pi}{\varepsilon kT}\sum\left(\frac{N_i}{V}\right)q_i^2\right)^{\frac{1}{2}}} = \frac{2\pi}{\varepsilon kT\kappa}\frac{q_i^2}{V}$$

$$\left(\frac{\partial A^{\text{el}}}{\partial N_i}\right)_{V,T,N_{j \neq i}} = \mu_i^{\text{el}} = -\frac{VkT}{4\pi a^3}\left(\frac{a}{1 + \kappa a}\frac{\partial \kappa}{\partial N_i} - a\frac{\partial \kappa}{\partial N_i} + a^2\kappa\frac{\partial \kappa}{\partial N_i}\right)$$

$$= -\frac{VkT}{4\pi a^3}\left(\frac{a}{1 + \kappa a} - a + a^2\kappa\right)\frac{\partial \kappa}{\partial N_i}$$

$$= -\frac{VkT(1 - 1 - \kappa a + \kappa a + \kappa^2 a^2)}{4\pi a^2(1 + \kappa a)}\frac{\partial \kappa}{\partial N_i}$$

$$= -\frac{VkT\kappa^2}{4\pi(1 + \kappa a)}\frac{\partial \kappa}{\partial N_i} = -\frac{VkT\kappa^2}{4\pi(1 + \kappa a)}\frac{2\pi}{\varepsilon kT\kappa}\frac{q_i^2}{V} = -\frac{\kappa q_i^2}{2\varepsilon(1 + \kappa a)}$$

$$(15.2\text{-}55)$$

Therefore, the electrostatic contribution to the solution nonidealities is

$$\mu_i^{\text{el}} = kT \ln \gamma_i^{\text{el}} = -\frac{\kappa q_i^2}{2\varepsilon(1+\kappa a)}$$

and

$$\gamma_i^{\text{el}} = e^{-\kappa q_i^2/2\varepsilon(1+\kappa a)kT} \qquad (15.2\text{-}56)$$

Notice that this model predicts that the electrostatic activity coefficient of an ion will always be less than unity.

15.3 THE MEAN IONIC ACTIVITY COEFFICIENT

In electrolyte solutions, the number of cations and the number of anions are not independently variable, since electrical neutrality must be maintained (Eq. 15.1-7). So it is not possible to measure the activity coefficient of one ionic species by varying only its concentration, as the concentration of the counterion must simultaneously change to maintain electrical neutrality. This is quite different than the thermodynamics of neutral molecules, where one can change the concentration of one species independent of all the others. Therefore, instead of using the activity coefficient of a single ionic species, common practice is to define the mean molar activity coefficient $\gamma_\pm^\nu$ for an electrolyte $A_{\nu_+}^{q_+} B_{\nu_-}^{q_-}$ as

$$\gamma_\pm^\nu = (\gamma_+)^{\nu_+}(\gamma_-)^{\nu_-} = \left(e^{-\kappa q_+^2/2\varepsilon(1+\kappa a)kT}\right)^{\nu_+} \left(e^{-\kappa q_-^2/2\varepsilon(1+\kappa a)kT}\right)^{\nu_-}$$

$$= e^{-\kappa(\nu_+ q_+^2 + \nu_- q_-^2)/2\varepsilon(1+\kappa a)kT}$$

or

$$\ln \gamma_\pm = -\frac{\kappa(\nu_+ q_+^2 + \nu_- q_-^2)}{2\varepsilon(1+\kappa a)kT(\nu_+ + \nu_-)} \qquad (15.3\text{-}1)$$

where $\nu = \nu_+ + \nu_-$. Now, noting that

$$\nu_+ q_+ + \nu_- q_- = 0; \quad \nu_+ q_+^2 = -\nu_- q_- q_+; \quad \text{and} \quad \nu_- q_-^2 = -\nu_+ q_- q_+$$

so that

$$\frac{\nu_+ q_+^2 + \nu_- q_-^2}{\nu_+ + \nu_-} = -\frac{\nu_+ q_+ q_- + \nu_- q_+ q_-}{\nu_+ + \nu_-} = -\frac{(\nu_+ + \nu_-)q_+ q_-}{\nu_+ + \nu_-} = -q_+ q_- = |q_+ q_-|$$

$$(15.3\text{-}2)$$

In writing the last form of the equation, we have noted that since the two charges are always of opposite sign, their product is negative and the term is positive. To avoid confusion, this is written as the absolute value of the product of charges. Consequently, we have

$$\ln \gamma_\pm = -\frac{|q_+ q_-|\kappa}{2\varepsilon(1+\kappa a)kT} = -\frac{|q_+ q_-|\left(\frac{4\pi}{\varepsilon kT}\sum\left(\frac{N_i}{V}\right)q_i^2\right)^{\frac{1}{2}}}{2\varepsilon(1+\kappa a)kT}$$

$$= -\frac{|q_+ q_-|\left(\frac{2\pi}{\varepsilon kT}I\right)^{\frac{1}{2}}}{2\varepsilon(1+\kappa a)kT} = -\frac{\alpha|q_+ q_-|\sqrt{I}}{1+\beta a\sqrt{I}}$$

or

$$\ln \gamma_{\pm} = -\frac{\alpha |q_+ q_-| \sqrt{I}}{1 + \beta a \sqrt{I}} \quad (15.3\text{-}3)$$

which is the Debye-Hückel expression for the activity coefficient of an electrolyte. In this equation

$$I = \text{ionic strength} = \sum \left(\frac{N_i}{V}\right) q_i^2, \alpha = \frac{1}{4\pi} \left(\frac{2\pi}{\varepsilon k T}\right)^{\frac{5}{2}} \text{ and } \beta = \left(\frac{4\pi}{\varepsilon k T}\right)^{\frac{1}{2}}.$$

Also note that in the limit of very low ionic strength ($\kappa \to 0$), we obtain

$$\ln \gamma_{\pm} = -\alpha |q_+ q_-| \sqrt{I} \quad (15.3\text{-}4)$$

which is the Debye-Hückel limiting law. Also, the activity coefficient of Eq. (15.3-3) corresponds to a mean ionic molar density of

$$\rho_{\pm}^{\nu} = \rho_{\pm}^{\nu_+ + \nu_-} = \rho_{A^{q_+}}^{\nu_+} + \rho_{B^{q_-}}^{\nu_-} \text{ or } \rho_{\pm} = (\rho_{A^{q_+}}^{\nu_+} + \rho_{B^{q_-}}^{\nu_-})^{\frac{1}{\nu_+ + \nu_-}}$$

The accuracy of Eqs. 15.3-3 and 15.3-4 in representing the $\gamma_{\pm}$ is illustrated in Fig. 15.3-1.

The equation above for the activity coefficient is a result of only the electrostatic contribution, and only in the limit of very dilute electrolytes. This equation

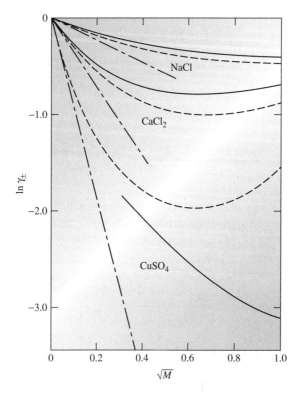

Figure 15.3-1 Activity coefficients of various salts in aqueous solution at 25°C as a function of salt molarity M. The solid line is the experimental data (from R. A. Robinson and R. H. Stokes, *Electrolyte Solutions*, 2nd ed., Butterworth, London, 1959). The dash-dot line is the result of the Debye–Hückel limiting law, Eq. 15.3-4, and the dashed line is the result of Eq. 15.3-5 with $\beta a = 1$. Note that for NaCl $I = M$; for CaCl$_2$ $I = 3M$; and for CuSO$_4$ $I = 4M$. Note that even for a simple salt like NaCl, significant departures from the Debye–Hückel limiting law occur at very low molalities $\sqrt{M} = 0.1$, or $M = 0.01$.

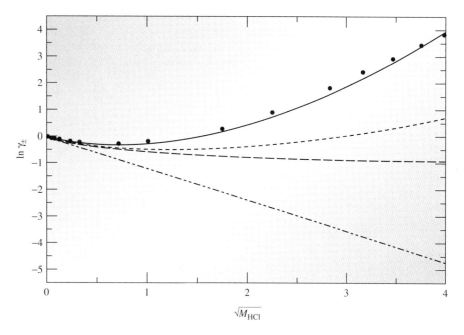

Figure 15.3-2 The mean molar activity coefficient for hydrogen chloride in water at 25°C as a function of the square root of the ionic strength $I = M_{HCl}$. The points are experimental data, the dash-dot line is the Debye–Hückel limiting law (Eq. 15.3-4), the large dashed line is the result of Eq. 15.3-3 with $\beta a = 1$, the small dashed line is from Eq. 15.3-5 with $\beta a = 1$ and $\delta = 0.1$, and the solid line is from Eq. 15.3-5 with $\beta a = 1$ and $\delta = 0.2$.

can be used as the basis for correlating data at much higher (up to several molal) concentrations using the following simple empirical modification

$$\ln \gamma_\pm = -\frac{\alpha |q_+ q_-| \sqrt{I}}{1 + \beta a \sqrt{I}} + \delta I \qquad (15.3\text{-}5)$$

where δ is an adjustable parameter fit to experimental data. This is shown in Fig. 15.3-2 for aqueous hydrogen chloride solutions up to $16M$. Consequently, we see that though the departures from the Debye-Hückel limiting law occur at very low molalities, the extended Debye-Hückel model of Eq. 15.3-5 can be used to reasonably high in molalities.

Another empirical correction that has been made is to recognize that since Debye-Hückel theory only provides the long-range electrostatic contribution to the activity coefficient, a short-range nonelectrostatic part contribution can be added, using one of the applied thermodynamic model developed for neutral molecules.

It is interesting to note that the Debye-Hückel theory combines statistical mechanics and a microscopic view for the ions (i.e., radial distribution functions and ensemble averages) with a macroscopic or continuum view (electrostatic theory and a solvent that appears only through its dielectric constant). The Debye-Hückel theory is an application of molecular modeling to develop an applied thermodynamic model, and perhaps should be considered molecular thermodynamics. The unusual form of the final result, the Debye-Hückel expression for the activity coefficient, is what has been

Chapter 15: A Theory of Dilute Electrolyte Solutions and Ionized Gases

obtained from theory. It would have been difficult to develop empirically solely by trying to correlate experimental data.

The contribution of the electrostatic charge to other thermodynamic properties can also be computed. For example, starting from

$$\frac{A^{el}}{kT} = -\frac{V}{4\pi a^3}\left[\ln(1+\kappa a) - \kappa a + \frac{\kappa^2 a^2}{2}\right] = -\frac{V\kappa^3}{12\pi}\tau(\kappa a) \quad (15.2\text{-}53)$$

where $\tau(\kappa a) = \frac{3}{\kappa^3 a^3}\left[\ln(1+\kappa a) - \kappa a + \frac{\kappa^2 a^2}{2}\right]$ and using

$$P = -\left(\frac{\partial A}{\partial V}\right)_{N,T} \quad \text{and} \quad \left(\frac{\partial \kappa}{\partial V}\right)_{N,T} = -\frac{\kappa}{2V} \quad (15.3\text{-}6)$$

one obtains

$$\frac{P^{el}}{kT} = -\frac{kT}{4\pi a^3}\left(\ln(1+\kappa a) - \kappa a + \frac{\kappa^2 a^2}{2} - \frac{\kappa^3 a^3}{2(1+\kappa a)}\right)$$

$$= \frac{\kappa^3}{24\pi}\left(\frac{3}{1+\kappa a} - 2\tau(\kappa a)\right) \quad (15.3\text{-}7)$$

The expression we have obtained for the activity coefficient of the electrolyte

$$\ln \gamma_\pm = -\frac{\alpha |q_+ q_-|\sqrt{I}}{1 + \beta a \sqrt{I}} \quad (15.3\text{-}3)$$

is only the electrostatic contribution. The total chemical potential on a molar basis for the electrolyte using the hypothetical ideal 1–molal solution as the standard state is

$$\overline{G}_e(T, P, M) = \overline{G}_e^\circ(T, P, M=1) + RT \ln\left(\frac{M_\pm \gamma_\pm}{M=1}\right)^\nu \quad (15.3\text{-}8)$$

where

$$M_\pm^\nu = (M_+)^{\nu_+}(M_-)^{\nu_-} = (\nu_+ M)^{\nu_+}(\nu_- M)^{\nu_-} = (\nu_+)^{\nu_+}(\nu_-)^{\nu_-} M^{\nu_+ + \nu_-} = M^\nu \Delta$$

and $\Delta = (\nu_+)^{\nu_+}(\nu_-)^{\nu_-}$, so that

$$\overline{G}_e(T, P, M) = \overline{G}_e^\circ(T, P, M=1) + \nu RT \ln\left(\frac{M \Delta \gamma_\pm}{M=1}\right) \quad (15.3\text{-}9)$$

In contrast, the partial molar Gibbs free energy of the solvent, water, is written as

$$\overline{G}_W(T, P, M) = \underline{G}_W^\circ(T, P, M=0) + RT \ln(x_W \gamma_W)$$
$$= \underline{G}_W^\circ(T, P, M=0) + RT \ln a_W \quad (15.3\text{-}10)$$

using the pure component standard state, where M is the molality of the salt, and $a_W = x_W \gamma_W$ to indicate the activity of water.

15.3 The Mean Ionic Activity Coefficient

Finally, even though the solvent (for example, water) does not have a charge, there is an electrostatic contribution to its activity coefficient. This comes about as follows. The Gibbs-Duhem equation written in terms of moles of solvent and electrolyte per kilogram of water is

$$n_W \, d\overline{G}_W = -n_e \, d\overline{G}_e$$

$$n_W RT \, d \ln a_W = -n_e \nu RT \, d \ln (M \Delta \gamma_\pm)$$

$$55.1 \times d \ln a_W = -M\nu \, d \ln (M \Delta \gamma_\pm) = -M\nu \, d \ln (M) - M\nu \, d \ln (\gamma_\pm) \quad \textbf{(15.3-11)}$$

where 55.1 is approximately the number of moles of water per kilogram, so that

$$55.1 \times d \ln a_W = -\nu \, dM - M\nu \, d \ln (\gamma_\pm) \quad \textbf{(15.3-12)}$$

Now for simplicity, using only the limiting law

$$\ln \gamma_\pm = -\alpha |q_+ q_-| \sqrt{I} = -\alpha |q_+ q_-| \sqrt{\frac{1}{2} \sum_{+,-} q_i^2 M_i}$$

$$= -\alpha |q_+ q_-| \sqrt{\frac{1}{2} \sum_{+,-} q_i^2 \nu_i M} = K M^{\frac{1}{2}}$$

$$\text{where } K = -\alpha |q_+ q_-| \sqrt{\frac{1}{2} \sum_{+,-} q_i^2 \nu_i} \quad \textbf{(15.3-13)}$$

gives

$$55.1 \times d \ln a_W = -\nu \, dM - M\nu K \, dM^{\frac{1}{2}}$$

and on integration

$$55.1 \times \ln \frac{a_W(M)}{a_W(M=0)} = 55.1 \times \ln a_W(M) = -\nu M - \frac{\nu K}{3} M^{\frac{3}{2}}$$

$$= 55.1 \times \ln (x_W \gamma_W)$$

or

$$x_W \gamma_W = a_W = e^{-\left(\nu M + \frac{\nu K}{3} M^{\frac{3}{2}}\right)/55.1} \quad \text{and} \quad \ln a_W = -\left(\nu M + \frac{\nu K}{3} M^{\frac{3}{2}}\right)/55.1$$

$$\textbf{(15.3-14)}$$

A somewhat more complicated expression is obtained if the extended expression for the Debye-Hückel model of Eq. 15.3-5 is used. The interesting result is that even though water is a neutral molecule, its activity and activity coefficient are affected by the electrostatics of the ions, and the coupling is through the Gibbs-Duhem equation. However, since the exponent is attenuated by a factor of 55.1—the number of moles of water per kilogram—the effect will not be large.

The most-used model for concentrated electrolyte solutions, and also for mixed solvent systems, is that of Pitzer[3] in the form of an excess Gibbs energy expression for all (charged and uncharged) species present in solution

$$\frac{G^{ex}}{RT} = f(I) + \sum_i \sum_j \lambda_{ij}(I) M_i M_j + \sum_i \sum_j \sum_k \delta_{ijk} M_i M_j M_k \quad (15.3\text{-}15)$$

where G^{ex} is the excess Gibbs energy of the mixture per kilogram of solvent, M_i is the molality of species i (either ions or neutral molecules), the $\lambda_{ij}(I)$ are ionic strength adjustable binary parameters, the δ_{ijk} are constant ternary parameters, and $f(I)$ is the Debye-Hückel contribution to the excess Gibbs energy. Consequently, this equation takes the Debye-Hückel expression as its leading term, and then adds terms that resemble the virial expansion to account for higher concentrations, as well as solvent-solvent and other interactions.

CHAPTER 15 PROBLEMS

15.1 Compute the Debye lengths in aqueous solutions at 25°C for 1:1 electrolyte (i.e., NaCl) at various ionic strengths. Repeat the calculation for a 2:2 electrolyte. (A charge closer to a central charge than $1/\kappa$ sees the charge directly and interacts with it; a charge that is further than $1/\kappa$ is shielded from the central charge by the ions of the salt solution.)

15.2 The following data are available for the mean ionic activity coefficient of sodium chloride in water at 25°C.

M	0.1	0.25	0.5	0.75	1.0
$\gamma_\pm$	0.778	0.720	0.681	0.665	0.657

M	2.0	3.0	4.0	5.0	6.0
$\gamma_\pm$	0.669	0.714	0.782	0.873	0.987

Assuming the value of $\beta a = 1$, determine how well Eqs. 15.3-3–15.3-5 correlate these data.

15.3 Compute the activity coefficient of water in each of the solutions in Problem 15.2.

15.4 The following data are available for the mean ionic activity coefficient of $Cr_2(SO_4)_3$ in water at 25°C.

M	0.1	0.2	0.3	0.5	0.6	0.8	1.0
$\gamma_\pm$	0.0458	0.0300	0.0238	0.0190	0.0182	0.0185	0.0208

Assuming the value of $\beta a = 1$, determine how well Eqs. 15.3-3, 4 and 5 correlate these data.

15.5 Compute the activity coefficient of water in each of the solutions in Problem 15.4.

---

[3] See, for example, K. S. Pitzer, in "Thermodynamics of Aqueous Systems with Industrial Applications," *ACS Symposium Series*, **133**, 451 (1980).

Chapter 16

The Derivation of Thermodynamic Models from the Generalized van der Waals Partition Function

The goals of the previous chapters have been to develop a rather detailed theory of statistical mechanics with a minimum of assumptions. Hopefully, this has led to some understanding of the subject and how it can be used. Further, we have obtained some results that are exact in certain limiting cases. For example, the exact expressions for the thermodynamic properties of the ideal gas and the virial equation of state, and with some assumptions, the heat capacity of a monatomic crystal, the Debye-Hückel model for very dilute electrolytes, the Flory-Huggins model for polymer solutions, and others. The underlying justification for this is that thermodynamic models that are based on theory should be better, and have a broader range of applicability, than models based solely on empirical functions obtained by data correlation.

In this concluding chapter, the Generalized van der Waals model is introduced. It provides a general framework that can be used for understanding the assumptions contained in thermodynamic models commonly used in applied thermodynamics, and can also be used for developing new models. By understanding what these assumptions are, they can then be tested by, for example, computer simulation, and perhaps then be improved upon, leading to new and better thermodynamic models.

INSTRUCTIONAL OBJECTIVES FOR CHAPTER 16

The goals of this chapter are for the student to:

- Understand the concept of the Generalized van der Waals partition function
- Be able to use this partition function to understand the assumptions that underlie commonly used equations of state and activity coefficient models, and be able to derive new ones
- Understand how computer simulation (discussed in Chapter 13) can be used to test some of these assumptions

298 Chapter 16: The Derivation of Thermodynamic Models

16.1 THE STATISTICAL-MECHANICAL BACKGROUND

The starting point of the analysis is the canonical partition function for a pure fluid of N identical molecules in a volume V and at a temperature T

$$Q(N, V, T) = \frac{q_{int}^N}{\Lambda^{3N} N!} \int \int e^{-u(\underline{r}_1, \ldots \underline{r}_N)/kT} \, d\underline{r}_1 \ldots d\underline{r}_N = \frac{q_{int}^N}{\Lambda^{3N} N!} Z(N, V, T) \tag{16.1-1}$$

where $\Lambda = \left(\frac{2\pi m kT}{h^2}\right)^{\frac{1}{2}}$ is the De Broglie wave length and $Z(N, V, T)$ is the configuration integral of Eq. (7.1-3):

$$Z(N, V, T) = \int_V \ldots \int_V e^{-u(\underline{r}_1, \ldots \underline{r}_N)/kT} \, d\underline{r}_1 \ldots d\underline{r}_N \tag{7.1-3}$$

In this equation, $\underline{r}_i$ is the position vector of molecule i—for nonspherical molecules, the description of position and the integration also include orientation. For the present, only pure fluids of simple molecules are considered (mixtures are discussed in Sections 16.3 and 16.4, chain molecules and polymers in Section 16.5, and associating fluids in Section 16.6). The first term in the product on the right-hand side of Eq. 16.1-1 contains the translational partition function Λ^{-3N}, q_{int} is the contribution from all the internal modes (rotational, vibrational, electronic, and nuclear, though these last two are unimportant here), and the configuration integral is the result of the interactions between all the molecules.

As pointed out in previous chapters, once the partition function is known, the thermodynamic properties can be computed. For example

$$A(N, V, T) = -kT \ln Q(N, V, T)$$
$$U(N, V, T) = kT^2 \left(\frac{\partial \ln Q}{\partial T}\right)_{N,V} \quad \text{and} \tag{16.1-2}$$
$$P(N, V, T) = kT \left(\frac{\partial \ln Q}{\partial V}\right)_{N,T}$$

Of somewhat more interest here are the residual properties (denoted by the superscript res) over that of an ideal gas (indicated by the superscript IG) at the same temperature and density $\rho = N/V$, which are the differences between the real fluid and ideal gas properties:

$$A^{res}(N, V, T) = A(N, V, T) - A^{IG}(N, V, T) = -kT \ln \frac{Z(N, V, T)}{V^N}$$
$$U^{res}(N, V, T) = U(N, V, T) - U^{IG}(N, V, T) = kT^2 \left(\frac{\partial \ln Z}{\partial T}\right)_{N,V} \tag{16.1-3}$$

and

$$P^{res}(N, V, T) = P(N, V, T) - P^{IG}(N, V, T) = kT \left(\frac{\partial \ln(Z/V^N)}{\partial V}\right)_{N,T}$$

This form of the equation focuses attention on the configuration integral, which cannot be evaluated exactly; the ideal gas contributions are known from Chapters 3 and 4.

16.1 The Statistical-Mechanical Background

Throughout the discussion here, it will be assumed that the interaction energy for an assembly of N molecules is pairwise additive:

$$u(\underline{r}_1,\ldots,\underline{r}_N) = \sum_i \sum_{\substack{j \\ 1 \le i < j \le N}} u(r_{ij}) \qquad (7.3\text{-}3)$$

where r_{ij} is the separation distance between molecules i and j. With this assumption, we have that the total interaction energy among the N molecules, based on Eq. 11.2-4, is

$$U(N,V,T) = \frac{N^2}{2V} \int u(r_{12}) g(r_{12}; \rho, T) \, d\underline{r}_{12}$$
$$= \frac{N^2}{2V} \int u(r) g(r; \rho, T) \, d\underline{r} \qquad (11.2\text{-}4)$$

To proceed, we note that the interactions between real molecules have a hard core for which $u(r)$ is infinite and the radial distribution function $g(r)$ is 0. Also, at infinite temperature—conditions at which the molecules have a very high (infinite) kinetic energy—only the hard-core part of the potential is important, not the weakly attractive or repulsive energies from the dispersive interactions, since kT is so much larger than such energies that the Boltzmann factor is 1 except in the hard-core region, where its value is 0. Now, integrating the expression for the configurational energy, Eq. 16.1-3 from infinite temperature to the temperature of interest is

$$\ln Z(N,V,T) - \ln Z(N,V,T=\infty) = \int_{T=\infty}^{T} \frac{U^{\text{res}}(N,V,T)}{kT^2} \, dT$$
$$= \frac{N^2}{2V} \int_{T=\infty}^{T} \frac{1}{kT^2} \left[\int u^S(r) g(r; \rho, T) \, d\underline{r} \right] dT \qquad (16.1\text{-}4)$$

where $U^{\text{res}}(N,V,T)$ is the residual or non-hard-core part of the interaction energy (the hard-core part is taken care of by $Z(N,V,T=\infty)$ term) and $u^S(r)$ is the two-body non-hard core part of the interaction energy. Defining the mean potential per atom Φ over the temperature range (which is not the potential of mean force defined in Section 12.7) to be

$$\Phi = -\frac{kT}{N} \int_{T=\infty}^{T} \frac{U^{\text{res}}(N,V,T)}{kT^2} \, dT = -\frac{NkT}{2V} \int_{T=\infty}^{T} \frac{1}{kT^2} \left[\int u^S(r) g(r; \rho, T) \, d\underline{r} \right] dT$$
$$(16.1\text{-}5)$$

and noting that at infinite temperature where only the hard-core forces are important $Z(N,V,T=\infty) = Z^{\text{HC}}(\rho)$ so that

$$Z(N,V,T) = Z(N,V,T=\infty) \exp\left(-\frac{N\Phi}{kT}\right) = Z^{\text{HC}}(\rho, T=\infty) \exp\left(-\frac{N\Phi}{kT}\right)$$
$$(16.1\text{-}6)$$

300 Chapter 16: The Derivation of Thermodynamic Models

The conceptual basis for obtaining the hard-core contribution to the configuration integral is by considering a single hard-sphere test particle of diameter σ moving throughout a volume V, with the locations of the other hard-sphere molecules fixed. The center of this test particle can be anywhere in the volume V except where it would overlap another hard sphere. Also, note that if the hard sphere is not in direct contact with another hard sphere, its interaction energy is 0; in which case the Boltzmann factor in the configuration integral is unity. However, if the hard sphere does contact another hard sphere, its interaction energy is infinite; and the Boltzmann factor is 0, as is its contribution to the configuration integral. Therefore, the contribution of this molecule to the configuration integral is the volume that is accessible to it, which is referred to as the free volume V_f—that is, the total volume less the volume excluded from the test particle by the presence of other hard spheres. The free volume for hard spheres is only a function of the total volume and density. If the molecules were of zero diameter, $Z^{HC}(\rho) = V^N$, while for molecules of finite size, $Z^{HC}(\rho) = V_f^N$. Therefore, we have

$$Z(N, V, T) = V_f^N \exp\left(-\frac{N\Phi}{kT}\right) \quad \text{and}$$

$$Q(N, V, T) = \frac{q_{\text{int}}^N}{\Lambda^{3N} N!} V_f^N \exp\left(-\frac{N\Phi}{kT}\right) \quad (16.1\text{-}7)$$

This last expression is the Generalized van der Waals partition function.[1,2,3]

The Helmholtz energy is then

$$-\frac{A(N, V, T)}{kT} = \ln Q(N, V, T) = \ln \frac{q_{\text{int}}^N}{\Lambda^{3N} N!} + N \ln V_f - \frac{N\Phi}{kT} \quad (16.1\text{-}8)$$

and the equation of state is obtained as follows:

$$P(N, V, T) = kT \left(\frac{\partial \ln Q}{\partial V}\right)_{N,T} = kT \left.\frac{\partial}{\partial V}\right|_{N,T} \left(\ln \frac{q_{\text{int}}^N}{\Lambda^{3N} N!} + N \ln V_f - \frac{N\Phi}{kT}\right)$$

$$= NkT \left(\frac{\partial \ln V_f}{\partial V}\right)_{N,T} - N \left(\frac{\partial \Phi}{\partial V}\right)_{N,T} = P^{\text{ent}} + P^{\text{eng}} \quad (16.1\text{-}9)$$

where P^{ent} is the entropic contribution to the equation of state resulting from the hard-core part of the interactions, and P^{eng} is referred to as the *energetic contribution*—though it also contains an entropic contribution, since from Eq. 16.1-5 the mean potential Φ depends on the distribution of molecules in the fluid (i.e., the radial distribution function).

The equations above and in Table 16.1-1 are exact; however, to proceed further we need explicit expressions for the free volume V_f and the mean potential Φ. To do this exactly requires solving the statistical mechanics of fluids composed of interacting molecules—something that cannot be done exactly, and which we are trying to avoid by using the Generalized van der Waals partition function. Instead, approximations will be made for the free volume, the average soft interaction energy,

---

[1] J. H. Vera and J. M. Prausnitz, *Chem. Eng. J.* **3**, 1 (1972).
[2] S. I. Sandler, *Fluid Phase Equilibria* **19**, 233 (1985).
[3] K. H. Lee, M. Lombardo, and S. I. Sandler, *Fluid Phase Equilibria* **21**, 177 (1985).

16.2 Application of the Generalized van der Waals Partition Function 301

Table 16.1-1 The Generalized van der Waals Partition Function and Thermodynamic Properties for a Pure Fluid

$$Q(N, V, T) = \frac{q_{\text{int}}^N}{\Lambda^{3N} N!} Z(N, V, T)$$

$$Z(N, V, T) = Z^{\text{HC}}(N, V, T = \infty) \exp\left(-\frac{N\Phi}{kT}\right) = V_f^N \exp\left(-\frac{N\Phi}{kT}\right)$$

$$U^{\text{res}}(N, V, T) = \frac{N^2}{2V} \int_{R^*} u^S(r) g(r; N, V, T) \, d\underline{r} = \frac{2\pi N^2}{V} \int_{R^*} u^S(r) g(r; N, V, T) r^2 \, dr$$

$$\Phi(N, V, T) = -\frac{kT}{N} \int_{T=\infty}^{T} \frac{U^{\text{res}}(N, V, T)}{kT^2} \, dT$$

$$= -\frac{NkT}{2V} \int_{T=\infty}^{T} \frac{1}{kT^2} \left[\int u^S(r) g(r; \rho, T) d\underline{r}\right] dT$$

$$A(N, V, T) = -kT \ln Q(N, V, T) = -kT \ln \frac{q_{\text{int}}^N}{\Lambda^{3N} N!} - NkT \ln V_f + N\Phi$$

$$P(N, V, T) = -\left(\frac{\partial A(N, V, T)}{\partial V}\right)_{N,T} = kT \left(\frac{\partial \ln Z(N, V, T)}{\partial V}\right)_{N,T}$$

$$= NkT \left(\frac{\partial \ln V_f}{\partial V}\right)_{N,T} - N\left(\frac{\partial \Phi}{\partial V}\right)_{N,T} = P^{\text{ent}} + P^{\text{eng}}$$

and the coordination number, and then use these approximations in this partition function to obtain pure fluid equations of state (Section 16.2), mixture equations of state and their mixing rules (Section 16.3), and activity coefficient models (Section 16.4).

For later reference, note that the coordination number for an atom (or molecule) in the Generalized van der Waals model, N_c, which is the number of molecules in the range of the soft part of interaction of a central molecule, is

$$N_c(\rho, T) = \frac{N}{V} \int_{R^*} g(r; \rho, T) \, d\underline{r} = \rho \int_{R^*} g(r; \rho, T) \, d\underline{r} \quad (16.1\text{-}10)$$

where R^* is the range of the potential.

16.2 APPLICATION OF THE GENERALIZED VAN DER WAALS PARTITION FUNCTION TO PURE FLUIDS

Equations of state for pure fluids can be developed by making approximations for the free volume V_f of the hard-core fluid and the mean potential Φ. We start with the free volume.

The Free Volume

The simplest model for the excluded volume, as used by van der Waals, is that around each molecule there is a volume of $4\pi\sigma^3/3$ from which other molecules are excluded

302 Chapter 16: The Derivation of Thermodynamic Models

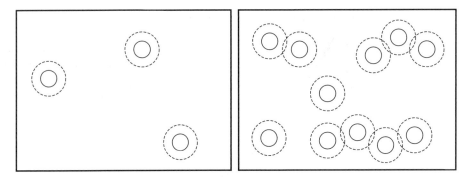

Figure 16.2-1 The free volume and excluded volume (within dotted region) for hard spheres at (a) low density (left) and (b) moderate density (right). Note the overlapping of excluded volumes at moderate density. Reprinted from Fluid Phase Equilibria, Vol. 34, "The Generalized van der Waals Partition Function. I. Basic Theory" by S. I. Sandler, pp 233–257, Copyright 1987, with permission from Elsevier. Similar to figure in Vera and Prausnitz[1].

(see Fig. 16.2-1a). However, we must attribute half of this excluded volume V_{exc} to each molecule of the interacting pair to avoid overcounting, so that

$$V_{\text{exc,vdW}} = \frac{2\pi(N-1)\sigma^3}{3} \approx \frac{2\pi N\sigma^3}{3} = N\beta \quad \text{and}$$

$$V_{\text{f,vdW}} = V - V_{\text{exc,vdW}} \approx V - N\beta = V(1-\rho\beta) \quad \text{so that}$$

$$P^{\text{ent}} = NkT\left(\frac{\partial \ln V_f}{\partial V}\right)_{N,T} = \frac{NkT}{V - N\beta} \tag{16.2-1}$$

where $\beta = 2\pi\sigma^3/3$. This simple expression for the free volume is accurate at low density, but not very accurate at moderate and high densities because it overestimates the excluded volume (underestimates the free volume). The reason for this can be seen in Fig. 16.2-1 in which the excluded volume for the test particle around each fixed molecule is shown. At low densities, where the molecules are widely separated, the excluded volume is just—as van der Waals assumed—the sum of the excluded volumes of each molecule as shown in Fig. 16.2-1a. However, at higher densities (Fig. 16.2-1b), the excluded volumes around neighboring molecules overlap, so the actual excluded volume is less than the number of molecules times the excluded volume around each isolated molecule. In fact, the simple van der Waals expression for the free volume of the hard-sphere fluid is only correct at very low density (it is also exact for a one-dimensional fluid; can you reason why that is so?).

There are more accurate expressions for the free volume of hard spheres (and some other hard geometric shapes). For example, the more accurate free-volume expression for the hard-sphere fluid corresponding to the Carnahan-Starling equation of state for hard spheres (Eq. 12.5-2) based on the Percus-Yevick solution of the Ornstein-Zernike equation is

$$V_f = V\exp\left[\frac{\eta(3\eta-4)}{(1-\eta)^3}\right] \quad \text{and} \quad P^{\text{ent}} = \frac{NkT}{V}\left[\frac{1+\eta+\eta^2-\eta^3}{(1-\eta)^3}\right] \tag{16.2-2}$$

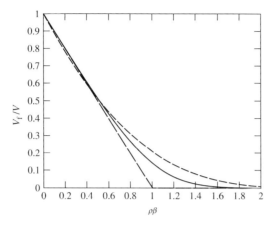

Figure 16.2-2 The free volume as a function of reduced density; solid line is the results of simulation and the Carnahan-Starling expression (which are indistinguishable on the scale of the figure), large dashed line results from the van der Waals expression, and the small dashed line is the Kim-Lin-Chao result. Reprinted with permission from S. I. Sandler, Chem. Eng. Educ., Winter 1990, page 12 et. seq.

where $\eta = \pi\rho\sigma^3/6$. A reasonably simple, and only somewhat less accurate, free-volume expression is that due to Kim, Lin, and Chao:[4]

$$V_f = V(1 - 0.42\rho\beta)^{17/6} = V(1 - 1.68\eta)^{17/6} \quad \text{and} \quad P^{\text{ent}} = \frac{NkT}{V}\left(\frac{1 + 3.08\eta}{1 - 1.68\eta}\right)$$

(16.2-3)

Figure 16.2-2 shows the ratio of the free volume to the total volume as a function of reduced density $\rho\beta$ for these models[5] together with computer simulation results. The figure also shows the value of computer simulation in establishing the accuracy of various assumptions that have been proposed for the free volume of hard-spheres.

The Mean Potential

To proceed in the simplest manner, first consider the interaction potential between the molecules to be the square-well potential to introduce the ideas involved, and then these will be generalized to more realistic interaction potentials. The square-well potential is

$$u(r) = \begin{cases} \infty & r \leq \sigma \\ -\varepsilon & \sigma < r < R_{\text{sw}}\sigma \\ 0 & r \geq R_{\text{sw}}\sigma \end{cases} \quad (8.1\text{-}5)$$

which has the advantage of a well-defined hard core and attractive regions. In this case u^S is exactly $-\varepsilon$, and R^* is the region from σ to $R_{\text{sw}}\sigma$. To compute the mean potential in the Generalized van der Waals partition function, note that the configurational energy for this fluid is

$$U^{\text{res}}(N,V,T) = \frac{N^2}{2V}\int u^S(r)g(r;\rho,T)\,d\underline{r} = -\frac{N^2\varepsilon}{2V}\int_{R^*} g(r;\rho,T)\,d\underline{r} = -\frac{N\varepsilon}{2}N_c(\rho,T)$$

(16.2-4)

---

[4]H. Kim, H. M. Lin, and K. C. Chao, *IEC Fund*. **25**, 75 (1986).

[5]S. I. Sandler, *Chem. Eng. Educ.*, **Winter 1990**, p. 12 et seq.

304 Chapter 16: The Derivation of Thermodynamic Models

and from Eq. 16.1-5

$$\Phi = -\frac{NkT}{2V} \int_{T=\infty}^{T} \frac{1}{kT^2} \left[\int u^S(r) g(r; \rho, T) \, d\underline{r} \right] dT$$

$$= \frac{kT\varepsilon}{2} \int_{T=\infty}^{T} \frac{1}{kT^2} \left[\int \frac{N}{V} g(r; \rho, T) \, d\underline{r} \right] dT$$

$$= \frac{kT\varepsilon}{2} \int_{T=\infty}^{T} \frac{N_c(\rho, T)}{kT^2} \, dT \quad (16.2\text{-}5)$$

where the coordination number N_c is the number of molecules in the well of a central molecule.

Therefore, to continue, we need to make an assumption about the coordination number and its dependence on temperature and density. If we were to make the assumption that the radial distribution function was a constant equal to 1 over the range of the interaction, then

$$N_c = 4\pi\rho\sigma^3(R_{sw}^3 - 1)/3 = C\rho \quad \text{where} \quad C = 4\pi\sigma^3(R_{sw}^3 - 1)/3 \quad (16.2\text{-}6)$$

From which it follows that

$$U^{res}(N, V, T) = -\frac{CN\varepsilon}{2}\rho \quad \text{and}$$

$$\Phi = -\frac{kT}{2} \int_{T=\infty}^{T} \frac{(-C\varepsilon\rho)}{kT^2} dT = \frac{TC\varepsilon\rho}{2} \int_{T=\infty}^{T} \frac{dT}{T^2} dT = -\frac{C\rho\varepsilon}{2} = -\frac{CN\varepsilon}{2V} \quad (16.2\text{-}7)$$

Using this result together with the van der Waals free volume term gives us

$$Q(N, V, T) = \frac{q_{int}^N}{\Lambda^{3N} N!} V_f^N \exp\left(-\frac{N\Phi}{kT}\right)$$

$$= \frac{q_{int}^N}{\Lambda^{3N} N!} (V - N\beta)^N \exp\left(\frac{CN^2\varepsilon}{2kTV}\right) \quad (16.2\text{-}8)$$

and

$$P(N, V, T) = kT \left(\frac{\partial \ln Q}{\partial V}\right)_{N,T} = kT \left(\frac{\partial}{\partial V}\right)_{N,T} \left[N \ln(V - N\beta) + \frac{CN^2\varepsilon}{2kTV}\right]$$

$$= NkT \left[\frac{1}{V - N\beta} - \frac{CN\varepsilon}{2kTV^2}\right] = \frac{NkT}{V - N\beta} - \frac{CN^2\varepsilon}{2V^2} \quad (16.2\text{-}9)$$

Now using that $n = N/N_{Av}$ where N_{Av} is Avogadro's number, n is the number of moles, $\underline{V} = V/n$ is the molar volume, and $R = N_{Av}k$ is the gas constant, one obtains

16.2 Application of the Generalized van der Waals Partition Function

$$P(N, V, T) = \frac{RT}{\underline{V} - b} - \frac{a}{\underline{V}^2} \quad (16.2\text{-}10)$$

which, with $b = N_{Av}\beta$ and $a = CN_{Av}^2\varepsilon/2$, is the familiar van der Waals equation with temperature-independent parameters.

An alternative assumption is to start with the low density radial distribution of Eq. 11.3-5:

$$g(r; \rho = 0, T) = e^{-u(r)/kT} \quad (11.3\text{-}5)$$

which, for the square-well fluid at zero density, is

$$g(r) = \begin{cases} 0 & r < \sigma \\ e^{\varepsilon/kT} & \sigma \leq r \leq R_{sw}\sigma \\ 1 & r > R_{sw}\sigma \end{cases} \quad (16.2\text{-}11)$$

and to use this simple model at all densities. This leads to Problem 16.1

$$N_c = 4\pi\rho \int_\sigma^{R_{sw}\sigma} e^{\varepsilon/kT} r^2 \, dr = \frac{4\pi\rho\sigma^3}{3}(R_{sw}^3 - 1)e^{\varepsilon/kT} = C\rho e^{\varepsilon/kT}$$

$$U^{res} = -\frac{N\varepsilon}{2}N_c = -\frac{N\varepsilon}{2}C\rho e^{\varepsilon/kT}; \quad \text{and} \quad \Phi = -\frac{kTC\rho(e^{\varepsilon/kT} - 1)}{2} \quad (16.2\text{-}12)$$

The resulting equation of state, Eq. 16.2-10, is again that of van der Waals, but now, $a(T) = CN_{Av}^2 kT(e^{\varepsilon/kT} - 1)/2$ and $b = N_{Av}\beta$. That is, we again obtain the van der Waals equation, but now with a temperature-dependent a parameter.

More generally, at any density, for the square-well fluid

$$U^{res}(N, V, T) = \frac{N^2}{2V} \int u(r) g(r; \rho, T) \, d\underline{r}$$

$$= -\frac{N^2\varepsilon}{2V} \int_{R^*} g(r; \rho, T) \, d\underline{r} = -\frac{N\varepsilon}{2} N_c(\rho, T) \quad (16.2\text{-}13)$$

Then, the generalized van der Waals partition function for the square-well fluid, Q_{GvdW}, is

$$Q_{GvdW}(N, V, T) = \frac{q_{int}^N}{\Lambda^{3N} N!} V_f^N \exp\left(-\frac{N\varepsilon}{2} \int_{T=\infty}^T \frac{N_c(\rho, T)}{kT^2} \, dT\right) \quad (16.2\text{-}14)$$

and, more generally, for any potential (not only the square-well)

$$Q_{GvdW}(N, V, T) = \frac{q_{int}^N}{\Lambda^{3N} N!} V_f^N \exp\left(-\frac{N\Phi(\rho, T)}{kT}\right) \quad (16.2\text{-}15)$$

Note that Eq. 16.2-15 is to be compared to the partition function if we had made the van der Waals assumption for the free volume and coordination number:

$$Q_{vdW}(N, V, T) = \frac{q_{int}^N}{\Lambda^{3N} N!} \left(V - \frac{2\pi N\sigma^3}{3}\right)^N \exp\left(\frac{N\varepsilon N_c}{2kT}\right) \quad (16.2\text{-}16)$$

Chapter 16: The Derivation of Thermodynamic Models

As shown above, in order to use the Generalized van der Waals partition function to develop new thermodynamic models, assumptions had to be made about the density (and also possibly temperature) dependence of the free volume, and the density and temperature dependence of the coordination number or the mean potential Φ. In this way, a direct relationship is obtained between the parameters in the thermodynamic models and molecular parameters. An alternative is to use the Generalized van der Waals model in reverse and work backward from commonly used macroscopic thermodynamic models to understand the assumptions on which they are based. This latter procedure has been used to understand the assumptions that underlie some of the commonly used engineering equations of state that are shown in Table 16.2-1. Notice that some use the simple van der Waals free-volume term, while others use the free-volume term that comes from the Carnahan-Starling expression derived from the Percus-Yevick equation, which is a better representation of hard-core free volume.

Table 16.2-1 Examples of Square-Well Based Equations of State

van der Waals

$$V_f = \underline{V} - b; \; N_c = \frac{4\pi}{3}\sigma^3(R_{sw}^3 - 1)\rho = C\rho \to \Phi = -\frac{2\pi}{3}\sigma^3\varepsilon(R_{sw}^3 - 1)\rho = -\frac{\varepsilon C\rho}{s};$$

which leads to $P = \dfrac{RT}{\underline{V} - b} - \dfrac{a}{\underline{V}^2}$ with $a = N_{Av}^2 \dfrac{C\varepsilon}{2} = bN_{Av}\varepsilon(R_{sw}^3 - 1)$

Temperature-dependent van der Waals

$$V_f = \underline{V} - b; \; N_c = \frac{4\pi}{3}\sigma^3(R_{sw}^3 - 1)e^{\varepsilon/kT} \to \Phi = -\frac{4\pi\sigma^3}{3}kT(R_{sw}^3 - 1)\rho(e^{\varepsilon/kT} - 1)$$

which leads to $P = \dfrac{RT}{\underline{V} - b} - \dfrac{a(T)}{\underline{V}^2}$ with $a(T) = \dfrac{C}{2}kT(e^{\varepsilon/kT} - 1) = bRT(e^{\varepsilon/kT} - 1)$

Redich-Kwong

$$V_f = \underline{V} - b; \; N_c = C\sqrt{\frac{T_o}{T}}\ln(1 + \beta\rho) \to \Phi = -\frac{C\varepsilon}{3}\sqrt{\frac{T_o}{T}}\ln\left(1 + \frac{b}{\underline{V}}\right)$$

which leads to $P = \dfrac{RT}{\underline{V} - b} - \dfrac{a/\sqrt{T}}{\underline{V}(\underline{V} + b)}$ with $a = \dfrac{N_{Av}C\varepsilon}{3}b\sqrt{T_o}$

Soave-Redlich-Kwong

$V_f = \underline{V} - b; \; N_c = C_2(T)\ln(1 + \beta\rho)$ which leads to $P = \dfrac{RT}{\underline{V} - b} - \dfrac{a(T)}{\underline{V}(\underline{V} + b)}$ with a given above

Peng-Robinson

$$V_f = \underline{V} - b; \; N_c = C_3(T)\rho\frac{\mathrm{atan}\left(\dfrac{N}{\sqrt{2Nb\rho - b^2\rho^2}}\right)}{\sqrt{2Nb\rho - b^2\rho^2}} = C_3(T)\frac{\mathrm{atan}\left(\dfrac{V}{\sqrt{2NbV - b^2}}\right)}{\sqrt{2NbV - b^2}}$$

which leads to $P = \dfrac{RT}{\underline{V} - b} - \dfrac{a(T)}{\underline{V}(\underline{V} + b) + b(\underline{V} - b)}$

with $a(T)$ (or $C_3(T)$) empirically fit to vapor pressure data

Note: $C = \dfrac{4\pi}{3}\sigma^3(R_{sw}^3 - 1)$ and $b = \dfrac{2\pi N_{Av}}{3}\sigma^3$

16.2 Application of the Generalized van der Waals Partition Function

One question that the reader may ask is, why don't all commonly used engineering equations of state use the more accurate Carnahan-Starling free-volume expression rather than the less accurate van der Waals expression? There are several reasons for this. First, if the excluded volume parameter b (or β) is treated as an adjustable parameter, the calculated free volume can be made to be closer to that obtained from computer simulation, at least over some ranges of density. Second, and perhaps more importantly, equations of state commonly used in engineering are applied to non-spherical molecules, and analogues of the Carnahan-Starling free-volume expression are not available for every molecular shape that occurs in chemical processing. Furthermore, even if such expressions were available, their use would require that the forms of the equations of state for molecules of different geometric shapes be different. Since engineers deal almost exclusively with mixtures, not pure components, it then would not be clear how to formulate an equation of state for a mixture. (It should be remembered that the way mixtures are treated now is to use the same equation of state for the mixture and all of the pure components, and obtain the parameters for the mixture from a set of mixing and combining rules.) A final reason that the simple van der Waals free-volume term is used is that calculations involving simple cubic equations of state are computationally very quick. At first glance, this may not seem important given the speed of computers. However, in the analysis and design of a chemical process using computer simulation software (for example Aspen),[6] or in petroleum reservoir simulation, thermodynamic calculations frequently take up to 90 percent of the computer time, and may be used hundreds of thousands or even millions of times in the iterations while the simulation is converging. Therefore, it is advantageous to have a simple, reasonably accurate equation of state rather than one that is more complicated but only slightly more accurate.

Molecular-level computer simulation, as described in Chapter 13, can be used to test the coordination number models discussed above. A comparison of computer simulation results with some of the models are shown in Fig. 16.2-3 for the square-well fluid for $R_{sw} = 1.5$ and for three values of the reduced well depth (or, equivalently, at different reduced temperatures for the same well depth). Also shown as the solid line is the result of the following relatively simple model[3]

$$N_c = \frac{N_m V_o e^{\varepsilon/2kT}}{V + V_o(e^{\varepsilon/2kT} - 1)} \qquad (16.2\text{-}17)$$

where $V_o = N\sigma^3/\sqrt{2}$ is the close-packed volume of hard-spheres and N_m is the coordination number at close-packing (18 for $R_{sw} = 1.5$). This equation was derived from a simple argument that the likelihood of the occurrence of two neighboring molecules is proportional to $e^{\varepsilon/2kT}$. Another test of this model is shown in Fig. 16.2-4.

Based on the accurate expression for the free volume (Eq. 16.2-3) and for the coordination number model of Eq. 16.1-17, the following equation of state is obtained for the square-well fluid from the generalized van der Waals partition function (Problem 16.13)

$$\frac{P\underline{V}}{RT} = \frac{1 + \eta + \eta^2 + \eta^3}{(1-\eta)^3} - \frac{N_m V_o(e^{\varepsilon/2kT} - 1)}{V + V_o(e^{\varepsilon/2kT} - 1)} \qquad (16.2\text{-}18)$$

---

[6]Aspen Plus® process simulator, Aspen Technology, Inc, Burlington, MA.

308 Chapter 16: The Derivation of Thermodynamic Models

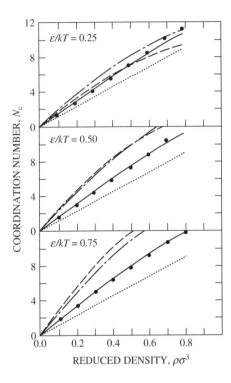

Figure 16.2-3 Coordination number for the square-well fluid ($R_{sw} = 1.5$) for various values of the reduced inverse temperature ε/kT. The points are the result of Monte Carlo simulation,[1] the dotted line is the van der Waals model (Eq. 16.2-7a), the dashed line is the Redlich-Kwong result, the dash-dot line is the Peng-Robinson result, and the solid line results from Eq. 16.1-17. Reprinted with permission from S. I. Sandler, Chem. Eng. Educ., Winter 1990, page 12 et. seq.

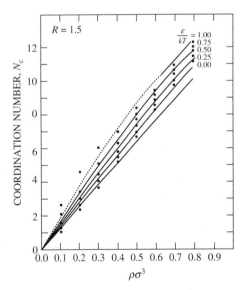

Figure 16.2-4 Coordination number for the square-well fluid ($R_{sw} = 1.5$) for various values of the reduced temperature ε/kT. The points are the result of Monte Carlo simulation,[1] and the solid line results from Eq. 16.1-17. Dotted portion of the $\varepsilon/kT = 1$ is in the two-phase region. Reprinted with permission from S. I. Sandler, Chem. Eng. Educ., Winter 1990, page 12 et. seq.

While this equation, for the reasons described earlier, is not used in engineering applications, it has been shown to be reasonably accurate for the description of the vapor-liquid equilibrium of simple, spherical real fluids by adjusting the values of its parameters, as shown in Fig. 16.2-5.

For interaction potential models other than the square-well potential, the general procedure is to keep the same structure as above by using the mean-value theorem of calculus to obtain

16.2 Application of the Generalized van der Waals Partition Function

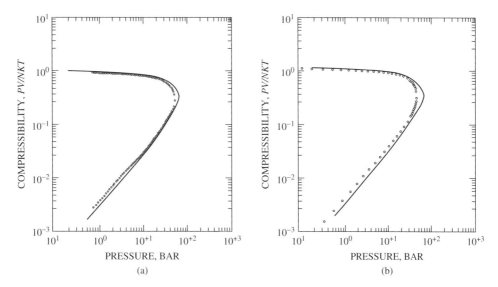

Figure 16.2-5 The compressibility $Z = PV/RT$ along the vapor-liquid equilibrium envelope for (a) argon and (b) methane; the points are the experimental data and the line results from Eq. 16.1-18. Reprinted with permission from S. I. Sandler, Chem. Eng. Educ., Winter 1990, page 12 et. seq.

$$U^{\text{res}}(N, V, T) = \frac{N^2}{2V} \int_{R^*} u^S(r) g(r; N, V, T) \, d\mathbf{r} = \frac{N^2 \langle u^S \rangle}{2V} \int_{R^*} g(r; N, V, T) \, d\mathbf{r}$$

$$= \frac{N \cdot N_c(N, V, T) \langle u^S \rangle}{2} \tag{16.2-19}$$

where $u^S(r)$ is the soft-core part of the interaction potential, $\langle u^S \rangle$ is an appropriately chosen average value of this part of the potential, N_c is an estimate of the coordination number defined by Eq. (16.1-10), and R^* is the range of the soft interaction. So a number of assumptions need to be made. One is to choose an effective hard-core diameter, for which in principle the equations of Section 14.4 can be used. Next is to pick an average value for the soft part of the interaction potential, $\langle u^S \rangle$. Finally, one also has to make an assumption about the temperature and density dependence about the coordination number. Alternatively, one can dispense with the idea of the coordination number, and make estimates directly for the temperature and density dependence of the integral

$$\int_{R^*} u^S(r) g(r; N, V, T) \, d\mathbf{r} = 4\pi \int_{R^*} u^S(r) g(r; N, V, T) r^2 \, dr$$

A less commonly used equation of state (in engineering) is that of Alder et al.,[7] that uses the following expression for the free volume:

$$V_f = V \exp\left[\frac{\eta(3\eta - 4)}{(1 - \eta)^3}\right]$$

---

[7] B. J. Alder, D. A. Young and M. A. Mark, *J. Chem. Phys.* **56**, 3013 (1972).

and, based on the results of computer simulation, the following coordination number model:

$$N_c = 2 \sum_m \sum_m m A_{mn} \left(\frac{\varepsilon}{kT}\right)^{m-1} \left(\frac{\rho\sigma^3}{\sqrt{2}}\right)^n$$

which leads to

$$P = \frac{RT}{\underline{V}}\left[\frac{1+\eta+\eta^2+\eta^3}{(1-\eta)^3}\right] - \sum_m \sum_n A_{mn}\left(\frac{\varepsilon}{kT}\right)^m (\rho\sigma^3)^n \qquad (16.2\text{-}20)$$

16.3 EQUATION OF STATE FOR MIXTURES FROM THE GENERALIZED VAN DER WAALS PARTITION FUNCTION

The straightforward extension of the Generalized van der Waals partition function to mixtures is

$$Q_{\text{Gvdw}}(N_1, N_2, \ldots, V, T) = \left[\prod_i \left(\frac{q_{\text{int},i}^{N_i}}{\Lambda_i^{3N_i} N_i!}\right)\right][V_f(N_1, N_2, \ldots, V, T)]^N$$

$$\times \exp\left(-\frac{N\Phi_{\text{mix}}(N_1, N_2, \ldots, V, T)}{kT}\right) \qquad (16.3\text{-}1)$$

with

$$U_{\text{mix}}^{\text{res}}(N_1, N_2, \ldots, V, T) = \sum_i \sum_j U_{ij}^{\text{res}}(N_1, N_2, \ldots, V, T) \qquad (16.3\text{-}2)$$

and

$$U_{ij}^{\text{res}}(N_1, N_2, \ldots, V, T) = \frac{N_i N_j}{2V} \int_{R_{ij}^*} u_{ij}^S(r) g_{ij}(r; N_1, N_2, \ldots, V, T)\, d\underline{r}$$

$$= x_i x_j \frac{N^2}{2V} \int_{R_{ij}^*} u_{ij}^S(r) g_{ij}(r; N_1, N_2, \ldots, V, T)\, d\underline{r}$$

$$(16.3\text{-}3)$$

where R_{ij}^* is the range of the $i-j$ interaction (the well region for the square-well potential), and

$$\Phi_{\text{mix}}(N_1, N_2, \ldots, V, T) = -\frac{kT}{N} \int_{T=\infty}^{T} \frac{U_{\text{mix}}^{\text{res}}(N_1, N_2, \ldots, V, T)}{kT^2}\, dT \qquad (16.3\text{-}4)$$

In these equations, x_i is the mole fractions of species i, $x_i = N_i/\sum_j N_j = N_i/N$, $u_{ij}^S(r) = u_{ji}^S(r)$, and $g_{ij}(r) = g_{ji}(r)$ is so defined that $N_i g_{ij}(r)\, d\underline{r}$ is the number of species i molecules in a volume element $d\underline{r}$ at a distance r from a central j molecule.

There are two different paths for using the Generalized van der Waals partition function for mixtures. The first is where the interest is in equations of state and their mixing rules, in which the density, temperature, and composition dependence of the coordination number is important. This is discussed in this section. The second way of proceeding is the use of the Generalized van der Waals partition function for

16.3 Equation of State for Mixtures

liquid mixtures to obtain activity coefficient models. The species-species coordination numbers are also of central importance here to, but since liquids (especially below their critical point) are not very compressible, a simplifying assumption is made that the density dependence of the coordination number can be neglected. This will be considered in Section 16.4.

To develop equations of state for mixtures—and the mixing rules to be used with such equations—the starting point is

$$P(N, V, T) = kT \left(\frac{\partial \ln Q}{\partial V}\right)_{N,T}$$

$$= kT \left.\frac{\partial}{\partial V}\right|_{N_i, T} \ln \left[\prod_i \left(\frac{q_{\text{int},i}^{N_i}}{\Lambda_i^{3N_i} N_i!}\right)\right] [V_f(N_1, N_2, \ldots, V, T)]^N$$

$$\times \exp\left(-\frac{N\Phi_{\text{mix}}(N_1, N_2, \ldots, V, T)}{kT}\right)$$

$$= kT \left.\frac{\partial}{\partial V}\right|_{N_i, T} \left[\ln \left[\prod_i \left(\frac{q_{\text{int},i}^{N_i}}{\Lambda_i^{3N_i} N_i!}\right)\right] + N \ln V_f(N_1, N_2, \ldots, V, T) \right.$$

$$\left. - \frac{N\Phi_{\text{mix}}(N_1, N_2, \ldots, V, T)}{kT}\right]$$

$$= NkT \left(\frac{\partial \ln V_f(N_1, N_2, \ldots, V, T)}{\partial V}\right)_{N,T} - N \left(\frac{\partial \Phi_{\text{mix}}(N_1, N_2, \ldots, V, T)}{\partial V}\right)_{N,T}$$

(16.3-5)

Here, the subscripts on the derivative indicate that the volume derivative is at fixed numbers of molecules of all species and temperature. It is useful to separate the equation of state into an entropic part P^{ent} that arises from the free-volume part of the Generalized van der Waals partition function, and an energetic part P^{eng} coming from the mean potential—that is, as before, using $P(N, V, T) = P^{\text{ent}}(N, V, T) + P^{\text{eng}}(N, V, T)$ with

$$P^{\text{ent}}(N, V, T) = NkT \left(\frac{\partial \ln V_f(N_1, N_2, \ldots, V, T)}{\partial V}\right)_{N,T} \quad \text{and} \quad (16.3\text{-}5a)$$

$$P^{\text{eng}}(N, V, T) = -N \left(\frac{\partial \Phi_{\text{mix}}(N_1, N_2, \ldots, V, T)}{\partial V}\right)_{N,T} \quad (16.3\text{-}5b)$$

To begin, we again consider the simplest case of the square-well potential, so that

$$U_{ij}^{\text{res}}(N_1, N_2, \ldots, V, T) = -\varepsilon_{ij} \frac{N_j N_i}{2V} \int_{R_{ij}^*} g_{ij}(r; N_1, N_2, \ldots, V, T) \, d\mathbf{r}$$

$$= -\varepsilon_{ij} \frac{N_j N_{ij}}{2} \quad (16.3\text{-}6a)$$

where $\varepsilon_{ij} = \varepsilon_{ji}$ is the well depth of an i-j interaction and N_{ij} is the average number of i molecules around a central j molecule—that is, the coordination number of i molecules around a j molecule, defined by

$$N_{ij}(N_1, N_2, \ldots, V, T) = \frac{N_i}{V} \int_{R_{ij}^*} g_{ij}(r; N_1, N_2, \ldots, V, T)\, d\underline{r} \qquad (16.3\text{-}7)$$

To proceed, one has to specify either each of the species-species coordination numbers (or, directly, the configurational energies U_{ij}^{res}) as a function of temperature and the number densities of each species in the mixture. It should be noted that, as before, for interaction potential models that are different than the square-well potential, we can keep the same structure as above using

$$\begin{aligned}
U_{ij}^{\text{res}}(N_1, N_2, \ldots, V, T) &= \frac{N_j N_i}{2V} \int_{R_{ij}^*} u_{ij}^{\text{S}}(r) g_{ij}(r; N_1, N_2, \ldots, V, T)\, d\underline{r} \\
&= \frac{N_j N_i \langle u_{ij}^{\text{S}} \rangle}{2V} \int_{R_{ij}^*} g_{ij}(r; N_1, N_2, \ldots, V, T)\, d\underline{r} \\
&= \frac{N_j N_{ij} \langle u_{ij}^{\text{S}} \rangle}{2} \qquad (16.3\text{-}6b)
\end{aligned}$$

where $u_{ij}^{\text{S}}(r)$ is the soft-core part of the interaction potential, and $\langle u_{ij}^{\text{S}} \rangle$ is an appropriately chosen average value for the i-j interaction. For the square-well potential, $\langle u_{ij}^{\text{S}} \rangle = -\varepsilon_{ij}$. In what follows, we will use ε_{ij}, but the results are easily generalized by replacing it with $-\langle u_{ij}^{\text{S}} \rangle$.

Table 16.3-1 summarizes the Generalized van der Waals partition function and its relationship to the thermodynamic properties of a mixture. In what follows, we convert this general formalism into equation of state mixing rules by considering specific choices for the free volume and mean potential.

Free Volume in Mixtures

For mixtures of hard spheres (or the hard-core part of the square-well fluid), the simple van der Waals approximation for the free volume is

$$V_{\text{f}}(N_1, N_2, \ldots, V, T) = V(N_1, N_2, \ldots, V, T) - b \sum_i N_i \quad \text{with} \quad b = \sum_i \sum_j x_i x_j b_{ij}.$$

For hard spheres, there is an exact combining rule

$$b_{ij} = \frac{1}{2}(b_{ii} + b_{jj}) \quad \text{so that} \quad b = \sum_i \sum_j x_i x_j b_{ij} = \sum_i x_i b_{ii}.$$

In this case,

$$\begin{aligned}
P^{\text{ent}}(N, V, T) &= NkT \left(\frac{\partial \ln V_{\text{f}}(N_1, N_2, \ldots, V, T)}{\partial V} \right)_{N_i, T} \\
&= NkT \left(\frac{\partial \ln(V - Nb)}{\partial V} \right)_{N_i, T} = \frac{NkT}{V - Nb} \qquad (16.3\text{-}8)
\end{aligned}$$

Of course, expressions other than the van der Waals equation could be used for the free-volume term. For example, the Carnahan-Starling expression discussed earlier and extended to mixtures could be used, resulting in greater complexity.

16.3 Equation of State for Mixtures

Table 16.3-1 The Generalized van der Waals Partition Function and Thermodynamic Properties for a Mixture

$$Q_{\text{GvdW}}(N_1, N_2, \ldots, V, T) = \left[\prod_i \left(\frac{q_{\text{int},i}^{N_i}}{\Lambda_i^{3N_i} N_i!}\right)\right] Z_{\text{mix}}(N_1, N_2, \ldots, V, T)$$

$$Z_{\text{mix}}(N_1, N_2, \ldots, V, T) = Z_{\text{mix}}^{\text{HC}}(N_1, N_2, \ldots, V, T) \exp\left(-\frac{N\Phi_{\text{mix}}}{kT}\right)$$

$$= V_{\text{f,mix}}^N \exp\left(-\frac{N\Phi_{\text{mix}}}{kT}\right)$$

$$\Phi_{\text{mix}}(N_1, N_2, \ldots, V, T) = -\frac{kT}{N} \int_{T=\infty}^{T} \frac{U_{\text{mix}}^{\text{res}}(N_1, N_2, \ldots, V, T)}{kT^2} dT$$

$$U_{\text{mix}}^{\text{res}}(N_1, N_2, \ldots, V, T) = \sum_i \sum_j U_{ij}^{\text{res}}(N_1, N_2, \ldots, V, T)$$

$$= \sum_i \sum_j \frac{N_i N_j}{2V} \int_{R_{ij}^*} u_{ij}^S(r) g_{ij}(r; N_1, N_2, \ldots, V, T) d\underline{r}$$

$$= \frac{N^2}{2V} \sum_i \sum_j x_i x_j \int_{R_{ij}^*} u_{ij}^S(r) g_{ij}(r; N_1, N_2, \ldots, V, T) d\underline{r}$$

$$A_{\text{mix}}(N_1, N_2, \ldots, V, T) = -kT \ln Q_{\text{mix}}(N_1, N_2, \ldots, V, T)$$

$$= -kT \ln \left[\prod_i \frac{q_{\text{int},i}^{N,i}}{\Lambda_i^{3N_i} N_i!}\right] \frac{Z(N_1, N_2, \ldots, V, T)}{V^N}$$

$$= -kT \ln \left[\prod_i \frac{q_{\text{int},i}^{N,i}}{\Lambda_i^{3N_i} N_i!}\right] - NkT \ln V_{\text{f,mix}}(N_1, N_2, \ldots, V, T)$$
$$+ N\Phi_{\text{mix}}(N_1, N_2, \ldots, V, T)$$

$$P(N_1, N_2, \ldots, V, T) = -\left(\frac{\partial A_{\text{mix}}}{\partial V}\right)_{N_1, N_2, \ldots T}$$

$$= kT \left(\frac{\partial \ln Q(N_1, N_2, \ldots, V, T)}{\partial V}\right)_{N_1, N_2, \ldots T}$$

$$= NkT \left(\frac{\partial \ln V_{\text{f,mix}}(N_1, N_2, \ldots, V, T)}{\partial V}\right)_{N_1, N_2, \ldots T}$$

$$- N \left(\frac{\partial \Phi_{\text{mix}}(N_1, N_2, \ldots, V, T)}{\partial V}\right)_{N_1, N_2, \ldots T}$$

Species-Species Coordination Number

The low-density limit of species-species coordination number for the square-well potential is

$$\lim_{\rho \to 0} N_{ij}(N_1, N_2, \ldots, V, T) = \frac{N_i}{V} \frac{4\pi}{3} \sigma_{ij}^3 (R_{ij}^3 - 1) e^{\varepsilon_{ij}/kT}$$

$$= \frac{N_i}{V} C_{ij} e^{\varepsilon_{ij}/kT} \quad \text{with} \quad C_{ij} = \frac{4\pi}{3} \sigma_{ij}^3 (R_{ij}^3 - 1) \quad \textbf{(16.3-9)}$$

314 Chapter 16: The Derivation of Thermodynamic Models

Of somewhat more interest in formulating models is to look at the ratio of i molecules to j molecules around a central j molecule—that is, the ratio of local compositions that can be different from the ratio of bulk compositions that would be obtained in a completely random mixture. For the square-well fluid, this local composition ratio at low density is

$$\lim_{\rho \to 0} \frac{N_{ij}(N_1, N_2, \ldots, V, T)}{N_{jj}(N_1, N_2, \ldots, V, T)} = \frac{\frac{N_i}{V}\frac{4\pi}{3}\sigma_{ij}^3(R_{ij}^3 - 1)e^{\varepsilon_{ij}/kT}}{\frac{N_j}{V}\frac{4\pi}{3}\sigma_{jj}^3(R_{jj}^3 - 1)e^{\varepsilon_{jj}/kT}}$$

$$= \frac{N_i\sigma_{ij}^3(R_{ij}^3 - 1)e^{(\varepsilon_{ij}-\varepsilon_{jj})/kT}}{N_j\sigma_{jj}^3(R_{jj}^3 - 1)}$$

$$= \frac{x_i\sigma_{ij}^3(R_{ij}^3 - 1)e^{(\varepsilon_{ij}-\varepsilon_{jj})/kT}}{x_j\sigma_{jj}^3(R_{jj}^3 - 1)} \quad (16.3\text{-}10)$$

There are several interesting observations from this result. First, the molecules in a low-density mixture are not distributed randomly—that is, the local composition ratio is not the same as the ratio of the bulk mole fractions

$$\lim_{\rho \to 0} \frac{N_{ij}(N_1, N_2, \ldots, V, T)}{N_{jj}(N_1, N_2, \ldots, V, T)} \neq \frac{x_i}{x_j} \quad (16.3\text{-}11)$$

Second, at high temperatures ($kT \gg \varepsilon$), the exponential terms are close to unity and cancel so that

$$\lim_{\rho \to 0} \frac{N_{ij}(N_1, N_2, \ldots, V, T)}{N_{jj}(N_1, N_2, \ldots, V, T)} = \frac{x_i\sigma_{ij}^3(R_{ij}^3 - 1)}{x_j\sigma_{jj}^3(R_{jj}^3 - 1)} \quad (16.3\text{-}12)$$

which indicates that the ratio of local compositions is essentially equal to the ratio of species volume fractions (this would be exactly true if $R_{ij} = R_{jj}$, which may be the case, as R_{sw} is frequently chosen to be about 1.5). Finally, at moderate temperatures, the ratio of local compositions is determined by both volume fractions of the species and the differences in the attractive parts of their interaction potentials.

Another interesting observation is that the total coordination number of a species, for example species 1 in a binary mixture, again using the low-density result, is

$$N_{11} + N_{21} = \frac{N_1}{V}\frac{4\pi}{3}\sigma_{11}^3(R_{11}^3 - 1)e^{\varepsilon_{11}/kT} + \frac{N_2}{V}\frac{4\pi}{3}\sigma_{21}^3(R_{21}^3 - 1)e^{\varepsilon_{21}/kT}$$

$$= \frac{N}{V}\frac{4\pi}{3}[x_1\sigma_{11}^3(R_{11}^3 - 1)e^{\varepsilon_{11}/kT} + x_2\sigma_{21}^3(R_{21}^3 - 1)e^{\varepsilon_{21}/kT}] \quad (16.3\text{-}13)$$

which indicates that, unless the molecules are of the same size and have the same interaction energies, the total coordination number of a species in a low-density mixture is a function of composition. This may also be true, but to a lesser extent, at liquid densities; this is different from the lattice theory assumption used in Chapter 10 that each lattice site has a fixed coordination number.

The simplest model species-species coordination number at other densities is the completely random mixture; in this case, $N_{ij} = \frac{N_i}{V}C$, where C is a constant and the same for all species. In this case

$$U_{ij}^{\text{res}}(N_1, N_2, \ldots, V, T) = -\frac{1}{2}\varepsilon_{ij} N_j N_{ij} = -\frac{1}{2}\varepsilon_{ij} C N_j \frac{N_i}{V} = -\frac{1}{2}\varepsilon_{ij} C N_j \rho_i \tag{16.3-14}$$

and

$$\Phi_{\text{mix}}(N_1, N_2, \ldots, V, T) = -\frac{1}{2}\frac{C}{N}\sum_i\sum_j \varepsilon_{ij} \frac{N_j N_i}{V} \quad \text{and}$$

$$\Phi_i(N_i, V_i, T) = -\frac{1}{2} C \varepsilon_{ii} \frac{N_i}{V}.$$

Using these expressions, we obtain

$$P^{\text{eng}}(N, V, T) = \frac{N^2}{V^2} \frac{C}{2} \sum_i \sum_j x_i x_j \varepsilon_{ij}$$

$$= -\frac{N^2}{V^2} \sum_i \sum_j x_i x_j a_{ij} \quad \text{with} \quad a_{ij} = \frac{C}{2}\varepsilon_{ij} \tag{16.3-15}$$

So then, the equation of state is

$$P(N, V, T) = P^{\text{ent}}(N, V, T) + P^{\text{eng}}(N, V, T)$$

$$= \frac{NkT}{V - N\beta} - \frac{aN^2}{V^2} \tag{16.3-16}$$

which is the van der Waals equation with the van der Waals one-fluid mixing rules

$$\beta = \sum_i \sum_j x_i x_j \beta_{ij} \quad \text{and} \quad a = \sum_i \sum_j x_i x_j a_{ij} \tag{16.3-17}$$

Other choices for the species-species (or local) coordination numbers result in other equations of state, as shown in Table 16.3-2. Of course, choices other than those shown in the table can be made, leading to different equations of state and activity coefficient models. Among those other choices are the model of Wilson[8]

$$\frac{N_{ij}}{N_{jj}} = \frac{N_i}{N_j} e^{(\varepsilon_{ij} - \varepsilon_{jj})/kT} \tag{16.3-18a}$$

of Whiting and Prausnitz[9]

$$\frac{N_{ij}}{N_{jj}} = \frac{N_i}{N_j} e^{(\varepsilon_{ij} - \varepsilon_{jj})N_{cj}/kT} \quad \text{with} \quad N_{ij} + N_{jj} = N_{cj} = C_j \rho \tag{16.3-18b}$$

of Hu et al.[10]

---

[8]G. M. Wilson, *J. Am. Chem. Soc.*, **86**, 127 (1964).
[9]W. B. Whiting and J. M. Prausnitz, *Fluid Phase Equilibria*, **9**, 119 (1982).
[10]Y. Hu, D. Ludecke and J. M. Prausnitz, *Fluid Phase Equilibria*, **17**, 217 (1984).

Table 16.3-2 Local Coordination Number Models and the Equations of State Mixing Rules that Result

Local coordination number	Equation of state mixing rule
$N_{ij} = \dfrac{N_i}{V} C$	van der Waals (vdW) 1-fluid
$N_{ij} = \dfrac{N_i}{V} C_{ij}$	vdW 1-fluid
$N_{ij} = \dfrac{N_i}{V} C_{ij} e^{\varepsilon_{ij}/kT}$	vdW 1-fluid
$\dfrac{N_{ij}}{N_{jj}} = \dfrac{N_i v_i}{N_j v_j}$ with $N_{ij} + N_{jj} = N_{cj} = \dfrac{N}{V} C_j$	nonquadratic mixing rule; $a = \dfrac{\sum_i \sum_j x_i x_j v_i a_{ij}}{\sum_i x_i v_i}$
$\dfrac{N_{ij}}{N_{jj}} = \dfrac{N_i v_i}{N_j v_j} e^{(\varepsilon_{ij}-\varepsilon_{jj})/kT}$ with $N_{ij} + N_{jj} = N_{cj} = \dfrac{N}{V} C_j$	As above
Surface area fractions	Nonquadratic mixing rule

$$N_{ij} = \frac{N_i}{V} \frac{4\pi}{3} \sigma_{ij}^3 (R_{ij}^3 - 1) e^{\alpha \varepsilon_{ij}/kT} \quad \text{with} \quad \alpha = 0.60 - 0.58 \rho \left(\sum_i x_i \sigma_{ij}^3 \right) \tag{16.3-18c}$$

and a model based on the simple extension of Eq. 16.1-1

$$N_{ij} = \frac{x_i N_m V_{o,ij} e^{\varepsilon_{ij}/2kT}}{V + V_{o,ij}(e^{\varepsilon_{ij}/2kT} - 1)} \quad \text{with} \quad V_{o,ij} = N\sigma_{ij}^3/\sqrt{2} \tag{16.3-18d}$$

with an effective coordination number $N_m = 18$ (for $R_{sw} = 1.5$).

Computer simulation[11] can be used to test the accuracy of these approximations. Figure 16.3-1 shows the deviations of the local composition ratios from the bulk composition ratios in terms of the quantities $\theta_{12} = \dfrac{N_{12}}{N_{22}} \dfrac{N_2}{N_1}$ and $\theta_{21} = \dfrac{N_{21}}{N_{11}} \dfrac{N_1}{N_2}$. Each of these ratios would be unity in a completely random mixture. For the square-well potential at low density it would be

$$\lim_{\rho \to 0} \theta_{ij} = \frac{\sigma_{ij}^3 (R_{ij}^3 - 1)}{\sigma_{jj}^3 (R_{jj}^3 - 1)} e^{(\varepsilon_{ij}-\varepsilon_{jj})/kT} \tag{16.3-19}$$

Another test of the models is to look at the total coordination number for the square-well system with $\varepsilon_{22} = 1.2kT = 2\varepsilon_{11}$, but varying diameter ratios and densities as shown in Fig. 16.3-2.

The main conclusion from the simulation results, especially Fig. 16.3-1, is that none of the local composition models are completely accurate; so if one can devise

---

[11] S. I. Sandler, *Chem. Eng. Educ.*, **Spring 1990**, p. 80.

16.3 Equation of State for Mixtures 317

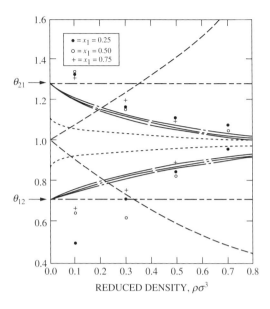

Figure 16.3-1 The ratios of the local compositions to the bulk compositions θ_{12} and θ_{21} as a function of reduced density for the square-well fluid with $\varepsilon_{22} = 1.2kT = 2\varepsilon_{11}$, $\sigma_{11} = \sigma_{22}$ and $R_{11} = R_{22} = 1.5$ obtained from Monte Carlo simulation. The arrows indicate the low density results, the points are the simulation results at different bulk mole fraction. The dash-double dotted line is the model of Eq. 16.3-18a small dotted line is the model of Eq. 16.3-18b, the dashed line is the model of Eq. 16.3-18c, and the solid lines result from Eq. 16.3-18d. Reprinted with permission from S. I. Sandler, Chem. Eng. Educ., Spring 1990, page 80 et. seq.

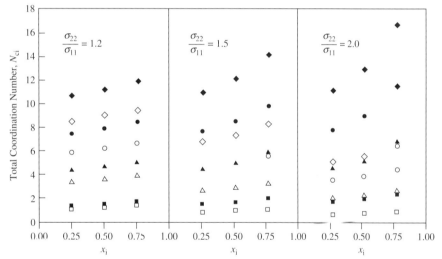

Figure 16.3-2 The total coordination numbers N_{c1} (unfilled points) and N_{c2} (filled points) as a function of density and mole fraction from simulation.[12] The squares, triangles, circles, and diamonds are for reduced densities of $\rho(x_1\sigma_{11}^3 + x_2\sigma_{22}^3) = 0.1$, 0.3, 0.5 and 0.7, respectively, for $R_{11} = R_{22} = 1.5$ Reprinted from Fluid Phase Equilibria, Vol. 34, K.-H. Lee and S. I. Sandler, "The Generalized van der Waals Partition Function. IV. Local Composition Models for Mixtures of Unequal Size Molecules" by K.-H. Lee and S. I. Sandler, pp.113–147, Copyright 1987, with permission from Elsevier.

improvements, the result may be mixture equations of state (and also activity coefficient models as will be seen later) with a better theoretical basis and greater accuracy. Nonetheless, the models presently used in engineering are effective, largely as a result of treating the parameters in the model as adjustable, and fitting to experimental data.

---

[12] K.-H. Lee and S. I. Sandler, *Fluid Phase Equilibria*, **34**, 113 (1987).

318 Chapter 16: The Derivation of Thermodynamic Models

16.4 ACTIVITY COEFFICIENT MODELS FROM THE GENERALIZED VAN DER WAALS PARTITION FUNCTION

The second way to use the Generalized van der Waals partition function is to develop an expression for the excess Helmholtz energy of mixing, and from that derive activity coefficient models. The main difference from the analysis used for equations of state discussed in Sec. 16.3 is that here, as is typical in formulating activity coefficient models, because of the limited range of liquid densities of interest, the total coordination number z of a molecule is not considered to be a function of density, though it will be a function of composition and (to a lesser extent) of temperature.

The starting point for the analysis in terms of the Generalized van der Waals partition function is to note that for mixing at constant temperature and total volume (so that $V(N_1, N_2, \ldots, T) = \sum_i V_i(N_i, T)$), the excess Helmholtz energy of mixing over that of forming an ideal mixture of the same components at the same temperature and total volume is

$$A^{ex}_{T,V}(N_1, N_2, \ldots, V, T) = A(N_1, N_2, \ldots, V, T) - \sum_i A_i(N_i, V_i, T) - kT \sum_i N_i \ln x_i$$

$$= -kT \ln \left[\frac{Q(N_1, N_2, \ldots, V, T)}{\prod_i Q_i(N_i, V_i, T)} \right] - kT \sum_i \ln \left(\frac{N_i}{N} \right)^{N_i}$$

$$= -kT \ln \left[\frac{Z(N_1, N_2, \ldots, V, T=\infty) \prod_i \left(\frac{N_i}{N} \right)^{N_i}}{\prod_i Z_i(N_i, V_i, T=\infty)} \right]$$

$$+ \left[N\Phi_{mix}(N_1, N_2, \ldots, V, T) - \sum_i N_i \Phi_i(N_i, V_i, T) \right]$$

$$= -kT \ln \left[\frac{V_{f,mix}(N_1, N_2, \ldots, V)^N \prod_i \left(\frac{N_i}{N} \right)^{N_i}}{\prod_i V_{f,i}(N_i, V_i, T)^{N_i}} \right]$$

$$- kT \int_{T=\infty}^{T} \left[\frac{U^{res}_{mix}(N_1, N_2, \ldots, V, T) - \sum_i U^{res}_i(N_i, V_i, T)}{kT^2} \right] dT$$

$$= A^{ex,ent}_{T,V}(N_1, N_2, \ldots, V) + A^{ex,eng}_{T,V}(N_1, N_2, \ldots, V, T) \quad \text{(16.4-1)}$$

The first term on the right-hand side of this last equation is the athermal contribution to the excess Helmholtz energy due to the hard-core molecular size and shape differences between the species in the mixture (since all soft-core forces are unimportant at high temperatures); it is entirely entropic. The second term is the contribution from other than the hard-core interaction, resulting from the attractive and soft-core

16.4 Activity Coefficient Models

repulsive forces. Also, note that

$$\left(\frac{\partial A^{\text{ex}}(N_1, N_2, \ldots, V, T)}{\partial N_i}\right)_{T,V,N_{j \neq i}} = \left(\frac{\partial G^{\text{ex}}_{T,P}(N_1, N_2, \ldots, P, T)}{\partial N_i}\right)_{T,P,N_{j \neq i}}$$

$$= kT \ln \gamma_i(x_1, x_2, \ldots, T, P) \quad \textbf{(16.4-2)}$$

So, starting with the Generalized van der Waals partition function, we can develop activity coefficients models or "reverse engineer" models commonly used to understand the assumptions inherent in such models. In this way, the Generalized van der Waals partition function can provide a theoretically-based platform for improving current activity coefficient models—or developing new ones. Also, once the model assumptions have been identified, they can be tested using computer simulation.

As an aside, it is useful to note that at high densities, and especially in liquids, the molecules are so close to each other that they are always subject a background attractive interaction field determined by their proximity to all the surrounding molecules. In this case, the local compositions are determined primarily by the hard-core molecular volumes. It is for this reason that some activity coefficient models separate the contribution to the excess Helmholtz energy into an entropic (sometimes referred to as the *configurational*) part depending on volume fractions (as in the Flory-Huggins[13] and UNIQUAC[14] models, though the latter also includes surface area fractions), and an enthalpic or energetic part (sometimes referred to as the *residual* part, though the term *residual* is used differently in this chapter) arising for the soft part of the interactions. However, when the attractive forces are very strong and specific to certain orientations, as in hydrogen bonding, different types of models are required, as discussed in Section 16.6.

Free Volume Term

As in the discussion of equations of state, the free volume (entropic) and energetic contributions will be considered separately. Therefore, the starting point is

$$A^{\text{ex,ent}}_{T,V}(N_1, N_2, \ldots, V) = A^{\text{conf}}(N_1, N_2, \ldots, V) - \sum_i A^{\text{conf}}_i(N_i, V_i)$$

$$= -kT \left[N \ln V_{f,\text{mix}} + \sum_i N_i \ln x_i - \sum_i N_i \ln V_{f,i} \right]$$

$$= -kT \ln \left[\frac{(V - N\beta)^N \prod_i x_i^{N_i}}{\prod_i (V_i - N_i \beta_i)^{N_i}} \right]$$

$$= kT \sum_i N_i \ln \frac{\phi_{f,i}}{x_i} \quad \textbf{(16.4-3a)}$$

where $\phi_{f,i} = V_{f,i}(N_i, V_i, T)/V_{f,\text{mix}}(N, V, T)$ is the free volume of the pure species divided by the total free volume in the mixture. (Note that if the simple van der Waals free-volume term is used, then $\phi_{f,i} = (V_i - N_i \beta_i)/(V - N\beta)$.) If the fractional free

---

[13] See Section 10.4.
[14] D. S. Abrams and J. M. Prausnitz, *AIChE J.*, **21**, 116 (1975).

320 Chapter 16: The Derivation of Thermodynamic Models

volume is the same for all components and in the mixture—that is, $V_{f,i}(N_i, V_i, T) = x_i V_{f,\text{mix}}(N, V, T)$—then the free-volume ratio becomes the mole fraction ratio, and $A_{T,V}^{\text{ex,ent}} = 0$. If, however, $\phi_{f,i}$ of each species is assumed to be equal to its volume fraction $\phi_i = x_i v_i / \sum_i x_i v_i$ where v_i is some measure of the volume of species i, then Eq. 16.4-3a is the Flory expression for the configurational contribution to the free energy of mixing used in Eq. 10.4-12. Alternatively, other assumptions can be made for the entropic term; for example, those based on lattice theories, leading to the Guggenheim-Staverman[15] expression that includes both surface area and volume fractions

$$\frac{A_{T,V}^{\text{ex,ent}}(N_1, N_2, \ldots, V)}{kT} = \sum_i N_i \ln \frac{\phi_i}{x_i} + \frac{z}{2} \sum_i N_i q_i \ln \frac{\theta_i}{\phi_i} \qquad (16.4\text{-}3b)$$

where q_i is proportional to the surface area of molecule i, ϕ_i is its volume fraction, θ_i is its surface area fraction and z is the (same) coordination number for all the species in the mixture.

To see how the activity coefficient is obtained from the expression for $A_{T,V}^{\text{ex,ent}}$, we will use as an example $\phi_{f,i} = \phi_i$, so that for a binary mixture

$$A_{T,V}^{\text{ex,ent}}(N_1, N_2, T, V) = kT \sum_i N_i \ln \frac{\phi_{f,i}}{x_i} = kT \sum_i N_i \ln \frac{v_i}{\sum_j x_j v_j}$$

$$= kT \left[N_1 \ln \frac{v_1}{x_1 v_1 + x_2 v_2} + N_2 \ln \frac{v_2}{x_1 v_1 + x_2 v_2} \right]$$

$$= kT \left[N_1 \ln \frac{N v_1}{N_1 v_1 + N_2 v_2} + N_2 \ln \frac{N v_2}{N_1 v_1 + N_2 v_2} \right] \qquad (16.4\text{-}4)$$

Therefore, to obtain the activity coefficient[16] of species 1

$$\ln \gamma_1^{\text{ent}}(x_1, x_2, T, P) = \frac{1}{kT} \left(\frac{\partial A^{\text{ex,ent}}(N_1, N_2, V, T)}{\partial N_i} \right)_{T,V,N_{j \neq 1}}$$

$$= \frac{\partial}{\partial N_1}\bigg|_{T,V,N_{j \neq 1}} \left[N_1 \ln \frac{N v_1}{N_1 v_1 + N_2 v_2} + N_2 \ln \frac{N v_2}{N_1 v_1 + N_2 v_2} \right]$$

$$= \ln \frac{N v_1}{N_1 v_1 + N_2 v_2} + \frac{N_1}{N} - \frac{N_1 v_1}{N_1 v_1 + N_2 v_2} + \frac{N_2}{N} - \frac{N_2 v_1}{N_1 v_1 + N_2 v_2}$$

$$= \ln \frac{v_1}{x_1 v_1 + x_2 v_2} + x_1 - \frac{x_1 v_1}{x_1 v_1 + x_2 v_2} + x_2 - \frac{x_2}{x_1} \frac{x_1 v_1}{x_1 v_1 + x_2 v_2}$$

$$= \ln \frac{\phi_1}{x_1} + 1 - \phi_1 - \frac{x_2}{x_1} \phi_1$$

$$= \ln \frac{\phi_1}{x_1} + 1 - \frac{\phi_1}{x_1} = \ln \frac{\phi_1}{x_1} + \phi_2 \left(1 - \frac{v_1}{v_2} \right) = \ln \gamma_1^{\text{ent}}(x_1, x_2, T, P) \qquad (16.4\text{-}5)$$

This entropic or free-volume contribution to the activity coefficient can be added to the energetic contribution from the mean potential that is considered next.

---

[15] This model was developed independently by E. A. Guggenheim (see "Mixtures", Claredon Press, Oxford, 1952) and A. J. Staverman (Rev. Trav. Chim. *Pays-Bas*, **69**, 163 (1950).

[16] See footnote 8 of Chapter 10 for the proper way of taking compositional derivatives.

16.4 Activity Coefficient Models   321

The Mean Potential

To obtain an expression for the mean potential, models for the species-species coordination number, N_{ij}, must be developed. These are then used in the following equations:

$$U_{ij}^{\text{res}}(N_1, N_2, \ldots, V, T) = -\frac{1}{2}\varepsilon_{ij} N_j N_{ij}$$

$$\int_{T=\infty}^{T} \frac{U_{ij}^{\text{res}}(N_1, N_2, \ldots, V, T)}{kT^2} dT = -\frac{1}{2}\varepsilon_{ij} N_j \int_{T=\infty}^{T} \frac{N_{ij}}{kT^2} dT \quad (16.4\text{-}6)$$

$$\Phi_{\text{mix}}(N_1, N_2, \ldots, V, T) = \frac{kT}{2N} \sum_i \sum_j \varepsilon_{ij} N_j \int_{T=\infty}^{T} \frac{N_{ij}}{kT^2} dT$$

$$\text{and} \quad \Phi_i(N_i, V_i, T) = \frac{kT\varepsilon_{ii}}{z} \int_{T=\infty}^{T} \frac{z_i}{kT^2} dT \quad (16.4\text{-}7a)$$

Then

$$A^{\text{ex,eng}} = N\left[\Phi_{\text{mix}}(N_1, N_2, \ldots, V, T) - \sum_i x_i \Phi_i(N_i, V_i, T)\right]$$

$$= \frac{1}{2} NkT \left[\sum_i \sum_j \varepsilon_{ij} x_j \int_{T=\infty}^{T} \frac{N_{ij}}{kT^2} dT - \sum_i x_i \varepsilon_{ii} \int_{T=\infty}^{T} \frac{z_i}{kT^2} dT\right] \quad (16.4\text{-}7a)$$

If the species-species and total coordination numbers are independent of temperature, this becomes

$$A^{\text{ex,eng}} = -\frac{1}{2}N\left[\sum_i \sum_j \varepsilon_{ij} x_j N_{ij} - \sum_i x_i \varepsilon_{ii} z_i\right] \quad (16.4\text{-}8a)$$

For the case of a binary mixture with temperature-dependent species coordination numbers, Eq. 16.4-6a becomes

$$A^{\text{ex,eng}} = \frac{NkT}{2}\left[\begin{array}{l} \varepsilon_{11}x_1 \int_{T=\infty}^{T} \frac{N_{11}}{kT^2} dT + \varepsilon_{12}x_2 \int_{T=\infty}^{T} \frac{N_{12}}{kT^2} dT + \varepsilon_{12}x_1 \int_{T=\infty}^{T} \frac{N_{21}}{kT^2} dT \\ + \varepsilon_{22}x_2 \int_{T=\infty}^{T} \frac{N_{22}}{kT^2} dT - \varepsilon_{11}x_1 \int_{T=\infty}^{T} \frac{z_1}{kT^2} dT - \varepsilon_{22}x_2 \int_{T=\infty}^{T} \frac{z_2}{kT^2} dT \end{array}\right]$$

$$= \frac{NkT}{2}\left[\begin{array}{l} \varepsilon_{11}x_1 \int_{T=\infty}^{T} \frac{N_{11}-z_1}{kT^2} dT + \varepsilon_{12}x_2 \int_{T=\infty}^{T} \frac{N_{12}}{kT^2} dT \\ + \varepsilon_{12}x_1 \int_{T=\infty}^{T} \frac{N_{21}}{kT^2} dT + \varepsilon_{22}x_2 \int_{T=\infty}^{T} \frac{N_{22}-z_2}{kT^2} dT \end{array}\right]$$

$$= \frac{NkT}{2} \left[\begin{array}{l} -\varepsilon_{11}x_1 \int_{T=\infty}^{T} \frac{N_{21}}{kT^2} dT + \varepsilon_{12}x_2 \int_{T=\infty}^{T} \frac{N_{12}}{kT^2} dT \\ + \varepsilon_{12}x_1 \int_{T=\infty}^{T} \frac{N_{21}}{kT^2} dT - \varepsilon_{22}x_2 \int_{T=\infty}^{T} \frac{N_{12}}{kT^2} dT \end{array} \right]$$

$$= \frac{NkT}{2} \left[x_2(\varepsilon_{12} - \varepsilon_{22}) \int_{T=\infty}^{T} \frac{N_{12}}{kT^2} dT + x_1(\varepsilon_{12} - \varepsilon_{11}) \int_{T=\infty}^{T} \frac{N_{21}}{kT^2} dT \right] \quad \text{(16.4-7b)}$$

since $z_1 = N_{11} + N_{21}$ and $z_2 = N_{12} + N_{22}$, where z_i is the total coordination number of species i. When all the local coordination numbers N_{ij} are independent of temperature, this last equation becomes

$$A^{\text{ex,eng}} = -\frac{N}{2}[x_2(\varepsilon_{12} - \varepsilon_{22})N_{12} + x_1(\varepsilon_{12} - \varepsilon_{11})N_{21}]$$

$$= \frac{N}{2}[x_2(\varepsilon_{22} - \varepsilon_{12})N_{12} + x_1(\varepsilon_{11} - \varepsilon_{12})N_{21}] \quad \text{(16.4-8b)}$$

Equations 16.4-7 and 16.4-8 for temperature-dependent and temperature-independent coordination numbers, respectively, will be used for all that follows.

The simplest species-species coordination number model is that each species has the same temperature-independent total coordination number z, and that the molecules are randomly distributed—that is, that the number of i molecules around a j molecule is just $N_{ij} = x_i z$. In this case, starting from Eq. 16.4-8b, we obtain

$$A^{\text{ex,eng}} = \frac{N}{2}[x_2(\varepsilon_{22} - \varepsilon_{12})x_1 + x_1(\varepsilon_{11} - \varepsilon_{12})x_2] = x_1 x_2 \frac{zN}{2}[\varepsilon_{11} + \varepsilon_{22} - 2\varepsilon_{12}]$$

$$= N\alpha x_1 x_2$$

with for the square-well potential

$$\alpha = \frac{z}{2}[\varepsilon_{11} + \varepsilon_{22} - 2\varepsilon_{12}] \quad \text{or} \quad \alpha = z[\sqrt{\varepsilon_{11}} - \sqrt{\varepsilon_{22}}]^2 \quad \text{if it is assumed that}$$

$$\varepsilon_{12} = \sqrt{\varepsilon_{11}\varepsilon_{22}}$$

or for other interaction potentials

$$\alpha = \frac{z}{2}[2\langle u_{12}^S \rangle - \langle u_{11}^S \rangle - \langle u_{22}^S \rangle] \quad \text{or} \quad \alpha = -\frac{z}{2}\left[\sqrt{\langle u_{11}^S \rangle} - \sqrt{\langle u_{22}^S \rangle}\right]^2$$

if it is assumed that $\langle u_{12}^S \rangle = \sqrt{\langle u_{11}^S \rangle \langle u_{22}^S \rangle}$ \quad (16.4-9)

So if we assume that $A_{T,V}^{\text{ex,ent}} = 0$ and write $A_{T,V}^{\text{ex,eng}}/N = \alpha x_1 x_2$, the one-constant Margules equation is obtained with an explicit expression for α, and the following

16.4 Activity Coefficient Models

expression for the activity coefficient:

$$kT \ln \gamma_1^{\text{eng}}(N_1, N_2, T, P) = \left(\frac{\partial A_{T,V}^{\text{ex,eng}}}{\partial N_1}\right)_{T,V,N_2} = \left(\frac{\partial}{\partial N_1} \alpha N_1 x_2\right)_{T,V,N_2}$$

$$= \left(\frac{\partial}{\partial N_1} \frac{\alpha N_1 N_2}{N_1 + N_2}\right)_{T,V,N_2} = \frac{\alpha N_2}{N_1 + N_2} - \frac{\alpha N_1 N_2}{(N_1 + N_2)^2} = \alpha(x_2 - x_1 x_2)$$

$$= \alpha x_2(1 - x_1) = \alpha x_2^2 \tag{16.4-10}$$

Note that the expression for the excess Helmholtz energy above is symmetric in composition, and consequently leads to activity coefficients that are mirror images of each other. That is, $kT \ln \gamma_1^{\text{eng}} = \alpha x_2^2$ and $kT \ln \gamma_2^{\text{eng}} = \alpha x_1^2$. Such behavior is not usually found in experimental data unless the species are very similar.

Equation 16.4-10 can be used as the complete expression for the excess Helmholtz energy of mixing, or it can be combined with Eq. 16.4-5 to give an augmented one-constant Margules equation:

$$A_{T,V}^{\text{ex}}(N_1, N_2, T, V) = A_{T,V}^{\text{ex,ent}} + A_{T,V}^{\text{ex,eng}} = kT\left(N_1 \ln \frac{\phi_{\text{f},1}}{x_1} + N_2 \ln \frac{\phi_{\text{f},2}}{x_2}\right) + \alpha N x_1 x_2$$

and

$$kT \ln \gamma_1(x_1, x_2, T, P) = \ln \frac{\phi_1}{x_1} + \phi_2\left(1 - \frac{v_1}{v_2}\right) + \alpha x_2^2 \tag{16.4-11}$$

which is similar to (but not exactly the same as) the Flory-Huggins model (see Eq. 10.4-12). Note that by adding Eq. 16.4-5 to Eq. 16.4-10, the mirror image symmetry is broken, unless coincidentally the molecules are of the same size.

This same procedure of adding together the separate entropic and energetic contributions to the excess Helmholtz energy and the activity coefficient can be used with any of the mean potential models that are discussed below.

A slightly more realistic model for the mean potential is to recognize that the molecules in a mixture can be of different sizes (and therefore have different coordination numbers), but the mixture is still random. That is, $N_{ij} = x_i z_j$, where $z_j \neq z_i$. Then, following the analysis above

$$A_{T,V}^{\text{ex,eng}} = -\frac{N}{2}\left[\sum_i \sum_j \varepsilon_{ij} x_j x_i z_j - \sum_i x_i \varepsilon_{ii} z_i\right] \tag{16.4-12a}$$

which, for a binary mixture, becomes

$$A_{T,V}^{\text{ex,eng}} = \frac{N}{2} x_1 x_2 [\varepsilon_{11} z_1 + \varepsilon_{22} z_2 - \varepsilon_{12}(z_1 + z_2)] = \alpha N x_1 x_2 \tag{16.4-12b}$$

This again leads to the one-constant Margules equation, but with

$$\alpha = \frac{1}{2}[\varepsilon_{11} z_1 + \varepsilon_{22} z_2 - \varepsilon_{12}(z_1 + z_2)].$$

If, instead, one uses for the species coordination numbers $z_i = x_i z_i^{\text{o}} + x_j z_{ji}^{\text{o}}$ where z_i^{o} is the coordination number of species i in a pure fluid, and z_{ji}^{o} is the number of j molecules around an infinitely dilute i molecule, then a two-constant Margules equation is obtained (Problem 16.7):

$$A_{T,V}^{\text{ex,eng}} = Nx_1x_2(\alpha + \beta x_1) \qquad (16.4\text{-}12c)$$

This expression for the excess Helmholtz energy is also not symmetric in composition, so that the resulting activity coefficients are not mirror images of each other.

Another alternative is to assume that the local coordination number of one species around another depends on the volume fractions rather than on mole fractions, that is

$$N_{ij} = \frac{x_i v_i}{\sum_i x_i v_i} z = \phi_i z \quad \text{where } \phi_i = \frac{x_i v_i}{\sum_i x_i v_i} \text{ is the volume fraction} \qquad (16.4\text{-}13a)$$

and v_i is some measure of the molecular volume[17] of species i. Using this in Eq. 16.4-8a, the following is obtained for a binary mixture:

$$\frac{A^{\text{ex,eng}}}{N} = \frac{1}{2}[x_2(\varepsilon_{22} - \varepsilon_{12})N_{12} + x_1(\varepsilon_{11} - \varepsilon_{12})N_{21}]$$

$$= \frac{z}{2}\left[x_2(\varepsilon_{22} - \varepsilon_{12})\frac{x_1 v_1}{x_1 v_1 + x_2 v_2} + x_1(\varepsilon_{11} - \varepsilon_{12})\frac{x_2 v_2}{x_1 v_1 + x_2 v_2}\right]$$

$$= \frac{z}{2}\frac{x_1 v_1 x_2 v_v}{x_1 v_1 + x_2 v_2}\left(\frac{\varepsilon_{12} - \varepsilon_{11}}{v_1} + \frac{\varepsilon_{21} - \varepsilon_{22}}{v_2}\right) = a\frac{x_1 v_1 x_2 v_2}{x_1 v_1 + x_2 v_2} \qquad (16.4\text{-}13b)$$

which is the Van Laar equation with $a = \dfrac{z}{2}\left(\dfrac{\varepsilon_{12} - \varepsilon_{11}}{v_1} + \dfrac{\varepsilon_{21} - \varepsilon_{22}}{v_2}\right)$. Once again, this contribution to the excess Helmholtz energy can be augmented with an entropic contribution—for example, of Eq. 16.4-3a.

Yet another choice is

$$\frac{N_{ij}}{N_{jj}} = \frac{N_i v_i}{N_j v_j} \quad \text{with the restriction that } N_{ij} + N_{jj} = z_j \qquad (16.4\text{-}14a)$$

which leads to (Problem 16.5)

$$N_{ij} = \frac{x_i v_i}{\sum_i x_i v_i} z_j \qquad (16.4\text{-}14b)$$

and for a binary mixture is

$$\frac{A^{\text{ex,eng}}}{N} = \frac{1}{2(x_1 v_1 + x_2 v_2)}[x_2(\varepsilon_{22} - \varepsilon_{12})x_1 v_1 z_2 + x_1(\varepsilon_{11} - \varepsilon_{21})x_2 v_2 z_1]$$

$$= \frac{x_1 x_2}{(x_1 v_1 + x_2 v_2)}\frac{[(\varepsilon_{22} - \varepsilon_{12})v_1 z_2 + (\varepsilon_{11} - \varepsilon_{21})v_2 z_1]}{2} = \alpha\frac{x_1 x_2}{x_1 v_1 + x_2 v_2}$$
$$(16.4\text{-}14c)$$

with $\alpha = \dfrac{[(\varepsilon_{22} - \varepsilon_{12})v_1 z_2 + (\varepsilon_{11} - \varepsilon_{21})v_2 z_1]}{2}$

which looks like a combination of the Margules and Van Laar equations.

---

[17] An alternative assumption is that v is instead a measure of surface area. All the equations here would be unchanged, except for the interpretation of v as a surface area instead of a volume.

16.4 Activity Coefficient Models

As a more complicated example, consider the following species-species coordination number model:

$$\frac{N_{ij}}{N_{jj}} = \frac{N_i v_i}{N_j v_j} e^{(\varepsilon_{ij}-\varepsilon_{jj})/kT} \quad \text{with the restriction that } N_{ij} + N_{jj} = z_j \quad \text{(16.4-15a)}$$

The ideas underlying this model are that the local composition of one species over another is favored if it has a more attractive interaction with the central molecule, and/or it occupies more space by having a larger value of the product $N_i v_i$. Solving Eq. 16.4-15a gives

$$N_{ij} = z_j \frac{N_i v_i e^{(\varepsilon_{ij}-\varepsilon_{jj})/kT}}{N_j v_j + N_i v_i e^{(\varepsilon_{ij}-\varepsilon_{jj})/kT}} \quad \text{(16.4-15b)}$$

Now

$$\int_{T=\infty}^{T} \frac{N_{ij}}{kT^2} dT = z_j \int_{T=\infty}^{T} \frac{N_i v_i e^{(\varepsilon_{ij}-\varepsilon_{jj})/kT}}{N_j v_j + N_i v_i e^{(\varepsilon_{ij}-\varepsilon_{jj})/kT}} \frac{1}{kT^2} dT$$

$$= z_j \int_{T=\infty}^{T} \frac{e^{(\varepsilon_{ij}-\varepsilon_{jj})/kT}}{\frac{N_j v_j}{N_i v_i} + e^{(\varepsilon_{ij}-\varepsilon_{jj})/kT}} \frac{1}{kT^2} dT \quad \text{(16.4-15c)}$$

To proceed with the integration, the following substitution is made:

$$x = \frac{\varepsilon_{ij} - \varepsilon_{jj}}{kT} \quad \text{so that} \quad dx = -\frac{\varepsilon_{ij} - \varepsilon_{jj}}{kT^2} dT \quad \text{or} \quad \frac{dx}{\varepsilon_{ij} - \varepsilon_{jj}} = -\frac{dT}{kT^2} \quad \text{and}$$

$$\int_{T=\infty}^{T} \frac{e^{(\varepsilon_{ij}-\varepsilon_{jj})/kT}}{\frac{N_j v_j}{N_i v_i} + e^{(\varepsilon_{ij}-\varepsilon_{jj})/kT}} \frac{1}{kT^2} dT = -\frac{1}{\varepsilon_{ij} - \varepsilon_{jj}} \int_{x=0}^{x=\frac{\varepsilon_{ij}-\varepsilon_{jj}}{kT}} \frac{e^x}{\frac{N_j v_j}{N_i v_i} + e^x} dx$$

$$= -\frac{1}{\varepsilon_{ij} - \varepsilon_{jj}} \left[\ln\left(e^{(\varepsilon_{ij}-\varepsilon_{jj})/kT} + \frac{N_j v_j}{N_i v_i}\right) - \ln\left(1 + \frac{N_j v_j}{N_i v_i}\right) \right]$$

$$= -\frac{1}{\varepsilon_{ij} - \varepsilon_{jj}} \left[\ln\left(\frac{e^{(\varepsilon_{ij}-\varepsilon_{jj})/kT} + \frac{N_j v_j}{N_i v_i}}{1 + \frac{N_j v_j}{N_i v_i}}\right) \right]$$

$$= -\frac{1}{\varepsilon_{ij} - \varepsilon_{jj}} \left[\ln\left(\frac{N_i v_i e^{(\varepsilon_{ij}-\varepsilon_{jj})/kT} + N_j v_j}{N_i v_i + N_j v_j}\right) \right]$$

Now using

$$\int_{T=\infty}^{T} \frac{N_{ij}}{kT^2} dT = -\frac{z_j}{\varepsilon_{ij} - \varepsilon_{jj}} \ln\left(\frac{N_i v_i e^{(\varepsilon_{ij}-\varepsilon_{jj})/kT} + N_j v_j}{N_i v_i + N_j v_j}\right)$$

$$= -\frac{z_j}{\varepsilon_{ij} - \varepsilon_{jj}} \ln(\phi_i e^{(\varepsilon_{ij}-\varepsilon_{jj})/kT} + \phi_j) \quad \text{(16.4-15d)}$$

Chapter 16: The Derivation of Thermodynamic Models

so that finally

$$A^{ex,eng} = -\frac{NkT}{2}\left[\begin{array}{c} x_2(\varepsilon_{12} - \varepsilon_{22})\dfrac{z_2}{\varepsilon_{12} - \varepsilon_{22}}\ln(\phi_1 e^{(\varepsilon_{12}-\varepsilon_{22})/kT} + \phi_2) \\ + x_1(\varepsilon_{12} - \varepsilon_{11})\dfrac{z_1}{\varepsilon_{21} - \varepsilon_{22}}\ln(\phi_2 e^{(\varepsilon_{21}-\varepsilon_{11})/kT} + \phi_1) \end{array}\right]$$

$$= -\frac{NkT}{2}[x_2 z_2 \ln(\phi_1 e^{(\varepsilon_{12}-\varepsilon_{22})/kT} + \phi_2) + x_1 z_1 \ln(\phi_2 e^{(\varepsilon_{21}-\varepsilon_{11})/kT} + \phi_1)] \quad \text{and}$$

$$\frac{A^{ex,eng}}{NkT} = -\frac{1}{2}[x_2 z_2 \ln(\phi_1 e^{(\varepsilon_{12}-\varepsilon_{22})/kT} + \phi_2) + x_1 z_1 \ln(\phi_2 e^{(\varepsilon_{21}-\varepsilon_{11})/kT} + \phi_1)]$$

(16.4-15e)

This looks like the Wilson activity coefficient model[8]

$$\frac{G^{ex,eng}}{NkT} = \frac{A^{ex,eng}}{NkT} = -x_1 \ln(x_1 + x_2 \Lambda_{21}) - x_2 \ln(x_2 + x_1 \Lambda_{12}) \quad (16.4\text{-}16)$$

In fact, Eq. 16.4-15e can be made to look more like the Wilson model by setting $v_1 = v_2$ (so $\phi_i = x_i$) and defining $\Lambda_{12} = e^{(\varepsilon_{12}-\varepsilon_{22})/kT}$ and $\Lambda_{21} = e^{(\varepsilon_{21}-\varepsilon_{22})/kT}$ to obtain

$$\frac{A^{ex,eng}}{NkT} = -\frac{1}{2}[x_1 z_1 \ln(x_1 + x_2 \Lambda_{21}) + x_2 z_2 \ln(x_2 + x_1 \Lambda_{12})] \quad (16.4\text{-}17)$$

Equation 16.4-17 is an augmented Wilson model with four adjustable parameters (z_1, z_2, Λ_{12}, and Λ_{21}). It reduces to the usual Wilson model when the coordination numbers of both species are the same (equal to 2). However, if $z_1 = z_2 = z$ but not equal to 2, and instead is treated as an adjustable parameter, a three-parameter Wilson equation[18,19] results.

Had surface areas a been used in the local coordination model of Eq. 16.4-15a instead of volumes v, we would instead obtain

$$\frac{A^{ex,eng}}{NkT} = -\frac{z}{2}\left[x_2 \frac{z_2}{z}\ln(\theta_1 \Lambda_{12} + \theta_2) + x_1 \frac{z_1}{z}\ln(\theta_2 \Lambda_{21} + \theta_1)\right]$$

$$= -\frac{z}{2}[x_2 q_2 \ln(\theta_1 \Lambda_{12} + \theta_2) + x_1 q_1 \ln(\theta_2 \Lambda_{21} + \theta_1)] \quad (16.4\text{-}18)$$

where θ_i is the surface area fraction of species i and it is reasonable to assume that the coordination number ratios q_1 and q_2 are proportional to the species surface areas. By combining Eq. 16.4-3b with the Guggenheim-Staverman equation for the entropic term (Eq. 19a), we obtain (written in multicomponent form for brevity)

$$\frac{G^{ex}}{NkT} = \frac{A^{ex,ent} + A^{ex,eng}}{NkT}$$

$$= \sum_i x_i \ln \frac{\phi_i}{x_i} + \frac{z}{2}\sum_i x_i q_i \ln \frac{\theta_i}{\phi_i} - \sum_i x_i q_i \ln\left(\sum_j \theta_j \Lambda_{ij}\right)$$

---

[18] G. Scatchard and G. M. Wilson, *J. Amer. Chem. Soc.* **86**, 133 (1964).
[19] H. Renon and J. M. Prausnitz, *AIChE J.* **15**, 785 (1969).

16.4 Activity Coefficient Models

which is the well-known UNIQUAC equation.[14] It is also the basis for the predictive UNIFAC[20,21] method when molecules are considered to be collections of functional groups.

In a similar manner, other assumptions about the species-species coordination number will lead to other expressions for the energetic contribution to the excess Helmholtz energy of mixing. Activity coefficient models are then obtained from the sum of $A^{\text{ex,eng}}$ and $A^{\text{ex,ent}}$.

The final example to be considered is a simplification of Eq. 16.4-15b by assuming the volume parameter of each species v is the same, but introducing a nonrandom parameter α, so that

$$N_{ij} = z_j \frac{N_i e^{\alpha(\varepsilon_{ij}-\varepsilon_{jj})/kT}}{N_j + N_i e^{\alpha(\varepsilon_{ij}-\varepsilon_{jj})/kT}} \qquad (16.4\text{-}19)$$

which, if we use in Eq. 16.4-8b but neglecting the temperature dependence of the coordination number, gives

$$A^{\text{ex,eng}} = \frac{N}{2}\left[\begin{array}{c} x_2(\varepsilon_{22}-\varepsilon_{12})z_2 \dfrac{N_1 e^{\alpha(\varepsilon_{12}-\varepsilon_{22})/kT}}{N_2 + N_1 e^{\alpha(\varepsilon_{12}-\varepsilon_{22})/kT}} \\ + x_1 z_1(\varepsilon_{11}-\varepsilon_{12}) \dfrac{N_2 e^{\alpha(\varepsilon_{21}-\varepsilon_{11})/kT}}{N_1 + N_2 e^{\alpha(\varepsilon_{21}-\varepsilon_{11})/kT}} \end{array}\right]$$

or

$$\frac{A^{\text{ex,eng}}}{NkT} = \frac{1}{2}\left[\begin{array}{c} x_2\dfrac{(\varepsilon_{22}-\varepsilon_{12})}{kT}z_2 \dfrac{x_1 e^{\alpha(\varepsilon_{12}-\varepsilon_{22})/kT}}{x_2 + x_1 e^{\alpha(\varepsilon_{12}-\varepsilon_{22})/kT}} \\ + x_1 z_1(\varepsilon_{11}-\varepsilon_{12}) \dfrac{x_2 e^{\alpha(\varepsilon_{21}-\varepsilon_{11})/kT}}{x_1 + x_2 e^{\alpha(\varepsilon_{21}-\varepsilon_{11})/kT}} \end{array}\right]$$

$$= \frac{x_1 x_2}{2}\left[z_2 \frac{\tau_{12} G_{12}}{x_2 + x_1 G_{12}} + z_1 \frac{\tau_{21} G_{21}}{x_1 + x_2 G_{21}}\right] \qquad (16.4\text{-}20)$$

where the following notation has been used: $\tau_{ij} = \frac{\varepsilon_{jj}-\varepsilon_{ij}}{kT}$ and $G_{ij} = e^{-\alpha(\varepsilon_{ij}-\varepsilon_{jj})/kT}$. If we take $z_1 = z_2 = 2$, the last form of Eq. 16.4-20 is the Nonrandom Two Liquid model (or NRTL model, as it is usually referred to) of Renon and Prausnitz,[13] though with a slight reinterpretation of the meaning of the parameters (which arises from assuming the temperature independence of N_{ij}). Instead, using the temperature dependence in the integral of Eq. 16.4-7b, which would result in a new and different form of the excess free-energy-of-mixing equation, similar to the Wilson equation. Also, different choices for the total species coordination numbers, z_1 and z_2, would result in either one (if $z_1 = z_2 \neq 2$) or two (if $z_1 \neq z_2$) additional parameters in the model.

Once again, the expression above for $A^{\text{ex,eng}}$ could be combined with the $A^{\text{ex,ent}}$ expressions discussed earlier to produce a new, as-yet unused activity coefficient

---

[20]A. Fredenslund, J. Gmehling and P. Rasmussen, *Vapor-Liquid Equilibrium Using UNIFAC*, Elsevier, Amsterdam, (1977).

[21]A. Fredenslund, R. L. Jones and J. M. Prausnitz, *AIChE J*. **21**, 1086 (1975), and numerous other papers by the Fredenslung group and the Gmehling group.

model. In a similar manner, different assumptions about the species-species coordination number lead to other expressions for the energetic contribution to the excess Helmholtz energy of mixing. Activity coefficient models are then obtained by taking the derivative with respect to the mole number of the sum of $A^{\text{ex,eng}}$ and $A^{\text{ex,ent}}$.

The equations developed above were all based on the square-well potential. For more realistic potentials, $-\varepsilon_{ij}$ of the square-well potential is replaced by $\langle u^s_{ij} \rangle$, which is the ensemble averaged non-hard-core part of the interaction potential between species i and j. The problem that arises is that $\langle u^s_{ij} \rangle$ is, in principle, a function of temperature and composition. If each $\langle u^s_{ij} \rangle$ is taken as a function of composition, the resulting excess Helmholtz (Gibbs) energy functions will have a higher-order mole fraction dependence than the comparable models in which it was taken to be a constant.

Table 16.4-1 summarizes the activity coefficient models that result from the Generalized van der Waals partition function for different choices for the local coordination number.

ILLUSTRATION 16.4-1

Show for a binary mixture of species with molecular volumes of v_1 and v_2 that, with the assumptions that $N_{21} = \phi_2 z_1$, $N_{12} = \phi_1 z_2$, and $z_2 = \frac{v_2}{v_1} z_1$ where $\phi_i = \frac{x_i v_i}{x_1 v_1 + x_2 v_2}$, the Flory-Huggins excess Helmholtz expression of Eq. 10.4–11 is obtained.

SOLUTION

Since the species-species coordination number is independent of temperature, we can start from

$$A^{\text{ex,eng}} = N[x_2(\varepsilon_{22} - \varepsilon_{12})N_{12} + x_1(\varepsilon_{11} - \varepsilon_{12})N_{21}]$$

$$= N[x_2(\varepsilon_{22} - \varepsilon_{12})\phi_1 z_2 + x_1(\varepsilon_{11} - \varepsilon_{12})\phi_2 z_1]$$

$$= N\left[x_2(\varepsilon_{22} - \varepsilon_{12})\phi_1 \frac{v_2}{v_1} z_1 + x_1(\varepsilon_{11} - \varepsilon_{12})\phi_2 z_1\right]$$

$$= Nz_1 \left[\frac{x_2(\varepsilon_{22} - \varepsilon_{12})x_1 v_1 \frac{v_2}{v_1}}{x_1 v_1 + x_2 v_2} + \frac{x_1(\varepsilon_{11} - \varepsilon_{12})x_2 v_2}{x_1 v_1 + x_2 v_2}\right] = Nz_1 x_1 \phi_2 [\varepsilon_{11} + \varepsilon_{22} - 2\varepsilon_{12}]$$

$$= \frac{Nz_1 x_1 v_1 \phi_2}{x_1 v_1 + x_2 v_2}[\varepsilon_{11} + \varepsilon_{22} - 2\varepsilon_{12}] \frac{1}{v_1}(x_1 v_1 + x_2 v_2)$$

$$= \phi_1 \phi_2 \left(x_1 + x_2 \frac{v_2}{v_1}\right) z_1 [\varepsilon_{11} + \varepsilon_{22} - 2\varepsilon_{12}]$$

$$= \chi \phi_1 \phi_2 \left(x_1 + x_2 \frac{v_2}{v_1}\right)$$

with the following for the Flory parameter $\chi = z_1[\varepsilon_{11} + \varepsilon_{22} - 2\varepsilon_{12}]$

Table 16.4-1 Local Coordination Number Models and the Activity Coefficient Models that Result

Local coordination number	Activity coefficient model
$N_{ij} = x_i z$	1-constant Margules
$N_{ij} = x_i z_j$	1-constant Margules
$N_{ij} = \dfrac{x_i v_i}{\sum_i x_i v_i} z = \phi_i z$	Van Laar equation
$\dfrac{N_{ij}}{N_{jj}} = \dfrac{N_i v_i}{N_j v_j}$ with $N_{ij} + N_{jj} = z_j$	a Van Laar-like equation
$\dfrac{N_{ij}}{N_{jj}} = \dfrac{N_i v_i}{N_j v_j} e^{(\varepsilon_{ij} - \varepsilon_{jj})/kT}$ with $N_{ij} + N_{jj} = z_j$	A family of local compositions models, such as the Wilson and UNIQUAC equations, depending on the other assumptions made.
Surface area fractions	As above with v_i being the surface area rather than volume
$N_{21} = \phi_2 z_1$, $N_{12} = \phi_1 z_2$, and $z_2 = \dfrac{v_2}{v_1} z_1$	Flory term in the Flory-Huggins equation

16.5 CHAIN MOLECULES AND POLYMERS

Developing equations of state by considering only simple spheres and proceeding as in the previous sections is not very productive for polymers and other chain molecules. Instead, a beads-on-a-chain model has been used for modeling long chain molecules. In such molecules, there are high-frequency (short wavelength) vibrational and rotational motions that are similar in behavior to those energy modes of small molecules—for example, bond vibrations and some rotations around bonds. But there are also quite different hindered rotations, chain entanglement motions, and other motions that in principle could be identified by normal mode analysis (see Chapter 4) involving many of the "beads" on the chain simultaneously. These are large wavelength (that is, extending over many bead diameters) motions, that are therefore low-frequency motions; they are very different from the internal modes of small molecules.

Prigogine[22] suggested that the totality of rotational and vibrational motions in a chain molecule be separated into internal (high-frequency) and external (low-frequency) energy modes. The high-frequency rotational and vibrational modes (referred to as *external modes*, designated by the subscript ext below) are then treated as was the case for small molecules (see Chapters 3 and 4), and therefore are density independent. However, the long wavelength modes (referred to as *internal modes*, indicated by the subscript int) may depend on chain entanglements between neighboring molecules, and are assumed to be a function of density.

---

[22] I. Prigogine, *The Molecular Theory of Solutions*, North Holland Pub. Co., Amsterdam, 1957, Chap. XVI.

330 Chapter 16: The Derivation of Thermodynamic Models

The Generalized van der Waals partition function for chain molecules is

$$Q(N, V, T) = \frac{1}{N!} \left(\frac{V}{\Lambda^3}\right)^N (q_{r,\text{int}} q_{v,\text{int}} q_{\text{elect}})^N (q_{r,\text{ext}} q_{v,\text{ext}})^N \left[\frac{V_f}{V} \exp\left(-\frac{\Phi}{kT}\right)\right]^N \tag{16.5-1}$$

The analogous partition function for an ideal gas (subscript IG) of these chain molecules is

$$Q^{\text{IG}}(N, V, T) = \frac{1}{N!} \left(\frac{V}{\Lambda^3}\right)^N (q_{r,\text{int}} q_{v,\text{int}} q_{\text{elect}})^N (q_{r,\text{ext}} q_{v,\text{ext}})^N_{\text{IG}} \tag{16.5-2}$$

where the external degrees of freedom may behave differently in a dense fluid and in an ideal gas. In fact, this occurs for the three translational degrees of freedom in the single-particle partition function in that

$$q_{\text{trans,IG}} = \frac{V}{\Lambda^3} \text{ in the ideal gas and } q_{\text{trans}} = \frac{V_f}{\Lambda^3} \exp\left(-\frac{\Phi}{kT}\right) \text{ in a dense fluid}$$

so that

$$\frac{q_{\text{trans}}}{q_{\text{trans,IG}}} = \frac{V_f}{V} \exp\left(-\frac{\Phi}{kT}\right) \tag{16.5-3}$$

Therefore

$$\frac{Q(N, V, T)}{Q^{\text{IG}}(N, V, T)} = \frac{(q_{r,\text{ext}} q_{v,\text{ext}})^N}{(q_{r,\text{ext}} q_{v,\text{ext}})^N_{\text{IG}}} \left[\frac{V_f}{V} \exp\left(-\frac{\Phi}{kT}\right)\right]^N \tag{16.5-4}$$

The further suggestion was that the ratio of the dense fluid to ideal gas external partition function for each mode of these low-frequency external motions have the same temperature and density dependence as a one-dimensional translational partition function of Eq. 16.5-3. That is, if there is a total of κ external models, then

$$\frac{q_{r,\text{ext}} q_{v,\text{ext}}}{(q_{r,\text{ext}} q_{v,\text{ext}})_{\text{IG}}} = \left[\frac{V_f}{V} \exp\left(-\frac{\Phi}{kT}\right)\right]^{\kappa/3} \tag{16.5-5}$$

Therefore

$$\frac{Q(N, V, T)}{Q^{\text{IG}}(N, V, T)} = \frac{(q_{r,\text{ext}} q_{v,\text{ext}})^N}{(q_{r,\text{ext}} q_{v,\text{ext}})^N_{\text{IG}}} \left[\frac{V_f}{V} \exp\left(-\frac{\Phi}{kT}\right)\right]^N$$

$$= \left[\left[\frac{V_f}{V} \exp\left(-\frac{\Phi}{kT}\right)\right]^{\kappa/3}\right]^N \left[\frac{V_f}{V} \exp\left(-\frac{\Phi}{kT}\right)\right]^N$$

$$= \left[\frac{V_f}{V} \exp\left(-\frac{\Phi}{kT}\right)\right]^{N(1+\kappa/3)} \tag{16.5-6}$$

and

$$A(N, V, T) - A^{\text{IG}}(N, V, T) = A^{\text{res}}(N, V, T)$$

$$= -N\left(1 + \frac{\kappa}{3}\right)\left[kT \ln \frac{V_f}{V} - \Phi\right] \tag{16.5-7}$$

16.5 Chain Molecules and Polymers

where $\kappa = 0$ for the simple molecules considered previously, and κ is equal to the number of rotations and vibrations to be treated as external degrees of freedom for the chain molecules being considered here. In application, κ is treated as an adjustable parameter.

The partition function of Eq. 16.5-6 is the basis for the perturbed hard chain (PHC) equation of state (including its variants, such as the simplified perturbed hard chain, the perturbed soft chain, etc.). In the PHC equation,[23] the free volume is modeled using the Carnahan-Starling expression extended to chain molecules:

$$V_f = V \exp\left[\frac{\eta'(3\eta' - 4)}{(1 - \eta')^3}\right] \quad \text{with} \quad \eta' = \frac{s\pi\rho\sigma^3}{3} = s\eta \quad (16.5\text{-}8)$$

where s is the effective number of segments (beads) on the chain, σ is the diameter of a bead, and ρ is the number density of chains (so that $s\rho$ is the number density of beads). Based on the Alder et al.[7] equation, the following expression was used for the mean potential:

$$\frac{\Phi}{kT} = \sum_m \sum_n A_{mn} \left(\frac{\varepsilon r}{kT}\right)^m (s\rho\sigma^3)^n \quad (16.5\text{-}9)$$

where r is a normalized surface area parameter defined so that $r = 1$ for a single segment (bead) of the molecule. With these choices, the perturbed hard-chain equation of state is

$$P = \frac{RT}{\underline{V}}\left[\frac{1 + \eta' + (\eta')^2 + (\eta')^3}{(1 - \eta')^3}\right] - \sum_m \sum_n A_{mn}\left(\frac{\varepsilon r}{kT}\right)^m (s\rho\sigma^3)^{n+1} \quad (16.5\text{-}10)$$

A simplified version of the perturbed hard-chain equation uses the Kim, Lin, and Chao[4] free-volume term of Eq. 16.2-3:

$$P = \frac{RT}{\underline{V}}\left[\frac{1 + 3.08\eta'}{1 - 1.68\eta'}\right] - \sum_m \sum_n A_{mn}\left(\frac{\varepsilon r}{kT}\right)^m (s\rho\sigma^3)^{n+1} \quad (16.5\text{-}11)$$

Other versions of the perturbed hard-chain equation have been obtained by replacing the mean potential term with some of the other expressions discussed in the previous section. Also, more complicated equations can be obtained by replacing the free-volume term for hard-spheres with those obtained from correlating simulation data for soft spheres, such as Lennard-Jones 12-6 potentials.

There are a larger number of adjustable parameters in models for chain molecules than in the simpler cubic equations of state. If the molecule is a hydrocarbon or a homopolymer (a polymer consisting of only a single type of functional group—for example CH_2, as in polyethylene), the number of segments s is obvious. However, for chain molecules consisting of several types of functional groups of different sizes and interaction energies, s is taken to be an adjustable parameter. In both cases, the number of external rotational and vibrational modes κ is also considered to be an adjustable parameter. These are in addition to the free volume and energy parameters in the model.

---

[23] S. Beret and J. M. Prausnitz, *AIChE J.*, **21**, 1123 (1975).

332 Chapter 16: The Derivation of Thermodynamic Models

A useful way of thinking about this model is that it is the summation of two contributions to the residual free energy. The first contribution is from the unattached segments, and the second is from the formation of chains from these free segments. Defining $\Phi' = \Phi/s$ to be the mean potential per segment (bead) of the molecule, Eq. 16.5-7 can be rewritten as

$$A^{\text{res}}(N, V, T) = -Ns \left[kT \ln \frac{V_f}{V} - \Phi' \right] - N \left[\left(1 + \frac{\kappa}{3} - s \right) kT \ln \frac{V_f}{V} \right]$$

$$= A^{\text{seg}} + A^{\text{chain}} \tag{16.5-12}$$

where the first term on the right is the contribution to the residual free energy from the segments (beads) if they were not connected to form a chain, and the second term is the additional contribution due to chain formation. Note that if $\kappa = 0$ (no external rotation or vibration motions), this expression reduces to that for a simple molecule (i.e., for diatomic, triatomic, or other small molecules for which chain entanglements are unimportant). Furthermore, if the number of segments s is 1, the equation reduces to that for an atomic fluid. More sophisticated expressions can be used for A^{seg} and A^{chain}; one set of approximations is discussed in the next section.

16.6 HYDROGEN-BONDING AND ASSOCIATING FLUIDS

Hydrogen-bonding and other types of associations in fluids involve very directional and strong interactions. These interactions are stronger than the forces considered in the previous sections, but weaker than a covalent chemical bond. In such cases, temporary chemical complexes form and then dissociate. While the lifetimes of these complexes (which may be as simple as dimers or as complex as the large hydrogen-bond network that occurs in water) are short on a macroscopic time scale, they are much longer than interaction collision times in nonassociating fluids. In such cases, the effects of association must be included. Chapman et al.[24,25] have, in a series of papers, developed the SAFT (statistical associating fluid theory) equation of state by writing the residual Helmholtz energy as

$$A^{\text{res}}(N, V, T) = A^{\text{seg}} + A^{\text{chain}} + A^{\text{assoc}} \tag{16.6-1}$$

which corresponds to a product of partial partition functions

$$\frac{Q(N, V, T)}{Q^{\text{IG}}(N, V, T)} = Q^{\text{seg}}(N, V, T) Q^{\text{chain}}(N, V, T) Q^{\text{assoc}}(N, V, T) \tag{16.6-2}$$

The extent of association is of special importance, since this determines the number of associated and unassociated molecules in the system—that is, the composition of the mixture. Based on the statistical mechanical perturbation theory of Wertheim,[26]

---

[24] W. G. Chapman et al., *Fluid Phase Eq.* **52**, 31 (1982).

[25] W. G. Chapman et al., *Ind. Eng. Chem. Res.* **29**, 1709 (1990).

[26] M. S. Wertheim, *J. Stat. Phys.* **35**, 35 (1984) and **42**, 459 (1986) and *J. Chem. Phys.* **85**, 2929 (1986) and **87**, 7323 (1987).

16.6 Hydrogen-Bonding and Associating Fluids

the association term used is

$$\frac{A^{\text{assoc}}}{NkT} = \sum_{A} \left[\ln X^A - \frac{X^A}{2} \right] + \frac{M}{2} \qquad (16.6\text{-}3)$$

In this equation, M is the total number of association sites on a molecule, X^A is the mole fraction of molecules that are not bonded at association site A, and the sum is over all association sites A. The calculation of X^A requires a number of assumptions, and these will not be considered here. However, the parameters in the calculation include site-site association strength parameters and bonding-site volumes. The approximate chain term used in the original SAFT equation based on simulation data is

$$\frac{A^{\text{chain}}}{NkT} = X(1-m) \ln g^{\text{hs}}(d) \qquad (16.6\text{-}4)$$

where m is the effective segment length and $g^{\text{hs}}(d)$ is the hard-sphere radial distribution function at contact for spheres of diameter d at the reduced density η, evaluated using Carnahan-Starling result

$$g^{\text{hs}}(d) = \frac{2-\eta}{2(1-\eta)^3} \qquad (12.5\text{-}3)$$

Finally, the segment term is

$$\frac{A^{\text{seg}}}{NkT} = \left[\frac{4\eta - 3\eta^2}{(1-\eta)^2} + \frac{\varepsilon}{kT} \rho\sigma^3 \left(a_0(\rho\sigma^3) + \frac{\varepsilon}{kT} a_1(\rho\sigma^3) \right) \right] Xm \qquad (16.6\text{-}5)$$

Here, the first term is the Carnahan-Starling expression for the Helmholtz energy of the hard-sphere fluid, and ε and σ are the Lennard-Jones parameters for segment-segment interactions; these last two parameters, together with m, were obtained from fitting vapor pressure and saturated liquid density data. The functions $a_0(\rho\sigma^3)$ and $a_1(\rho\sigma^3)$, which depend on reduced density, were obtained from correlations of computer simulation results for the Lennard-Jones 12-6 fluid.

The resulting equation of state for this model is obtained from the volume derivative of the Helmholtz energy, the internal energy is obtained from its temperature derivative, and the chemical potential from its particle number derivative. Each of these results in rather complex equations. For example, a change in particle number needed to determine the chemical potential changes the density, which then changes the hard-sphere radial distribution at contact and the extent of association; so the segment, chain, and association terms all change, and these contributions must be included in the derivative.

Only the original SAFT equation has been discussed here; there are many later versions and improvements, and these have been shown to be very effective for describing the properties and phase behavior of both pure fluids and mixtures of nonassociating chain fluids (i.e., hydrocarbons) and associating fluids (i.e., acetic acid, methanol, and water). This might be expected, given the theoretical basis and the number of parameters in the model, which includes the number of association sites on each molecule. Also, the resulting equation of state is in good agreement with computer simulation. This is not unexpected, since with the exception of the form of the association term, computer simulation data were used either directly or indirectly in the formulation of the model.

Chapter 16: The Derivation of Thermodynamic Models

CHAPTER 16 PROBLEMS

16.1 Show that using Eq. 16.2-12 leads to the van der Waals equation of state.

16.2 Show that the expression for the coordination number in Table 16.2-1 leads to the Redlich-Kwong equation of state.

16.3 Show that the expression for the coordination number in Table 16.2-1 leads to the temperature-dependent van der Waals equation of state.

16.4 Show that the expression for the coordination number in Table 16.2-1 leads to the Peng-Robinson equation of state, and develop the relationship between $C_3(T)$ and $a(T)$ in that equation.

16.5 Show that the assumption $N_{ij} = \frac{N_i}{V} C_{ij} e^{\varepsilon_{ij}/kT}$ leads to the van der Waals one-fluid mixing rule for equations of state.

16.6 Derive Eqs. 16.4-14b and 16.4-14c.

16.7 Derive the two-constant Margules equation, Eq. 16.4-12c, using the coordination number described in the text.

16.8 Show that the assumption $N_{ij} = x_i z_j e^{\varepsilon_{ij}/kT}$, with the restriction that $N_{ij} + N_{jj} = z_j$, leads to the one-constant Margules equation with a temperature-dependent parameter.

16.9 Derive the expression for the excess Helmholtz energy of mixing at constant T and V that arises from assuming $\frac{N_{ij}}{N_{jj}} = \frac{N_i}{N_j} e^{(\varepsilon_{ij}-\varepsilon_{jj})/kT}$ with the restriction that $N_{ij} + N_{jj} = z_j$.

16.10 Derive the activity coefficient model corresponding to the excess Helmholtz free energy of mixing of Problem 16.9.

16.11 Using the simple van der Waals model for the free volume and coordination number, develop explicit expressions for A^{seg} and A^{chain} from Eq. 16.5-12 and the residual contribution to the equation of state.

16.12 Show that by assuming that $\langle u_{ij}^s \rangle = u_{ij,0}^s + x_j u_{ij,1}^s$ and using the random mixing model with each species having the same coordination number, $N_{ij} = x_i z$, leads to the two-constant Margules equation

$$A^{\text{ex,eng}} = N x_1 x_2 (\alpha + \beta x_1)$$

16.13 Show that using the coordination number model of Eq. 16.2-17 leads to the equation of state of Eq. 16.2-18.

Index

activity
  absolute, 119, 122–123
  coefficient, 164, 168–178, 279, 291–295
    models, 171–172, 291–296, 318–334
      Debye-Huckel, 291–296
      Flory-Huggins, 172–178, 184, 279, 297, 319, 323, 328–329
      Guggenheim-Staverman, 177, 320, 326
      Lattice, 169–178
      Margules, 322–329, 334
      NRTL, 327
      regular solution, 171, 184, 324
      UNIFAC, UNIQUAC, 178, 327
      Wilson, 172, 239, 315, 326–329
adsorption, 82, 95, 97, 162, 252
amino acid, 182–183
association reaction, 86, 123, 332–333

Barker-Henderson perturbation theory, 262–272, 275
binomial expansion, 67, 83–84, 97
Boltzmann
  energy distribution, 6, 18–19
  factor, 18–19, 22, 29, 102, 139, 164, 167, 172, 177, 182, 186, 189, 196, 198, 220, 224, 227–228, 239, 249–252, 263, 299–300
bond
  bending, 61
  chemical, 48, 161, 213
  dissociation energy, 46, 49, 55, 78
  energy, 49
  hydrogen, 123, 161, 211, 319, 332
  stretching, 44, 53, 61
    asymmetric, symmetric, 61
Born-Oppenheimer approximation, 22, 44
Boyle temperature, 126–129

canonical partition function, 8–14, 16–32, 64, 83, 89, 94, 99–100, 116, 119, 178, 298
cell model, 163, 184–185
chain molecule, 172–182, 242–245, 274, 298, 329–333
chemical
  equilibrium, ideal gas mixture, 64–84

equilibrium constant, 67–86
extent of reaction, 74, 77–78
potential, 21, 26–27, 50, 82, 89, 93–96, 159, 215, 237–238, 252, 279, 290, 294, 333
reaction, ideal gas mixture, 64–86
stoichiometric coefficient, 73
cluster diagram, function, 102, 110–114, 118, 194, 197, 218
collective motions, 151, 329–331
colloids, 137–139, 213, 227, 232, 235–236, 274
compressibility, 96, 123–124, 144, 214, 239–240, 256, 269, 272–274, 309
  equation, 201–202, 223, 227
  isothermal, 200, 207–208
computer simulation, 163, 198–199, 210, 213, 219, 228, 228, 232, 241–257, 269–271, 297, 303, 307–310, 316–319, 331–333
  entropy, 249
  Gibbs energy, 249
  Helmholtz energy, 249
  internal energy
  molecular-dynamics, 198, 210, 241–256, 270
  Monte Carlo, 198, 241–256, 270, 308, 317
    transition probability, matrix, 250–251
configurational
  distribution functions, 187–189
  energy, 190–191, 298–334
  integral, 100, 102, 186–196, 298–334
cooperative behavior, 181
coordination
  number, 164, 169, 172–173, 211, 301–329, 334
  shell, 162, 191, 197, 208, 212, 225, 232
    first, 208, 211–212, 225
    second, 212, 232
correlation
  direct, 197, 220–221
  function
    n-body, 188–189, 218, 269
    pair, 185–275
    total, 207, 220–221
    two-body, 185–275
  indirect, 197, 221
corresponding states, 131, 145–146

335

crystal, 147–162
  Debye, 150–162
  Einstein, 147–162
  entropy, 149, 153, 161

De Broglie wave length, 23, 47, 100, 298
Debye
  crystal, 150–162
  -Hückel, 244, 280–298
  length, 235, 288, 296
  limiting law, 276, 280, 292–295
degeneracy, 4–6, 9–11, 19, 43, 48, 87–88, 161, 164, 172
  electronic, 6, 24–49, 54, 79, 85–88
  nuclear, 25, 45
  rotational, 45–47
  translational, 4, 29, 35–36
  vibrational, 47–48
departure functions
  enthalpy, enthalpy, internal energy, 116, 145–146
diatomic molecule
  partition function, 44–49
  rotational energy, 12–13, 44–49, 53, 63, 99, 115
  vibrational energy, 23, 44–45, 47, 63, 147–156
dielectric constant, 235–236, 277–281, 293
dimerization, 86
dipole, 126, 130, 138–139, 234–235, 275
dissociation, 78, 80, 85–86
  energy, 46, 49, 55, 78
distribution function
    three-particle, 218–220
    two-particle, see radial distribution function
  Gaussian probability distribution, 29–32, 72–73, 199–200
    standard deviation, 29–30, 72, 199
  probability, 2, 8, 10–11, 30–32, 151, 186, 250–251
  random, 167, 171, 196
DLVO model, 236

Einstein crystal, 147–162
electrical neutrality, 285
electrolyte solutions, 276–295
electronic
  energy states, 23–28, 42, 48–49, 54, 76–78, 99
  partition function, 23–28, 42, 48–49, 54, 76–78, 99
electrostatic, 281–291
  energy, 285–287, 290–291
  potential theory, 281, 284
energy
  dissociation, 46, 49, 55, 78
  fluctuations, 10, 29–33
  interaction, see interaction energy
  kinetic, 2, 10, 42, 53, 59, 63, 98–99, 126, 129, 132, 247, 254–256, 299
  landscape, 147–148
  level, 3–10
  modes
    electronic, 23–28, 42, 48–49, 54, 76–78, 99
    independent, 9, 12–13, 22
    potential, 13, 59, 98–101, 122, 126, 130, 282; see also interaction energy
    rotational, 45–46, 52–53, 63
    translational, 3–4, 11–14, 17, 22–23, 29, 35, 44–45, 52–53, 61–62, 80, 98–99, 262, 298, 330
    vibrational, 44–49, 51–55, 58–60, 148–159
  states, 23–28, 42, 48–49, 54, 76–78, 99
  accessible, 13, 66
ensemble
  canonical, 8–14, 16–32, 64, 83, 89, 94, 99–100, 116, 119, 178, 298
  grand canonical, 2, 87, 89–92, 95–98, 118–122, 199–201, 207, 252
    restricted, 93
    semi, 93
  isobaric-isothermal, 92–93
  microcanonical, 2, 87–88, 94–95, 164, 172, 248, 255
enthalpy, 27, 50, 118, 131, 160, 176–177
  of mixing, 176–177
    excess, 176–77
entropy, 20, 27, 33–35, 37, 50, 92–95, 150, 155, 161, 164, 169–170, 172–177, 249
  at absolute zero, 63, 149–150, 155, 161
  departure, 50, 116, 145–146
  excess of mixing, 168–177
  Gibbs equation, 16, 33–34, 92
  of mixing, 168–177, 318–329
    excess, 168–177, 318–329
  residual, 161
  third law, 20, 63, 147, 150, 155, 161
equal *a priori* probability, 5–10
equation of state
  Alder, 331
  Bender, 141
  Benedict-Webb-Rubin, 141

Carnahan-Starling, 227–228, 239–240, 302–307, 312, 331–333
combining rules, 136–138, 143–144, 307, 312
cubic, 137, 143, 166–168, 273, 307, 331
hard sphere, 227–228, 239–240, 302–307, 312, 331–333
mixing rules, 142–144, 301, 310–312, 315–316, 334
    one-fluid, 143–144, 315–316
    nonquadratic, 316
Peng-Robinson, 142, 167–168, 184, 273, 306–308, 311, 334
perturbed
    hard-chain, 274, 331
    hard-sphere, 274
    soft-chain, 274, 331
    soft-sphere, 174
Redlich-Kwong, 142, 167–168, 184, 273, 306–308, 334
SAFT, 332–333
Soave, 146, 167–168, 184, 273, 306, 308, 334
van der Waals, 142–144, 166–168, 184, 273–274, 306, 316–316, 334
    extended, 274
virial, 87, 95, 98–146, 166, 168, 222, 224, 237–238, 245, 247, 273, 278, 297
    extended, 141
    polyatomic molecules, 114
    range of applicability, 123
equipartition of energy, 35
equilibrium constant, 67–86
ergodic, 5, 18, 205, 242, 251, 255
    hypothesis, 5, 18, 205, 242, 255
excluded volume, 211, 232, 273, 301–302

fiber, 86
Flory-Huggins model, 172–178, 184
fluctuations, 6, 29–33, 94, 96, 130, 198–201, 242, 247, 255–256
    chemical reactions, 70–73
    dependence on system size, 29–33, 94–96
    energy, 10, 29–33
    number of particles, 70–73, 94
Fourier transform, 197, 203–208, 223–224
free volume, 272, 300–312, 319–320, 331–334
fugacity, 117, 146

gas, 18–86, 301–317
    ideal, 16–86
        diatomic, 44–86

monatomic, 16–42
polyatomic, 44–86
nonideal, *see* virial and other equations of state
Gauss' theorem, 284–285
Gaussian distribution, 29–32, 72–73, 199–200
Gibbs
    -Duhem equation, 295
    energy, 21, 26, 87, 95–96, 169, 171, 279, 290, 296, 328
        excess, of mixing, 169, 171, 296, 328
    entropy equation, 16, 33–34, 92
    paradox, 16, 37–39
graph theory, 98, 102–114, 197, 221–222, 223
graphite, 82, 97, 161
ground state
    electronic, 24
    nuclear, 25
Guggenhein-Staverman model, 177, 320, 326

Hamaker constant, 234–237
hard-sphere fluid, 125–127, 136, 139, 196–197, 211–213, 222–228, 231–235, 252, 257–265, 274, 277, 284, 300–302
harmonic oscillator, 45, 47–48, 63, 148, 161
heat capacity, 131, 200, 254
    crystal
        Debye, 153–159
        Einstein, 149–151, 157–159
        DuLong-Petit Law, 150
    ideal gas
        diatomic, 49, 51
        monatomic, 21, 25, 27, 30–31, 42
        polyatomic, 56–58
    reaction, 77–81
    Ising model, 181
Heisenberg uncertainty principle, 3, 13, 47, 98
Helmholtz energy, 21, 7, 49, 55, 67, 95–96, 165, 169, 228, 231, 261, 281, 290, 300, 318–319, 323–328, 332–333
    excess, 318–319, 323–324, 327–328
Henry's law, 279
helix-coil, 43, 86, 181–184
hydrogen, 45–46, 63, 85–90, 137, 197, 203, 211, 213
    bonding, 123, 161, 211, 319, 332
    deuterium exchange, 85, 211
    ortho, para, 45, 85
hypernetted chain equation, 197, 222

ideal
    gas, 14, 16–82, 85–86
    mixture, 14, 64–82, 85–86
indistinguishability, 3–4, 14–15, 37–39, 45–46, 65, 82, 97, 119, 162–165, 169, 172, 174, 187, 303
infinite dilution, 279–280
information theory, 35
integral
    configuration, 98–115, 119, 122, 185–187, 191, 194, 231, 282, 298–300
    equation theory, 163, 197–198, 210, 213, 216–228
    irreducible, 102–111
interacting molecules, 98–334
interaction energy, 98–102, 110, 115, 137, 148–149, 166–171, 198, 186, 189–197, 243, 245, 248, 270, 278, 287, 299–300, see also potential energy
interaction potential, 125–144
    atom-atom, see specific models
    attractive, 125–144, 270–271
    binary interaction parameter, 136–137
    charge-charge, see Coulomb interaction
    charge-dipole, 234
    charge-induced dipole, 234
    combining rules, 136–138, 143–144, 307, 312
        Lorentz-Berthelot, 136, 215
    Coulomb, 133, 235–237, 287
        shielded, 133, 235–237, 287
    dipole-dipole, 235
    exponential-6, 132
    hard-sphere, 125–127, 136, 139, 196–197, 211–213, 222–228, 231–235, 252, 257–265, 284
        diameter, 212–213
        packing fraction, 224–225, 272
        Lennard-Jones, 130–139, 196, 211, 229, 233, 252–255, 270, 274, 277, 331–333
    mean, 299–303, 306, 311–312, 321–323, 331–332
    mean force, potential, 213, 230–239, 274, 278, 284–284
    mixture, 136–137
    modified Buckingham, 132, 137
    multiatom molecules, see nonspherical molecules
    nonpairwise additivity, 101
    nonspherical molecules, 137–138, 197, 210–211, 272, 298, 307
    pairwise additivity, 101–102, 118, 122, 190–194, 197–198, 210, 217–219, 229, 241, 245, 253, 262–263, 299
    perturbation, 258, 262–270
    point centers of repulsion, 125
    reference, 257–258, 262–272
    repulsive, 125–128, 270–272
    shielded Coulomb, 133, 235–237, 287
    site-site, 137–140, 333
    square-well, 128–131, 134–135, 140, 144–145, 253, 256–265, 274–275, 303–307, 313, 317, 322, 328
    Stockmayer, 139
    Sutherland, 146
    triangular well, 146, 240, 275
    two-body see specific models
    united atom, 137–138
    Yukawa, 133
internal
    energy, 9–11, 18, 25–27, 49–52, 53–58, 76–77, 94, 144–145, 214–215, 229, 246, 275, 333
        diatomic ideal gas, 49–52
        monatomic ideal gas, 25–27, 41–42
        polyatomic ideal gas, 53–58
    modes
        bond bending, stretching 44, 53, 63
            asymmetric, symmetric, 61
        rotational, 45–46, 52–53, 63
        vibrational, 44–49, 51–55, 58–60, 148–159
integral
    configuration, 98–115, 119, 122, 185–187, 191, 194, 231, 282, 298–300
    equation theory, 163, 197–198, 210, 213, 216–228
    irreducible, 102–111
ion cloud, 244, 287
ionic strength, 276, 288, 292–293, 296
ionization, 28, 64, 78–81, 276–280
    degree, 28, 78–80
    energy, 78–81
    potential lowering, 79
ionized gas, 28, 64, 78–81, 276–280
ions, 64, 78–81, 276–280
Ising model, 178–184
isoelectric point, 234
isothermal compressibility, 200

Kirkwood superposition approximation, 219–220, 265

lattice
  crystal, 64, 78–81, 147–161, 163–183
  hexagonal closed packed, 164
  models, 163–183
  simple cubic, 64, 78–81, 212, 276–280, 164, 212,
  spacing, 169, 172
  vacancies, 164–165
liquids, 185–334
  molecular, 210–211
local
  composition, 172, 314–319, 325, 329
  mole fraction, 172
Lorentz-Berthelot combining rules, 136, 215

macroscopic state, definition, 2
magnetization, 178
MATLAB®, 222–226, 252–256, 274–275
maximum term method, 69–71, 94
Maxwell-Boltzmann distribution of velocities, 39–41
Maxwell distribution of speed, 40–41
Mayer cluster function, integral, 102–109, 194–195, 218
McMillan-Mayer standard state, 238
mean
  field approximation, 265
  ionic activity coefficient, 276, 291–296
  potential, 299–306, 311–312, 320–322, 331–332
  spherical approximation, 198, 222
Metropolis algorithm, 243, 250–251
microscopic
  reversibility, 251
  state, definition, 2
mixing rules, 142–144, 215, 301, 310–316
  one-fluid, 143–144, 315, 334
mixtures, 14, 21, 64–80, 83, 93, 99, 136–138, 142–144, 164, 168–172, 175–177, 228–229, 252, 279, 296, 310–329, 334
  ideal, 168, 175, 318
  ideal gas, 16–64, 64–86
  multicomponent, 65, 93, 99, 142–144, 168–169, 177, 228, 279, 296, 310–329
  one-fluid model, 215, 229
  random, 229, 314, 316
  virial equation for, 136–137, 142–144
molality, 279, 294–296
molecular simulation, 198, 241–256, 270, 308, 317

molecular dynamics, 198, 210, 241–256, 270
Monte Carlo, 198, 241–256, 270, 308, 317
molecule
  heteronuclear, 44–61, 64–80
    linear, 45, 53–54, 57–58, 61–62, 115, 138
    nonlinear, 52–57, 62, 115
  homonuclear, 45, 47, 50, 210
muscle, 86

neutron scattering, 197, 202–213
nonideal
  gas, 87, 95, 98–146, 166, 168, 222, 224, 237–238, 245, 247, 273, 278, 297
  solution, 164, 168–178, 279, 291–295
normal mode analysis, 53, 59–63, 151, 329
nuclear partition function, 24–26, 33, 45

occupation number, 4, 13–14, 65
one-fluid model, 143–144, 215, 229, 315–316
Ornstein-Zernike equation, 197, 220–224, 227, 303
osmotic
  equilibrium, 93, 235–240, 252
  pressure, 232, 237–238
  virial coefficient, 238–240

packing fraction, 224–225, 272
pair distribution function, 185–275
pairwise additivity, 101–102, 118, 122, 190–194, 197–198, 210
partially ionized gas, 28, 64, 78–81, 276–280
partition function
  canonical, 8–14, 16–32, 64, 83, 89, 94, 99–100, 116, 119, 178, 298
  electronic, 23–28, 42, 48–49, 54, 76–78, 99
  Generalized van der Waals, 297–334
  grand canonical, 2, 87, 89–92, 95–98, 118–122, 199–201, 207, 252
    restricted, 93
    semi, 93
  isobaric-isothermal, 92–93
  microcanonical, 2, 87–88, 94–94, 164, 172, 248, 255
  mixture, 14, 21, 64–80, 83, 93, 99, 136–138, 142–144, 164, 168–172, 175–177, 228–229, 252, 279, 296, 310–329, 334
  nuclear, 24–26, 33, 45
  rotational, 45–46, 52–53, 63

partition function (*continued*)
  translational, 3–4, 11–14, 17, 22–23, 29, 35, 44–45, 52–53, 61–62, 80, 98–99, 262, 298, 330
  vibrational, 48, 52–53, 55, 58, 61–63, 149, 152, 156
Percus-Yevick equation, 197–198, 216, 222–228, 240, 256, 274–275, 302, 302
periodic boundary conditions, 243–244
  lattice, 147
perturbation
  potential, 258, 270–271
  theory, 257–274, 332
    first order, 257–267, 269–272
    second order, 265–269
Planck's constant, 3, 17
plasma, *see* ionized gas
Poisson equation, 281–284
  -Boltzmann equation, 285
    linearized, 285–286
polyatomic molecules, 23, 49, 52–63, 100, 110, 114–116, 252, 297–334
  chain molecules, 172–182, 242–245, 274, 298, 329–333
  linear, 45, 53, 57, 59, 61–62, 115, 139
  nonlinear, 53–55, 59, 115
polymer, 43, 62, 86, 146, 164, 172–178, 181–184, 202, 213, 232–238, 261, 274, 329–332
postulates, 5
  equal a priori probability, 5–6, 10
  ergodic hypothesis, 5, 18, 205, 242, 255
potential, *see* interaction energy
  of mean force, 213, 230–239, 274, 278, 284–284
pressure
  energetic contribution, 165–6, 300, 311
  entropic contribution, 165–167, 300, 311, 318–320
  equation, 160, 193, 201, 227
probability
  distribution, 2, 5, 8, 10–11, 30–32, 151, 186, 250–251
  of occurrence, 5–8, 10–11, 28, 33–34, 40, 71–73, 87–88, 92, 96, 180, 182–184, 251
protein, 86, 133, 139–140, 172, 182, 213, 232–239, 256, 274

quantum
  mechanics, 3, 16–17, 24, 36, 39–40, 45, 47, 203, 213
  particle in a box, 3, 16–17
  numbers
    rotational, 46, 53
    translational, 3, 16–17
    vibrational, 47, 53

radial distribution function, 185–275
  mixtures, 211, 228–230
random numbers, 249, 251, 255
  pseudo, 249
reactions, chemical, 64–86
reduced
  temperature, 129–132, 255, 257, 260, 307–308
  virial coefficient, 129–132
regular solution model, 171, 184, 324
residual properties, 161, 298–299, 319, 332–334
rigid rotator, 45–46
rotational
  degeneracy, 45–46
  energy, 45–46, 52–53, 63
  hindered, 63
  partition function, 45–46, 52–53, 63
  temperature, 46, 53, 55

Sackur-Tetrode equation, 26
salt, 232, 239, 279–280, 292
scattering
  coherent, 202–203
  factor, 203–205, 213
  neutron, 197, 202–213
    small angle (SANS), 213
  x-ray, 197, 202–213
series reversion, 120
solvation forces, 232
solvent activity, 295–296
spatial probability density function, *see* radial distribution function
spectroscopy
  infrared, 48, 53
  microwave, 46, 202
  Raman, 48, 53
  ultraviolet, 24
spinodal decomposition, 184
spin state, 24, 45, 85, 178–181, 184
standard state, 26, 117, 238, 294–295
statistical average, 2, 5
steric hindrance, 190, 211–212
Stirling's approximation, 25, 69, 71
structure factor, 210–211
sublimation pressure, 159–162

sum over states, 8, 12–15, 48
superposition
    approximation, Kirkwood, 219–220, 265
    principle, 281
surface area fraction, 178, 316, 319–320, 326, 329
symmetry number (factor), 47, 53

Taylor series expansion, 101, 227, 257–261, 272, 277, 285
temperature
    absolute zero, 18, 63, 150, 155, 161
    rotational, 46, 53, 55
    vibrational, 48, 52–53, 55, 58, 61–63, 149, 152, 156
thermal expansion, 161, 239
thermodynamic properties, *see* individual properties
third law of thermodynamics, 20, 63, 147, 150, 155, 161
total correlation function, 207, 220
translational
    degeneracy, 4, 29, 35–36
    energy, 3, 16–17, 42, 53, 59, 63, 98–99, 126, 129, 132, 247, 254–256, 299
    partition function, 3–4, 11–14, 17, 22–23, 29, 35, 44–45, 52–53, 61–62, 80, 98–99, 262, 298, 330

van der Waals, 142–144, 166–168, 184, 273–274, 306, 316–316, 334
    extended, 274
    Generalized partition function, 297–334
    one theory, 143–144, 315

radius, 212–213
vibrational
    anharmonicity, 63
    energy, 44–49, 51–55, 58–60, 148–159
    frequency, 44–49, 51–55, 58–60, 148–159
    modes, 44–49, 51–55, 58–60, 148–159
    partition function, 48, 52–53, 55, 58, 61–63, 149, 152, 156
    temperature, 48, 52–53, 55, 58, 61–63, 149, 152, 156
virial
    coefficient, 98–146
    equation of state, 87, 95, 98–146, 166, 168, 222, 224, 237–238, 245, 247, 273, 278, 297
        range of applicability, 123–124, 141
    osmotic coefficient, 238–240
    theorem, 247
volume
    free, 166, 272, 300–320, 331–334
    fraction, 175–178, 184, 239, 314, 319–320, 324

wavelength, 23–24, 43, 100, 148, 151–152, 197, 202–204, 298, 329
Weeks-Chandler-Andersen perturbation theory, 270–272

x-ray scattering, 197, 202–213

Yvon-Born-Green (YBG) equation, 216–218

zero-point energy, 149, 153, 160
Zimm-Bragg model, 182–184